Transport and Chemical Transformation of Pollutants in the Troposphere

Series editors: Peter Borrell, Patricia M. Borrell, Tomislav Cvitaš, Kerry Kelly and Wolfgang Seiler

Springer
Berlin
Heidelberg
New York
Barcelona
Budapest
Hong Kong
London
Milan
Paris
Santa Clara
Singapore
Tokyo

Transport and Chemical Transformation
of Pollutants in the Troposphere

Volume 4

Biosphere-Atmosphere Exchange of Pollutants and Trace Substances

Experimental and Theoretical Studies of Biogenic Emissions and of Pollutant Deposition

Sjaak Slanina (Petten)
Editor and Coordinator

BIATEX Steering Group

Jan H. Duyzer	Delft
David Fowler	Edinburgh
Günter Helas	Mainz
Øystein Hov	Bergen
Franz X. Meixner	Mainz
Sten Struwe	Copenhagn

Springer

Editor
Prof. Dr. SJAAK SLANINA
Netherlands Energy Research Foundation
P.O. Box 1
NL-1755 ZG Petten
The Netherlands

With 211 Figures and 45 Tables

The cover photograph shows Han Mols (ECN, Petten) servicing a wet annular denuder for the measurement of atmospheric concentrations of HNO_3, HNO_2, HCl and SO_2 above the canopy of the Speulder Forest site in the Netherlands.

ISBN 3-540-61711-6 Springer-Verlag Berlin Heidelberg New York

Library of Congress Cataloging-in-Publication Data. Biosphere-atmosphere exchange of pollutants and trace substances: experimental and theoretical studies of biogenic emissions and of pollutant deposition / Sjaak Slanina, editor and coordinator. p. cm. – (Transport and chemical transformation of pollutants in the troposphere; v. 4) Includes index. ISBN 3-540-61711-6 (hardcover: alk. paper) 1. Environmental chemistry. 2. Man – Influence on nature. I. Slanina, Sjaak. II. Series. TD193.B55 1996, 628.5′3–DC20 96-36462 CIP

Cover Design: Struve & Partner, Heidelberg

Coverpicture from Sjaak Slanina (BIATEX)

SPIN 10514718 30/3136-5 4 3 2 1 0 – Printed on acid-free paper

Transport and Chemical Transformation of Pollutants in the Troposphere

Series editors: Peter Borrell, Patricia M. Borrell, Tomislav Cvitaš, Kerry Kelly and Wolfgang Seiler

Volume 1	Transport and Chemical Transformation of Pollutants in the Troposphere *An overview of the work of EUROTRAC*
Volume 2	Heterogeneous and Liquid Phase Processes *Laboratory studies related to aerosols and clouds*
Volume 3	Chemical Processes in Atmospheric Oxidation *Laboratory studies of chemistry related to troposphere ozone*
Volume 4	Biosphere-Atmosphere Exchange of Pollutants and Trace Substances *Experimental and theoretical studies of biogenic emission and of pollutant deposition*
Volume 5	Cloud Multi Phase Processes and High Alpine Air and Snow Chemistry *Ground-based cloud experiments and pollutant deposition in the high Alps*
Volume 6	Tropospheric Ozone Research *Tropospheric ozone in the regional and sub-regional context*
Volume 7	Tropospheric Modelling and Emission Estimation *Chemical transport and emission modelling on regional, global and urban scales*
Volume 8	Instrument Development for Atmospheric Research and Monitoring *Lidar profiling, DOAS and tunable diode laser spectroscopy*
Volume 9	Exchange and Transport of Air Pollutants over Complex Terrain and the Sea *Field measurements and numerical modelling; aircraft, ship, ocean platform and laboratory measurements*
Volume 10	Photo-oxidants, Acidification and Tools: Policy Applications of EUROTRAC Results

Foreword by the Series Editors

EUROTRAC is the European co-ordinated research project, within the EUREKA initiative, studying the transport and chemical transformation of pollutants in the troposphere. The project has achieved a remarkable scientific success since its start in 1988, contributing substantially both to the scientific progress in this field and to the improvement of the scientific basis for environmental management in Europe. EUROTRAC, which at its peak comprised some 250 research groups organised into 14 subprojects, brought together international groups of scientists to work on problems directly related to the transport and chemical transformation of trace substances in the troposphere. In doing so, it helped to harness the resources of the participating countries to gain a better understanding of the trans-boundary, interdisciplinary environmental problems which beset us in Europe.

The scientific results of EUROTRAC are summarised in this report which consists of ten volumes.

Volume 1 provides a general overview of the scientific results, prepared by the Scientific Steering Committee (SSC) and the International Scientific Secretariat (ISS) of EUROTRAC, together with brief summaries of the work of the fourteen individual subprojects prepared by the respective subproject coordinators.

Volumes 2 to 9 comprise detailed overviews of the subproject achievements, each prepared by the respective subproject coordinator and steering group, together with summaries of the work of the participating research groups prepared by the principal investigators. Each volume also includes a full list of the scientific publications from the subproject.

The final volume, 10, is the complete report of the Application Project, which was set up in 1993 to assimilate the scientific results from EUROTRAC and present them in a condensed form so that they are suitable for use by those responsible for environmental planning and management in Europe. It illustrates how a scientific project such as EUROTRAC can contribute practically to providing the scientific consensus necessary for the development of a coherent atmospheric environmental policy for Europe.

A multi-volume work such as this has many contributors and we, as general editors, would like to express our thanks to all of them: to the subproject coordinators who have borne the brunt of the scientific co-ordination and who have contributed so much to the success of the project and the quality of this report; to the principal investigators who have carried out so much high-quality scientific work; to the members of the International Executive Committee (IEC) and the SSC for their enthusiastic encouragement and support of EUROTRAC; to the participating governments in EUROTRAC, and in particular the German Government (BMBF) for funding, not only the research, but also the ISS publication activities; and finally to Mr. Christian Witschell and his colleagues at Springer Verlag for providing the opportunity to publish the results in a way which will bring them to the notice of a large audience.

Peter Borrell (Scientific Secretary, ISS)
Patricia May Borrell
Tomislav Cvitaš
Kerry Kelly
Wolfgang Seiler (Director, ISS)

EUROTRAC ISS
Fraunhofer Institute (IFU)
Garmisch-Partenkirchen

Table of Contents

Individual Reports from BIATEX Contributors

List of Contributors

Dr. M. Ammann
Universität Bern
Pflanzenphysiologisches Institut
Altenbergrain 21
CH-3013 Bern
Switzerland

Dr. C. Ammann
Max Planck Institut für Chemie
Postfach 3060
D-55020 Mainz
Saarstraße 23
D-55122 Mainz
Germany

Prof. Dr. M. O. Andreae
Max Planck Institut für Chemie
Postfach 3060
D-55020 Mainz
Saarstraße 23
D-55122 Mainz
Germany

Dr. A. Anghelus
Agentia Pt. Supravegherea
Si Protectia Mediuli
Timisora
Str. Cozia Nr. 7
RM-1900 Timisoara
Rumania

Dr. J. Baak
TNO MEP
P. O. Box 6011
NL-2600 JA Delft
Netherlands

Dr. F.-W. Badeck
Institut für Physikalische u.
Theoretische Chemie der
J. W. Goethe-Universität
Niederurseler Hang
D-60339 Frankfurt
Germany

Dr. P. von Ballmoos
Universität Bern
Pflanzenphysiologisches Institut
Altenbergrain 21
CH-3013 Bern
Switzerland

Dr. U. Bartell
Max Planck Institut für Chemie
Postfach 3060
D-55020 Mainz
Saarstraße 23
D-55122 Mainz
Germany

Mr. J. Beck
Max Planck Institut für Chemie
Postfach 3060
D-55020 Mainz
Saarstraße 23
D-55122 Mainz
Germany

Ms. S. Blezinger
Max Planck Institut für Chemie
Postfach 3060
D-55020 Mainz
Saarstraße 23
D-55122 Mainz
Germany

Dr. M. Bliefernicht
Max Planck Institut für Chemie
Postfach 3060
D-55020 Mainz
Saarstraße 23
D-55122 Mainz
Germany

Dr. K. Bode
Max Planck Institut für Chemie
Postfach 3060
D-55020 Mainz
Saarstraße 23
D-55122 Mainz
Germany

Dr. P. Boeckx
University of Ghent
Laboratory for Physical and
Radiobiol. Chemistry
Coupure Links 653
B-9000 Gent
Belgium

Dr. C. Boissard
Environmental Science Division
Lancaster University
GB-Lancaster LA1 4YQ
United Kingdom

Dr. M. Borgarello
CISE S.p.A.
Via Reggio Emilia 39
I-20124 Milano
Italy

Dr. C. Brunold
Universität Bern
Pflanzenphysiologisches Institut
Altenbergrain 21
CH-3013 Bern
Switzerland

Dr. G. Brusasca
ENEL-CRAM
Via Rubattino 54
I-20134 Milano
Italy

Mr. F. Böswald
Fraunhofer Institut für
Atmosphärische Umweltforschung
Kreuzeckbahnstraße 19
D-82467 Garmisch-Partenkirchen
Germany

Dr. T. Camaroschi
Agentia Pt. Supravegherea
Si Protectia Mediuli
Timisora
BD. Mihai Viteazu Nr. 32
RM-1900 Timisoara
Rumania

Dr. H. Casado
Universidad del Pais Vasco
Facultad de Farmacia
Portal de Lasarte s/n
E-01007 Vitoria
Spain

Dr. G. Catenacci
CISE S.p.A.
Via Reggio Emilia 39
I-20124 Milano
Italy

Dr. C. Cavicchioli
CISE S.p.A.
Via Reggio Emilia 39
I-20124 Milano
Italy

Dr. T. W. Choularton
UMIST
Department of Atmospheric Physics
P. O. Box 88
GB-Manchester M60 1QD
United Kingdom

Dr. S. Cieslik
CEC Joint Research Centre
Environmental Research Institute
I-21020 Ispra (VA)
Italy

Dr. O. van Cleemput
University of Ghent
Laboratory for Physical and
Radiobiol. Chemistry
Coupure Links 653
B-9000 Gent
Belgium

Dr. B. Clement
Ecole Nationale Supérieure de Chimie –
Laboratoire Chimie et Environnement
118, route de Narbonne
F-31077 Toulouse-Cedex
France

Prof. Dr. R. Conrad
Max-Planck Institut für
Terrestrische Mikrobiologie
Karl-von-Frisch-Straße
D-35043 Marburg/Lahn
Germany

Dr. W. Conze
Max Planck Institut für Chemie
Abt. Biogeochemie
Postfach 3060
D-55020 Mainz
Saarstraße 23
D-55122 Mainz
Germany

Dr. G. Deinum
TNO MEP
P. O. Box 6011
NL-2600 JA Delft
Netherlands

Dipl. Chem. T. Dietrich
Zentrum für Umweltforschung (ZUF)
J. W. Goethe-Universität Frankfurt
Georg-Voigt-Straße 14
D-60325 Frankfurt
Germany

Dr. R. Dlugi
Universität München
Meteorologisches Institut
Theresienstraße 37
D-80333 München
Germany

Ms. J. Dominguez Lorenzo
Universidad de Valladolid
Lab. de Contaminacion Atmosferica
E-Valladoid
Spain

Dr. G. P. J. Draaijers
RIVM-Air Research Laboratory
P. O. Box 1
NL-3720 BA Bilthoven
Netherlands

Dr. A. Druilhet
Université Paul Sabatier
Laboratoire d'Aerologie URA/CNRS 354
118, route de Narbonne
F-31062 Toulouse Cedex
France

Dr. S. C. Duckham
Lancaster University
Institute of Environmental &
Biological Sciences
GB-Lancaster LA1 4YQ
United Kingdom

Ing. J. H. Duyzer
TNO MEP
P. O. Box 6011
NL-2600 JA Delft
Netherlands

Dipl. Biol. G. Eichstädter
Universität München
Inst. für Botanik und Mikrobiologie
Arcisstraße 21
D-80333 München
Germany

Dr. J. W. Erisman
RIVM
Antonie van Leewenhoeklaan 9
P. O. Box 1
NL-3720 BA Bilthoven
Netherlands

Dr. U. Fehsenfeld
Fraunhofer Institut für
Atmosphärische Umweltforschung
Kreuzeckbahnstraße 19
D-82467 Garmisch-Partenkirchen
Germany

Dr. M. Ferm
Swedish Environmental
Research Institute (IVL)
Box 47086
S-402 58 Göteborg
Sweden

Dr. Th. Foken
Deutscher Wetterdienst
Meteorolog. Observatorium Lindenberg
D-15864 Lindenberg
Germany

Prof. Dr. J. Fontan
Université Paul Sabatier
Laboratoire d'Aerologie
118, route de Narbonne
F-31062 Toulouse Cedex
France

Dr. R. Forkel
Fraunhofer Institut für
Atmosphärische Umweltforschung
Kreuzeckbahnstraße 19
D-82467 Garmisch-Partenkirchen
Germany

Dr. D. Fowler
Institute of Terrestrial Ecology
Edinburgh Research Station
Bush Estate, Penicuik
GB-Midlothian EH32 ORA
United Kingdom

Dr. J. L. Fugit
Ecole Nationale Supérieure de Chimie et
Laboratoire Environnemental
118, route de Narbonne
F-31077 Toulouse-Cedex
France

Dr. E. Führer
Forest Research Institute
Frankel Leó u. 44
H-1023 Budapest
Hungary

Dr. R. Gabriel
Max Planck Institut für Chemie
Postfach 3060
D-55020 Mainz
Saarstraße 23
D-55122 Mainz
Germany

Dr. M. W. Gallagher
UMIST
Department of Atmosheric Physics
P. O. Box 88
GB-Manchester M60 1 QD
United Kingdom

Dr. J. A. Garcia
Universidad de Extremadura
Departamento de Fisica
E-06071 Badajoz
Spain

Dr. H. Gfeller
Institut für Organische Chemie
Universität Bern
Freiestraße 3
CH-3012 Bern
Switzerland

Dr. D. Goedhart
RIVM
Antonie van Leewenhoeklaan 9
P. O. Box 1
NL-3720 BA Bilthoven
Netherlands

Dr. T. Gori
Universität München
Meteorologisches Institut
Thersienstraße 37
D-80333 München
Germany

Dr. L. Granat
University of Stockholm
Department of Meteorology
S-106 91 Stockholm
Sweden

Dr. C. Gries
Arizona State University
Dept. of Botany
USA-Tempe AZ 85287-1601
USA

Dr. H. Güsten
KfK Karlsruhe
Institut für Meteorologie
und Klimaforschung (IMK)
Postfach 3640
D-76021 Karlsruhe
Germany

Dr. J. Hahn
Fraunhofer Institut für
Atmosphärische Umweltforschung
Kreuzeckbahnstraße 19
D-82467 Garmisch-Partenkirchen
Germany

Dr. F. J. M. Harren
University of Nijmegen
Molecular and Laser Physics
Toernooiveld
NL-6525 ED Nijmegen
Netherlands

Dr. M. Haziza
Ecole Nationale Supérieure de Chimie
Chimie-Energie-Environnement
118, route de Narbonne
F-31077 Toulouse Cedex
France

Dr. G. Heinrich
Kernforschungszentrum Karlsruhe
Institut f. Meteorologie u. Klimaforsch.
Postfach 3640
D-76021 Karlsruhe
Germany

Dr. G. Helas
Max Planck Institut für Chemie
Abt. Biogeochemie
Postfach 3060
D-55020 Mainz
Saarstraße 23
D-55122 Mainz
Germany

Dr. C. N. Hewitt
University of Lancaster
Institute of Environmental and
Biological Sciences
GB-Lancaster LA1 4YW
United Kingdom

Dr. C. Hilse
University Mainz
Botanical Institute
D-55099 Mainz
Germany

Dr. R. Hofmann
Max Planck Institut für Chemie
Abt. Biogeochemie
Postfach 3060
D-55020 Mainz
Saarstraße 23
D-55122 Mainz
Germany

Dr. U. Hofmann
Max Planck Institut für Chemie
Abt. Biogeochemie
Postfach 3060
D-55020 Mainz
Saarstraße 23
D-55122 Mainz
Germany

Dr. J. Hogenkamp
RIVM
Antonie van Leewenhoeklaan 9
P. O. Box 1
NL-3720 BA Bilthoven
Netherlands

Dr. L. Horváth
Institute for Atmospheric Physics
PO Box 39
H-1675 Budapest
Hungary

Dr. O. Hov
University of Bergen
Geophysics Institute
Allégaten 70
N-5007 Bergen
Norway

Dr. A. Hubert
Max Planck Institut für Chemie
Abt. Biogeochemie
Postfach 3060
D-55020 Mainz
Saarstraße 23
D-55122 Mainz
Germany

Dr. P. Hummelshoj
Riso National Laboratory
Section of Chemistry
Department of Environmental Science and Technology
Postbox 49
DK-4000 Roskilde
Denmark

Dr. G. Inclan
Universität München
Meteorologisches Institut
Theresienstraße 37
D-80333 München
Germany

Dr. W. Jaeschke
J. W. Goethe-Universität
Zentrum für Umweltforschung
Postfach 111932-196
D-60054 Frankfurt
Georg-Voigt-Straße 14
D-60325 Frankfurt
Germany

Mr. R. Janson
Department of Meteorology
Stockholm University
Arrhenius Laboratory
S-106 91 Stockholm
Sweden

Dr. N. O. Jensen
Risö National Laboratory
Dept. of Meteorology & Wind Energie
Postbox 49
DK-4000 Roskilde
Denmark

Dr. C. Johansson
ITM Air Pollution Laboratory
Stockholm University
University Stockholm
Frescati Hagväg 16 B
S-106 91 Stockholm
Sweden

Dr. P. A. C. Jongejan
Netherlands Energy Research Foundation
P. O. Box 1
NL-1755 Petten
Netherlands

Dr. A. Kasper
Technische Universität Wien
Inst. für Analytische Chemie
Getreidemarkt 9/151
A-1060 Wien
Austria

Dr. E. Kemkers
RIVM
Anatomie van Leewenhoeklaan 9
P. O. Box 1
NL-3720 BA Bilthoven
Netherlands

Dr. J. Kesselmeier
Max Planck Institut für Chemie
Abt. Biogeochemie
Postfach 3060
D-55020 Mainz
Saarstraße 23
D-55122 Mainz
Germany

Dr. A. Khlystov
Netherlands Energy Research Foundation
P. O. Box 1
NL-1755 ZG Petten
Netherlands

Dr. L. Kins
Meteorologisches Institut
Universität München
Barbarastraße 16
D-80797 München
Germany

Dr. M. Klauer
J. W. Goethe-Universität
Zentrum für Umweltforschung (ZUF)
Postfach 111932-196
D-60054 Frankfurt
Georg-Voigt-Straße 14
D-60325 Frankfurt
Germany

Prof. Dr. G. H. Kohlmaier
J. W. v. Goethe Universität Frankfurt
Institut für Physikalische und
Theoretische Chemie
Marie-Curie-Straße 11
D-60439 Frankfurt/Main
Germany

Dr. Ch. Kolb
Fraunhofer Institut für
Atmosphärische Umweltforschung
Kreuzeckbahnstraße 19
D-82467 Garmisch-Partenkirchen
Germany

Dr. G. Kramm
Fraunhofer Institut für
Atmosphärische Umweltforschung
Kreuzeckbahnstraße 19
D-82467 Garmisch-Partenkirchen
Germany

Dr. L. Kristensen
Risö National Laboratory
Dept. of Meteorology & Wind Energy
Postbox 49
DK-4000 Roskilde
Denmark

Mr. U. Kuhn
Max Planck Institut für Chemie
Postfach 3060
D-55020 Mainz
Saarstraße 23
D-55122 Mainz
Germany

Dr. E. Köhler
Universität München
Meteorologisches Institut
Thersienstraße 37
D-80333 München
Germany

Dr. D. Köhler
Universität München
Meteorologisches Institut
Thersienstraße 37
D-80333 München
Germany

Dr. A. Labatut
CEC Joint Research Centre
Environmental Research Institute
I-21020 Ispra (VA)
Italy

Dr. E. Lamaud
Université Paul Sabatier
Lab. d'Aerologie URA/CNRS 354
118, route de Narbonne
F-31062 Toulouse Cedex
France

Dr. K. Lederle
Fraunhofer Institut für
Atmosphärische Umweltforschung
Kreuzeckbahnstraße 19
D-82467 Garmisch-Partenkirchen
Germany

Dr. A. Lopez
Université Paul Sabatier
Lab. d'Aerologie URA/CNRS 354
118, route de Narbonne
F-31062 Toulouse Cedex
France

Dr. J. Ludwig
Max Planck Institut für Chemie
Abteilung Biogeochemie
Postfach 3060
D-55020 Mainz
Saarstraße 23
D-55122 Mainz
Germany

Dr. E. A. Lörinczi
Agentia Pt. Supravegherea
Si Protectia Mediuli
Timisora
Str. Cozia Nr. 7
RM-1900 Timisoara
Rumania

Mrs. M. Lörinczi
Agentia Pt. Supravegherea
Si Protectia Mediuli
Timisora
Str. Cozia Nr. 7
RM-1900 Timisoara
Rumania

Dr. G. Manzi
ENEL-CRAM
Via Rubattino 54
I-20134 Milano
Italy

Dr. F. X. Meixner
Max Planck Institut für Chemie
Postfach 3060
D-55020 Mainz
Saarstraße 23
D-55122 Mainz
Germany

Dr. M. Mennen
RIVM
Antonie van Leewenhoeklaan 9
P. O. Box 1
NL-3720 BA Bilthoven
Netherlands

Dr. L. Merk
Max Planck Institut für Chemie
Postfach 3060
D-55020 Mainz
Saarstraße 23
D-55122 Mainz
Germany

Dr. H. Müller
Fraunhofer Institut für
Atmosphärische Umweltforschung
Kreuzeckbahnstraße 19
D-82467 Garmisch-Partenkirchen
Germany

Dr. J. J. Möls
Netherlands Energy Research
Foundation
P. O. Box 1
NL-1755 ZG Petten
Netherlands

Dr. E. Mönnich
Institut f. Meteorologie
u. Klimaforschung
Postfach 3540
D-76021 Karlsruhe
Germany

Dr. Z. Nagy
Institute for Atmospheric Physics
PO Box 39
H-1675 Budapest
Hungary

Dr. T. H. Nash III
Arizona State University
Dept. of Botany
USA-Tempe AZ 85287-1601
USA

Dr. T. Nunes
Universidade de Aveiro
Departemento de Ambiento
e Ordenamento
P-3800 Aveiro
Portugal

Dr. S. Nussbaum
Universität Bern
Pflanzenphysiologisches Institut
Altenbergrain 21
CH-3013 Bern
Switzerland

Dr. R. P. Otjes
ECN Stichting Energieonderzoek
Postbus 1
NL-1755 ZG Petten
Netherlands

Dr. U. Özer
Uludag University
Fen-Edebiyat Faculty
Kimya Bölümü
TR-16059 Görükle, Bursa
Turkey

Dr. K. Pilegaard
Riso National Laboratory
Environmental Science
P. O. Box 49
DK-4000 Roskilde
Denmark

Dr. C. Pio
Universidade de Aveiro
Departamento de Ambiente
e Ordenamento
P-3810 Aveiro
Portugal

Dr. M. A. H. G. Plantaz
ECN Stichting Energieonderzoek
Postbus 1
NL-1755 ZG Petten
Netherlands

Dr. M. Plöchl
Potsdam Institut für
Klimafolgenforschung (PIK)
Telegrafenberg
D-14473 Potsdam
Germany

Dr. G. Protoschill-Krebs
Max Planck Institut für Chemie
Postfach 3060
D-55020 Mainz
Saarstraße 23
D-55122 Mainz
Germany

Dr. A. van Pul
RIVM
Antonie van Leewenhoeklaan 9
P. O. Box 1
NL-3720 BA Bilthoven
Netherlands

Dr. R. Rabong
Fraunhofer Institut für
Atmosphärische Umweltforschung
Kreuzeckbahnstraße 19
D-82467 Garmisch-Partenkirchen
Germany

Dr. P. Ramge
Institut für Physikalische und
Theoretische Chemie
J. W. v. Goethe Universität Frankfurt
Marie-Curie-Straße 11
D-60439 Frankfurt/Main
Germany

Dr. Th. Rausch
Max Planck Institut für Chemie
Postfach 3060
D-55020 Mainz
Saarstraße 23
D-55122 Mainz
Germany

Dr. J. Reuder
Universität München
Meteorologisches Institut
Theresienstraße 37
D-80333 München
Germany

Dr. J. Reuss
University of Nijmegen
Molecular and Laser Physics
Toernooiveld
NL-6525 ED Nijmegen
Netherlands

Dr. K. Reußwig
Universität München
Meteorologisches Institut
Theresienstraße 37
D-80333 München
Germany

Dr. M. L. Riba
Ecole Nationale Supérieure de Chimie –
Laboratoire Chimie et Environnement
118, route de Narbonne
F-31077 Toulouse-Cedex
France

Dr. A. Ritivoiu
Agentia Pt. Supravegherea
Si Protectia Mediuli
Timisora
Bd. Mihai Viteazu Nr. 32
RM-1900 Timisoara
Rumania

Mr. R. Rodriguez de la Torre
Universidad de Valladolid
Lab. de Contaminacion Atmosferica
E-Valladolid
Spain

Dr. G. Roider
Universität München
Meteorologisches Institut
Theresienstraße 37
D-80333 München
Germany

Dr. A. Rondon
University of Stockholm
Department of Meteorology
S-106 91 Stockholm
Sweden

Dr. A. Ruggaber
Universität München
Meteorologisches Institut
Theresienstraße 37
D-80333 München
Germany

Dr. K. Ruoss
Universität München
Meteorologisches Institut
Theresienstraße 37
D-80333 München
Germany

Prof. Dr. R. San José
Universidad Politécnica de Madrid
Facultad de Informática
Environmental Modelling & Software
Campus de Montegancedo
E-28660 Boadilla del Monte (Madrid)
Spain

Dr. M. L. Sanchez Gomez
Universidad de Valladolid
Lab. de Contaminacion Atmosferica
E-Valladolid
Spain

Dr. U. Schickedanz
Zentrum für Umweltforschung (ZUF)
J. W. v. Goethe-Universität Frankfurt
Georg-Voigt-Straße 14
D-6035 Frankfurt
Germany

Dr. J. K. Schjörring
Royal Veterinary & Agricultural University
Centre of Ecology & Environment
Thorvaldsenvej 40
DK-1871 Frederiksberg C
Denmark

Dr. U. P. Schlunegger
Institut für Organische Chemie
Universität Bern
Freiestraße 3
CH-3012 Bern
Switzerland

Dr. A.-M. Schreiner
Fraunhofer Institut für
Atmosphärische Umweltforschung
Kreuzeckbahnstraße 19
D-82467 Garmisch-Partenkirchen
Germany

Dr. P. Schröder
GSF Forschungszentrum
Institut für Biochemische
Pflanzenpathologie
D-85758 Oberschleißheim
Germany

Dr. J. Schween
Universität München
Meteorologisches Institut
Theresienstraße 37
D-80333 München
Germany

Dr. W. Schürmann
Fraunhofer Institut für
Atmosphärische Umweltforschung
Kreuzeckbahnstraße 19
D-82467 Garmisch-Partenkirchen
Germany

Dr. L. Schäfer
Max Planck Institut für Chemie
Abt. Biogeochemie
Postfach 3060
D-55020 Mainz
Saarstraße 23
D-55122 Mainz
Germany

Dr. R. Schönwitz
Fraunhofer Institut für
Atmosphärische Umweltforschung
Kreuzeckbahnstraße 19
D-82467 Garmisch-Partenkirchen
Germany

Dr. W. Seidl
Fraunhofer Institut für
Atmosphärische Umweltforschung
Kreuzeckbahnstraße 19
D-82467 Garmisch-Partenkirchen
Germany

Dr. P. Seifert
Universität München
Meteorologisches Institut
Theresienstraße 37
D-80333 München
Germany

Chapter 1

Assessment and Achievements of BIATEX

Sjaak Slanina

ECN, Petten, The Netherlands

Introduction

An understanding of the exchange of pollutants and trace substances between the atmosphere and the biosphere is essential to a full appreciation of the transport and chemical transformation of pollutants in the troposphere, and to the implementation of any policy for pollutant abatement. The biosphere is the ultimate sink for all pollutants or their oxidation products, and so the rates of deposition play a critical role in determining pollutant lifetimes in the atmosphere, their transport range and their concentration.

The biosphere is also a source of many compounds which are precursors for photo-oxidants, and so the contribution of biogenic emissions must be taken into account when determining the anthropogenic contribution to pollutant concentrations and deposition.

When BIATEX (BIosphere/ATmosphere EXchange of pollutants) was established as a subproject of EUROTRAC, the general notion prevailed that the absence of data on dry deposition and biogenic emissions would be an important barrier to obtaining insights into transport, export and import of pollution in Europe, as well as estimates on acid deposition and oxidant formation.

The project description reflects this state of affairs and was primarily oriented towards these problems.

A number of factors influenced the development of BIATEX, as will be explained in the next sections.

- At the time it was still unclear that eutrophication, exposure of ecosystems to too large amounts of nutrients such as phosphate and nitrate, was becoming a grave environmental problem in many parts of Europe, as important as acidification or exposure to increasing oxidant concentrations.
- In the initial phase of BIATEX, the critical load concept as a device for formulating environmental policy in Europe was developed. This critical load

concept created a need to provide more accurate estimates of deposition loads of pollutants in Europe, in particular of the contribution of dry deposition, the most uncertain component.

- An environmental policy to limit oxidant formation in Europe became more important. A requirement for the policy development are assessments of biogenic, as well as anthropogenic, emissions of reactive organic compounds. The cheapest environmental policy is based on limiting volatile organic compounds, but this policy is doomed if sufficient organic compounds are emitted by the biosphere.
- The progress of environmental measures made it necessary to provide better estimates of pollution loads. Uncertainties in pollution loads of several hundred percent can be quite acceptable if very large exceedences (transgressions) of critical loads or critical concentrations are encountered. But environmental policy in many countries in Europe has now progressed to the point that quantitative decisions must be taken, for example on. whether emissions must be reduced by 60 or 90 %. The increase in costs is exponential for reductions over 50 %, so hard evidence is necessary to support large expenditures for abatement.

Many of the research groups engaged in BIATEX were already involved the application of results to the environment, and they, at least, had some experience in supporting national environmental policy. However, no mechanism existed for transferring scientific data in a form suitable for the development of environmental policy on a European scale.

So the BIATEX community was not only confronted with a number of scientific problems regarding measurement and modelling of exchange fluxes, but also with the necessity of adapting BIATEX to the changing circumstances, and to developing suitable instrumentation to contribute to environmental policy development.

The very open and fruitful cooperation within BIATEX has been a great asset which has enabled us to reach many of the ambitious gaols described in the original BIATEX proposal. Frequent contacts within EUROTRAC, the SSC, the ISS and contacts with coordinators and investigators from other EUROTRAC subprojects helped in obtaining a good overall picture of priorities and possibilities.

Collaboration with national and international agencies, especially EMEP, should also be mentioned, which has been essential in developing the application of BIATEX results to a description of the exchange of pollutants on a European scale.

BIATEX was also supported by the European Communities at a later stage. This support was not only important financially for some activities within BIATEX but also paved the way to an integration of European research in this area.

Promises versus final products of BIATEX

The collaboration of some 30 European groups in research devoted to assessment of the exchange of pollutants between atmosphere and biosphere has been successful and has promoted coherence between European researchers.

This last point is not trivial: The development of an European Environmental Policy is dependent on a European consensus and hence on a coherent scientific data base for Europe. The contribution of EUROTRAC to European collaboration in environmental research has been very important in general and BIATEX is certainly not an exception in this.

The next question in a final report of such a large programme as BIATEX must, by necessity, be the evaluation of the results in comparison to the original objectives, as laid out in the description of BIATEX during the starting phase of the subproject.

The final report of BIATEX should provide answers to the following questions:

- Has BIATEX answered the questions posed as the original objectives at the beginning of the subproject?
- If these original objectives have not been met, what are the reasons for not delivering the promised results?
- Which important gaps exist in the present state of knowledge?

The original description of BIATEX formulated the following objectives (the brackets indicate omitted sentences or parts of sentences in the original text) :

A (...) to quantify fluxes of pollutants between atmosphere and biosphere, to study the mechanisms which are responsible for the observed exchange, and to provide **regional fluxes on seasonal and annual scales**. (...) Priorities are: nitrogen compounds, volatile organic compounds and sulphur compounds.

B (...) data on spatial and temporal variations of some biogenic emission rates of some important trace constituents, (...) partial insight into mechanisms controlling emission fluxes and first order estimates of **regional fluxes on seasonal and annual scales**. (...)Priorities are: volatile organic compounds, sulphur compounds, nitrogen compounds.

C (...) secondary products will be developed: Method development for eddy-correlation and gradient methods. Quality assurance and quality control for the whole measurement systems for exchange measurements. Development of mechanistic studies to generalise observations. Model development up to 3-D models describing deposition, and simpler models for biogenic emission. Assessment of results of individual and joint experiments.

Summarising the conclusions of the 1995 BIATEX workshop in Madrid the following evaluation of the achievements in these areas can be made:

A Fluxes of pollutants between the atmosphere and the biosphere have been quantified, the mechanisms which are responsible for the observed exchange are now partially characterised, and **regional fluxes on seasonal and annual scales** have been provided.

The priorities are, according to the BIATEX results: nitrogen compounds, volatile organic compounds and sulphur compounds.

B Data on spatial and temporal variations of biogenic emission rates of some important trace constituents have been obtained. Partial insight into mechanisms controlling emission fluxes has been provided and first order estimates of **regional fluxes on seasonal and annual scales** for biogenic emitted volatile organic compounds will be obtained before the end of 1995.

Priorities within BIATEX are: volatile organic compounds, nitrogen compounds and sulphur compounds.

C Method development for eddy correlation and gradient methods has taken place,
important progress has been achieved for reduced nitrogen compounds, some oxidants, VOCs and reduced sulfur compounds.

Quality assurance and quality control for the integral measurement systems for exchange measurements has not been performed in a rigorous way but first order qualifications for the precision and accuracy of a number of procedures are available.

Development of mechanistic studies to generalise observations has been partially successful *e.g.* in regarding the influence of a water layer on deposition, and systematic approaches on isoprene emissions, but have not been developed for other aspects.

Model development has been less successful than expected:
No 3-D models for deposition have been developed, but models are now available based on resistance analogy, which can provide better estimates of dry deposition fluxes.

No approaches for inhomogeneous situations have been developed.
A simple model for biogenic emission is currently under development and will be finished before the end of 1995.

The assessment of results of individual and joint experiments has been performed in a qualitative way however the integration of individual results in communal assessments is still a weak point.

To summarise the confrontation between promises and results, BIATEX has achieved or will achieve in the near future a number of the main objectives as promised in 1989, such as regional and seasonal deposition fluxes and emission fluxes of biogenic emitted organic compounds. Estimates of biogenic emissions of other compounds, for example NO or reduced sulfur, are not yet possible. Progress has been made in terms of additional information, but this information is still insufficient for estimates of emission fluxes on a European scale.

At the onset of BIATEX, far reaching plans were developed for model development in the area of deposition and biogenic emissions. The need for advanced models in this area has been clearly demonstrated, but the gap proved to be so large between the possibility of a systematic meteorological approach, and the quality of data for exchange measurements and available data sets of important parameters, that different approaches had to be chosen. In fact it was decided to employ resistance modelling to provide new estimates for the deposition of important compounds in Europe.

The ambitious goal to use biological mechanistic studies as a basis for extrapolation of exchange data in time and space, and under different biological and meteorological conditions has only partially been met. However, good cooperation between different disciplines, biology, biochemistry and atmospheric chemistry plus micrometeorology, has emerged within BIATEX. This cooperation will form the basis for future developments in this area.

International cooperation has been strengthened by EUROTRAC and certainly within BIATEX. One of the important developments in BIATEX is the increased coherence in the scientific community engaged in assessments of exchange between atmosphere and biosphere.

Principal results obtained in BIATEX

The reviews of the different areas of BIATEX which are published in this volume, provide a more detailed account of the results of BIATEX in the context of the general developments in the area of measurement and assessments of exchange of compounds between atmosphere and biosphere. But a number of principal achievements will be mentioned here:

1. The methodology to derive better deposition estimates for Europe, has been established. A model, based on a resistance analogy approach, has been derived by RIVM (BIATEX annual report 1993). This method provides estimates of the resistances for turbulent transport in the atmosphere (R_a), based on parameters commonly available such as wind speed, estimates of heat fluxes *etc*. Detailed land use descriptions are used to derive the roughness height and the resistance for deposition due to the stagnant layer near the surface (R_b). Estimates for the surface resistances (R_c) are provided for Europe, based on the detailed land use descriptions mentioned earlier and on better surface resistance parameterisation, which have been developed by BIATEX participants. The deposition is calculated as passing through a layer at an

altitude of 50 meters, to minimise effects of non-homogeneous terrains. Depositions of SO_2 and NH_3, calculated by means of this improved approach are presented in Figs. 1.1 and 1.2. The surface resistances and hence deposition fluxes at constant concentration vary over a much wider range as assumed in older models. The influence of surface wetness and variation in turbulence leads to much higher deposition in countries like Netherlands, Denmark, Belgium and northern Germany compared to much dryer areas in Spain or Italy. The first results of deposition measurements in Portugal confirm the important role of water layers for the deposition of these compounds.

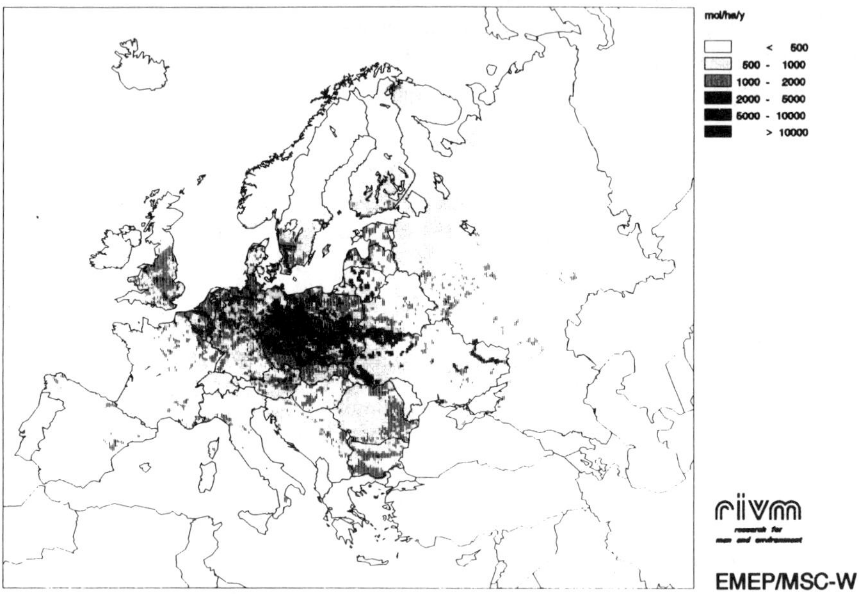

Fig. 1.1: Deposition of acidifying compounds over Europe (mol ha^{-1} year^{-1}). The sum of wet and dry deposition is based on new estimates of dry deposition velocities.

Collaboration between BIATEX models and EMEP has been established, the parameterisation of deposition as developed in BIATEX is currently implemented in EMEP modelling. The change in deposition patterns is not only of importance in establishing, with better precision, exceedence of critical loads in Europe, but will also influence export and import of pollutants for different countries. The collaboration between BIATEX and EMEP provides an ideal vehicle to transfer scientific findings and conclusions into a format which is suitable as a basis for European environmental policy.

2. The development of generalised descriptions of biogenic emissions in Europe has started. With the help of financial support from the BMFT, Germany, a good start has been made in this area. During the course of BIATEX joint experiments organised in the Bavarian forest, and individual experiments have provided more and better data on biogenic emissions of volatile organic compounds (BEVOC).

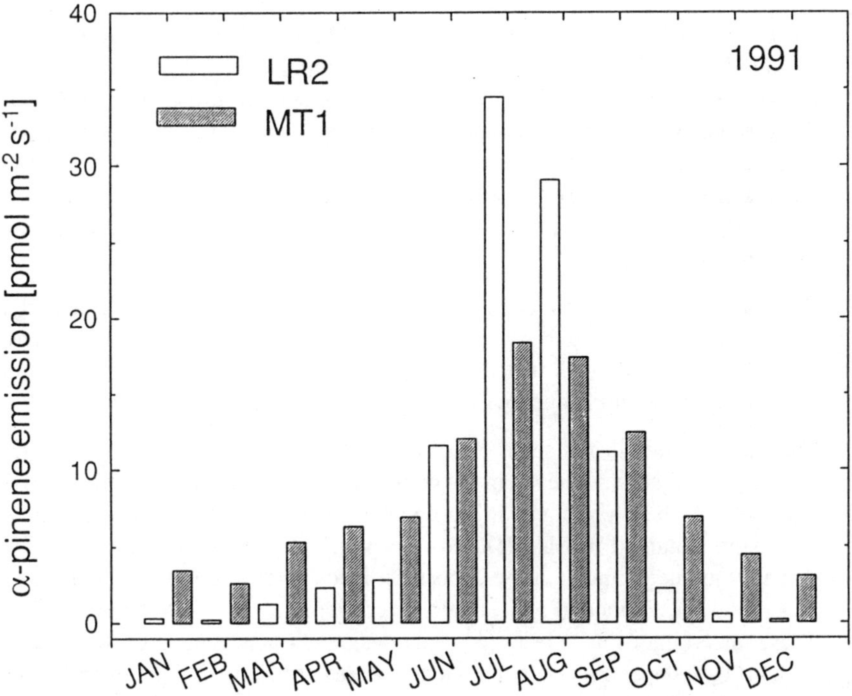

Fig. 1.2: Comparison between results from the US models (MT1) and model calculations (LR2) performed at IFU and based on observations in the Bavarian forest over a period from 1987 to 1991.

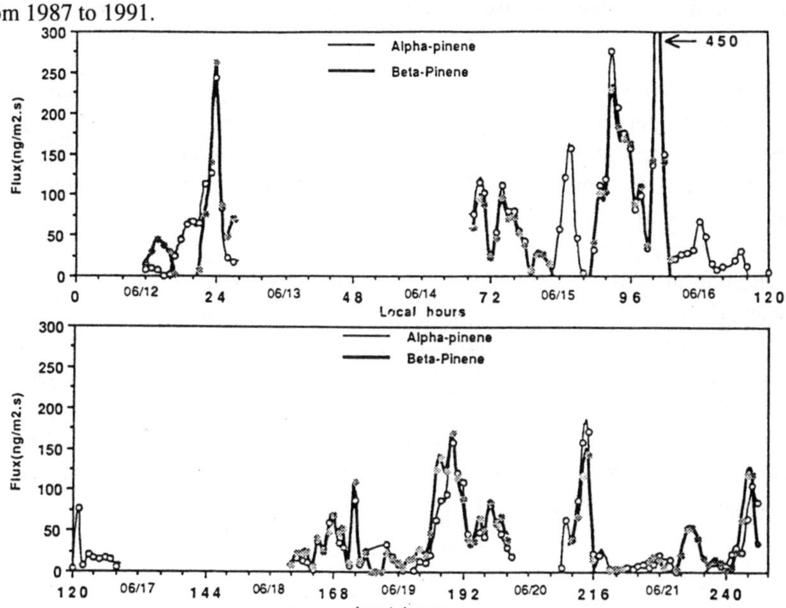

Fig. 1.3: Pinene emissions measured in the Bavarian forest using an automated GC system developed at the University of Toulouse.

Over the last two years, Günter Helas, MPI Mainz, has put together efforts to provide better descriptions of BEVOC in Europe. At the workshop organised by BIATEX and the European Community, the state of affairs was evaluated. An analysis of European research in this area showed that extrapolation of US data would result in large errors and that emission descriptions must be parameterised and validated with European data. An example of studies to derive a parameterisation for isoprene emissions under European conditions is given in Fig. 1.3.

The best possible approaches for estimates of BEVOC in Europe were discussed at this workshop. This approach will be implemented by IFU (Garmisch, Germany) in close collaboration with the GENEMIS subproject and EMEP. The first estimates of BEVOC for Europe will be available at the end of 1995.

3. The uncertainties in the surface resistances for deposition of gases like SO_2, NH_3, NO_2 and NO and hence in the contribution of dry deposition processes to the loads of these pollutants in Europe has been reduced considerably due to the results obtained in BIATEX. But a warning is necessary: When BIATEX started, the uncertainties in the deposition loads of different compounds were, depending on compound and location, very large, even up to 200 to 300 %. These uncertainties have been reduced considerably but remain of the order of 50 to 100 % depending on compound and area. Abatement policies in Europe will need a further decrease in these uncertainties in order to obtain good cost/benefit analyses for steadily increasing expenditure in this area. This will require a considerable effort in fundamental mechanistic research, in monitoring of exchange fluxes and in the development of models.

4. The first results on dry deposition of particles indicate that the deposition velocity of particles is higher than expected. Research on deposition of particles was incorporated at a late stage in BIATEX, according to recommendations of the SSC. Measurements of deposition velocities for aerosols has been carried out at different locations using different techniques. The University of Toulouse has applied eddy correlation techniques in combination with particle counting, the same approach has been used by RISØ. ECN and TNO approached the problem by means of gradient measurements of sulfate in aerosol and ECN has studied the distribution of lead isotopes as these isotopes are deposited on particles and by way of the particles on vegetation. All these studies have indicated that the deposition velocity of particles over low vegetation as well as forests is much larger than formerly assumed on the basis of wind tunnel experiments, deposition modelling and the use of surrogate surfaces. Deposition velocities of 1 to 2 cm/sec have been observed over forests.

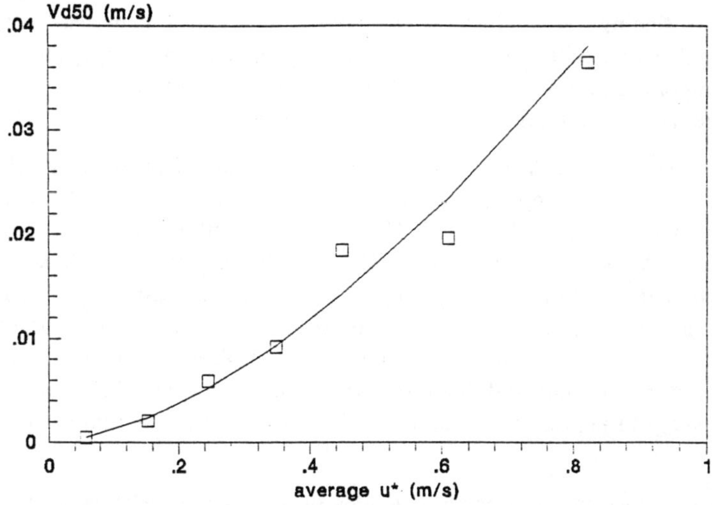

Fig. 1.4: Deposition velocity of sulfate aerosol measured in the Speulder forest, as a function of wind friction velocity, u_*.

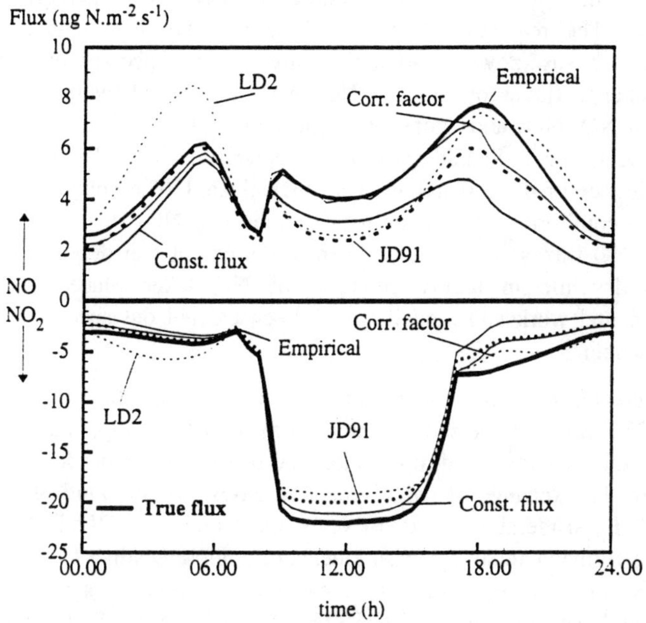

Fig. 1. 5: Influence of photochemical reactions on NO and NO_2 fluxes. Constant flux indicates the exchange fluxes without correction for reactions, true flux is the value after optimal correction for chemical reactions.

These findings (Fig. 1.4) are not only important for the assessment of exceedence of critical loads for different compounds in Europe, but play also an important role in estimates of life time and transport of aerosols and hence in the radiative forcing caused by small particles.

5. In the cadre of BIATEX, the first reliable estimates of exchange fluxes of reduced sulfur compounds, other than SO_2, have been made (Fig. 1.5). Exchange of COS, CS_2, DMS, H_2S and other sulfur compounds have been studied. After the initial phase of BIATEX the decision was taken that these compounds would not have the highest priority in the second part of BIATEX, but recent developments, for example, results of reduced sulfur emissions in Europe, have shown that the contribution of these compounds cannot be ignored in the future and that research on the exchange of these compounds is necessary to provide information on the sulfur balance in Europe in the near future.

6. The influence of atmospheric reactions on the exchange between atmosphere and biosphere of different compounds has been shown to be quite important. The influence of the decomposition of ammonium nitrate on the deposition fluxes of nitric acid is a good example, another is the interaction of photo-oxidants. The reactions of NO_x and ozone have been characterised and modelled. It was known for some time already that especially measurements of the exchange fluxes of NO and NO_2 were influenced by the rapid reaction between NO, NO_2 and ozone under influence of solar radiation. In several experiments, *e.g.* at Speulderbos in the Netherlands and in locations in the UK, the influence of these reactions were investigated. The conclusion is that the shifts in equilibrium do not influence ozone and NO_2 fluxes very much, but modify NO fluxes considerably, even to the extent that deposition fluxes are measured while in reality emission of NO takes place. Integration of experimental work and modelling has been carried out successfully to solve this problem.

7. Research on exchange of radiatively active gases has increased within BIATEX. Basically the same techniques as used for compounds important for acidification problems, eutrophication and increased oxidant levels are used to measure the exchange of radiative active gases. As this kind of research has expanded considerably in the last years, more and more BIATEX participants have performed exchange measurements of compounds like CO_2 and methane. The activities cover a wide range, *e.g.* methane emissions from wetlands in Scotland have been reported (BIATEX annual report 1993), but biological mechanistic research has been carried out as well (Fig. 1.6) to characterise the formation and oxidation of methane in soils.

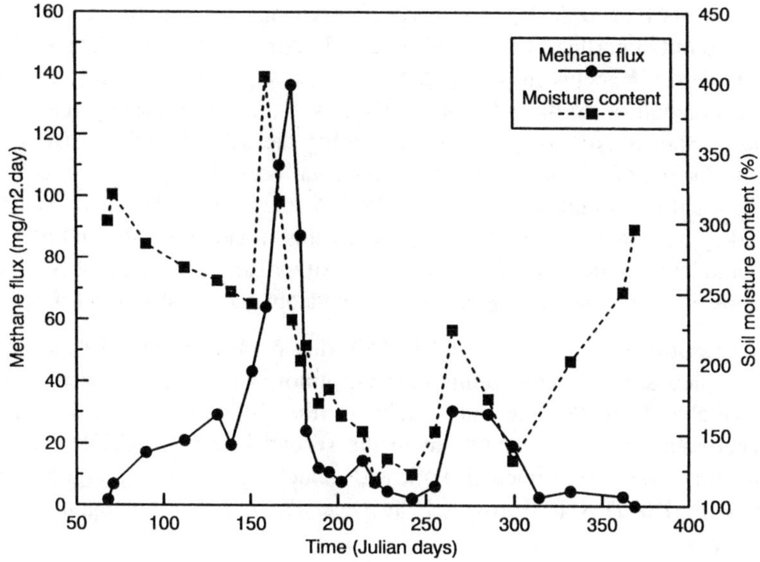

Fig. 1.6: Methane emissions as a function of soil humidity, determined in laboratory experiments.

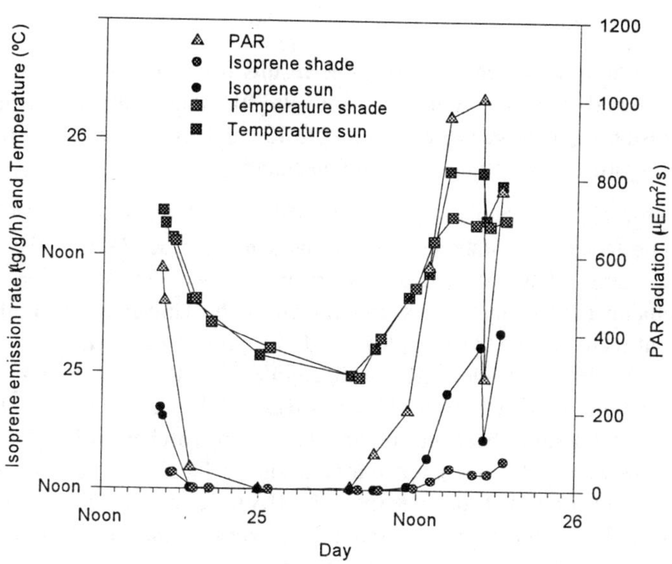

Fig. 1.7: Isoprene emissions from eucalyptus trees measured in Portugal.

8. At the start of BIATEX all work on exchange measurements was mainly carried out in northern Europe. European descriptions of the exchange of gases and aerosols between biosphere and atmosphere were all based on extrapolation of northern European data even though everybody realised that this extrapolation was questionable. During the second half of BIATEX the participation of southern European groups has been strengthened considerably. Important contributions from Italy (ISPRA and others), Portugal, Spain and Turkey have laid a proper basis to remedy this situation. By the end of 1995 it should be possible to verify a number of extrapolations made in assessments of depositions and biogenic emissions with data obtained in southern Europe.

9. The assumption at the start of BIATEX that modelling activities in different areas, deposition, biogenic emissions and atmospheric reactions, would provide a suitable focus to integrate results of the different joint and individual experiments has been confirmed in the second half of BIATEX. The close combination of experimental work and modelling that has emerged in many areas of BIATEX has provided the necessary means for integration of the results.

Why did BIATEX not succeed in some areas?

The BIATEX community made a proper diagnosis of a number of quite difficult problems in the initial phase of the subproject and formulated the research objectives accordingly.

Good examples are:

> To be able to draw general applicable results from exchange measurements at one time and at one location which are valid for other locations and under other circumstances, it is necessary to couple research on exchange of trace gases with investigations of the underlying mechanisms.

Specific biological mechanisms are mentioned in the description of BIATEX as of prime importance: this statement has proved to be true. During the course of BIATEX, attempts have been made to combine the two disciplines, for example in the systematic model approaches undertaken at the University of Frankfurt. As already outlined above, the complexity of the problem and the time needed to become acquainted with the different disciplines, together with a lack of suitable data, proved more of a barrier than foreseen at the start. On the other hand, effective collaboration between biologists and researchers who are active in exchange measurements has been initiated and is leading to new results. Good examples were provided during the EU/BIATEX workshop on biogenically emitted volatile organic carbon compounds, where biogenic emissions were linked to biological mechanisms as a first step in providing a more generalised approach to describe these emissions.

In other areas, for example on the influence of water layers on the deposition rates of SO_2 and NH_3, which is a simpler problem than some of the biological issues, acceptable generalised descriptions have been obtained which indeed are a great asset in generalising the results.

> The description of deposition in non-ideal conditions requires models of a total different character compared to the resistance modelling generally employed. It had been rightly pointed out at the start of BIATEX that large areas of Europe cannot be described in terms of ideal surfaces for classical micrometeorological approaches. For this reason emphasis was given to the development of 2-D and 3-D models for deposition. As already mentioned, the gap between the parameterisation required by modellers to describe these processes and available data sets, together with problems connected to finding compromises between fundamental meteorological modellers and the down-to-earth approach favoured by many scientists engaged in exchange measurements, brought these developments to a stand-still during the first half of BIATEX. On the other hand, the confrontation between modellers engaged in fundamental research in these areas, and those scientists who are responsible for providing estimates for deposition loads and biogenic emission fluxes, has made the position clear. Future work in this area now has a good basis and further possibilities can be envisaged.

These types of difficulties will be encountered in any scientific endeavour like BIATEX. The inclusion of ambitious goals in the initial phases should be defended even if it is generally accepted that there is a fair chance that they will not be attained.

Another problem was directly connected to the way EUROTRAC and consequently BIATEX was organised. From the beginning, the generalisation and application of the results from BIATEX in terms of improved descriptions for deposition and biogenic emissions had a high priority. The Steering Group had clear ideas on how to proceed in this matter, but it has been difficult for BIATEX participants, mainly financed by national organisations, to devote sufficient money and manpower to the project. During the last stages of BIATEX these problems were alleviated by important contributions from RIVM, supported by the Ministry of the Environment, VROM, in the Netherlands, and from MPI Mainz and IFU Garmisch-Partenkirchen, supported by the BMFT in Germany. An important conclusion of the Steering Group regarding this issue is that the proper means and pathways for generalisation and application must be in place at the start of any research; implementation of generalisation and application halfway through a project is a hazardous operation and can never operate optimally. The late start of this development blocked an extension to, for example, assessments of biogenic NO and reduced sulfur emissions. In all honesty, it must be said that less data were available in these areas than others, but a lack of communal funds was the main impediment.

Summa summarum, central funds for the generalisation and application of results, as well as very rigorous planning, are necessary in order to be able to transfer the results of scientific research into formats which are suitable for supporting the development of environmental policy.

Future developments in atmosphere/biosphere exchange

Research on atmosphere/biosphere exchange will be needed in the future to support the existing and future sulfur, nitrogen and oxidant protocols as well in the area of climate change. Important deficiencies exist in a number of areas:

- The problems regarding the description of deposition and biogenic emissions over non-ideal terrains have not been solved. It will be necessary to develop and validate new models to tackle this problem. Not only model development, but also validation of deposition over non-homogeneous terrains will require new approaches.
- Though good progress has been made in the last years in measuring deposition fluxes of aerosols, only very few reliable data are available at present. It will be necessary to expand the data base on deposition of aerosols considerably in the near future.
- Though participation of southern European groups has been considerably increased in the second half of BIATEX, insufficient data on atmosphere/biosphere exchange are available for Southern Europe. Much effort will be needed to acquire data for southern Europe of the same quality as are available for the northern part.
- Validation of deposition modelling especially for larger areas has not yet been carried out. The uncertainties in estimates of deposition and especially biogenic emissions are still very large (50 to 100 %) and will remain so until methods are developed to verify exchange fluxes over large areas.
- As abatement policy is implemented in Europe sources will become important which have more or less been ignored up to now. Examples are biogenic NO and reduced sulfur emissions. Careful examination is needed to map future needs as a function of changing emission patterns in Europe.
- Due to the abatement policy already mentioned, the contribution of compounds like nitric acid, nitrous acid, organic acids *etc.* in acidification and eutrophication will become much more important in the near future. Study of the exchange of these compounds will be necessary as little reliable data are available at present.

Chapter 2

Biosphere - Atmosphere Exchange of Ammonia

M.A. Sutton[1], G.P. Wyers[2], F.X. Meixner[3], J.K. Schjørring[4], J. Kesselmeier[3], G. Kramm[5] and J.H. Duyzer[6]

[1] Institute of Terrestrial Ecology (Edinburgh Research Station) Penicuik, Midlothian, EH26 0QB, UK
[2] Netherlands Energy Foundation (ECN), 1755-ZG Petten, the Netherlands
[3] Dept. Biogeochemistry, Max Planck Institute for Chemistry, Postfach 3060, 55020 Mainz, Germany
[4] Centre for Ecology & Environment, Royal Veterinary and Agricultural University, Copenhagen, Denmark
[5] Fraunhofer-Institut für Atmosphärische Umweltforschung (IFU), D-82467, Garmisch-Partenkirchen, Germany
[6] TNO Institute of Environmental Sciences, P.O. Box 6011, 2600 JA Delft, the Netherlands.

Summary

Substantial progress has been made in the last eight years in the understanding and quantification of ammonia exchange between the atmosphere and biosphere. Much of the work has been linked to the joint EC/EUROTRAC subproject BIATEX (BIosphere ATmosphere EXchange), which has served as the main European forum for work in this area. In the mid-1980s there was still much confusion and uncertainty over the rate and direction of ammonia fluxes with different ecosystems; although the results of isolated studies were available, there was no clear overview of the key factors affecting ammonia fluxes. Work since that time has highlighted the dominant effects of ecosystem type and management, as well as humidity and wetness, on ammonia exchange. Ammonia is a key component of plant metabolism, so that ammonia emission may occur from plants in relation to nitrogen nutrition and plant growth stage. In contrast, ammonia is highly soluble and may be efficiently captured by leaf cuticles and surface wetness allowing large deposition velocities. The consequence is that ammonia exchange is bi-directional over agricultural ecosystems, though for most semi-natural ecosystems dry deposition dominates, being a significant component of the total atmospheric nitrogen input. The work within BIATEX has focused in more detail on the processes controlling these differences and, using the results of both micrometeorological and controlled environment measurements, has developed new models that are able to provide the synthesis necessary to predict ammonia fluxes. Long term and regional estimates of ammonia net exchange are still uncertain, though the models developed now provide the necessary framework to guide future measurements.

2.1 Introduction

The exchange of ammonia between the atmosphere and plant communities has been a subject of interest for over a century. However, up until very recently, still very little was known about the magnitude or even the direction of ammonia fluxes with most ecosystem types. Interest was started during the last century as part of the debate of agricultural science on the source of nitrogen for plant nutrition. In these early studies, most attention was given to the contribution of ammonium (NH_4^+) inputs in wet deposition (*e.g.* Way 1855). However, it was also shown qualitatively that gaseous ammonia (NH_3) in the atmosphere could be absorbed by plant leaves (Ville 1850) - what we would now call dry deposition. Hall and Miller (1911) provided the first measurements of vertical concentration gradients of NH_3 above differently managed land (see Sutton *et al.* 1993e), which gave early evidence of the differences in NH_3 exchange between ecosystems, allowing both emission and dry deposition.

Interest in NH_3 in the atmosphere continued this century, with an appreciation of its importance as the major gaseous base in atmospheric chemistry (*e.g.* Junge and Ryan 1958). However, it has not been until the more recent interest in NH_3 as an atmospheric pollutant, that substantial progress has been made in understanding and quantifying NH_3 exchange with different components of the biosphere. Concern about ammonia as a pollutant started to grow in the mid 1980s as it was realized that both wet and dry deposition of reduced nitrogen (NH_x) contribute to ecosystem acidification, alongside deposition of SO_x and NO_y from SO_2 and NO_x emissions. Although NH_3 is a base in the atmosphere, once deposited, it may be oxidized (nitrified) by soil microflora to form NO_3^- liberating up to 1 H^+ per NH_3; where neutralization of atmospheric acids to form NH_4^+ has already taken place, inputs provide up to 2 H^+ per NH_4^+ (Gundersen and Rasmussen 1988; Sutton *et al.* 1993e). In addition to the acidification concerns, the old issue of atmospheric NH_x as a source of nitrogen to plants has returned, as inputs (along with deposited NO_y) may be substantial and cause significant changes in the structure of semi-natural plant communities. Ecological impacts include loss of semi-natural grass and heathlands, as well as decreases in forest vitality (Bobbink *et al.* 1992a; INDITE 1994).

The environmental concern about the impacts of ammonia highlighted the great lack of knowledge available to quantify inputs of NH_x to ecosystems. In non-complex terrain, quantification of wet deposition was reasonably sound, while for most ecosystems, inputs of ammonium in submicron aerosols were acknowledged to be small (Fowler *et al.* 1991). In contrast, little information was available to quantify the gaseous exchange (dry deposition, emission) of NH_3 with land surfaces. The national and international research programmes, including that of

BIATEX, that have been initiated in the past 10 years have provided much information towards remedying this situation, although it is recognized that there are still major uncertainties. The aim of the recent work has been closely linked to science policy on the impacts of regional air pollutants. With the rise of the 'critical loads' approach, which defines maximum no-effect deposition thresholds for sensitive ecosystems, it has become increasingly important to be able to quantify the nitrogen deposition to these different ecosystems (Grennfelt and Thörnelöf 1992; Hornung *et al.* 1995). Within the UN-ECE Convention on Long Range Transboundary Air Pollution (UN-ECE 1994), the critical loads approach has already been applied within the Second Sulfur Protocol (June 1994), and work is now underway for the Second NO_x (or nitrogen) Protocol to consider both NO_x and NH_3 using the critical loads approach. It is evident that work such as this needs to be underpinned by sound estimates of the deposition of ammonia.

2.2 BIATEX research on ammonia

Within the framework of the EC/EUROTRAC subproject BIATEX, the importance of quantifying inputs of ammonia to ecosystems was appreciated at an early stage (*e.g.* Duyzer *et al.* 1987; Sutton *et al.* 1989), and work has sought to provide the necessary scientific basis for such estimates. In particular, BIATEX has provided the main European forum for research groups in this area, which has enabled a much wider perspective to be taken than from individual national research projects.

One of the main limitations to understanding ammonia exchange processes in the past has been the lack of instrumentation capable of measuring NH_3 concentrations with sufficient precision and time resolution necessary for quantifying fluxes, and work in BIATEX has provided key developments here. Throughout the work of BIATEX the mainstay of research on NH_3 has been the application of micrometeorological techniques micrometeorological to quantify surface fluxes. These are particularly attractive since they quantify fluxes over the field scale, do not disturb the plant environment (*c.f.* chambers) and provide information for developing process based models to predict fluxes (Fowler and Duyzer 1989). Although the ecological concern is specifically focused on semi-natural ecosystems, work over agricultural ecosystems has also been an important element. This is relevant, since the models for describing the atmospheric transport and dispersion of pollutants (*e.g.* Sandnes and Styve 1992, Asman and van Jaarsveld 1992) need to integrate surface exchange over all land surface types. In addition, fluxes over agricultural crops are of interest in their own right as a gain or loss of N for farming systems (Schjørring 1991). In addition to micrometeorological measurements, controlled environment and chamber measurements of NH_3 fluxes have been an important contribution to the research, especially as they provide the only simple method to distinguish the relative contribution of soil and vegetation to net fluxes above the canopy. Furthermore, they provide useful means to identify in more detail the plant physiological processes processes controlling ammonia fluxes. The final part of the work has been the development of process based models of ammonia exchange, which may be used to predict fluxes with the surface. Although these are still extremely uncertain, substantial progress has been made within BIATEX towards a more general understanding and quantification of ammonia exchange. Each of these areas are considered in the following sections, with examples of some of the key research results, with the final section identifying the present achievements and uncertainties.

2.3 Development of instruments to measure exchange of ammonia

A great deal of progress has been made in ammonia measurement techniques recently, particularly within the frame of BIATEX. Ammonia has always been considered a difficult gas to measure, both because of its low Henry's coefficient (high solubility), resulting in potential problems with adsorption on inlet lines, and because it is very easy to contaminate atmospheric ammonia samples by contact with the skin, breath and ammonia in laboratory air. The main difficulty, however, is the need to separate gaseous NH_3 from aerosol NH_4^+. Up until the beginning of BIATEX, the main available techniques were simple batch methods relying on either multi-stage filter packs (Harrison *et al.* 1989; Sutton *et al.* 1993b) or simple and annular batch denuders (Ferm 1979; Allegrini *et al.* 1984). These methods were extremely labour intensive and limited to sampling periods of generally 1 hour or more.

New techniques developed at the beginning of BIATEX have included automated rotating batch wet denuders (Keuken *et al.* 1988), as well as ammonia thermodenuders (Keuken *et al.* 1989; Meixner *et al.* 1993), both typically run with 30–60 minutes time resolution. The denuder principle is attractive in that it relies on the rapid diffusion of the small NH_3 molecule to a reactive surface, while (under the condition of laminar flow) the submicron aerosol containing NH_4^+ passes through uncollected. More recently there has been an increased move towards continuous ammonia sampling techniques with the sensitivity required for atmospheric measurements. Within BIATEX a 'mist scrubber' system has also been used which collects by scavenging the ammonia in sampled air with a fine mist and analyzing the liquid. The system performed well in comparison with denuders and differential absorption spectrosocopy (DOAS), during a methods intercomparison at Rome in 1990 (Ernst 1990). Outside the frame of BIATEX progress has been made in the development of membrane diffusion scrubbers, which collect a fraction of the ammonia sampled by passing an acidic solution through a tube with semi-permeable membrane, into which the NH_3 diffuses (*e.g.* Dasgupta 1993). At the same time work within BIATEX has progressed rapidly on advancing the wet rotating denuders of Keuken *et al.* (1988) to measure NH_3 continuously (Wyers *et al.* 1993b). In this system, acid sampling solution is cycled continuously through the wet rotating annular denuder and analyzed on line as NH_4^+ in solution by a sensitive conductivity system. The advantage of this method is that the denuder collects all the NH_3 entering the system, avoiding the need to correct results for the fraction collected, while it is also extremely sensitive, with a detection limit of 0.02 μg m^{-3} and two minute time resolution. The availability of these new methods has allowed great advances in the measurement of ammonia biosphere-atmosphere exchange and many of the measurements made here represent the first application of these techniques.

2.4 Field measurements of ammonia exchange over different surfaces

2.4.1 Exchange of ammonia over semi-natural land - initial results

The first micrometeorological measurements of ammonia fluxes over semi-natural land were those of Duyzer *et al.* (1987), who measured exchange over dry heathlands in the Netherlands. The overall impression of this work as well as measurements of Sutton *et al.* (1989, 1993b), who measured fluxes over upland moorland, neutral and calcareous grasslands, was that gaseous ammonia is almost exclusively deposited to the land surface, and at rates close to the maximum permitted by atmospheric turbulence. This was explained on the basis of ammonia being an extremely soluble/reactive gas, with much of the deposition occurring on to leaf surfaces (cuticles) rather than being restricted to uptake through stomata.

These findings were derived on the basis of interpreting the measured fluxes (F_g) in relation to deposition velocities (V_d) and the resistance model of turbulent exchange often used in micrometeorology (*e.g.* Fowler and Duzyer 1989; Sutton *et al.* 1993b; Müller *et al.* 1993). Analogy is made with Ohms law, and it is assumed that the air concentration (χ) at the surface is zero, such that:

$$R_t\{z\} = [\chi_{surface} - \chi\{z\}]/F_g = -\chi\{z\}/F_g \qquad (1)$$

$$1/V_d\{z\} = R_t\{z\} = R_a\{z\} + R_b + R_c \qquad (2)$$

where z is a reference height above the surface for atmospheric terms, and the resistances are: $R_t\{z\}$, total resistance; $R_a\{z\}$, turbulent atmospheric resistance; R_b, quasi-laminar sublayer resistance; R_c, canopy resistance. The value of $1/(R_a + R_b)$ defines the maximum possible deposition velocity permissible by turbulence (V_{max}), and this is found from micrometeorological measurements. For ammonia, the flux is generally determined by vertical concentration profiles above a large uniform surface with associated turbulence measurements, which allows V_d to be calculated, and hence R_c found by difference. The measurements of Duyzer *et al.* (1987) and Sutton *et al.* (1989; 1993b) provided average estimates of R_c typically of 0–40 s m^{-1}, and the latter found a mean R_c for moorland of 5 s m^{-1}. These resistances are extremely small compared with resistances for uptake through stomata (R_s), which are typically 50–200 s m^{-1} for growing vegetation in the day, and much larger (>1000 s m^{-1}) at night. Thus these data indicate that leaf cuticles or deposits on them must provide the main site of uptake.

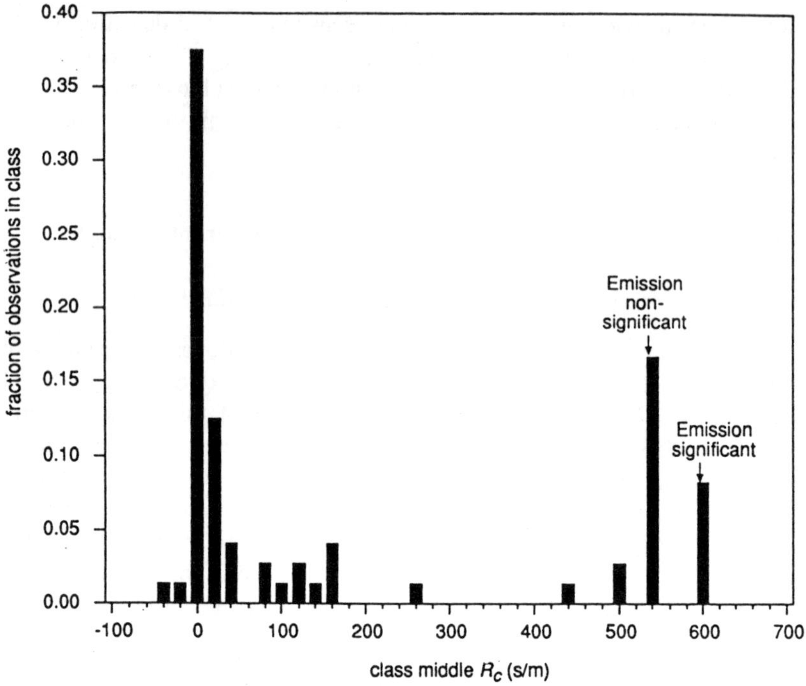

Fig. 2.1: Frequency distribution of observed canopy resistances (R_c) for NH_3 over a Douglas fir forest in the Netherlands. R_c is notionally > ∞ where emission occurs; significant and non-significant emission fluxes are shown here for convenience at 600 and 550 s m^{-1} (Duyzer *et al.* 1994).

An exception to this general pattern was found by Sutton *et al.* (1993b) for harrowed calcareous grassland, which showed a mean R_c > 100 s m^{-1}. They suggested that this was related to the disturbed basic soil (pH 8.4) being present on leaf surfaces, which would reduce NH_3 solubility and restrict uptake. A further exception is for frozen conditions and snow surfaces where R_c of 100–200 s m^{-1} is typical as adsorption is inefficient (Sutton *et al.* 1993b; Erisman and Wyers 1993). Nevertheless the general pattern of rapid deposition to semi-natural and forest vegetation in northern European conditions was supported by further measurements. For example, measurements of Andersen *et al.* (1993), Wyers *et al.* (1993a) and Duyzer *et al.* (1992, 1994) have each shown rapid rates of dry deposition to forest. The results of Duyzer *et al.* (1994) over Douglas fir in the Netherlands are presented in Fig. 2.1 to show this effect, with the most frequent value of R_c in the range –10 to + 10 s m^{-1}. In principle, a negative value of R_c is not possible (deposition faster than turbulence), though such values may occur because of random error in the results. However, in contrast to the measurements over moorlands and semi-natural grasslands noted above, these measurements showed a significant number of ammonia emission periods, generally when the air concentration was small (for this Dutch location) at

< 5 μg m^{-3}. Observations of this kind have been shown in a number of other recent measurements and this is discussed further in the following sections.

2.4.2 Exchange of ammonia over agricultural croplands

Micrometeorological measurements of ammonia fluxes have been made over croplands by a number of authors both within and outside BIATEX. Many of the earlier measurements focused on the emissions of ammonia from nitrogen fertilizers. Such measurements have often shown significant losses of nitrogen, particularly from urea (Fenn and Hossner 1985) (*e.g.* 10–20 % loss of fertilizer N), although much smaller emissions have been recorded for other fertilizers such as ammonium nitrate (Sutton *et al.* 1995a). Work in a dry continental area of north America (Harper *et al.* 1987) found larger rates of emission from wheat plants fertilized with ammonium nitrate in the weeks following application fertilization, (7 kg N ha^{-1}), however this emission was considered to have been mediated by changing the nitrogen metabolism of the plants so that emissions were larger. Prior to anthesis during a period when soil N was temporarily unavailable, NH_3 deposition was observed, while later on during senescence and ripening of the wheat emission was again observed. In the case of emission from senescing leaves the larger emission was related to the remobilization of nitrogen from leaves to allow transport to the grain.

The work within BIATEX has investigated both these broad scale seasonal / plant growth stage effects, and also investigated in more detail the processes regulating ammonia exchange in relation to environmental conditions and diurnal cycles. A case study of this may be provided by the measurements of Meixner *et al.* (1991, 1995) over a wheat crop at Manndorf in southern Germany. It may be noted that 30 % of the agricultural land in the mid European landscape consists of wheat fields (FAO 1985), so estimates of exchange over this crop are essential to quantifying net regional fluxes. Ammonia concentrations were determined using the thermodenuder system developed by Keuken *et al.* (1989) and fluxes determined by both the aerodynamic gradient and Bowen ratio methods. The measurements results were subjected to rigorous data screening criteria to ensure that the necessary micrometeorological and measurement conditions were met (Meixner 1994). A summary of the average diurnal course of the ammonia flux for the period 14–19 July 1990 from the screened measurements is shown in Fig. 2.2, together with the air concentrations for the period. The results show that during the entire period there is a strong diurnal cycle to the exchange process, with emission fluxes (up to +30 ng N m^{-2} s^{-1}) occurring in the day, and deposition fluxes (to –30 ng N m^{-2} s^{-1}) during night and early morning. It may be noted that the emission flux is largest during the late morning when the plants are most active and stomatal resistances the smallest, indicating the plant physiological control of the exchange process. As a result the flux shows a clear correlation with water vapour fluxes during daytime periods (Meixner *et al.* 1991). In addition to this, parallel measurement of CO_2 fluxes showed an anti-correlation between CO_2 and

NH_3 fluxes (Fig. 2.3). The largest CO_2 uptake at high light intensities around mid-day suggests a high stomatal conductance and, possibly a high leaf temperature, which would favour NH_3 emission. Links between ammonia fluxes and a range of plant physiological processes have been discussed by Schjørring (1991), Sutton *et al.* (1994; 1995b) and Meixner *et al.* (1995). For example, measurements of Schjørring *et al.* (1993a; b) have shown a strong link between ammonia emission fluxes and mobilization of leaf enzymes associated with protein breakdown for senescing wheat.

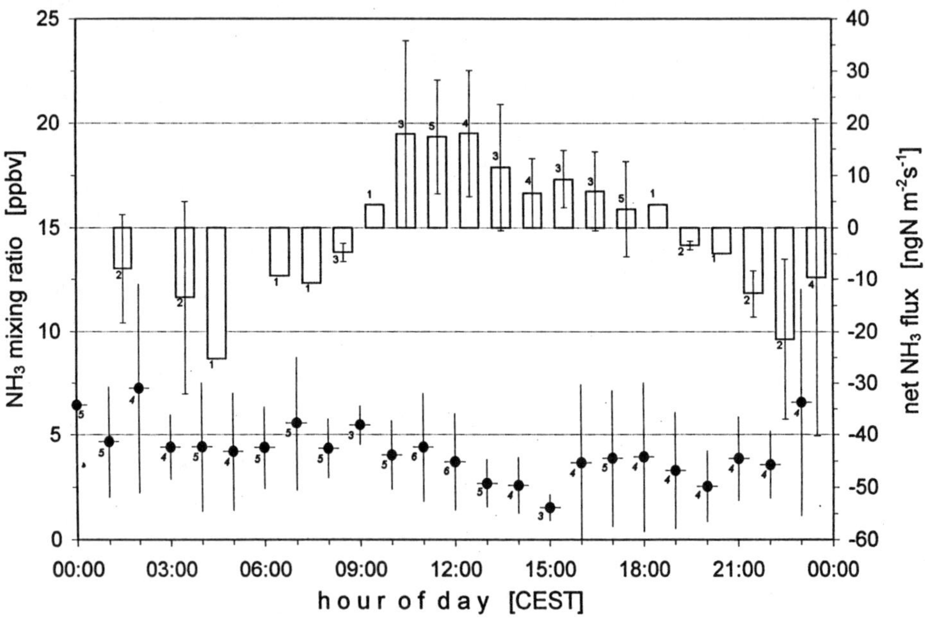

Fig. 2.2: Summary of the diurnal cycle of ammonia fluxes over a senescing (ripening) wheat crop measured during the Manndorf experiment, 14–19 July, 1990 (Meixner *et al.* 1995). Vertical bars show the average flux (positive values denote emission), while solid points show the air concentrations. Both are shown with the standard deviation for the number of days where measurements met the objective screening criteria.

A brief comment is necessary with regard to the micrometeorological requirements for the measurements reported in Fig. 2.2, as well as in the other diurnal courses of NH_3 fluxes (*e.g.* Figs. 2.3, 2.4, 2.6 and 2.7). The approach used is dependent on stationary (or little changing) conditions in the boundary layer, so that measurements around daybreak or dusk may be subject to storage errors in the flux. Such errors have been discussed in detail by Sutton *et al.* (1993a) and Meixner (1994), and in some cases such measurements may have to be excluded from a detailed resistance analysis. However, storage errors are largest for slowly exchanging gases and, as the absolute value of F_g/χ is generally large for NH_3, these errors are often trivial for measurements over short vegetation.

Similar bi-directional fluxes of ammonia over agricultural croplands have been observed by Sutton *et al.* (1993a; c; 1995a). Again diurnal cycles of ammonia fluxes were observed, though in those studies more emphasis was placed on the role of surface wetness in controlling the direction of ammonia flux. Thus these authors generally found deposition when the canopy was wet, and emission when the plants were dry. A helpful complement to these short-term detailed measurements has been the development of a new approach for providing long-term passive sampling of ammonia fluxes over cropland (Schjørring *et al.* 1992; Schjørring 1995). This technique is potentially extremely useful for estimating the net exchange over entire growing seasons of crops, and further work is needed to provide a verification of the approach by comparison with classical methods.

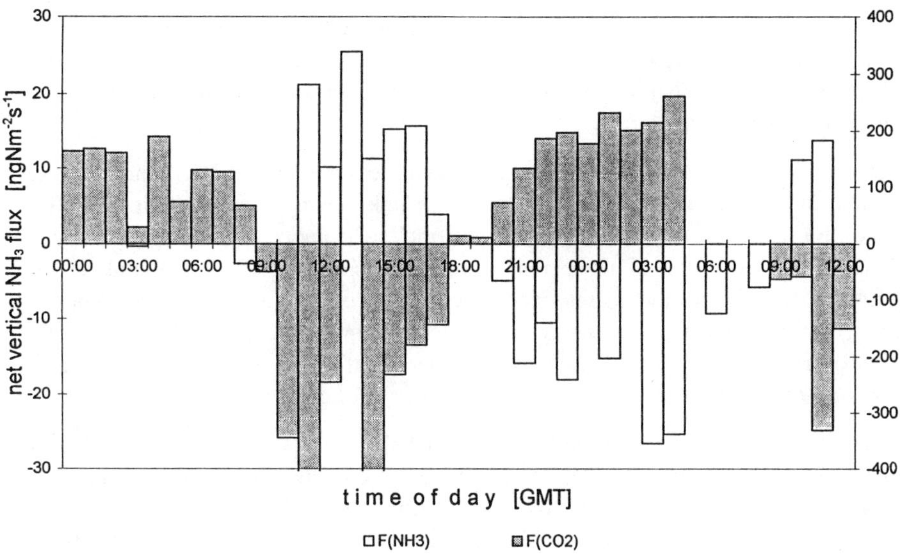

Fig. 2.3: Example of NH_3 and CO_2 fluxes over a ripening wheat canopy measured during the Manndorf experiment, 18–19 July, 1990. The results show a clear anti-correlation between the diurnal exchange cycles of these gases.

Comparing the exchange of ammonia between agricultural and semi-natural (unfertilized) vegetation, it seems clear that the reason for the differences is the altered biology of the plants in relation to nitrogen fertilization and species type. It is well known that fertilized agricultural plants generally have higher foliar N contents than unfertilized plants (Sutton *et al.* 1993e), and this difference is expected to be reflected in ammonium production and consumption in the plants. A key term that has often been used to describe the exchange of ammonia with leaf tissues is the ammonia 'compensation point' concentration (Farquhar *et al.* 1980; Lemon and van Houte 1980). This is the air concentration of NH_3 at equilibrium with ammonium in the leaf tissues, according to Henry's law, so that when air concentrations are larger than the compensation point (χ_{cp}), deposition is expected, and when they are smaller, emission is expected. The term is used since it reflects the compensation between ammonium consumption and production processes in the leaf tissues. It might then be expected that the fertilized agricultural plants have larger compensation points than the semi-natural plants, allowing for increased potential for emission. Since ammonia is a soluble gas, there is competition for deposition to the leaf cuticles, so that where the compensation point is small, net deposition dominates. Ammonia compensation point concentrations have been estimated by a number of authors (see sections 5 and 6). As an example, Meixner *et al.* (1995) estimated a net NH_3 compensation point of 2.7 $\mu g\ m^{-3}$ from their measurements over wheat.

2.4.3 Interactions of ammonia exchange with other acid gases

The possibility that ammonia fluxes may interact with concentrations, or parallel fluxes of other acid gases, such as SO_2, HNO_3 and HCl, is of substantial importance when quantifying budgets of the transport and deposition of the range of acidifying pollutants.

During the 1980s substantial information from measurements of NH_4^+ and SO_4^{2-} in forest throughfall suggested that there was a link between the deposition of NH_3 and SO_2, often referred to as 'co-deposition'. A number of authors found 1:1 equivalent deposition of these ions (van Breemen *et al.* 1982; Bobbink *et al.* 1992b), though in other cases excess deposition of NH_3 was found to enhance SO_4^{2-} inputs, but not necessarily at equivalent deposition rates (Draaijers 1993, Sutton *et al.* 1993d). The consequence of this effect would be to reduce the extent of long-range transport of these pollutants, as rates of dry deposition increase.

Micrometeorological measurements within BIATEX have addressed this question from a more mechanistic standpoint, by providing the first simultaneous flux measurements of NH_3 and SO_2 over both semi-natural land (Erisman and Wyers 1993) and agricultural croplands (Sutton *et al.* 1993a; 1995b). Erisman and Wyers provided an extended series of flux measurements over several months, but did not find a 1:1 equivalent deposition link as had been observed in the throughfall studies, although they suggested that at air humidity > 70–80 % a large air concentration of SO_2 could reduce the value of R_c for NH_3 and *vice versa*. The measurements of Sutton *et al.* (1993a, 1994) also showed that over agricultural cropland the situation is more complicated, as bi-directional ammonia fluxes are usual and enhanced SO_2 concentrations may (at least indirectly) result in increased NH_3 emission. An example of the measurements reported by Sutton *et al.* (1995b) is shown in Fig. 2.4. Rather than SO_2 providing an enhanced deposition of NH_3, the opposite is seen, as periods of very high SO_2 concentration are coupled to brief periods of NH_3 emission. The explanation is probably that elevated SO_2 concentrations are associated with formation of acidic SO_4^{2-} aerosol which react to deplete the NH_3 concentration. At concentrations of NH_3 smaller than the compensation point for this agricultural canopy, ammonia emission is then possible. From these different studies it appears therefore that SO_2 may either enhance NH_3 deposition (uptake by leaf cuticles) or enhance NH_4^+ aerosol formation and NH_3 emission from stomata. In agricultural areas it is therefore possible that the presence of SO_2 increases the mean transport distance of NH_3 (NH_3 emission and NH_4^+ formation are favoured, limiting removal by dry deposition), whereas NH_3 transport distances may be reduced by SO_2 in areas dominated by unfertilized ecosystems (rates of dry NH_3 deposition enhanced). The exact balance between these effects is still to be quantified.

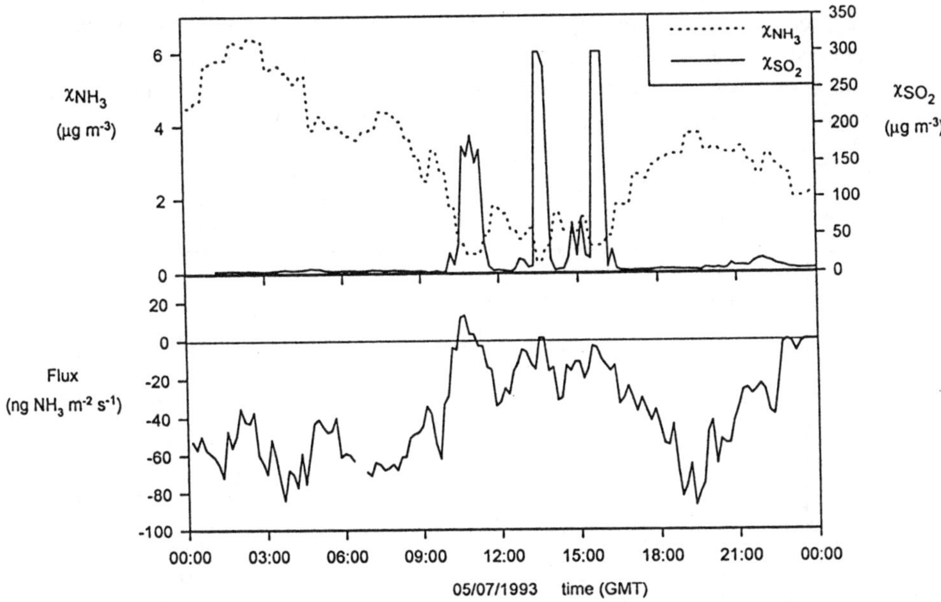

Fig. 2.4: Ammonia fluxes over a ripening wheat crop at Sutton Bonington, central England in relation to air concentrations of NH_3 and SO_2. Elevated concentrations of SO_2 are coupled with smaller concentrations of NH_3 and periods of NH_3 emission (Sutton *et al.* 1995b).

A further interaction between ammonia and other acid gases is the potential for reaction with HNO_3 and HCl to form ammonium nitrate and chloride respectively. It is well established that the atmospheric equilibrium between these species is governed by air concentrations, temperature and humidity (Stelson and Seinfeld 1982). However, there is also debate as to the rate of these reactions and the extent to which rapid reaction could modify the calculation and magnitude of surface fluxes of each of the species (Brost *et al.* 1988; Harrison *et al.* 1989; Pandis and Seinfeld 1990; Sutton *et al.* 1993e; Kramm and Dlugi 1994). The classical micrometeorological approach used to estimate surface fluxes assumes that the flux is constant with height in the measurement layer above the surface, so that any production or consumption provides an error in interpreting measurements. Secondly, should the rate of reaction be rapid, this may alter deposition budgets of net exchange, as well as atmospheric transport.

A substantial amount of modelling work within BIATEX has addressed this question, both from an analytical framework relevant to other gases, such as nitrogen oxides and ozone (Kramm *et al.* 1991; Duyzer 1992), and more specifically for the reaction of ammonia with nitric acid (Kramm and Dlugi 1994; Kramm 1995; Zhang *et al.* 1995). An example from the work of Kramm *et al.* (1993) is shown in Fig. 2.5 to illustrate this effect. Apart from the need to develop further the basis of the extensive calculation methods needed to treat this effect, the key uncertainty is the magnitude of the reaction rate which defines whether these effects are significant. With a reaction rate of 10 ppm^{-1} min^{-1}, the effects are substantial, as can be seen in the example of Fig. 2.5 by the apparent 'emission' gradients of NH_4NO_3, which are a consequence of aerosol formation, and from the variation in fluxes with height.

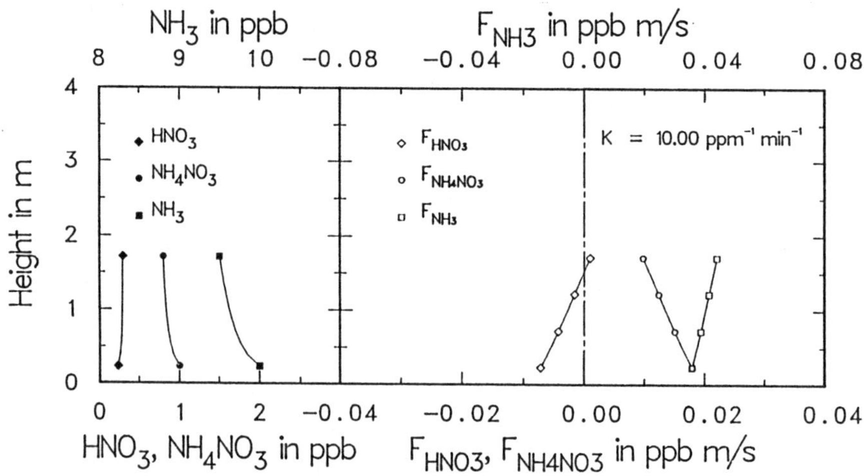

Fig. 2.5: Example of the effect of atmospheric reaction of HNO_3, NH_3 and NH_4NO_3 on the vertical profiles of concentrations and turbulent fluxes (from Kramm *et al.* 1993). Where reaction rates are negligible, fluxes (F) are conserved and constant with height above the surface. In this model example the reaction rate (K) for formation of NH_4NO_3 is set at 10 ppm^{-1} min^{-1} which provides a significant divergence of the fluxes.

The magnitude of the effect is also a consequence of the degree of perturbation of the equilibrium, so that near sources of ammonia (*e.g.* > 50 $\mu g\ m^{-3}$) there may be a large effect on HNO_3 gradients. From more typical background exchange measurements Harrison *et al.* (1989) concluded that the reaction rate was too slow to significantly affect fluxes, although a re-analysis of their measurements (Sutton *et al.* 1993e) showed a significant interaction between measured R_c (HNO_3, HCl) and ammonia fluxes, which was probably due to this effect. Recently Zhang *et al.* (1995) have provided an analysis of measured NH_3 and HNO_3 fluxes in the Netherlands. They observed apparent 'emission' of HNO_3, which they suggested was due to dissociation of NH_4NO_3 aerosol during warm daytime conditions, and estimated a first order correction to the NH_3 fluxes. It is evident that further work is required in this area to establish the extent to which classical micrometeorological estimates need to be corrected.

2.4.4 Evidence of ammonia emission from European semi-natural ecosystems.

As noted in section 2.4.1, the initial flux measurements over semi-natural ecosystems indicated rapid ammonia deposition, with only slim evidence of periods of ammonia emission. In general this picture is supported by the more recent measurements. However, there is now an increasing body of data collected within BIATEX to suggest that this distinction is not so sharp, and that ammonia emission over semi-natural land is also bi-directional, but with an increased tendency to deposition compared with agricultural land. The first evidence for this came from heathlands in the Netherlands, subject to intense atmospheric concentrations of ammonia (Sutton *et al.* 1992a; Erisman and Wyers 1993). Sutton *et al.* (1992a) postulated that a possible reason for this could be that the biology of these systems may be changed when subjected to such elevated concentrations of ammonia, resulting in an enhanced compensation point and ammonia emission. In this context, it should be noted that the measurements of Sutton *et al.* were made over a heather canopy in a state of decline relative to competing purple moor grass, and with a nitrogen content between 1.6–2.3 %, which is far in excess of typical values 0.8–1.5 % in cleaner areas (Sutton *et al.* 1993e).

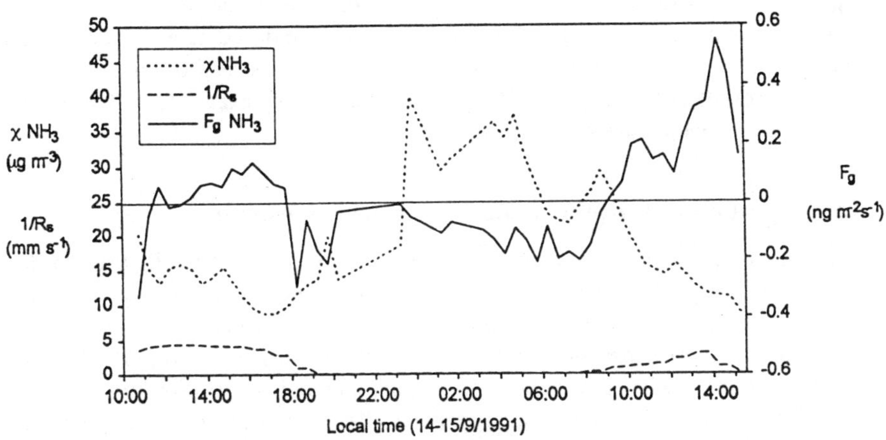

Fig. 2.6: Ammonia concentrations and fluxes over heathland in the Netherlands (Elspeet Veld). The area is subject to very large air concentrations of ammonia and shows significant periods of ammonia emission from the heathland during the day for this period.

An example of the measurements made using the sampling method of Wyers *et al.* (1993b) is shown in Fig. 2.6 for the Elspeet heathland in the Netherlands. As with the measurements over agricultural croplands, the emission fluxes occur during the daytime when stomata are open ($1/R_s$ large) and when the crop is typically subjected to smaller relative humidities.

Further evidence of ammonia emission from semi-natural vegetation has since come from acidic upland moorland systems. Using batch filter pack sampling, with 2 hour time resolution, earlier work of Sutton *et al.* (1993b) found rapid deposition with mean $R_c = 5$ s m^{-1}. However, using the more recent continuous sampling systems at the same site, a more detailed picture may be seen, as exemplified in Fig. 2.7. Overall the flux is still dominated by deposition, with a mean R_c (calculated from mean V_d and V_{max}) of 27 s m^{-1} for this period. Nevertheless, brief periods of emission are observed.

A possible constraint on this interpretation of the observed NH_3 emission from both semi-natural and agricultural ecosystems is the existence of rapid reactions between gaseous and particulate phases. For example, in warm dry conditions over vegetation under drought conditions a decrease in relative humidity near the surface may lead to evaporation of ammonium nitrate aerosol providing apparent ammonia emission (section 2.4.3). Nevertheless, it is clear from micrometeorological measurements where aerosol NH_4^+ gradients are insignificant, and from laboratory studies, that NH_3 emissions do occur in many situations.

The existence of a possible feed back between long term N deposition and the direction of the ammonia flux has important consequences for estimating pollutant inputs to semi-natural ecosystems, and knock on effects for the comparison of deposition with critical loads. For example, where reduced rates of deposition (and periods of emission) in the NH_3 polluted areas are observed, this may be an indication of altered biology with the critical load already exceeded. Clearly, this is an aspect which requires further investigation.

2.5 Laboratory and cuvette studies of ammonia exchange processes.

Controlled environment measurements provide a useful complement to the micrometeorological approach and a number of studies have been made within BIATEX to apply these to investigate the surface exchange of ammonia. An important series of experiments were conducted by van Hove *et al.* (1987, 1989, 1990). In these studies the authors found that ammonia deposition to plants was generally limited by uptake through stomata. They showed that adsorption onto leaf surfaces could occur, and also be enhanced in the presence of SO_2, but that the total contribution of leaf surface adsorption was very small compared with stomatal uptake. This is very different to the efficient uptake by leaf cuticles observed by the field measurement over semi-natural land, and is probably due to

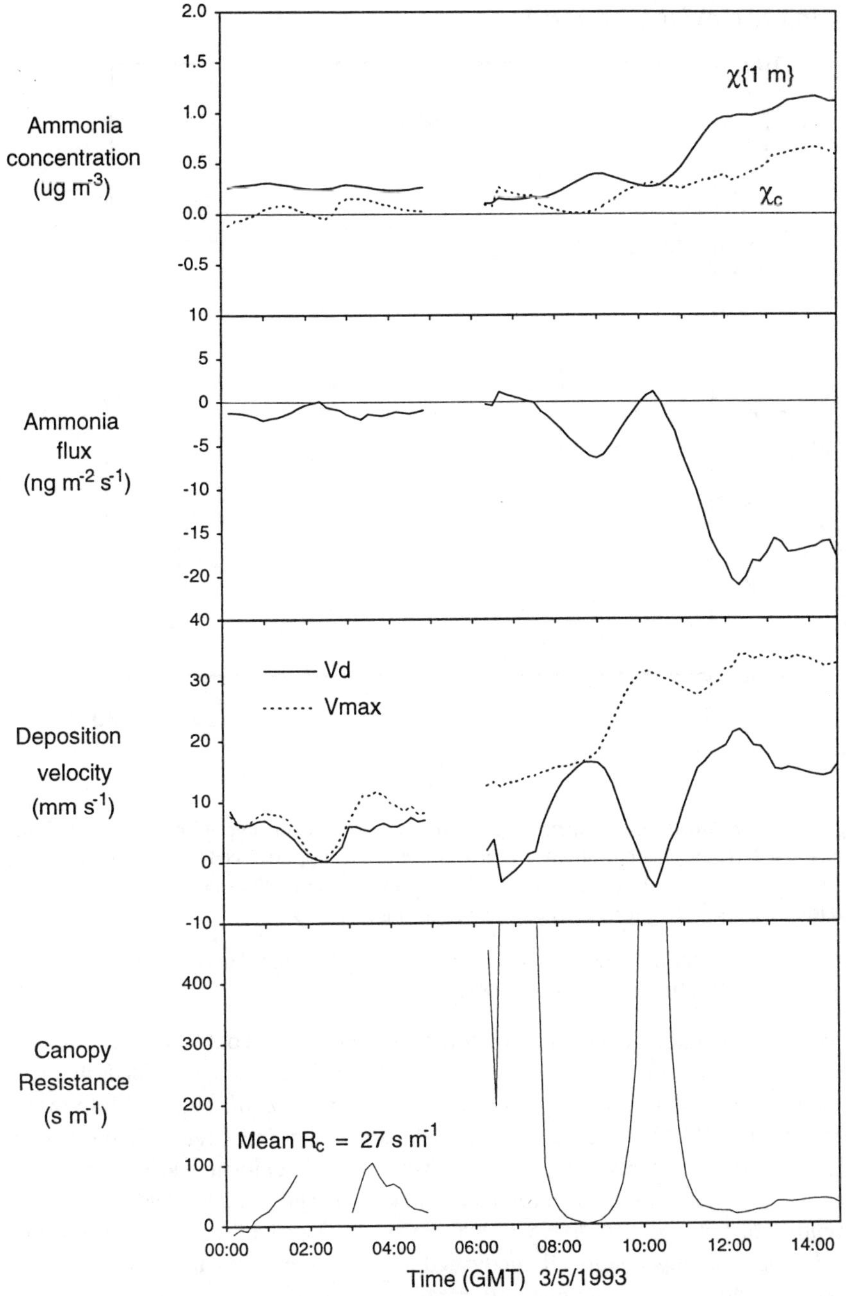

Fig. 2.7: Time course of ammonia exchange over moorland at Moorhouse, northern England. The term χ_c is the estimated concentration at the surface and represents the concentration analogue of R_c (Sutton and Fowler 1995).

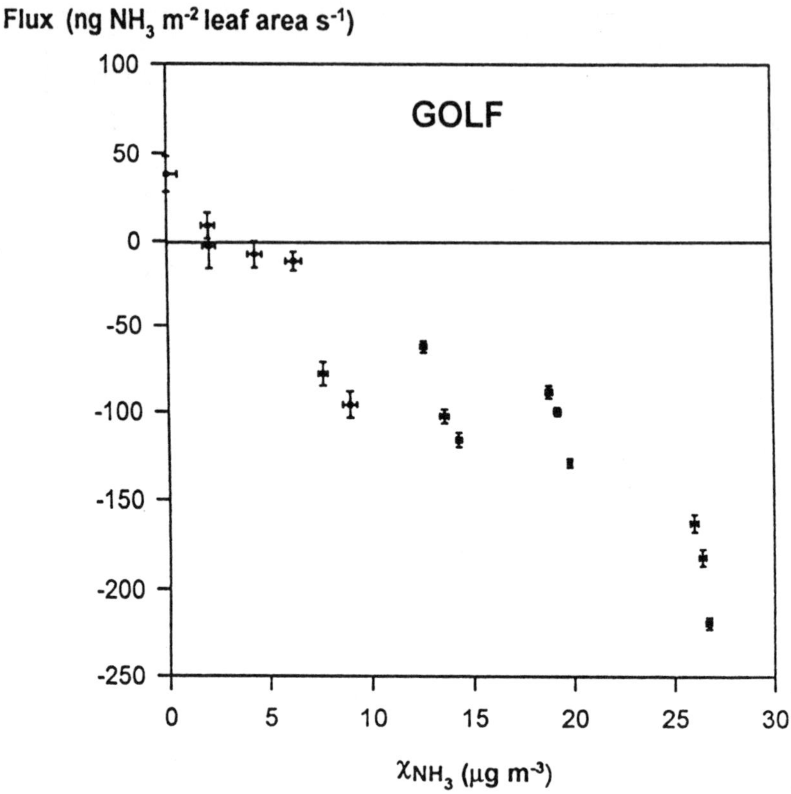

Fig. 2.8: Controlled environment measurements of NH_3 exchange with barley (*cv.* Golf) plants and NH_3 air concentrations. The measurements support a NH_3 compensation point of 2.8 μg m^{-3} for plants at 20 °C, 55 % relative humidity and light intensity of 520 mmol m^{-2} s^{-1} (data from Husted and Schjørring 1995).

the high air concentrations (50-100 μg m^{-3}) applied by van Hove *et al.* (see Sutton *et al.* 1992b, 1993b; Duyzer *et al.* 1994).

In contrast to studies using such high air concentrations, controlled environment measurements, at more typical smaller concentrations, also provided the first evidence of the ammonia compensation point (Farquhar *et al.* 1980). More recent work within BIATEX has extended such studies to investigate compensation points for selected fertilized and unfertilized plants. Schjørring (1991) provided laboratory measurements of ammonia fluxes from agricultural plants, and more recently Husted and Schjørring (1995) have developed an advanced controlled environment facility specifically dedicated to quantifying NH_3 and CO_2 fluxes. To determine the ammonia compensation point, environmental conditions are held constant to avoid confounding from variations due to temperature, humidity and light, and the plants exposed to a range of ammonia concentrations to identify the concentration at which a switch from emission to deposition occurs. An example of the measurements is shown in Fig. 2.8 (Husted and Schjørring 1995; Sutton *et al.* 1995b). This shows an NH_3 compensation point of 2.8 μg m^{-3} for barley (*cv.* Golf) plants at 20 °C.

Other compensation point determinations have been made in the field for tree foliage by Bingemer *et al.* (1991) and Kesselmeier *et al.* (1993) and as expected from the distinction noted from thc micrometeorological measurements, these show much smaller values than for fertilized agricultural plants. These measurements were made as part of the joint programme at the Schachtenau spruce forest in Bavaria (Enders *et al.* 1992) using a dynamic cuvette system at 23 m height within the canopy (Schäfer *et al.* 1992). Emission and deposition of ammonia was found in close correlation with ambient concentrations, suggesting a link with the ammonia compensation point. Overall, the results indicate a compensation point of 0.43 ppb (0.28–0.62; 95 % confidence level), equivalent to 0.33 (0.21–0.47) μg NH_3 m^{-3} (see Fig. 2.9). The authors noted a smaller value during night and low light conditions which they explained on the basis of a standstill of nitrate reduction to form ammonium in the cells, as this is a light

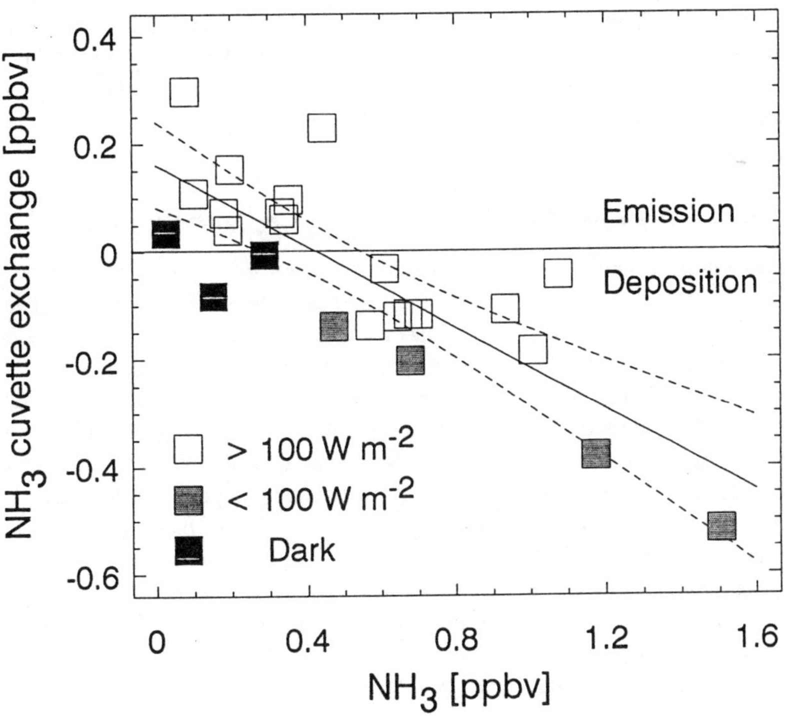

Fig. 2.9: Ammonia exchange measured with a dynamic chamber technique at 23 m height in a spruce forest. The overall regression provides a compensation point of 0.43 ppb (0.33 μg m^{-3}). Distinction may be made within the data in relation to light intensity, with smaller compensation points in low light conditions.

driven process. Such small values of the compensation point are consistent with the other limited data for forests (Gravenhorst and Breiding 1990; Langford and Fehsenfeld 1992; Sutton *et al.* 1993e), and confirm the general picture that forests act as sinks for ammonia emitted by agricultural ecosystems. The estimates also point to the need to provide a more mechanistic and quantitative basis for the observed differences in compensation points. In this respect, measurements to relate compensation points to physiological processes (*e.g.* nitrate reduction, protein degradation and photo-respiration) and alternative laboratory methods to determine compensation points from cell apoplast ammonium concentrations and pH are both essential.

2.6 Synthesis of ammonia exchange processes

The results presented above indicate that the exchange of ammonia between the biosphere and atmosphere is regulated by both ecosystem factors, such as plant community type, land management, nitrogen nutrition and plant physiology, as well as by environmental factors, such as surface wetness, humidity, temperature and atmospheric turbulence. Different components of the surface may act as both sources and sinks and there is a need to integrate these into a modelling framework for ammonia exchange. Hence leaves may both absorb or emit ammonia via the stomata in relation to the compensation point of the leaf tissues. Leaf cuticles, on the other hand, may capture ammonia from the atmosphere, or that emitted by stomata and soil in an internal cycle within the crop canopy (Denmead *et al.* 1976; Sutton *et al.* 1993e).

2.6.1 Pre-existing models of ammonia biosphere atmopshere exchange

Until recently, existing models of ammonia fluxes have failed to provide the necessary synthesis of these multiple exchange processes. Two main approaches have been used. Firstly, the canopy resistance (R_c) model (Eqs. 1 and 2) has been used to parametrize situations where ammonia deposition dominates. This provides a useful means to separate the surface processes from the atmospheric, and also to describe the effects of wetness and humidity on uptake by leaf cuticles and stomata (*e.g.* Fowler and Unsworth 1979; Sutton *et al.* 1993b, d; Duyzer *et al.* 1994; Erisman *et al.* 1994). However, it provides a poor description of bi-directional ammonia fluxes, as the model explicitly assumes that the concentration at the absorbing surface is zero (Eq. 1). The consequence is that a negative value of R_c can indicate either deposition faster than possible by turbulence or an emission flux. Bi-directional fluxes may be modelled using a modified R_c approach, but this requires external criteria (eg. a value of relative humidity) to manually switch between emission and deposition (Wyers *et al.* 1993a).

The second resistance model that has been applied to describe ammonia exchange is the simple compensation point model assuming transfer solely through plant stomata (Farquhar *et al.* 1980). This may be written as:

$$F_g = (\chi_s - \chi\{z\})/(R_a\{z\} + R_b + R_s) \qquad (3)$$

where χ_s is the compensation point for the leaf apoplastic solution, or 'stomatal compensation point' and R_s is stomatal resistance. This simple model provides the necessary condition for bi-directional modelling of ammonia with the leaves, but is unable to describe the situation where parallel deposition occurs onto leaf surfaces, as observed in many of the field studies.

2.6.2 Canopy compensation point model of ammonia exchange.

More recent work within BIATEX has developed a new resistance model that is able to describe both the uptake of ammonia by plant cuticles and the bi-directional exchange with stomata. The model described in detail elsewhere by Sutton and Fowler (1993) and Sutton *et al.* (1993a, 1995b) accounts for the competition for exchange within the canopy by defining and calculating the 'canopy compensation point' for ammonia (χ_c). This is the concentration analogue of R_c, and may be estimated to calculate the flux from:

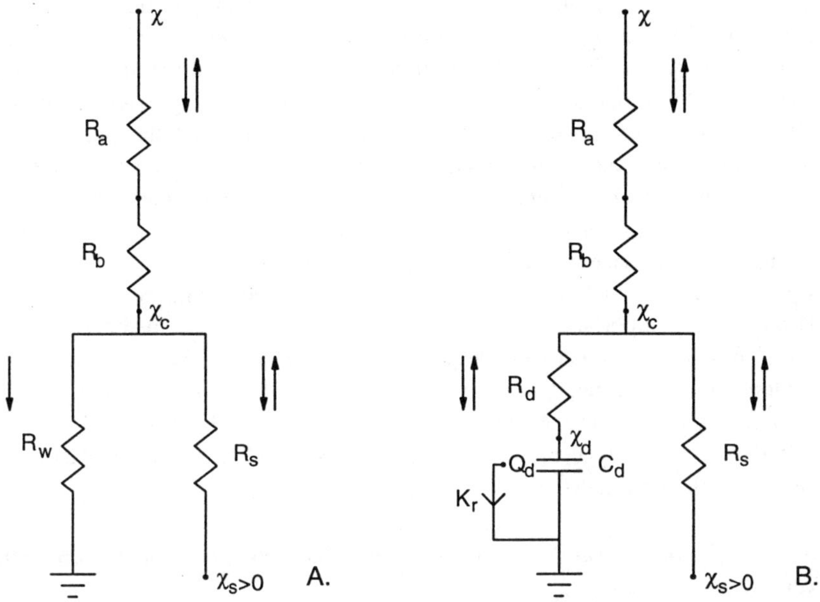

Fig. 2.10: Model diagrams to describe the bi-directional exchange of ammonia between a plant canopy and the atmosphere. The net flux is described as the result of the exchange between the air concentration (χ) and the 'canopy compensation point' (χ_c), which is the result of competition between exchange to the leaf surface and transfer through stomata (R_s) with a stomatal compensation point (χ_s). Deposition to the leaf surface is treated in different ways: **A**, as a simple resistance (R_w); **B**. as an adsorption capacitance (C_d) of given charge (Q_d) with adsorption resistance (R_d) and a reaction rate (K_r). R_a and R_b are the resistances to atmospheric turbulence.

$$F_g = (\chi_c - \chi\{z\})/(R_a\{z\} + R_b) \quad (4)$$

$$\chi_c = \frac{[\chi\{z\}/(R_a\{z\}+R_b)+\chi_s/R_s]}{[(R_a\{z\}+R_b)^{-1}+R_s^{-1}+R_w^{-1}]} \quad (5)$$

A schematic representation of this model is provided in Fig. 2.10A, and it may be seen that depending on the value of the leaf cuticle resistance (R_w), part or all of the ammonia emitted by stomata may be recaptured by deposition to the leaf surfaces. This model provides a substantial advance over the pre-existing R_c and simple compensation point models, since it provides the opportunity to separate the physiological factors affecting ammonia exchange (χ_s, R_s) from the environmental factors affecting capture by the leaf surface. In addition, it provides a single model concept for quantifying ammonia fluxes over both semi-natural and agricultural land. The model so far been applied over diurnal time-scales by Sutton and Fowler (1993, 1995) and Sutton *et al.* (1995b). An example of is shown in Fig. 2.11 for measurements of ammonia fluxes over Speulder forest in the Netherlands. Values of R_w were determined from average night time R_c for different relative humidity, which is possible since R_s is very large at night, and this showed good agreement with model humidity responses to R_w (Sutton *et al.* 1995b). The good agreement shown in Fig. 2.11 demonstrates the usefulness of the model, and there is now a need to refine the physiological basis of estimates of χ_s to allow the model to be applied over longer periods. It should be noted that the pH and NH_4^+ concentration applied in this model fit are much larger than noted elsewhere (Kesselmeier *et al.* 1993), and provide for example a stomatal compensation point of roughly 10 μg m^{-3} at 15 °C. This large value deserves further investigation and may be a consequence of the large long-term deposition of nitrogen to this forest as part of the feed-back discussed in section 2.4.4. The estimates of leaf apoplast pH are equally important to refining these estimates. The assumed apoplastic pH here is based on estimates of Farquhar *et al.* (1980), though more recent work (Schjørring, unpublished data) suggests that smaller values in the region pH 6 may be more typical. It is the apoplastic ratio of $[NH_4^+]/[H^+]$ that is important in defining χ_s, which suggests that the concentrations of NH_4^+ in the apoplast may be even larger than estimated here.

2.6.3 Leaf cuticle capacitance - canopy compensation point model of ammonia exchange.

For the present time the above model description may provide the most practical parametrization of the ammonia exchange process, by providing a useful balance between simplicity and treatment of the different processes. It should be recognized however, that this is still a simplification of the full set of interacting fluxes within the canopy. For example, no account is made in the equations noted above of competing soil emissions, which may be important from some fertilized arable ecosystems. Equally, the description of the flux to leaf cuticles is probably an over simplification as leaf cuticles may act more like a 'capacitor' for ammonia adsorption and desorption than a resistance for permanent capture. An example of this was provided by Sutton *et al.* (1995a) who in some cases could not explain mid-morning periods of ammonia emission from a wheat crop on the basis of the canopy compensation point model of Eqs. 4 and 5. A more detailed description treating the leaf surfaces as a capacitance for ammonia adsorption was developed by Sutton *et al.* (1995a) to attempt to explain these observations. The model necessarily becomes more complicated to parametrize, and especially to use to interpret measured fluxes, as the number of unknowns increases. In addition, whereas straightforward resistances models are independent of time, this treatment becomes dependent on the previous history of the canopy and follows the evolution of the magnitude of the cuticle capacitance.

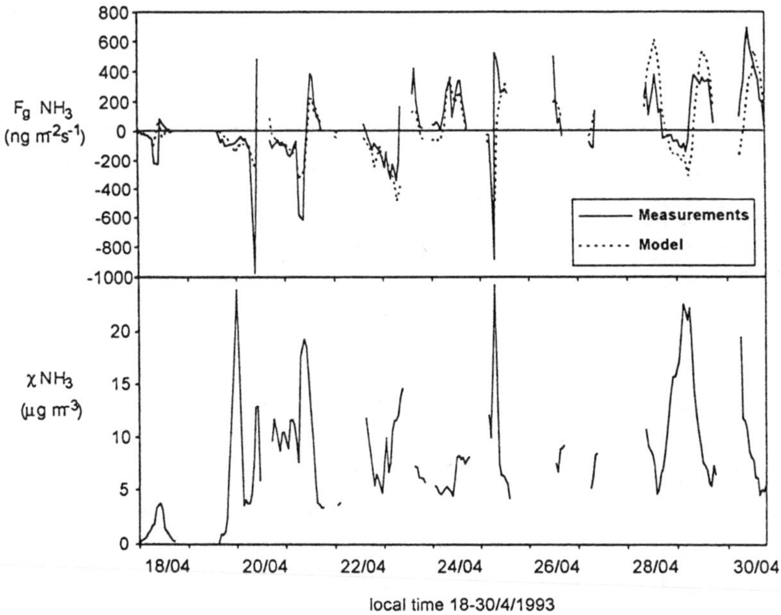

Fig. 2.11: Air concentrations of ammonia and comparison of the measured flux over Douglas fir (Speulder forest in the Netherlands) with that modelled according to Equations 4-5. The model assumes leaf apoplast pH 6.8 and 550 µmol l^{-1} NH_4^+; R_w is treated as a function of relative humidity from long-term average nocturnal R_c.

A schematic representation of the capacitance model and a description of the key parameters is shown in Fig. 2.10B. The charging resistance (R_d) may be defined in relation to the cuticular capacitance (C_d) which provides a time constant to charging and discharging which makes the model more stable in the first instance. The value of C_d was estimated on the basis of water film thickness for different relative humidities, leaf area index and Henry's law, given an estimated pH of the leaf surface water. At any given time step (i) the canopy compensation point is calculated according to the modified equation:

$$\chi_{c\{i\}} = \frac{\left[\chi\{z-d\}/(R_a\{z-d\}+R_b)+\chi_s/R_s+\chi_d/R_d\right]}{\left[(R_a\{z-d\}+R_b)^{-1}+R_s^{-1}+R_d^{-1}\right]} \tag{6}$$

The value of the adsorption concentration (χ_d) is either initiated or found according to:

$$\chi_{d\{i\}} = Q_{d\{i\}}/C_d \tag{7}$$

with the flux into or out of the cuticular capacitor (F_d) being:

$$F_d = (\chi_{d\{i\}}-\chi_{c\{i\}})/R_d \tag{8}$$

After *t* seconds, the new capacitance charge ($Q_{d\{i+t\}}$) is then found as:

$$Q_{d\{i+t\}} = Q_{d\{i\}} - (F_d.t) \tag{9}$$

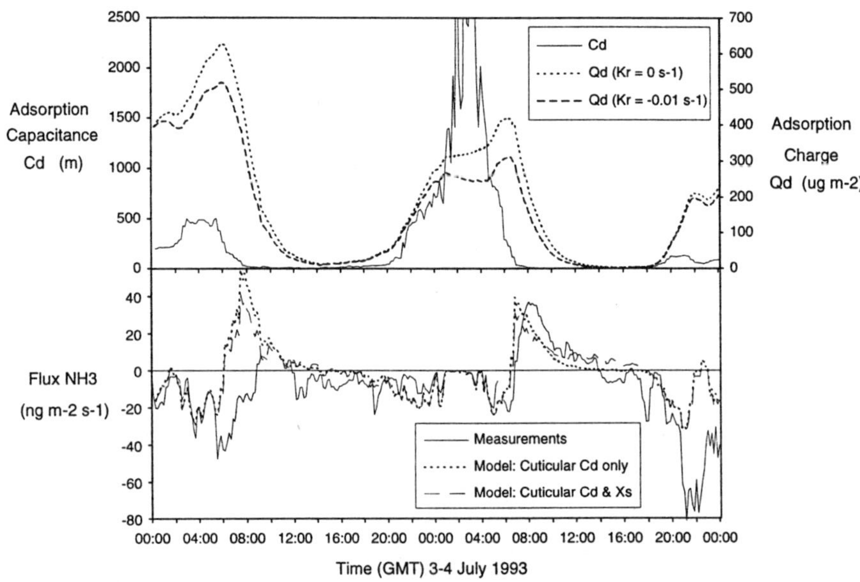

Fig. 2.12: Comparison of measured ammonia flux over a ripening wheat canopy at Sutton Bonington, central England (78 cm tall, early senescence) with model estimates parametrized using the capacitance only (C_d) description of NH_3 uptake onto leaf surfaces, and the full model allowing for both C_d and stomatal exchange (R_s, χ_s). The model flux assumes leaf intercellular pH 6.8 and 50 μmol NH_4^+ L^{-1}, leaf surface pH 4.5, with an equivalent water-film thickness and C_d according to Sutton *et al.* (1995a). The model fluxes are calculated applying an adsorption reaction rate (K_r) of -0.01 s^{-1}, the effect of which is shown on the adsorption charge (Q_d), alongside the modelled value of C_d.

The application of this model to a case study of bi-directional fluxes over a wheat canopy (Sutton *et al.* 1995a) is shown in Fig. 2.12. The top graph shows how the cuticle capacitance is estimated to fluctuate in relation to surface wetness of the canopy, and also provides two values of the adsorption charge (Q_d), in the presence and absence of a further removal reaction of rate K_r. For the second day of the measurement period, the model is able to explain the peak in ammonia emission in the morning as being due to evaporation from water-layers, however, the model fit is much poorer on the first day, indicating a need for improved parametrization of water-film thickness.

The above detailed model description of cuticular capacitance for ammonia provides a helpful means to integrate the ammonia exchange processes, and this needs to be further developed to make the results more widely applicable. For example, in the results of Fig. 2.11, enhanced peaks of morning ammonia emission (beyond that predicted by the canopy compensation point model), may be due to ammonia desorption from water-layers. Nevertheless, a balance is needed in model complexity, and the greater simplicity of the canopy compensation point model may make this more useful for more general parametrization of ammonia fluxes.

2.7 Current progress and future developments

Over the past eight years significant progress has been made in our understanding and quantification of the biosphere atmosphere exchange of ammonia, and much of this has been achieved within the framework of the joint EC/EUROTRAC project BIATEX. Both micrometeorological and controlled environment measurements have highlighted the difference of ammonia fluxes between semi-natural and agricultural ecosystems, and provided the necessary material for developing models of the ammonia exchange process. As a result it is now clear that ammonia fluxes between the atmosphere and ecosystems depend on a diverse set of environmental and ecological factors. As a key component in plant metabolism ammonia may be both emitted and absorbed by plant leaves, via stomata and this is dependent on both the physical equations of solubility (Henry's Law) and physiological and phenological (growth stage) factors affecting intercellular pH and NH_4^+ concentrations. In contrast, as a soluble compound ammonia may be deposited on to leaf surfaces, a process regulated by humidity and wetness at the leaf surface as well as by acidity of pre-existing leaf surface deposits. From here it may be either absorbed by the leaf (cuticle or stomata), washed to the ground in rain or desorbed as leaf wetness evaporates, providing a pulse of ammonia emission.

The quantification and modelling of these processes has undergone substantial progress within BIATEX and it is now possible for the first time to be able to integrate a large number of these factors into single models. The magnitude of many of the terms and prediction of ammonia fluxes over long periods and wide areas is still extremely uncertain. For example, the quantitative importance of gas-particle reactions on the measured fluxes is still to be established. However, the model frameworks that have been developed now allow a more focused examination and isolation of the individual terms. Thus studies in chambers at low humidity, where leaf surface adsorption is less important, may be applied to investigate the factors affecting the stomatal compensation point. In contrast, where the stomatal compensation point is inaccessible (*e.g.* night-time fluxes), measurements may be used to focus on leaf surface exchange processes.

Future work needs to provide the quantitative estimates of many of the terms that have now been identified and their temporal and spatial variability for different ecosystems. With this basis, it will then begin to be possible to provide a more realistic picture of the biosphere atmosphere exchange of ammonia over regional and continental scales for different ecosystems. At present, current scaling up estimates are extremely uncertain, being based either on simple deposition velocities, or at best canopy resistances. Given the importance of ammonia deposition to inputs of acidity and nitrogen deposition, and the need to provide sound estimates of net fluxes for critical loads comparisons, such improved estimates should play an important part of the assessment of air pollutant impacts in Europe.

Acknowledgements

Work on ammonia biosphere atmosphere exchange within BIATEX has been funded by a number of sources and the authors are particularly grateful to: the UK Department of Environment (Air Quality Division)(MAS); the Dutch Ministry of Spatial Planning and Environment and Ministry of Economic Affairs (GPW, JHD); the Bundesminister für Bildung und Forschung (FXM, JK); the Bundesminister für Forschung und Technology (GK); the Danish Agricultural and Veterinary Research Council (JKS); the EC for project EV4V-0083-C 'Development of automated chemical sensors for dry deposition measurements by the gradient method' (GPW, FXM). In addition, this work is currently supported within the framework of the EC Environment Programme project EXAMINE ('Exchange of Ammonia with European Ecosystems'), which provides an essential integration of the national expertise (MAS, GPW, FXM, JKS).

References

Allegrini, I., De Santis, F., Di Palo, V., Liberti, A., (1984) Measurement of particulate and gaseous ammonia at a rural site by means of newly developed high performance diffusion tubes. *J. Aerosol Sci.* **15**, 465–71.

Andersen, H.V., Hovmand, M.F., Hummelshøv, P., Jensen, N.O., (1993) Measurements of ammonia flux to a spruce stand in Denmark. *Atmos. Environ.* **27A**, 189–202.

Asman, W.A.H., Jaarsveld, J.A. van, (1992) A variable-resolution transport model applied for NH_x for Europe. *Atmos. Environ.* **26A**, 445–464.

Bingemer, H.G., Ernst, L., Fischer, M., Kesselmeier, J., Andreae M.O., Helas G., (1991) Volatile organic acids and ammonia measurements in a spruce forest. in: P. Borrell, P.M. Borrell, W. Seiler (eds), *Proc. EUROTRAC Symp. '90*, SPB Academic Publishing bv, The Hague, pp. 201–205.

Bobbink, R., Boxman, D., Fremstad, E., Heil, G., Houdijk, A., Roelofs J., (1992a) Critical loads for nitrogen eutrophication of terrestrial and wetland ecosystems based upon changes in vegetation and fauna. in: Grennfelt, P., Thörnelöf, E. (eds) *Critical loads for nitrogen*, Nord 1992: 41, Nordic Council of Ministers, Copenhagen, pp. 111–160.

Bobbink, R., Heil, G.W., Raessen, M.B.A.G., (1992b) Atmospheric deposition and canopy exchange processes in heathland ecosystems. *Environ. Pollut.* **75**, 29–37.

Breemen, N. van, Burrough, P.A., Velthorst, E.J. Dobben, H.F. van, de Wit, T., Ridder, T.B., Reijnders, H.F.R., (1982) Soil acidification from atmospheric ammonium sulfate in forest canopy throughfall. *Nature* **299**, 548–550.

Brost, R.A., Delany, A.C., Huebert B.J., (1988) Numerical modeling of concentrations and fluxes of HNO_3, NH_3, and NH_4NO_3 near the surface. *J. Geophys. Res.* **93**, 7137–7152.

Dasgupta, P.K. (1993) Automated measurement of atmospheric trace gases: diffusion based collection and analysis. in: L. Newman (ed) *Measurement challenges in atmospheric chemistry. Advances in Chemistry* **232** American Chemical Soc, Washington, pp. 41–90..

Denmead, O.T., Freney, J.R., Simpson J.R., (1976) A closed ammonia cycle within a plant canopy. *Soil Sci. Biochem.* **8**, 161–164.

Draaijers G. (1993) The variability of atmospheric deposition to forests; the effects of canopy structure and forest edges. Ph.D. Thesis, University of Utrecht, Utrecht, the Netherlands.

Duyzer, J.H. (1992) The influence of chemical reactions on surface exchange of NO, NO_2 and O_3: results of experiments and model calculations. in: Schwartz and Slinn (eds) *Precipitation Scavenging and atmosphere-surface exchange processes, 5th Int. Conf.* Richland, Washington, pp. 1105–1114.

Duyzer, J.H., Verhagen, H.L.M., Westrate, J.H., Bosveld, F.C., (1992) Measurement of the dry deposition flux of NH_3 on to coniferous forest. *Environ. Pollut.* **75**, 3–13.

Duyzer, J.H., Bouman, A.M.H., Diederen, H.S.M.A., Aalst R.M. van, (1987) Measurement of dry deposition velocities of NH_3 and NH_4^+ over natural terrains. *Report R 87/273* MT-TNO, Delft, the Netherlands.

Duyzer, J.H., Verhagen, H.L.M., Westrate, J.H., Bosveld, F.C., Vermetten, A.W.M., (1994) The dry deposition of ammonia onto a Douglas fir forest in the Netherlands. *Atmos. Environ.* **28**, 1241–1253.

Enders, G., Dlugi, R., Steinbrecher, R., Clement, B., Daiber, R., Eijk, J.v., Gaab, S., Haziza, M., Helas, G., Herrmann, U., Kessel, M., Kesselmeier, J., Kotzias, D., Kourtidis, K., Kurth, H.H., McMillen, R.T., Roider, G., Schuermann, W., Teichmann, U., Torres, L., (1992) Biosphere–atmosphere interactions: Integrated research in a European coniferous forest ecosystem. *Atmos. Environ.* **26A**, 171–189.

Ernst, L., (1990) Anwendung eines Sprühsammlers auf da Problem der Differenzierung von gas förmigem und partikel-gebundenem Ammoniak in der Atmosphäre. Diploma Thesis, University of Mainz, Germany.

Erisman, J.W., Wyers, G.P., (1993) Continuous measurements of surface exchange of SO_2 and NH_3: implications for their possible interaction in the deposition process. *Atmos. Environ.* **27A**, 1937–1949.

Erisman, J.W., van Pul, A., Wyers, G.P., (1994) Parametrization of surface resistance for the quantification of atmospheric deposition of acidifying pollutants and ozone. *Atmos. Environ.* **28**, 2595–2607.

FAO (1985) *Production Yearbook.* Food and Agriculture Organization, Rome.

Farquhar, G.D., Firth, P.M., Wetselaar, R., Wier B., (1980) On the gaseous exchange of ammonia between leaves and the environment: determination of the ammonia compensation point. *Plant Physiol.* **66**, 710–714.

Ferm, M., (1979) Method for the determination of atmospheric ammonia. *Atmos. Environ.* **13**, 1385–1393.

Fenn, L.B., Hossner L.R., (1985) Ammonia volatilization from ammonium or ammonium forming nitrogen fertilizers. *Adv. Soil Sci.* **1**, 123–169.

Fowler, D., Duyzer, J.H., Baldocchi D.D., (1991) Inputs of trace gases, particles and cloud droplets to terrestrial surfaces. *Proc. Roy. Soc. Edin.* **97B**, 35–59.

Fowler, D., Duyzer J.H., (1989) Micrometeorological techniques for the measurement of trace gas exchange. in: Andreae, M.O., Schimel, D.S. (eds) *Exchange of trace gases between terrestrial ecosystems and the atmosphere.* Wiley, Chichester, pp. 189–207.

Fowler, D., Unsworth, M.H. (1979) Turbulent transfer of sulfur dioxide to a wheat crop. *Quart. J. Roy. Meteor. Soc.* **105**, 767–783.

Gravenhorst, G., Breiding H., (1990) NH_3 transfer between the atmosphere and coniferous trees. in: Beilke, S., Millan M. (eds) Field measurements and interpretation of species derived from NO_x, NH_3 and VOC emissions in Europe. *Air Pollut. Res. Report. 25*, CEC, Luxembourg. pp. 118–146.

Grennfelt, P., Thörnelöf, E., (1992) (eds) *Critical loads for nitrogen - a workshop report.* Nord 1992:41, Nordic Council of Ministers, Copenhagen.

Gundersen, P., Rasmussen, L., (1988) Nitrification, acidification and aluminium release in forest soils. in: Nilsson, J., Grennfelt, P. (eds), *Critical loads for sulfur and nitrogen.* Nordic Council of Ministers, Copenhagen, pp. 225–268.

Harper, L.A., Sharpe, R.R., Langdale, G.W., Giddens, J.E., (1987) Nitrogen cycling in a wheat crop: soil, plant and aerial nitrogen transport. *Agron. J.* **79**, 965–973.

Hall, A.D., Miller, N.H.J., (1911) On the absorption of ammonia from the atmosphere. *J. Agric. Sci.* **4**, 56–68.

Harrison, R.M., Rapsomanikis, S., Turnbull, A., (1989) Land-surface exchange in a chemically reactive system; surface fluxes of HNO_3, HCl and NH_3. *Atmos. Environ.* **23**, 1795–1800.

Hornung, M., Sutton, M.A., Wilson, R.B., (1995) Mapping and modelling of critical loads for nitrogen - a workshop report. *Report of an UN-ECE workshop, Grange-over-Sands, 23-26 October 1994,* Institute of Terrestrial Ecology, Edinburgh.

Hove, L.W.A. van, Koops, A.J., Adema, E.H., Vredenberg, W.J., Pieters G.A., (1987) Analysis of the uptake of atmospheric ammonia by leaves of *Phaseolus vulgaris* L. *Atmos. Environ.* **21**, 1759–1763.

Hove, L.W.A. van, Adema, E.H., Vredenberg, W.J., Pieters, G.A., (1989) A study of the adsorption of NH_3 and SO_2 on leaf surfaces. *Atmos. Environ.* **23**, 1479–1486.

Hove, L.W.A. van, Vredenberg, W.J., Adema, E.H., (1990) The effect of wind velocity, air-temperature and humidity on NH_3 and SO_2 transfer into Bean-leaves (*Phaseolus vulgaris* L.). *Atmos. Environ.* **24**, 1263–1270.

Husted, S., Schjørring, J.K., (1995) A computer controlled system for studying ammonia exchange, photosynthesis and transpiration of plant canopies growing under controlled environmental conditions. *Plant, Cell & Environ.* **18**, 1070–1077

INDITE (1994) *Impacts of Nitrogen Deposition in Terrestrial Ecosystems.* Report of the INDITE review group. Dept. of the Environment, London.

Junge, C.E., Ryan T.G., (1958) Study of the SO_2 oxidation in solution and its role in atmospheric chemistry. *Quart. J. Roy. Meteor. Soc.* **84**, 46–55.

Kesselmeier, J., Merk, L., Bliefernicht, M, Helas G., (1993) Trace gas exchange between terrestrial plants and atmosphere: carbon dioxide, carbonyl sulfide and ammonia under the rule of compensation points. in: Slanina, J., Angeletti, G., Beilke, S. (eds) General Assessment of biogenic emissions and deposition of nitrogen compounds, sulfur compounds and oxidants in Europe., *Air Pollut. Res. Report* **47,** CEC, Brussels, pp. 71–80.

Keuken, M.P., Wayers-IJpelaan, A., Mols, J.J., Otjes, R.P., Slanina, J., (1989) The determination of ammonia in ambient air by an automated thermodenuder system. *Atmos. Environ.* **23**, 2177–2185.

Keuken, M.P., Schoonebeek, C.A.M., van Wensveen-Louter, A., Slanina, J., (1988) Simultaneous sampling of NH_3, HNO_3, HCl, SO_2 and H_2O_2 by a wet annular denuder system. Atmos. Environ. 22, 2541–2548.

Kramm, G. (1995) Zum Austausch von Ozon und reaktiven Stickstoffverbindungen zwischen Atmosphäre und Biosphäre. (Schriftenreihe des Fraunhofer -Instituts für Atmosphärische Umweltforschung, Garmisch-Partenkirchen; Bd 34) Zugl.: Berlin, Humboldt-Univ., Diss. 1995.

Kramm, G., Müller, H., Fowler, D., Höfken, K.D., Meixner, F.X., Schaller E., (1991) A modified profile method for determining the vertical fluxes of NO, NO_2, ozone and HNO_3 in the atmospheric surface layer. *J. Atmos. Chem.* **13**, 265–288.

Kramm, G., Dlugi R., (1994) Modelling of the vertical fluxes of nitric acid, ammonia and ammonium nitrate. *J. Atmos. Chem.* **18**, 319–357.

Kramm, G., Müller, H., Schröder, P., Seiler, W., Dlugi, R., Seiler, T., Foken, T., Mölders, N., Sievering, H., (1993) Estimating the dry deposition of ozone and reactive nitrogen compounds within the framework of the project SANA. *A and WMA Trans* (submitted).

Langford, A.O., Fehsenfeld, F.C., (1992) Natural vegetationas a source or sink for atmospheric ammonia: a case study. *Science* **255**, 581–583.

Lemon, E., Van Houtte, R., (1980) Ammonia exchange at the land surface. *Agron. J.* **72**, 876–883.

Meixner, F.X.., (1994) Surface exchange of ammonia. Three different micrometeorological procedures applied. in: Angeletti, G., Restelli, G.O. (eds) *Physico-chemical behaviour of atmopsheric pollutants.* EUR 15609. CEC, Luxembourg, pp. 749–754.

Meixner, F.X., Possanzini, M., Slanina J., (1993) Development of automated chemical sensors for dry deposition measurements by the gradient method. *Final Report Project: EV4V-0083-C.* CEC, DG-XII, Brussels.

Meixner, F.X., Bliefernicht, M., Wyers, G.P., Helas, G., Kesselmeie, J., (1995) Ammonia exchange between terrestrial plants and the atmosphere controlled by plant physiology: compensation point and CO_2 exchange, this volume.

Meixner,, F.X., Böswald F., Eiblmeier, K., Kramm, G., Ludwig, J., Müller, H., Nestlen, M., Terrey, B., Weber, B., (1991) Measurements of vertical fluxes of nitrogen compounds. in: *EUROTRAC Annual Report, Part 4.* EUROTRAC ISS, Garmisch-Partenkirchen, pp. 60–66.

Müller, H., Meixner, F.X., Kramm, G., Fowler D., Dollard, G.J., Possanzini, M., (1993) Determination of HNO_3 deposition by modified Bowen ratio and aerodynamic profile techniques. *Tellus* **45B**, 346–367.

Pandis, S.N., Seinfeld, J.H., (1990) On the interaction between equilibration processes and wet or dry deposition. *Atmos. Environ.* **24A**, 2313–2327.

Sandnes, H., Styve, H., (1992) Calculated budgets for airborne acidifying components in Europe, 1985, 1987, 1988, 1988, 1989, 1990 and 1991. *EMEP report 1/92* Norwegian Meteor. Inst. Oslo, Norway.

Schaefer, L., Kesselmeier, J., Helas, G., (1992) Formic and acetic acid emission from conifers measured with a cuvette technique. in: G. Angeletti, S. Beilke, J. Slanina (eds). Field measurements and interpretation of species related to photooxidants and acid deposition. *Air Pollution Res. Report.* **39**, CEC Brussels, pp. 319–323.

Schjørring, J.K. (1991) Ammonia emissions from the foliage of growing plants. in: T.D. Sharkey, E.A. Holland, H.A. Mooney (eds), *Trace gas emissions by plants.* Academic Press, San Diego, pp. 267–292..

Schjørring, J.K., Kyllingsbaek, A., Mortensen, J.V., Byskov-Nielsen, S., (1993a) Field investigations of ammonia exchange between barley plants and the atmosphere. I. Concentration profiles and flux densities of ammonia. *Plant, Cell and Environ.* **16**, 161–167.

Schjørring, J.K., Kyllingsbaek, A., Mortensen, J.V., Byskov-Nielsen, S. (1993b) Field investigations of ammonia exchange between barley plants and the atmosphere. II. Nitrogen remobilization, free ammonium content and activities of ammonium assimilating enzymes in different leaves. *Plant, Cell and Environment* **16**, 169–178.

Schjørring, J.K., Ferm, M., Sommer, S.G., (1992) Measurement of NH_3 emission and deposition by the gradient method: Can passive flux samplers be used to obtain the net exchange of NH_3 through periods of several days with varying windspeed and atmospheric NH_3 concentration? in: I. Allegrini (ed), Development of Analytical Techniques for Atmospheric Pollutants, *Air Pollution Res. Report* **41**, CEC Brussels, pp. 21–34..

Schjørring, J.K., (1995) Long-term quantification of ammonia exchange between agricultural cropland and the atmosphere. I. Evaluation of a new method based on passive flux samplers in gradient configuration. *Atmos. Environ.* **29A**, 855–893.

Stelson, A.W., Seinfeld, J.H., (1982) Relative humidity and temperature dependance of the ammonium nitrate dissociation constant. *Atmos. Environ.* **16**, 983–992.

Sutton, M.A., Asman, W.A.H., Schjørring, J.K., (1994) Dry deposition of reduced nitrogen.*Tellus* **46B**, 255–273.

Sutton, M.A., Burkhardt, J.K., Guerin, D., Fowler, D., (1995a) Measurement and modelling of ammonia exchange over arable croplands. in: G.J. Heij, J.W. Erisman (eds), *Acid rain research: Do we have enough answers?* Elsevier Science bv, pp. 71–80.

Sutton, M.A., Fowler, D., (1993) A model for inferring bi-directional fluxes of ammonia over plant canopies. in: *Proc. WMO Conf. on the measurement and modelling of atmospheric composition changes including pollutant transport.* WMO/GAW-91 WMO Geneva, pp. 179–182.

Sutton, M.A., Fowler, D., (1995) Continuous measurements of ammonia deposition to acidic moorland and application of a new inferential technique to describe ammonia exchange with the plant canopy. in: R.W. Battarbee (ed), *Acid rain and its impact: the critical loads debate.* 162-168. Ensis Publishing, 26 Bedford Way, London WC1H.

Sutton, M.A., Fowler, D., Hargreaves, K.J., Storeton-West, R.L., (1992a) Effects of land-use and environmental conditions on the exchange of ammonia between vegetation and the atmosphere. in: Angeletti, G., Beilke, S., Slanina, J. (eds), Field measurements and interpretation of species related to acid deposition. *Air Pollution report* **39**, CEC Brussels, pp. 211–217.

Sutton, M.A., Fowler, D., Hargreaves, K.J., Storeton-West, R.L., (1993a) Interactions of NH_3 and SO_2 exchange inferred from simultaneous flux measurements over a wheat canopy. in: Slanina, J., Angeletti, G., Beilke, S. (eds), General Assessment of biogenic emissions and deposition of nitrogen compounds, sulfur compounds and oxidants in Europe. *Proc.of the joint CEC/BIATEX workshop Air Pollut. Res. Report* **47**, CEC Brussels, pp. 165–182.

Sutton, M.A., Fowler, D., Moncrieff, J.B., (1993b) The exchange of atmospheric ammonia with vegetated surfaces. I: Unfertilized vegetation. *Quart. J. Roy. Meteor. Soc.* **119**, 1023–1045.

Sutton, M.A., Fowler, D., Moncrieff, J.B., Storeton-West, R.L., (1993c) The exchange of atmospheric ammonia with vegetated surfaces. II: Fertilized vegetation. *Quart. J. Roy. Meteor. Soc.* **119**, 1047–1070.

Sutton, M.A., Fowler, D., Smith, R.I., Eager, M., Place, C.J., Asman, W.A.H., (1993d) Modelling the net exchange of reduced nitrogen. in: Slanina, J., Angeletti, G., Beilke, S. (eds) General Assessment of biogenic emissions and deposition of nitrogen compounds, sulfur compounds and oxidants in Europe. *Proc. joint CEC/BIATEX workshop, Air Pollut. Res. Report* **47**, *CEC, Brussels,*pp. 117–131.

Sutton, M.A., Moncrieff, J.B., Fowler, D., (1989) Measurement of atmospheric ammonia and the assessment of is exchange with vegetated surfaces. in: Changing composition of the troposphere. *WMO conf. Special Environ. Report No.* **17**. WMO, Geneva, pp. 154–157.

Sutton, M.A., Moncrieff, J.B., Fowler, D., (1992b) Deposition of atmospheric ammonia to moorlands. *Environ. Pollut.* **75**, 15–24.

Sutton, M.A., Pitcairn, C.E.R., Fowler, D., (1993e) The exchange of ammonia between the atmosphere and plant communities. *Adv. Ecol. Research.* **24**, 301–393.

Sutton, M.A., Schjørring, J.K., Wyers G.P., (1995b) Plant - atmosphere exchange of ammonia. *Phil. Trans. Roy. Soc., London. A.* **351**, 261–278.

UNECE, (1994) *Convention on Long-Range Transboundary Air Pollution*, GE.94-31533 /ECE94 /14. Environment and Human Settlements Division, UNECE, Geneva.

Ville, G., (1850) Note sur l'assimilation de l'azote du l'air, par les plantes, et sur l'influence qu'excerce l'ammoniaque dans la végétation. *Compte Rendus Acad. Sci., Paris.* **31**, 578–580.

Way, J.T., (1855) The atmosphere as a source of nitrogen to plants being an account of recent researches into this subject. *J. Roy. Agr. Soc.* **16**, 249–267.

Wyers, G.P., Otjes, R.P., Erisman, J.W., (1993a) Dry deposition of ammonia onto a coniferous forest. in: J. Slanina, G. Angeletti, S. Beilke (eds) General assessment of biogenic emissions and deposition of nitrogen compounds, sulfur compounds and oxidants in Europe. *Air Pollution Research Report* **47**, EC. Brussels, pp. 139–145.

Wyers, G.P, Otjes, R.P., Slanina,. J., (1993b) A continuous flow denuder for the measurement of ambient concentrations and surface fluxes of ammonia. *Atmos. Environ.* **27A,** 2085–2090.

Zhang, Y., ten Brink, H., Slanina J., Wyers, P., (1995) The influence of ammonium nitrate equilibrium on the measurement of exchange fluxes of ammonia and nitric acid. in: G.J. Heij, J.W. Erisman (eds) *Acid Rain Research: Do we have enough answers?* Elsevier Science bv, pp. 103–112.

Chapter 3

Atmospheric Particles and their Interactions with Natural Surfaces

Martin Gallagher[1] and Jacques Fontan[2], Paul Wyers[3], Walter Ruijgrok[4], Jan Duyzer[5], Paul Hummelshøj[6] Kim Pilegaard[6] and David Fowler[7]

[1] Institute of Science and Technology, University of Manchester, UK
[2] Laboratoire d'Aerologie, Université Paul Sabatier, France
[3] ECN, Energy Research Foundation, the Netherlands
[4] KEMA Environmental Science, Arnhem, the Netherlands
[5] TNO Institute of Environmental Sciences, Delft, the Netherlands
[6] RISØ National Laboratory, Roskilde, Denmark
[7] Institute of Terrestrial Ecology, UK

Summary

The purpose of this short review will be to focus on those aspects of aerosol exchange which are of importance to the BIATEX community in order to aid in the assessment of acidic or alkaline deposition, from dry and wet aerosol (in the form of cloud or fog droplets) over Europe with an eye to the larger scale processes in which aerosols play a critical role. One of the important goals of the Biatex program, emphasised by its co-ordinator, has been the necessity to deliver information about these processes which are "useful" to wider interests, in particular the modelling community. In light of the co-ordinators plea for "immediately useful information" a detailed summary of parameterisations which have been forthcoming in BIATEX and related programmes for both dry and wet aerosol deposition are listed in the Appendix.

Within the BIATEX program particulate deposition has not received as much attention as has trace gas exchange and is an area which still requires much investigation. Encouragingly, recent work reported at the EUROTRAC BIATEX workshop in Madrid, 1995, has produced reliable parameterisations that are now well supported by a growing experimental database. The results reported were obtained using very different instrumentation and measurement techniques and this strengthened the final conclusions drawn. The oft held view that there is significant disagreement surrounding dry deposition measurements would appear to recede in the light of these results. In particular there has been pleasing agreement between measurements of dry deposition velocity, V_d, for aerosols to forest canopies by micrometeorological techniques and throughfall-stemflow or surface type measurements quoted in the literature. The eddy correlation (EC) technique employed for aerosol flux measurements which has often been plagued by susceptibility to low-frequency contamination and instrumental limitations has significantly benefited from the work of Lamaud *et al.* (1994) who show that aerosol fluxes cannot be "blindly" computed using the analysis techniques generally used for trace gas fluxes.

We are now confident that the dry deposition velocity is well defined by measurement for typical atmospheric aerosol loadings, probably better than to within a factor of 2. Gaps still exist in our understanding of the detailed particle size dependence for the various physical transport processes in the crucial size region 0.1 - 1.0 μm. The classical model picture handed down to us, and based on wind tunnel data does not take into account the influence of *e.g.* stability and related burst effects, and would appear to underestimate deposition velocities in this size range for rough vegetation. It still remains for the latest experimental results to be compared in detail with the latest models, which are now beginning to address this discrepancy by the inclusion of additional transport mechanisms.

Introduction: Particles in the atmosphere

Aerosols are generally categorised according to their physical state and the method by which they have been generated. Terminology for different types of aerosol follows no strict pattern, however, there are some rough guidelines. Aerosol is generally defined as a suspension of solid or liquid particles in a gas, with a full description requiring knowledge of the concentration, chemical composition, hygroscopicity, phase and morphology. This definition strictly includes both the particles and the suspending gas, usually air. Aerosols are ubiquitous in the atmosphere ranging in size from 0.001 μm to greater than several hundred microns and generally persist for seconds to a year or more. Particles which are less than 2.5 μm in size are generally referred to as "fine" aerosol and those greater than 2.5 μm are called "coarse". "Size" here refers to equivalent sphere diameter but this distinction can be lost particularly when measuring the optical as opposed to aerodynamic size of the aerosol.

Particles may be produced by a bewildering array of different processes, the most important of which are outlined in Fig. 3.1. These are nucleation, coagulation, chemical conversion and aeolian and volcanic processes. Already we may begin to see the difficulty that arises in quantifying individual sources since many are intimately coupled. Most of the particulate mass in the atmosphere is formed from gaseous emissions, although there are still large uncertainties in source magnitudes. Natural sources are generally well distributed over the globe whereas man made sources are concentrated in industrialised areas of the world where they generally exceed natural sources. Once in the atmosphere aerosols will grow either by promoting condensation or by coagulation with other particles. Particles which grow to sizes larger than a micron are generally removed by impaction or sedimentation onto surfaces. Smaller particles are generally removed by diffusion or act as nucleation sites for raindrops or ice crystals and are removed by washout

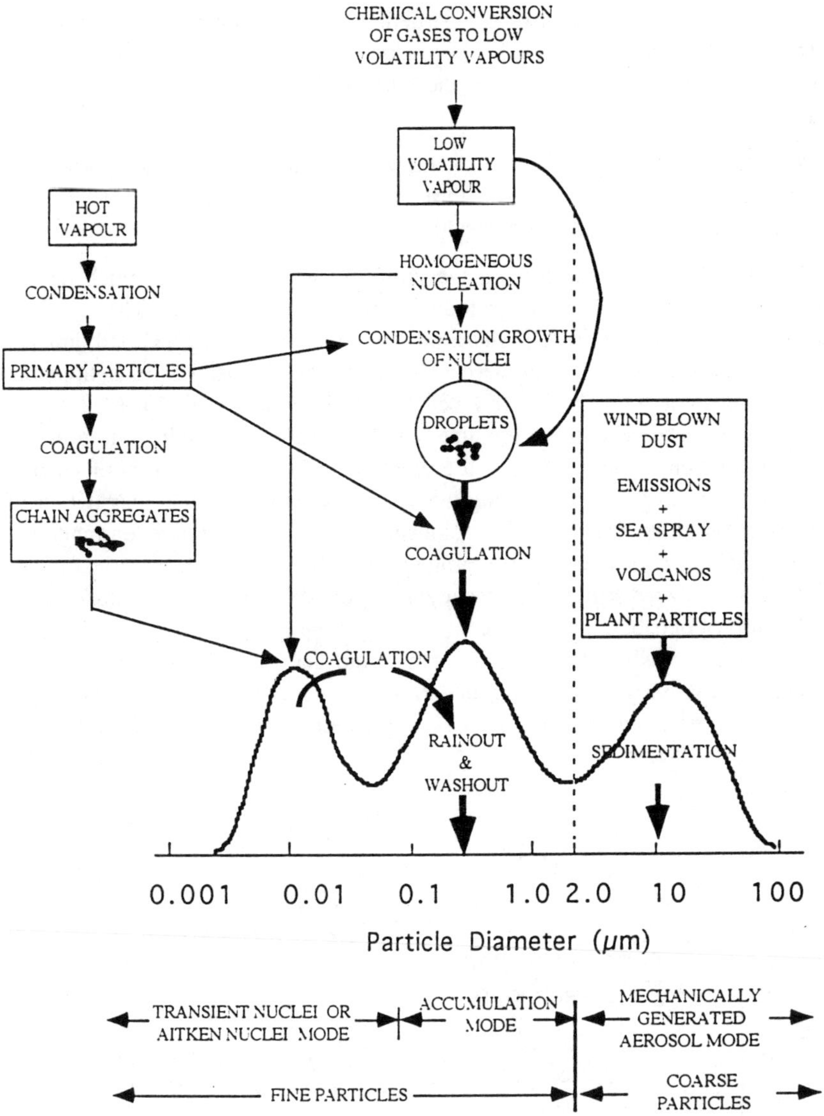

Fig. 3.1: Idealised particle surface area distribution of atmospheric aerosol. Adapted from Whitby and Cantrell (1976).

or scavenging processes. Most of the particulate mass is concentrated within the lowest 2 km of the atmosphere while above 5 km particle concentrations are little influenced by direct emissions from the surface (apart from episodic volcanic activity).

Boundary layer stability and depth near the earth's surface may also significantly affect local concentrations which in turn may have important consequences for environmental pollution and human health.

The size distribution of aerosol in the atmosphere depends very much on the primary source and the processes of growth, transport and removal. The so called nuclei mode, *i.e.* the size range where particles which are less than 0.1 μm are found, accounts for most of the aerosol in terms of number concentration, but rarely accounts for more than a few percent of the total aerosol mass, most of which is found in the 0.1 to 1.0 μm size range, the so called accumulation mode. Particles in this important size range are produced by the condensation of volatile vapours onto existing aerosol causing them to grow as well as the coagulation of nuclei mode particles. Particle removal mechanisms in the accumulation mode are generally the least efficient in the atmosphere. The chemical fractionation of typical aerosol size distributions also depends on these various processes with typical relative contributions of different species to the two modes as shown in Fig. 3.2. In the fine aerosol mode, sulfate, ammonium and nitrate ions are the major contributors with varying contributions from biogenic sources, whereas in the coarse mode ferrous and calcium ions as well as silicates from aeolian processes dominate.

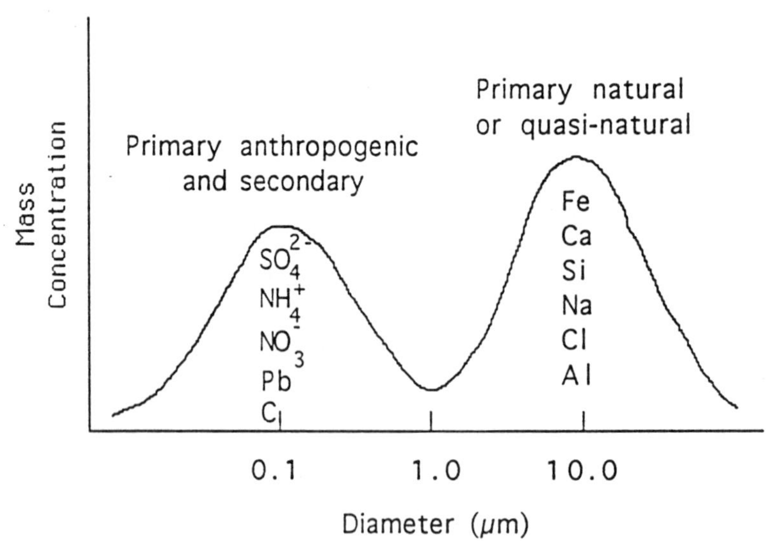

Fig. 3.2: Idealised aerosol mass distribution and fractionation into typical chemical species: Reproduced from Seinfeld (1986).

One of the ways of representing measured atmospheric aerosol size distributions is as a number concentration versus size on a log-log plot as in Fig. 3.3, which shows idealised composite plots for background or clean air aerosol size distributions compared to those from rural and urban sites, the latter being typified by an increase in number concentration in all size ranges, but particularly for particles

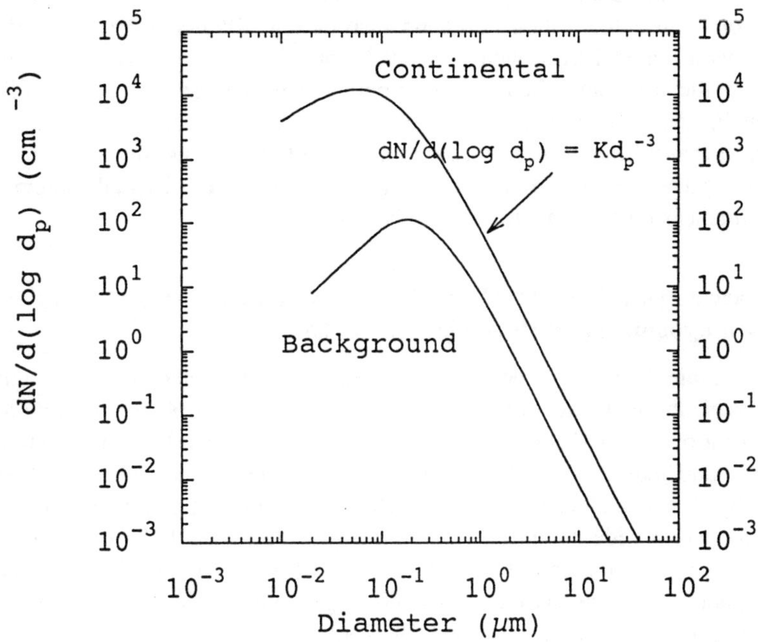

Fig. 3.3: $dN/d\log(d)$ aerosol size distributions for clean and urban air.

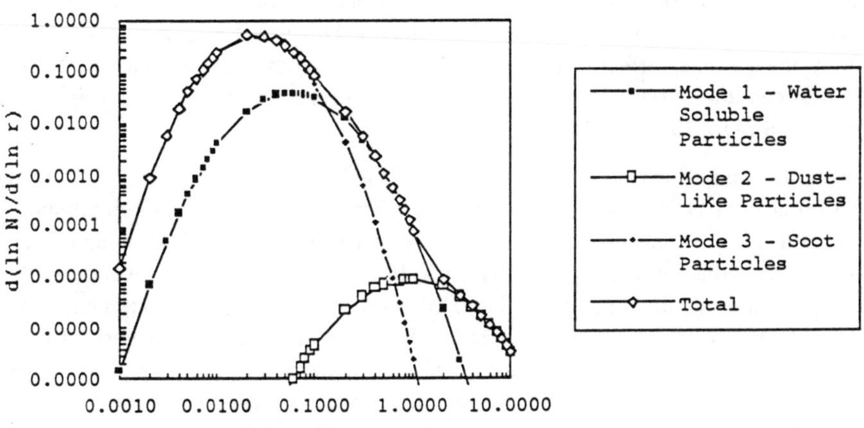

Fig. 3.4: Normalised aerosol distribution for urban air using the mode approach, Deepak and Gerber (1983)

smaller than 1 μm. The straight line portion of the graph has often been expressed in terms of an inverse power law distribution, $dN/d\log d_P = Kd^{-n}$, where $n \approx 3$ (the Junge-Power law distribution). The mass-size spectrum for aerosol at a given site is generally parameterised using the mode approach of *e.g.* Deepak and Gerber (1983), shown in Fig. 3.4, to describe the contributions from different sources. This type of presentation can, however, obscure important detail and so it is also common to use the volume or mass distribution, $dV/d\log d_P$, which generally produces a bimodal trace with a "saddle" point in the size range 1 to 3 μm.The ratio of the mass and chemical contribution between the two modes will vary depending on location and source processes as already mentioned, Fig. 3.2. As a result of these different processes fine mode particles are usually acidic in nature while coarse mode particles are generally alkaline. This dichotomy has consequences for pollutant transfer processes.

Why are aerosols important? Global reasons, radiative properties of aerosols, and some common particle production mechanisms

Aerosols are thought to be important regulators in global and local climatic processes through radiative feedback mechanisms such as absorption and scattering of solar radiation. Aerosol can therefore effectively cool the earth both directly and indirectly. In the lower atmosphere they are known to cool limited areas by direct shading from the sun, Dutton (1991), although there are no hard data for world-wide cooling by *e.g.* sulfur particles. Models have predicted that anthropogenic production of sulfur particles by fossil fuel burning in the northern hemisphere alone can cause sufficient cooling to almost offset the warming effect predicted as a result of increases in CO_2, Rodhe (1991).

Aerosols also have an indirect role in contributing to global cooling. Modification of ambient aerosol population causes a small but significant shift of the effective radius of cloud droplets and hence their albedo. This effect is, sometimes referred to as the Twomey effect. It has been shown that a 2 μm change in cloud effective radius can alter cloud albedo sufficiently to offset the warming due to greenhouse gases, (Slingo, 1990).

Within cloud droplets SO_2 can be transformed to sulfuric acid and sulfate by reaction with O_3 or H_2O_2. These droplets then evaporate leaving behind sulfate aerosol which contribute and modify the aerosol size distribution and hence indirectly radiative feedback mechanisms. In the marine boundary layer oxidation of SO_2 via biogenic dimethyl sulfide (DMS) arising from phytoplankton emissions may again reduce incoming solar radiation levels at the surface, Charlson *et al.* (1987). Bi-molecular nucleation of sulfuric acid and water vapour in clean maritime conditions can also occur, another source of new aerosol particles. The magnitude and interaction of many of these feedback mechanisms remain to be quantified.

Global atmospheric sulfur budget

At present the global sulfur budget estimate show anthropogenic sources (mainly SO_2 from fossil fuel combustion) at about 80 TgS annually, compared to the total budget of 147 TgS, (IPCC 92). This is at least as large as the natural or background emissions. In industrialised areas anthropogenic sources are about ten times larger than natural sourccs, Galloway *et al.* (1984). The atmospheric concentrations from these sources over the last century has accelerated, particularly in the northern hemisphere, although recently there has been a decline in North America and parts of Europe. These gaseous emissions have more than doubled since pre-industrial times and are important since SO_2 and oxides of nitrogen are the main precursors of acid rain. Sulfate aerosol are primarily formed from such gaseous emissions and these aerosol may have a profound influence on global climate through their role as cloud condensation nuclei and radiative scatterers and absorbers, and on local climate through acidic deposition.

Regional scale importance - current gaps in knowledge

The role of sulfur emissions on environmental pollution has been studied extensively since the 1970s where the focus of attention was on freshwater acidification. Since then research activities have focused on atmospheric pollutants and the observed widespread decline in forest and other ecosystems in Europe, Krause *et al.* (1986) and in North America, Johnson (1987). Causal links between particle and gaseous pollutant capture by natural vegetation especially forests, and their decline have been sought. As sulfur concentrations and (possibly) deposition has declined over large areas of western Europe attention has recently turned to include regional scale eutrophication of ecosystems in general by deposition and exchange of oxides of nitrogen and reduced nitrogen species such as gaseous ammonia (NH_3) and particulate ammonium (NH_4^+), Sutton *et al.* (1994). The acidifying effect of SO_4^{2-}, NO_3^- and NH_4^+ containing particles to an ecosystem may be significant. Routine monitoring of concentrations and size distributions of these species are available and used by *e.g.* the EMEP model for regional scale deposition estimates of sulfur and the TREND model (Asman and Van Jaarsveld, 1990) for NH_4^+, although the distribution of ammonia sources is not as well known. In the EMEP model (Iversen *et al.* 1990), deposition velocities are assumed to be uniform, and independent of surface type and atmospheric stability. Very different results have been obtained using these models as a result of the uncertainty in the schemes used to describe V_d.

As reported by Hummelshøj *et al.* (1993) the largest gap in our knowledge in the use of such models is the uncertainty in aerosol deposition velocity used for different types of surfaces. These uncertainties are particularly acute for large aerosol as concentration measurements of sufficient accuracy to study processes in the surface layer are difficult to make. These uncertainties were highlighted by the international workshop on the Deposition of Acidifying Substances, Lövblad *et al.* (1993);

"More research is needed to narrow the uncertainties associated with the deposition processes concerning small particles. This uncertainty is caused by the poor knowledge of the deposition velocity." The authors proceed with the following recommendations ;

> Experiments should be directed to give deposition velocities as a function of atmospheric stability, friction velocity and surface type.... inference model estimates.. can be improved with simple parameterisations of V_d which... should be verified for different surfaces. We should also add: as a function of particle size.

Similar problems exist at present with base cation deposition because of the difficulty in measuring the concentration of these particles in air accurately. These problems must be addressed in order to provide accurate critical loads assessments.

Secondary particle production mechanisms from biogenic emissions and their interaction with inorganic trace gases and aerosol

A further area of importance which has hardly been addressed in terms of field measurements is that of secondary organic aerosol and the contribution from biogenic sources. Organic particles may be formed from the photooxidiation of monoterpenes and other biogenic volatile organic compounds (VOCs) emitted from vegetation and this is thought to be an important source of secondary organic aerosol (SOA) in both rural environments and in urban areas with extensive natural vegetation, Spyros *et al.* (1991). It has been shown, *e.g.* Bidleman (1988), that significant fractions of aerosol *mass* in such areas can be attributed to organic material. Annual estimates vary from 27 to 38 %, Gray *et al.* (1986) and 9–26 % by Hildemann (1990). Typical summer midday average concentrations range between 1.6 and 5.1 $\mu g\ m^{-3}$, Larson *et al.* (1989). There are two main SOA production processes: new particle formation via gas-particle conversion resulting from the photo-oxidation of biogenic hydrocarbons in the presence of ozone, and new aerosols formed by condensation of organic gases onto existing inorganic aerosol. The main production pathways are shown schematically in Fig. 3.5.

Gas-to-particle conversion processes involving biogenic volatile organic compounds (VOCs) have been studied in the laboratory where it has been found that monoterpene is the main precursor for new aerosol formation, isoprene being about an order of magnitude less in its effectiveness as a precursor for new particle formation. β-pinene, *e.g.*, produces new aerosol by photo-oxidation in the presence of ozone, Hatakayama *et al.* (1989) and OH, Izumi *et al.* (1989). Photo-oxidation of anthropogenic hydrocarbons can also contribute to secondary aerosol, Wang *et*

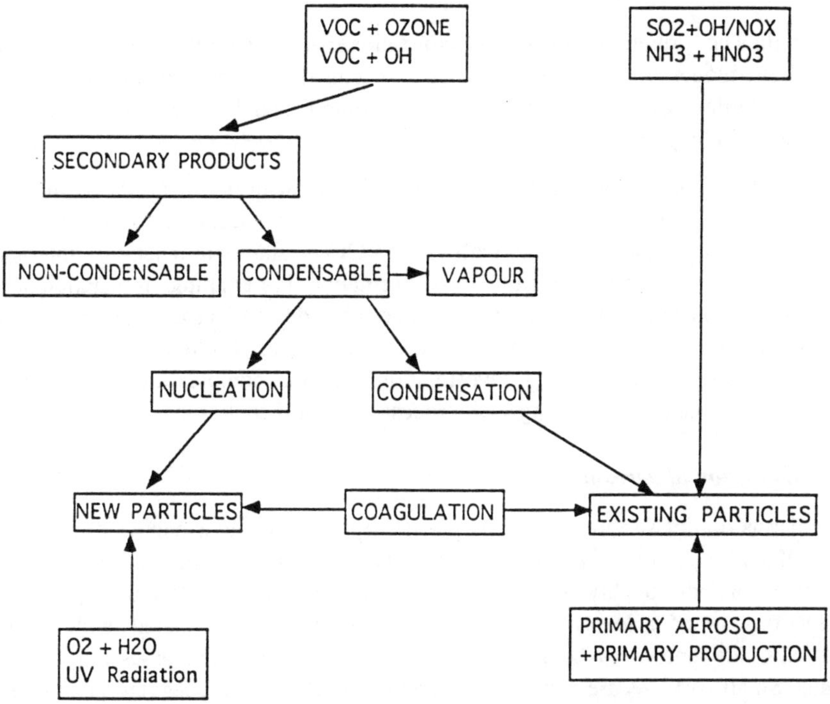

Fig. 3.5: Principal pathways for particle production from organic and inorganic trace gases.

al. (1992a). There are many uncertainties in the precise chemical pathways involved and at present only an approximate assessment of their contribution to ambient aerosol concentrations is available. Kinetic mechanisms for many of the precursor species involved in these reactions are still not known, however, Spyros *et al.*(1991) estimate that as much as 50 % of secondary, or new aerosol in "green" urban areas (such as Los Angeles) alone could be attributed to natural hydrocarbons.

Inorganic gas-particle equilibria

Gas to particle conversion can also occur with inorganic species such as SO_2 and NH_3. The presence of the latter species together with biogenic hydrocarbons produces varying effects on aerosol yields but generally leads to further increases, Wang *et al.* (1992a,b). Ambient aerosol will also be modified by the condensation of low vapour pressure products from the oxidation of organic gases, both biogenic and anthropogenic. Mechanisms and rates are still uncertain. As yet the contribution of this process to the total atmospheric aerosol loading, its spatial and temporal variability, its potential for participation in atmospheric photochemical processes in the boundary layer and the affect of anthropogenic NO_x and SO_2 emissions, Wang *et al.* (1992b), on these particle production mechanisms, as well as the subsequent affect on regional scale deposition and radiative budgets is very uncertain.The formation processes of atmospheric aerosol are thus closely coupled with both biogenic emissions of VOCs and with the presence and emissions of other volatile gases. These processes are complicated by the inherent spatial variability in biogenic sources.

Surface fluxes of ammonia will cause significant disturbance of the equilibria between ammonium nitrate and chloride aerosols. It is unclear as to the extent to which aerosol produced by both pathways (VOC or ammonia reactions) interact, and how these may affect rates of aerosol formation. For example, it is anticipated that the pre-existence of organic aerosol may provide condensation nuclei for the ammonia-nitrate-chloride reactions, and vice versa, (Sutton, 1994 private communication). Experimental investigation into the complex interactions and feedback mechanisms involving such processes are still continuing.

Dry deposition of aerosol

Dry deposition of sub-micron acidic particles provide a major contribution to the overall acidic input to ecosystems particularly for sulfate deposition. Sulfate levels have remained roughly constant over northern Europe despite significant reductions in SO_2 levels. Alkaline species, primarily associated with larger particles, are known to be important to nutrient cycling and assessing critical loads. Small particles are generally formed by condensation, coagulation and gas-particle conversion processes and include ammonium sulfate, nitrate, soot and a host of organic compounds. They also include many trace elements such as Lead and Cadmium. Typical idealised aerosol mass distribution and chemical fractionation is shown in Fig. 3.2, Seinfeld (1986).On average sulfate particles have long residence times in the atmospheric boundary layer and so the spatial variability in concentrations is reasonably well predicted by *e.g.* the EMEP model on scales of the order of 100 km. EMEP model estimates are also in line with measurements. The agreement between model and measurements for ammonium and nitrates is rather poor because of the equilibrium reactions that occur between ammonium nitrate aerosol, ammonia and nitric acid.

Size distributions of anthropogenically generated sulfate aerosol are relatively uniform with mass median diameters of typically 0.5 μm. This observation, applies to dry conditions since for relative humidities between 70 and 80 %, ammonium sulfate aerosol swell as water diffuses towards their surfaces, the rate of particle growth being predicted by standard theory. Ammonium nitrate particle size distributions are generally similar to that for ammonium sulfate although they deliquesce at lower relative humidities, > 60 %.

Particles of size greater than about 2 μm are generally produced by mechanical processes, including aeolian effects. This includes production from soils and sea spray. These large particles can be an important source of base cation deposition to sensitive ecosystems, Duyzer *et al.* (1994), due to transport from adjacent agricultural sources. Nitric acid can react with these large alkaline particles contributing to the nitrate fraction. The contribution of large particles to deposition loads is not well categorised due to limited measurements over relatively short periods and at few sites. The conclusions drawn from the Madrid BIATEX workshop, reported here, is that dry deposition velocities to forests are up to an order of magnitude larger than currently predicted by most models. If these large deposition velocities are correct, then the consequences for regional or even global scale sinks of species such as nitrogen and sulfate will be significant.

Aerosol deposition velocities

One of the crucial parameters needed to determine atmospheric inputs to ecosystems is the aerosol deposition velocity, V_d, and it is here that most uncertainty has arisen. Uncertainty in comparison of reported field measurements by different techniques has been exacerbated by an order of magnitude discrepancy between field measurements and values inferred from wind tunnel studies and theoretical models with apparently little success in unifying field and model investigations. This picture appears to be overly pessimistic and it is the contention that most of the discrepancy between field measurements can be accounted for.It is only recently that efforts have been made to compare model results with these factors taken into account. Ruijgrok *et al.* (1994) show that theoretical model deposition velocities integrated over a typical aerosol distribution with mass median diameter of 0.5 μm can be 3 to 6 times larger than the deposition velocity corresponding to an individual particle of that size. Throughflow type measurements *e.g.* by their very nature are integrating techniques providing mass fluxes with deposition velocities associated with a mass median size which depends on ambient aerosol size distributions. It is difficult to directly compare such measurements with model predictions because not enough detailed information is provided, *e.g.* friction speed, surface roughness and particle size distribution, in order to apply the necessary integrations in the models.

Uncertainties have also arisen because of the technical difficulties of performing measurements. As a result there are very few measurements available. In the past EC systems were used which incorporated optical particle counters, Seivering (1982; 1983), Neumann and Den Hartog (1985). For typical atmospheric particle size distributions these instruments have technical limitations to sampling volumes and are not able to provide statistically reliable fluxes for intermediate size particles but only for very small particles, < 0.3–0.5 μm. Conversely reliable fluxes for larger particle sizes can at present only be obtained with such instruments at the expense of information on the small particle fluxes due to electronic limitations. It is only with the advent of new fast response electrical particle counters, *e.g.* Lopez *et al.* (1994), Lamaud *et al.* (1994), that these problems have been addressed. It is important to emphasise the work by the latter since they show that filtering out of unwanted low frequency contributions to particle flux data by application of cospectral models, is vital for correct interpretation of particle fluxes. There are probably many data sets on particle fluxes which would benefit from this approach.

Mass transfer resistance analogy and nomenclature

Most predictive models and measurement techniques employ the deposition velocity concept, whereby the deposition velocity V_d is defined as

$$V_d = -\overline{w'\chi'}/\chi$$

where χ is the particle concentration, w is the vertical component of velocity attributed to the particle and $\overline{w'\chi'}$ is the vertical particle flux. The negative sign is used by convention to describe the deposition velocity as being positive to a surface *i.e.* downwards and negative as being away from the surface. The deposition velocity will depend on the characteristic vertical velocity component at the surface, which will be stability dependent, and the surface mass transfer resistance at the surface, or indeed within the surface for a porous, complex surface such as a forest canopy. The deposition velocity approach assumes that the removal rate or mass transfer at the surface is steady state and can be described by single point concentration measurements which arise in order to drive the mass transfer process by diffusion from high to low concentration regions. This approach is open to debate, particularly over forests, where mass transfer processes may not be adequately described by steady state relationships. Most models used in the micrometeorological field, however, adopt the one dimensional steady state continuity equation for simplicity viz.

$$F(d_P) = -K(z,d_P)\,\delta\chi(z,d_P)/\delta z - V_T(d_P)\chi(z,d_P) = V_d(z,d_P)\,\chi(z,d_P)$$

where $K(z)$ is the eddy diffusivity for the particles and V_T is the terminal or settling velocity for particles of size d_P. Assuming steady state conditions, *i.e.* a constant flux near the surface, then the solution to the equation is

$$V_d(z,d_P) = V_T(d_P)/[1-\exp[-(V_T(d_P)/(\kappa u_*)).\ln(z/z_o))]$$

this includes the differential effect of sedimentation on the concentration profile, *e.g.* Chamberlain (1966), Sehmel and Hodgson (1980) and Choularton and Carruthers (1986). The lower limit for deposition velocity is thus V_T. As diffusional resistance becomes less important the deposition velocity becomes increasingly greater than V_T. The most appropriate deposition velocity to use based on the resistance to mass transfer approach (for both particles and gases) is the surface deposition velocity, V_{ds},

$$V_{ds} = (V_d^{-1} - r_a)^{-1}$$

where r_a is the aerodynamic resistance between the measurement height and the surface in question, and is given by the following relation (Hicks *et al.*, 1989):

$$r_a = \frac{1}{ku_*}\left\{\ln(\frac{z-d}{z_0}) - \Psi_M(\frac{z}{L})\right\}$$

where z is the measurement height, d is the zero plane displacement, the height in the canopy where the wind is effectively zero (actually the wind is assumed to be zero at height $d + z_o$), z_o is the roughness length of the canopy and $\Psi_M(\zeta = z/L)$ is an empirical stability correction function after Businger *et al.* (1971) which depends on the measurement height and the Monin-Obukhov length scale which may be determined from local surface sensible heat and momentum fluxes. The stability correction function may also be subject to further correction terms which are themselves a function of height and site dependent, Westrate and Duyzer (1994). r_a may also be written as, r_a $\delta u/ u_*^2$ where u_* can be obtained from measurements of the local stress. Some authors use V_S instead of V_T, and V_G instead of V_d so that V_T may be used to describe the turbulent component of the deposition velocity (= $V_G - V_S$). The turbulent component of the deposition velocity is then compared with V_M the momentum deposition velocity which is an upper limit for mass transfer. This approach, suggested by Slinn (1982), is useful for comparison of a wide variety of field and laboratory results with model predictions since the surface type is essentially "normalised" (assuming that the diffusivity is adequately known).

The choice of reference height in quoting deposition velocities is also very important since the surface mass transfer resistance controls the overall deposition. The deposition velocity is relatively insensitive to reference height for particles smaller than 1 to 2 microns, however, for particles greater than about 5 μm the deposition velocity varies with reference height, with V_d declining significantly with height and approaching V_T as diffusion becomes less important. It is in selection of the reference height that discrepancies between model predictions for V_d will occur especially for particles greater than a few microns, or smaller than 0.1 μm, where again V_d increases as the reference height approaches the surface in question. Thus it is important when describing results from field measurements to provide details both of the particle size distribution as well as the reference concentration height with respect to the surface roughness and zero plane displacement as well as the friction speed and atmospheric stability under which the measurements were carried out. Deposition velocities should be proportional to friction speed for a uniform rough surface and will increase significantly with roughness length. Taking into account all these variables, model predictions suggest that particle deposition velocities in the atmospherically important size ranges may vary by up to several orders of magnitude.

A short review of aerosol deposition velocities by micrometeorological and related techniques

Appendix Table C summarises typical aerosol deposition velocities determined by various workers using micrometeorological methods for a variety of surfaces. Also included are several chemical throughfall and surface type measurements of deposition velocity obtained over forest canopies. For ease of comparison measurements have been listed which have some degree of size segregation, to avoid integration effects. Although there appears to be a variation of up to an order of magnitude in the results a large fraction of this may be due to the very different surfaces over which the measurements were made and under different atmospheric conditions. The results for snow surfaces, Duan *et al* (1988), might be expected to be very different to those for rough vegetated surfaces *e.g.* due to significant enhancement in collection efficiency, Slinn (1982). The rather small values obtained by Neumann *et al.* (1985) of around 0.5 to 1.0 mm s^{-1} may have been the result of statistical counting problems, however, their experimental arrangement was very similar to that of Sievering (1988) who used an identical optical particle counter. Most of the other results are reasonably consistent with values ranging from 1 to 6 mm s^{-1} . However, as noted by Sievering (1982, 1988), there is a significant influence on particle deposition velocity by the local stability resulting in smaller deposition velocities being observed during stable stratification and much larger velocities observed during unstable conditions. These influences are not always identified.

Fig. 3.6 shows a summary of surface deposition velocities, V_{ds}, (defined below), as a function of size, for tall vegetation. Included are the measurements reported by Waraghai and Gravenhorst (1989) using throughfall and related techniques for spruce canopies. Fig. 3.6 also compares this data with size segregated field measurements of super-micron particles (cloud droplets) reported by Beswick *et al.* (1991) and Gallagher *et al.* (1992) for a Sitka spruce canopy. Since the throughfall measurements are generally reported without associated micrometeorological information and are integrated over relatively long periods of time we can only compare un-normalised deposition velocities. Information is also not always provided to enable the surface deposition velocity to be distinguished. Despite this the overlap between the different data sets obtained using very different techniques is encouraging. The different throughfall data sets suggest a stability and wind speed and hence friction speed dependence is evident, but obviously not fully accounted for here. The magnitude and behaviour of the deposition velocities also appears partially consistent at higher wind speeds with the model predictions of Slinn (1982) and Peters and Eiden (1992). This is despite the fact that the data of Gallagher *et al.* (1992) was obtained over a snow covered canopy. This should result in a reduced collection efficiency for all particles since the snow pack would provide a smoother surface by covering micro-structure components such as needles. However, assuming the data from Waraghai and Gravenhorst (1989) was obtained at lower wind speeds, the data suggests a slight decrease with a secondary minimum before increasing again. This does not occur except at higher wind speeds in the various models mentioned earlier. There is also some evidence for a plateau or even yet another minimum between 10 and 20 μm. This latter may be due to inappropriate use of a suitable reference height as discussed earlier. The models do predict a plateau in deposition velocity for particles beyond 20 μm but the change is not as strong until one reaches particle sizes in excess of 200 μm. In general, as reported by Fowler *et al.* (1989), Beswick (1990) and Weyers *et al.* (1994), deposition velocities for these large particles closely follow that predicted by sedimentation and momentum. It would appear, however, that some of the detailed predictions by these models even for impaction processes still do not fully reproduce the field observations.

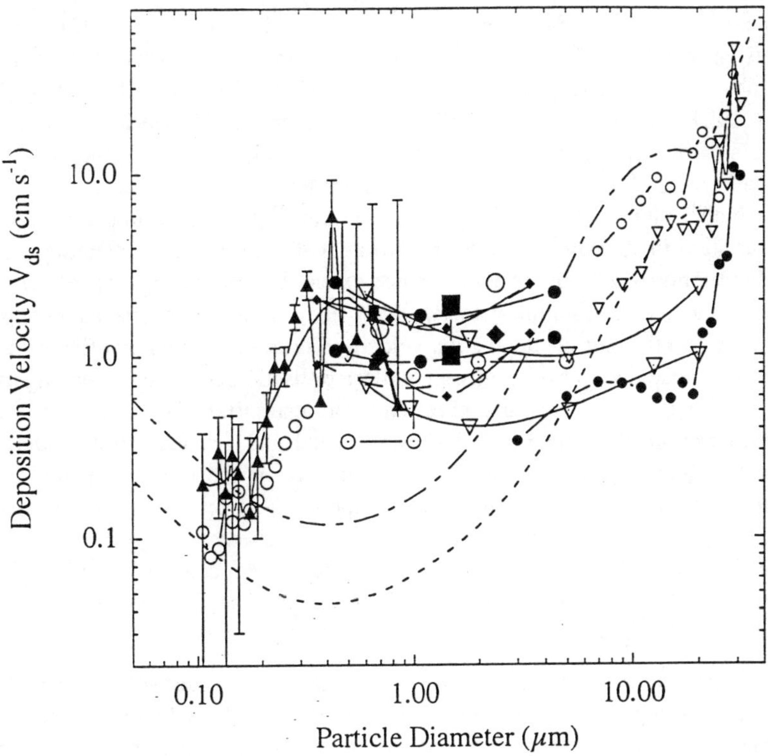

● Grosch and Schmitt (1988)

■ Brüeckman (1988)

◆ Hoefken and Gravenhorst (1983)

▽ Waraghai and Gravenhorst (1989)

● Bewsick et al. (1991) $u_* = 0.37$ ms^{-1} $z_o = 30.1$ cm

▽ Gallagher et al. (1992) $u_* = 0.71$ ms^{-1} $z_o = 15.7$ cm

o Gallagher et al.(1992) $u_* = 0.75$ ms^{-1} $z_o = 18.6$ cm

O Joutsenoja (1992) $u_* = 0.68$ m s^{-1} u=5.6 m s^{-1} $z_o = 0.08$-0.13 m

◆ Hofken et al. (1983) TF Beech

O Hofken et al. (1983) TF Spruce

⊙ Lorence (1989) Gradient Loblolly Pine, h=9 m

▲ Gallagher et al. (1995) $u_* = 0.5$ ms^{-1} $z_o = 3.5$ m

- - - Slinn (1982) Unmodified Model Result $u_* = 0.65$ m s^{-1}

Fig. 3.6: Summary of field measurements of deposition velocity, V_d (cm s^{-1}), to rough surfaces by micrometeorological and related techniques.

The greatest variability and uncertainty has aparently been found to occur for sub-micron particles. In general many experiments have shown V_{ds} for sub-micron particles to increase strongly with increasing u_* , or in conditions with strong fluctuations in wind speed, Sievering (1987), Duyzer and Bosveld, (1988), Hummelshøj (1992), Gallagher *et al.* (1992), and Joutsenoja (1992), which is consistent with theoretical predictions of impaction. Many workers, however, have also observed organised up-fluxes of particles in this size range, Sievering (1982; 1983; 1987), Katen and Hubbe (1985), and Gallagher *et al.* (1992), and this effect appears to be dependent on the nature of the underlying surface, the state of the vegetation, *e.g.* if it is senescent, on the atmospheric stability, Wesely *et al.* (1985), and whether ammonia emissions may be influencing the measurements. Some of the variation in V_{ds} will be due to the different structure of the collecting surfaces involved. This may be overcome, to some extent, by normalising the measured V_{ds} to the momentum deposition velocity, d_P $V_M \approx u_*^2/u$, which represents an apparent upper limit to the particle deposition velocity. Further confounding effects on local fluxes can arise due to re-suspension and aerosolisation processes which again can be very site specific. These processes can lead again to up-fluxes which may be a strong function of particle size and wind speed and can lead to a reduction in calculated mean deposition velocities, particularly if the instrumentation used is limited in terms of size discrimination. Some of these up fluxes are undoubtedly the result of unwanted contributions from low frequency fluxes (Fontan, private communication, Sievering, 1982). These effects can confuse interpretation of micrometeorological measurements and care should be taken to eliminate them before considering the basic underlying processes of deposition.

There are in general very few measurements of aerosol deposition by micrometeorological techniques reported for forests. Wesely *et al.* (1977) and Garland and Cox (1982) obtained results in broad agreement with model predictions for small particles and EC measurements were reported by Wesely *et al.* (1983, 1985) for a variety of surfaces including bare soil, grassland grasslandand a forest canopy forest canopies. However, Wesely *et al.* (1985), observed a diurnal variation in V_{ds} from 0.4 cm s^{-1} during the night to 1.4 cm s^{-1} during the day, much larger than predicted by models. Hicks *et al.* (1982) measured total particulate sulfur deposition rates of 0.7 cm s^{-1} to a forest using EC, again significantly larger than most model predictions but, assuming a typical mass size distribution, the results are compatible with the throughfall data presented in Fig. 3.6. Lamaud *et al* . (1994 a, b) used a new technique to determine the bulk aerosol particle flux to various surfaces. Typical deposition velocities for forest canopies were 0.8 cm s^{-1} with a strong dependence on stability, see Appendix. The influence of the ABL depth on v_{ds}, was also inferred from surface measurements of turbulent kinetic energy. Parameterisations of V_{ds} as a function of stability are given in the Appendix. These parameterisations represent the "state of the art" as far as our knowledge of aerosol dry deposition velocity is concerned and we are now confident that reliable parameterisations for the bulk aerosol V_{ds} exists for different vegetated surfaces under different stability regimes.

The variation of deposition velocity with aerosol size

Duyzer *et al.* (1994) made measurements of aerosol V_{ds} as a function of particle size by utilising the same instruments and techniques as Sievering (1988) to the Speulder Forest, in the Netherlands. Typical results are summarised in Fig. 3.6. The data apparently confirm the secondary peak or step in the behaviour of V_{ds} as suggested by the throughfall data sets but indicate a very sharp step change in behaviour below about 0.2 μm. Also shown are the EC measurements of Joutsenoja (1992) which were made over a wheat canopy, again by EC and again apparently confirming the behaviour obtained from the forest work. There also appears to be a roughness length dependence since the friction speeds for this data set are comparable to the forest data set but produced slightly smaller values of V_{ds} for all size particles when normalised by V_m. The observation that the microstructure of the collecting surfaces is a much stronger controlling factor than assumed for particles in this size range, is supported by the measurements of Duyzer and Bosveld(1986) and Neumann *et al.*(1986) which were conducted over grass. These latter measurements agree with those in Fig. 6 for the very smallest particles but not for intermediate size particles (0.2–0.5 μm). This is conclusively shown by the experimental results of Lamaud *et al.* (1994a,b,c).

The importance of stability was first suggested by Wesely *et al.* (1985), *viz* "..there seems to be considerable influence by convective motions with dimensions nearly as large as depth z_i of the mixed layer of the lower atmosphere, this is concomitant with the suggestion that rapid multi-directional flow near the fine structure in the grass canopies aids deposition processes such as particle impaction. Indeed, past work over surfaces without fine surface structure produces much smaller estimates of deposition velocity." Lamaud *et al.* (1994) highlight the additional influence of a canopy structure that is complex enough to allow for significant in-canopy flow. These factors have not been adequately addressed in wind tunnel studies upon which most model parameterisations of particle collection efficiency rely.

By overlaying the various data sets described above for rough surfaces and for similar friction speeds, even without normalisation by the usual controlling variables, a relatively clear picture of the behaviour of particle deposition is observed. The details of the observed behaviour are not, however, well replicated by the various models although the model of Peters and Eiden (1992), and the modified Slinn model used by Ruijgrok *et al.* (1994), described below, do come closest in this respect.

Some important model descriptions of particle deposition

There are many models which attempt to predict the dry deposition to a variety of surfaces, Bache (1979), Davidson *et al.* (1982), Slinn (1982), Wiman and Agren (1985), Peters and Eiden (1992). This variety may in part be accounted for the as yet unsatisfactory performance of such models. A detailed review of these models is beyond the scope of this paper, but a thorough review and intercomparison may be found in van Aalst (1986) and Ruijgrok *et al.* (1994). A qualitative review of predictive model results may also be found in Slinn (1986), where some discrepancies between various models was noted but not explained, particularly for supermicron particles where one might expect there to be good agreement. Ruijgrok *et al.* (1994) point out the significant difference between *e.g.* Wiman's formulation and that of Slinn's for sub-micron aerosol which results in a factor of 5 larger deposition velocities for these particles in the Wiman model. This discrepancy highlights some of the problems of comparing not only different models but also field measurements over different surfaces, that is the use of different parameterisations for collection efficiency and also the inconsistent use of an appropriate reference height and integrated deposition velocity.

Calculations of particle collection efficiency by vegetated surfaces reveal a strong dependence on particle size. The major ionic pollutant species, nitrate and sulfate, as discussed earlier, are generally found in the size range 0.1 to 1.0 µm where theoretical predictions initially indicated very inefficient transport to surfaces occurs, except over rough, complex canopies. Theoretical calculations have generally underestimated sub-micron aerosol deposition velocities to rough surfaces by a factor of four or five compared to most field studies *e.g.* Belot *et al.* (1976) used wind tunnel studies to predict deposition velocities of 0.5 µm aerosol to a 15 m tall pine forest. These predictions suggested that V_g (= $(r_a + r_b)^{-1}$) varied from 0.2 mm s^{-1} to 1.0 mm s^{-1} as the wind at the forest top changed from 1 to 5 m s^{-1}. More recent measurements in the field suggest that these values are significantly underestimated. Most models of particle deposition are only validated against wind-tunnel studies which may not be representative of field conditions. The general lack of investigation, until now, of the affect of thermal stability is probably the most important factor in the observed discrepancies.

Slinn model (1982)

One of the most commonly used models is that by Slinn (1982). This is a semi-analytical model which attempts to model in detail most of the processes of interception, impaction, rebound, retention, diffusion and sedimentation to and within a complex canopy taking into account the micro and macro scale structure of the canopy. Although very detailed the model is described in a functional form based on individual collection efficiencies for each of the different processes. The deposition velocity is given by

$$V_d = V_g + u_*^2/u_r \, [\, 1 + u_h/\, u_r \, \{ \, (1 - \mathrm{E})/(\, \mathrm{E} + (\mathrm{E}^{1/2} \tanh(\gamma\,(\mathrm{E}^{1/2}))\} \,]^{-1}$$

where $u_{r,h}$ are the wind speeds at the reference height and at the top of the canopy of height h, γ is a shape parameter describing the wind profile in the canopy and E is the total collection efficiency. V_g is the settling velocity which is calculated using Stoke's law. The total collection efficiency, E , is sub-divided into separate collection efficiencies for Brownian diffusion, E_B, interception, E_{IN}, and impaction, E_{IM},

$$E = R\,(E_B + E_{IN} + E_{IM})$$

where R is a rebound or bounce-off factor which reduces the overall collection efficiency. The parameterisations used for the individual collection processes are still uncertain as there is no detailed comparison with field measurements. A typical output from this model is shown in Fig. 3.7 which we have tailored to represent a particular forest canopy, Speulder forest, in the Netherlands. This is a 18 m tall Douglas fir canopy which has been the subject of several comprehensive field measurement programmes in the Dutch Priority Programme on Acidification, Duyzer *et al.* (1994), Weyers *et al.* (1994). Here we can see the predicted surface deposition velocity for different wind speeds for dry aerosol particles in the size range 0.05 to 30 μm.

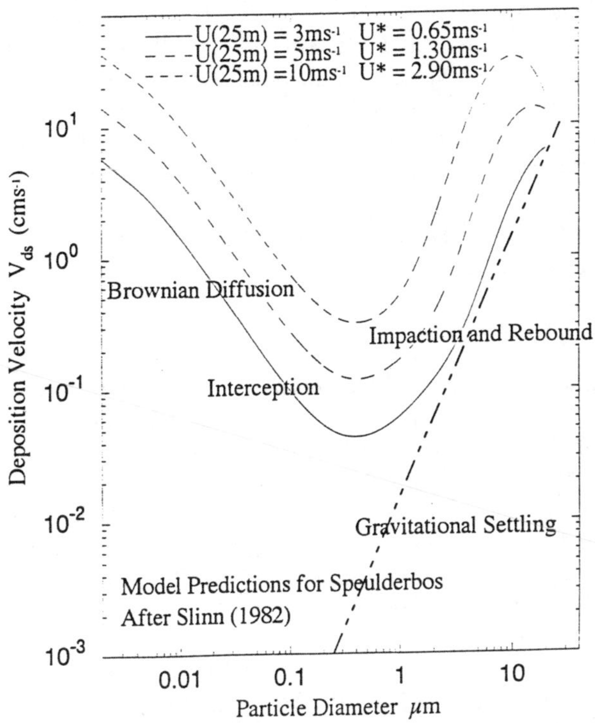

Fig. 3.7: Deposition velocity V_d (cm s^{-1}) as a function of particle size predicted by the the Slinn (1982) Model (unmodified) for Speulder Forest for different wind speeds.

The characteristic 'V' shape in the particle deposition velocity as a function of size is the result of combining the different processes of diffusion, impaction, interception and sedimentation. Beyond about 10 μm in size the deposition velocity is essentially equal to that predicted by momentum considerations alone, and which has been verified by a number of field measurements to date, primarily measurements of cloud and fog droplet deposition.

Peters and Eiden model (1992)

Peters and Eiden (1992) have modelled the size dependency of particle deposition velocity to a forest using a multi-layer analytical model. This model examines in detail the plant morphology and diffusion through the various plant sub-layers. This was achieved by introducing formulations for equivalent micro-scale roughness for needle surfaces. They also considered the phenomenon of burst effects in the sub-layer which could lead to accelerated particle transport across the viscous sublayer to the plant surface. Their results again show a minimum in the deposition velocity spectrum for particle sizes around 0.1–0.2 μm for typical wind speeds of 5 m s^{-1}. This minima moved to about 0.5 μm as wind speeds decreased to 0.5 m s^{-1}, consistent with some throughfall/surface measurements. Typical deposition velocities varied from 0.02 to 0.5 cm s^{-1} for 0.1 μm particles and from 0.02 to 2.0 cm s^{-1} for 1.0 μm particles as wind speeds increased from 0.5 to 5 m s^{-1}. The phenomenon of burst effects lead therefore to a significant enhancement in deposition velocity in the crucial 0.1–1.0 μm range and supports many results reported from field measurements. A detailed comparison between this model and measurements still remains to be carried out. The model of Peters and Eiden (1992), does seem however, to reproduce the magnitude of the observed v_{ds} values in the 0.2–0.5 μm size range at higher wind speeds observed by Lamaud *et al.* (1994) and Duyzer *et al* (1994). As yet the model is difficult to implement for quantitative comparison with current field measurements.

Hummelshøj model (1992)

Another useful model, is that by Hummelshøj *et al.* (1992). Although intended for use for smooth surfaces including water/sea surfaces, *e.g.* it includes features such as the effect of bubble bursting, (of interest to coastal and marine studies), it is analytically identical to Slinn's Model in its main features. However, it is described in terms of summation of the deposition velocities for individual processes, elucidating the physics and having the advantage that it can be much more easily modified for different surfaces. The dry deposition velocity, V_d, at a reference height h is then given by

$$V_d = (V_h + V_{g,d})(V_d + V_{g,w})/(V_h + V_d + V_{g,d})$$

where V_h is the turbulent transfer velocity from reference height, h, $V_{g,d}$ is the settling velocity for dry aerosol, $V_{g,w}$ is the settling velocity for wet aerosol at a given relative humidity, and V_d is the transfer velocity across the laminar sub-layer which depends on the diffusion velocity and the impaction velocity.

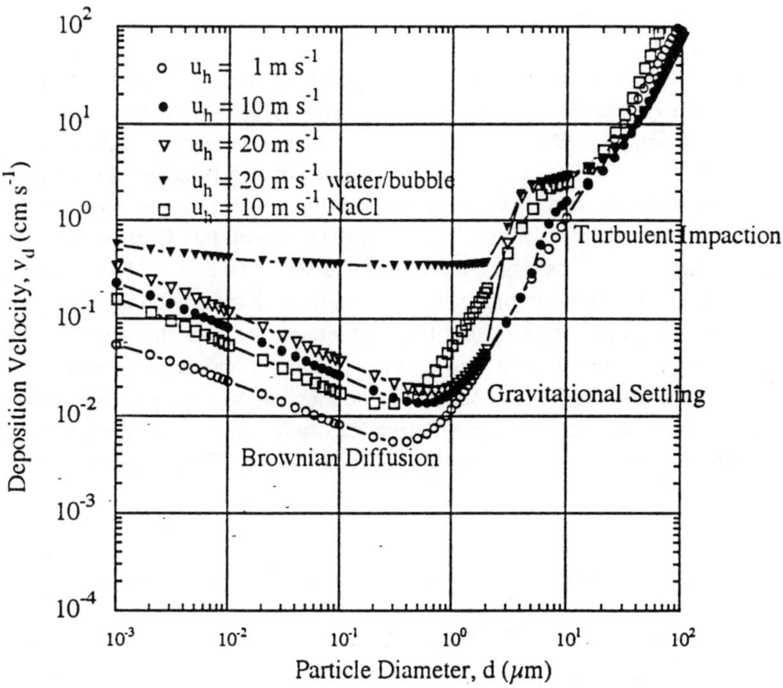

Fig. 3.8: Deposition velocity V_d (cm s^{-1}) as a function of particle size predicted by the the Hummelshøj (1992) Model for a mud/estuary site for ammonium sulfate and sodium chloride particles at different wind speeds at h = 10 m.

Fig. 3.8 shows the results of computer model calculations, based on this model, of the aerosol deposition velocity as a function of particle diameter to a smooth surface (a mud flat) for different wind speed cases and at a relative humidity of S = 98 %. The large increase in deposition velocity at high wind speeds for larger particles is due to the increasing importance of gravitational settling and impaction, *i.e.* the momentum deposition dominates at higher wind speeds. The model also includes the growth of wet particles as a function of relative humidity and the solute content of the particle, based on the parameterisation suggested by Fitzgerald (1975), something which most other models do not include. The inclusion of relative humidity effects is found to be important in the size range 1 to 10 μm, with the minimum value of V_d shifting towards smaller sizes as the humidity increases. The model calculations shown Fig. 3.8 are for ammonium sulfate aerosol particles but we have also shown the deposition velocity for NaCl particles for wind speeds at 10 m s^{-1}. As discussed by Hummelshøj *et al.* (1992) there is still uncertainty surrounding the correct Brownian diffusion parameterisation to use in such models even for an apparently simple surface such as water.

Ruijgrok et al. model (1994)

This model is essentially the Slinn Model (1982), however, in light of experimental evidence it has been modified. The usual Slinn parameterisation for collection efficiency by Brownian diffusion has been replaced by that suggested by Wiman and Agren (1985) which allows particles over a broader size range to be collected by diffusion and this results in a significantly larger deposition velocity. Fig. 3.9, reproduced from Ruijgrok *et al.* (1994), shows the effect of the different parameterisations on V_d as a function of particle size. The authors have included in their model the growth in particle size due to relative humidity changes for aerosol of different chemical composition. They integrated V_d over the size distribution function using typical values for mass median diameter and the geometric standard deviation of the distribution to obtain a "bulk" deposition velocity. They have also carried out the most complete quantitative sensitivity analysis of the uncertainties in the parameterisations used in the Slinn Model to date. They concluded that the

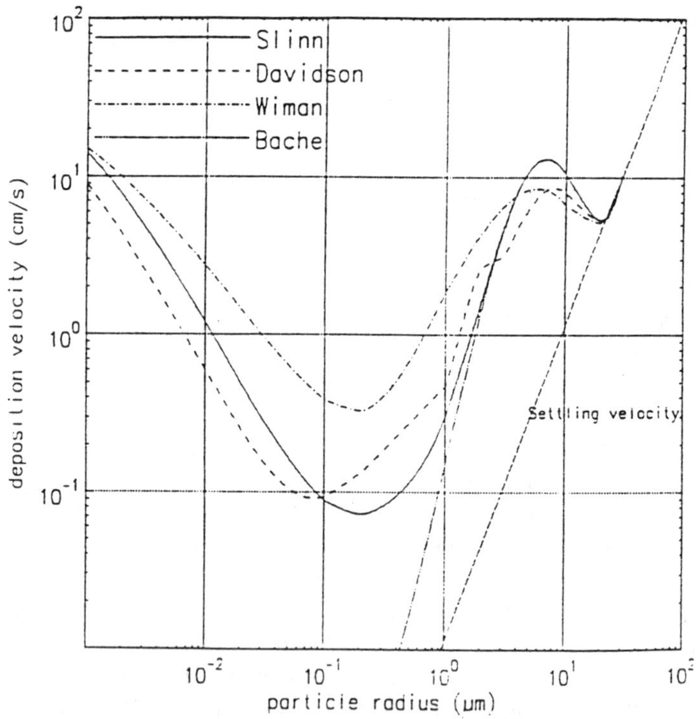

Fig. 3.9: Variation in predicted deposition velocity V_d (cm s^{-1}) as a function of particle size using the Slinn (1982) model for a forest canopy with different formulations for diffusion, interception and impaction based on models of Bache (1979), Davidson *et al.* (1982), Wiman and Agren (1985) and Slinn (1982). reproduced from Ruijgrok *et al.* (1994).

model was very sensitive, particularly in the 0.1 to 1.0 μm size range, to the input conditions used, such as wind speed, roughness length and zero plane displacement. Typical integrated values of V_d varying from 1 mm s^{-1} to 5 cm s^{-1} were obtained.

Using this model the authors successfully reproduced integrated deposition velocities measured at Speulder Forest for SO_4^{2-}, NO_3^-, ^{214}Pb and fog droplets as a function of friction speed. They found that on average the bulk deposition velocity for SO_4^{2-} was 1.3 ± 1.2 cm s^{-1} during daytime,(but still obtained large deposition velocities during stable conditions at night-time, 1.0 ± 1.4 cm s^{-1}).

Typical model flux estimates for the contribution by dry deposition to total inputs to this ecosystem were significant at 18 % for SO_4^{2-}, 23 % for NH_4^+ and 38 % for NO_3. The authors concluded that although useful the uncertainties in the model were such that simple parameterisations would provide just as accurate estimates for V_d.

Turbulent deposition of cloud water droplets

The interception of particles, dry aerosol, cloud or fog droplets and rain, by forests has long been considered an important mechanism providing large pollutant inputs to these canopies than to shorter vegetation *e.g.*, White and Turner, (1970) and Mayer and Ulrich (1974). Once aerosol particles become incorporated within cloud droplets the effective deposition velocity and pollutant flux is substantially enhanced. A causal link between cloud deposition and forest decline was deduced from observations that forest decline in North America (Johnson and Siccama, 1983) and northern Europe (Schut and Cowling, 1985) increased with altitude where the most likely contributory factor was interception of fog and orographic hill cloud containing high concentrations of pollutants. In fact the importance of the turbulent deposition and subsequent phytotoxic effects was first highlighted by Shuttleworth (1977), Dollard *et al.* (1983) and Unsworth *et al.* (1984) although original work on capture of fog to vegetation may be attributed to Hori (1953).

Aerosol composition within the polluted boundary layer is dominated by SO_4^{2-}, NH_4^+,NO_3^- and H^+ ions and these may act as cloud condensation nuclei if the surrounding air is raised to the appropriate supersaturation where activation can occur allowing growth by diffusion of water vapour to their surface to form cloud droplets. This readily occurs if the air is orographically raised over elevated terrain, the updraught speed controlling the peak supersaturation reached. On rising the air cools adiabatically and at a given altitude, called the lifting condensation level, an altitude governed by the lapse rate within the boundary layer, most of the aerosol mass will be activated, and cloud droplets will form. Once activated, the initial growth of the cloud droplets is very rapid so that within a short distance of cloud base the concentration of the major ionic pollutants will reach a maximum. As the air continues to rise, cooling at the reduced saturated adiabatic lapse rate, the cloud droplets continue to grow but at a reduced rate. The total cloud liquid water content increases with altitude roughly at about 0.15–0.20 g m^{-3} per 100 m,(depending on altitude), assuming no entrainment of air into the cloud occurs. The ionic concentration of the cloud water thus declines with increasing height.

This is graphically illustrated in Fig. 3.10, from Fowler *et al.* (1988), which shows the variation of concentration of the major ions in cloud water with height within an orographic hill cap cloud, at the Great Dun Fell site in the UK.Also shown is the corresponding concentrations in rainfall as a function of height over the hill. The enhancement in rainfall concentration is due to the seeder-feeder mechanism, Browning *et al.* (1974), whereby relatively clean rain from a frontal seeder cloud falls through the relatively polluted cap cloud. The rain drops efficiently sweep out the cloud droplets increasing the concentration within the rain and the total ionic flux. The enhancement factor typically varies from 2 to 6 depending on local meteorological conditions and the scale of the hill.

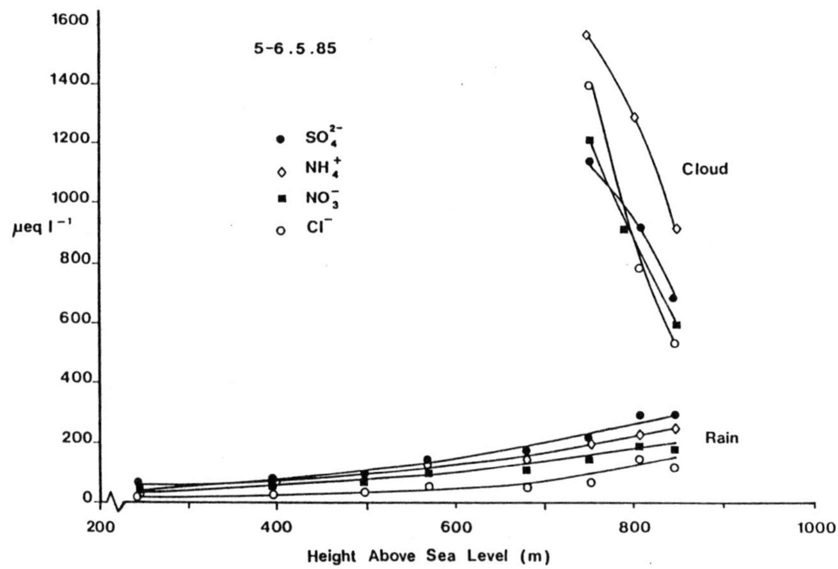

Fig. 3.10: Variation in ionic pollutant concentration in rain and cloud water as a function of elevation over a hill, Great Dun Fell (850 m a.s.l. Reproduced from Fowler *et al.* (1988).

The factors which determine cloud frequency, chemical composition and transfer rates at elevated sites have been investigated by a number of workers, *e.g.*, Lovett (1984), Milne *et al.* (1988), Kroll and Winkler (1989), Fowler *et al.* (1989,1990),Vong *et al.* (1990, 1995). The results show that at some coniferous forest sites at high altitudes the input of trace substances by cloud and fog interception cam be as high or higher than that from rain, particularly nitrate deposition. For lowland sites the contribution by fog deposition may however be as low as 5 %. Fowler *et al* (1988) suggest that for much of Western Britain at altitudes greater than 600 m cloud water deposition adds about 12 % to the total precipitation column but contributes 40 % of additional solutes to the wet deposition. This process significantly enhances both sulfur and nitrogen inputs to upland regions. Milne *et al.* (1988) estimated that ~ 33 % by volume of water input to a forest site, (Castlelaw hill, 480 m a.s.l.), may be due to cloud water deposition. A review of cloud water deposition estimates elevated forest regions in the eastern United States is given by Vong (1991) with typical estimated contribution to the total precipitation (rain + cloud + dry deposition) varying from 25–50 %, depending on the site elevation. Sites above 1500 m received between 45–80 % of total SO_4^{2-} deposition by cloud water.

There have been several detailed investigations into the turbulent transport of cloud droplets to vegetated surfaces using micrometeorological techniques. The first field work was performed by Dollard and Unsworth (1982) who used filter techniques to measure the total cloud water gradient above tussock grass tussock at an elevated moor, Great Dun Fell, in the UK. They showed that the cloud water flux was close to that predicted from momentum considerations alone, *i.e.* $V_d \approx V_m$. Further work at the same site, Gallagher *et al.* (1988), Morse (1988) used optical particle counters at different heights to determine cloud droplet V_{ds} as a function of droplet size using the flux gradient method. Care must be taken when interpreting such data as the effect of sedimentation on the cloud water flux profile should be taken into account, although this is generally small for particles less than 20 μm in size, (Chamberlain, private communication). The above work again showed that V_d the mass median droplet sizes corresponded closely to that for momentum, however, a strong variation in V_{ds} was observed for particles less than 10 μm in size. Since the contribution to the total cloud water flux from these particles sizes was small due to the observed cloud drop size distribution, they concluded that it was adequate to use the momentum deposition velocity in most models of cloud water deposition. An interesting and novel approach to cloud water deposition measurement was employed by Fowler *et al.* (1988) whereby a lysimeter was used in reverse to "weigh" the depositing cloud water. The rate of increase in weight of the lysimeter with time was used to estimate the total cloud water flux.

Beswick (1990), and Beswick *et al.* (1992) used the EC technique to measure V_{ds} as a function of particle size to a coniferous forest. Very similar results were obtained compared to those measured over tussock grass by Morse (1988) and Gallagher *et al.* (1988) with the same variation of V_{ds} with particle size observed for particles less than 10 μm in size, see Appendix and Fig. 3.6. Similar measurements to a snow covered forest, Gallagher *et al.* (1992) revealed the influence of collection efficiency on cloud water deposition. Ibrahim *et al.* (1983) measured the dry deposition to snow and showed that for supermicron particles the deposition velocity was smaller than for grass while for sub-micron particles it was greater.

Wyers *et al.* (1994) as part of BIATEX and Dutch Priority Programme on Acidification measured the bulk cloud water flux to a lowland forest canopy during fog events, Speulder Forest. The techniques used were similar to those of Beswick *et al.* (1992). They found a strong correlation of the turbulent deposition velocity with friction speed, confirming that transport of the larger droplets, which dominate the cloud water flux, was close to that for momentum. The *turbulent* deposition velocity, V_T, was found to be about half that compared with V_M and consistent with the measurements by Beswick *et al.* (1992). Vong *et al.* (1995) showed that evaporative effects significantly influence the fluxes for small cloud droplets and lead to upward fluxes for these particles. This phenomenon has been qualitatively observed over a number of forests canopies subsequent to wetting by rainfall or dew formation. This phenomenon may also cause significant leaf surface phytotoxic effects, Unsworth (1984).

The concentration of pollutants in cloud droplets depends strongly on the droplet size, with smaller droplets being more concentrated due to the large surface to volume ratio. This variation is important for a range of processes including cloud droplet deposition to vegetation, human health as a result of inhalation of water droplets, in-cloud chemical transformation processes, accretional growth of precipitation drops and ice crystals and precipitation chemistry, Lovett *et al.* (1982), Seinfeld (1986) and Borys *et al.* (1988). Model calculations, *e.g.* Coe *et al.* (1991), Bower *et al.* (1991), comparing droplet deposition fluxes of sulfate with bulk aqueous phase chemical and drop size specific chemical descriptions suggest that in general this will not significantly affect deposition patterns over hills. In contrast conversion processes involving ammonia/ammonium will be critically dependant on the chemical composition and pH of cloud droplets as a function of drop size. Ammonia is the principal basic gas in the atmosphere and tends to be taken up by acidic particles and outgassed by alkaline particles. Comparison of bulk aqueous phase cloud chemistry models and explicit drop size physico-chemical models, Bower *et al.* (1995) indicate that reduced nitrogen fluxes may be significantly influenced by in cloud processing. In-cloud sulfate production by the oxidation of SO_2 with O_3 (and H_2O_2) is directly linked with ammonium aerosol production. (in the absence of other salt forming species). Various scenarios dependent on summer winter variations in the concentration of available oxidants of ozone and hydrogen peroxide, and ammonium compensation points for the terrain downstream of cloud processing events are discussed in detail by Bower *et al.* (1995). On average a large fraction of the ammonia entering a cloud will be converted to ammonium aerosol. This air, depleted in NH_3, will have the potential to collect additional ammonia from new sources, including ammonia emitted from fertilised ground downwind of the cloud, effectively increasing the transport range of reduced nitrogen species altering the deposition pattern for NH_x.

The presence of nitric acid will also influence in-cloud conversion processes by forming ammonium nitrate in the aerosol stream leaving the cloud which will then be deposited. These in-cloud processes could provide important pathways for large inputs of NH_x to terrestrial ecosystems. Such models are currently being tested by field experiments, however, these are hampered due to the difficulty in constructing instruments which can provide high resolution measurements of drop size chemical composition, Collett *et al.* (1994). In addition long term measurements of cloud water deposition with instrumentation to measure cloud liquid water and droplet size are rare because of the difficulty and cost of maintaining the instrumentation for such a programme. One approach would be to estimate cloud liquid water content from available visibility networks, however, since the parameterisation used to compute liquid water content from visibility is sensitive to the drop size distribution, which may vary considerably in time and space, uncertainties using this method may be large.

Caveat to cloud water fluxes

Recent multi-height eddy correlation measurements of cloud droplet deposition to forests, (Vong and Kowalski, private communication) suggest '...that the additional terms in the cloud water flux budget may be important', *viz* LWC 'storage', d LWC/dt = advection + source (condensational growth) + flux divergence in the vertical. They have observed that the 'net flux' may change with height which may be taken account of. In other words, when liquid water is produced during condensational growth simultaneous to droplet removal by turbulent transfer and impaction, the net flux at one level is smaller than the surface flux and can be partly or totally offset by an upward flux due to condensational growth and/or evaporation. This will, however, depend on the type of cloud. (Kowalski *et al.* (1996) have, to date, provided the most detacted and complete evaluation of cloud water flux measurements using the eddy correlation technique).

Relative importance of deposition processes

Table 3.1 shows the typical contributions to annual deposition by dry and wet deposition processes for a forest site, 300 m a.s.l., estimated by Fowler *et al.* (1989), and depicted in Fig. 3.11, in order to assess the effect of increased aforestation on sinks of sulfur and nitrogen. They found that the magnitude of the increase in deposition was very site specific with low elevation sites showing little effect of aforestation on sulfur inputs whereas NH_3 and HNO_3 may be considerably enhanced. Table 3.2, from Fowler *et al.* (1990) shows the relative contributions to wet deposition by rainfall and cloud at a “high” elevation site in the UK, Great Dun Fell, 850 m a.s.l.

Estimates of Atmospheric Inputs to Vegetated Ecosystems

The effect of aforestation of moorland sites at these altitudes is to increase the cloud water deposition contribution for various ionic species from 12 to 44 %. Typical variations in the contribution to total cloud water deposition by turbulent deposition and sedimentation is shown in Table 3.3, from Wyers *et al.* (1994), for a lowland forest, Speuld, in the Netherlands. This will be very site specific with the turbulent component dominating the sedimentation component at high elevation sites where wind speeds are generally greater. At lowland sites where fogs will be more frequent, will be less turbulent and which have much larger drop size distributions the sedimentation flux may be comparable to or greater than the turbulent flux.

Table 3.1: Atmospheric inputs to Kielder forest (300 m a.s.l.) in kg/ha/yr; from Fowler *et al.* (1989).

	Wet deposition		Cloud deposition		Dry deposition SO_2,NO_2, HNO_3,NH_3		Total	
	S	N	S	N	S	N	S	N
Moorland	13.0	8.0	1.3	0.4	3.1	4.0	17.5	12.4
Forest	13.0	8.0	6.5	1.9	3.1	13.5	22.7	23.4

Table 3.2: How important is cloudwater deposition ? from Fowler *et al.* (1990)
Great Dun Fell, 850 m a.s.l., 2000 hr yr^{-1} in cloud, Average LWC 0.3 g m^{-3}.

	z_o m	V_g cm s^{-1}	PPT mm yr^{-1}	%PPT	%Annual wet deposition [SO_4, NO_3, H^+, NH_4^+]
Moorland	0.01	5	116	6	12
Forest	1.00	20	432	22	44

Table 3.3: Calculated turbulent cloud water fluxes (F_t) and measured cloud/fog water fluxes (Fg) in mg/m²/s; from Weyers *et al.* (1994)

Period	F_t	F_g	F_{total}
December	1.47	1.97	3.44
February	1.76	0.26	2.02
Total	1.72	0.51	2.23

Table 3.4: Estimates of total atmospheric inputs at Speulder Forest (mol /ha/yr) from Ruijgrok *et al.* (1994). Measurements from Erisman *et al.* (1993) and Wyers *et al.* (1994)

	SO_x	NO_y	NH_x	Na	Cl	Ca	K	Mg	Total
Gas	663	356	1443	0	0	0	0	0	2462
Aerosol	216	414	645	599	855	101	34	118	2982
Fog	34	23	96	2	4	1	1	0	161
Total Input	913	793	2184	601	859	102	35	118	5605
Uncertainty(%)	40	40	50	50	50	50	50	50	
Net Throughfall	924	394	1728	692	802	159	305	138	
Net TF-canopy ex.	924	394	1983	692	802	86	35	98	
Uncertainty(%)	30	40	40	30	30	40	50	40	
Average V_d (cm s^{-1})	1.3	1.9	1.0	5.0	5.0	5.7	2.8	5.8	
σV_d	1.5	2.4	1.1	4.1	4.1	4.1	4.4	2.9	

Table 3.4 shows total atmospheric inputs to Speulder forest estimated by Ruijgrok *et al.* (1994) based primarily on gradient and throughfall measurements and inferential modelling using the aerosol deposition velocity parameterisations and collection efficiencies in Table 3.2, together with equations 10–14 and results from Erisman *et al.* (1993), Vermeulen *et al.* (1994) and Wyers *et al.* (1994). The atmospheric deposition of SO_x is comparable to the net throughfall flux that was measured. Taking into account the influence of the canopy on the throughfall fluxes, Ruijgrok *et al.* (1994) show these fluxes compare reasonably well. A detailed examination of the errors associated with these estimates suggest that the model of Ruijgrok *et al.* (1994) does not significantly underestimate the deposition velocity for the different atmospheric components provided the Wiman parameterisation for collection efficiency is used. They also concluded from their error analysis of the available measurements and model comparisons that the filter technique was generally unreliable for flux measurements. Their overall conclusion was that deposition velocities to rough surfaces are much higher than previously thought and confirm the modifications made by Erisman (1993) to the Wesely *et al.* (1985) parameterisation in order to obtain higher values of V_{ds} for acidifying particles. The total uncertainties in the deposition velocities used are about a factor of two. These were primarily obtained from gradient measurements. Use of the parameterisations obtained by Lamaud *et al.* (1994) by EC measurements may improve on this. The final deposition velocities obtained by Ruijgrok *et al.* (1994) are in good agreement with the measurements reported by Lamaud *et al.* (1994), Beswick *et al.* (1994) and Duyzer *et al.* (1994).

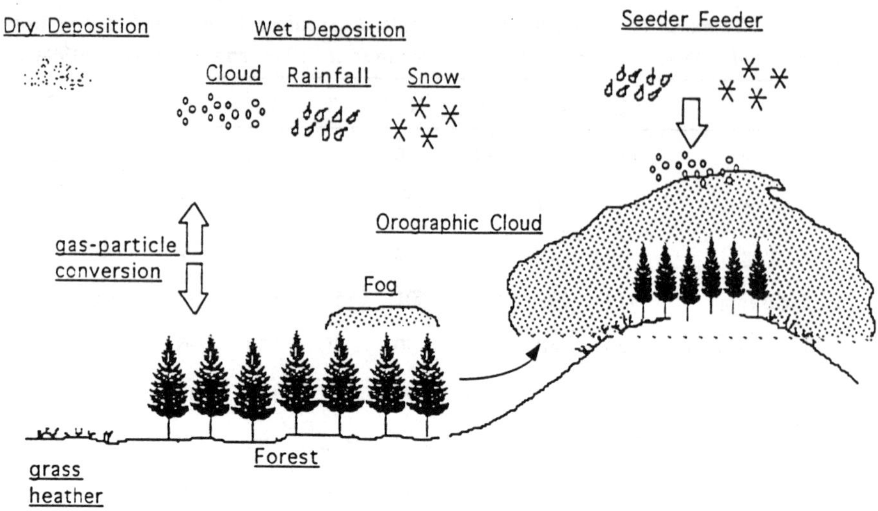

Fig. 3.11: Main deposition and removal pathways for atmospheric aerosol. Adapted from Fowler *et al.* (1989).

Conclusions

We now know that the deposition for sub-micron aerosol to forest canopies is large ≈ 1 cm s^{-1} or more for typical friction speeds and daytime stability conditions. We know that deposition velocity depends strongly on friction speed and stability and we know what that dependence is. We can identify differences in deposition velocity between forest canopies and short grass or soil surfaces. What do these results mean for estimates of total inputs of sulfur and nitrogen to forested regions? Have we been underestimating them? Probably. This means that critical load assessments and long range transport contributions to various atmospheric inputs may be in error, simply because a single parameter has not been adequately researched in the field. The technology has certainly been available to us to address most of the uncertainties associated with this problem, yet they have remained a major stumbling block for the last decade or more. A concerted effort is required to integrate the diverse and disparate measurements and modelling exercises. BIATEX has gone some way towards achieving this.

More field experiments are needed. Most of the uncertainties in current models appear to stem from parameterisations of the impaction and interception efficiency for sub-micron particles at leaf and needle surfaces under unstable conditions. Current wind tunnel derived parameterisations of collection efficiencies are inadequate. Field experiments or better wind tunnel studies are needed. In field experiments particle size information is sometimes inadequately known and relative the importance between concentration and mass fluxes are not adequately resolved by different measurement techniques. Canopy architecture and meteorological and micrometeorological information is not reported in sufficient detail to generalise results for comparison with predictive models. Measurements of the efficiencies of the individual transport mechanisms of diffusion, impaction and interception for aerosol as a function of size to and within complex canopies are required to address these uncertainties. Techniques for high temporal resolution measurements of aerosol fluxes as a function of their chemical composition are also desperately needed, traditional impactors being inadequate for such process investigations.

Addressing the directive from the BIATEX co-ordinator we have summarised the most important micrometeorological deposition measurements made to date for vegetated surfaces, see the Appendix, and listed the recommended parameterisations for aerosol and cloud water deposition. Regrettably it is not within the scope of this simple review to extend our discussions in detail to other types of surfaces. We hope that the results reported here may aid in stimulating further research into the interaction of particulates from the atmosphere with complex surfaces.

Acknowledgements

The authors would like to thank the following for useful contributions, discussions and the making available of recent reports, particularly those involved in the Dutch Priority Programme on Acidification, M. Sutton, Karl Beswick, Niels Otto Jensen, H. Westrate, S. Slarina, Rick Vong, Dave Carruthes, Tom Choularton and Dave Inglis.

References

Aalst, R.M. van (1986). Dry deposition of aerosol particles. In: S.D. Lee, T.Scheider, L.D. Grant, P.J.Verkerk (eds), *Aerosols.* Lewis Publ., Chelsea (MI), pp. 933–949.

ApSimon, H.M., Kruse,M., Bell,J.N.B (1987) Ammonia emissions and their role in acid deposition. *Atmos.Environ.* **21**, 1947–1987.

Asman W.A.H., Jaarsveld, H.A. van (1991). A variable-resolutio statistical transport model applied to ammonia and ammonium. RIVM, Bilthoven, report 228471007

Bache, D.H. (1979). Particle transpor within plant canopies. II prediction of deposition velocities. *Atmos. Environ.* **13**, 1681–1687.

Belot, Y., Baille, A., Delmas, J.L., (1976), Modéle numérique de dispersion des pollutants atmosphérique en présence de couverts végétaux. *Atmos.Environ.* **10**, 89–98.

Bache,D.H. (1979) Particle Transport Within Plant canopies - I. A Framework for Analysis. *Atmos. Environ.* **13**, 1257–1262.

Bache, D.H.(1979) Particle Transport Within Plant canopies - II. Prediction of Deposition Velocities. *Atmos. Environ.* **13**, 1681–1687.

Beswick,K.M., Hargreaves, K.J., Gallagher, M.W., Choularton,T.W., Fowler,D., (1991) Size-resolved measurements of cloud droplet deposition velocity to a canopy using an eddy correlation technique. *Quart. J. Roy. Met. Soc.* **117**, 623–645.

Beswick,K.M., M.W.Gallagher, P.Hummelshøj, K.Pilegard, N.O.Jensen, J.Duyzer (1994) Aerosol exchange to Speulder Forest. in: P.M. Borrell, P. Borrell, T. Cvitaš, W. Seiler (eds), *Proc. EUROTRAC Symp. '94*, SPB Academic Publishing bv, The Hague 1994, pp. 683–688.

Beswick,K.M. (1990) Measurements of Cloud Water Deposition to Forests and Moorland. PhD Thesis. University of Manchester Institute of Science and Technology.

Bower, K.N., Wells, M.,Choularton,T.W.,Sutton,M.A. (1995) A Model of Ammonia/Ammonium Conversion and Deposition in a Hill Cap Cloud.*Quart. J. Roy. Met. Soc.* **121**, 569–591.

van Breeman,N., Burrough, P.A.,Velthorst, E.J.,van Dobben, H.F., de Wit, T., Ridder, T.B. Reijnders, H.F.R. (1982) Soil acidification from atmospheric ammonium sulfate in forest canopy throughfall. *Nature* **299**, 548–550.

Brueckmann,A. (1988). Radionuklidbilanz von 4 Waldoekosystemen nach dem Reaktorunfall in Tschernobyl und eine Bestimmung der trockenen Deposition. Diplomarbeit Forstwissenschaftlicher Fachbereich, Göttingen.

Chamberlain, A.C., (1960), Aspects of the deposition of radioactive and other gases and particles. Int. *J. Air. Pollut.* **3**, 63–68.

Chamberlain, A.C. (1966), Transport of Gases to and from Grass and Grass-Like Surfaces, *Proc. Roy. Soc. London* **A**, 236–265.

Chamberlain, A.C. (1975), The moverment of particles in plant communities. In: .J.L.Monteith (ed), *Vegetation and the Atmosphere,* Academic Press, London, Vol.1, pp. 155–201.

Chamberlain, A.C.(1983), Deposition and Resuspension. In: Prupacher *et al* (eds), *Precipitation Scavenging, Dry Deposition and Resuspension.* Elsevier Science Publishing Co.Inc.

Coe, H., Choularton, T.W., Carruthers, D.J., Gallagher, M.W., Bower, K.N. (1991) A model off occult deposition applicable to complex terrain. *Quart. J. Roy .Met. Soc.* **117**, 803–823.

Collett, J.Jr., Lovinelli,R., Demoz,B.(1995) A Three-stage cloud impactor for size resolved measurements of cloud drop chemistry. (1995), *Atmos.Environ.* in press.

Collett, J.Jr., Oberholzer, B., Staehelin, J. (1993) Cloud Chemistry at Mt. Rigi, Switzerland: dependence on drop size and relationship to precipitation chemistry, *Atmos. Environ..* **27A**, 3–42.

Collett, J.L.Jr., Bator, A., Rao, X., Demoz, B.B. (1994) Acidity variations across the cloud drop size spectrum and their influence on rates of atmospheric sulfate production. *Geophys. Res. Lett.* **21**, 2393–2396.

Davidson,C.I., Miller, J.M., Pleskow, M.A. (1982) The influence of surface structure on predicted particle dry deposition to natural grass canopies. *Water Air and Soil Pollut.* .**18**, 25–44.

Dollard, J.G., Unsworth, M.H., Harvey, M.J. (1983), Pollutant transfer in upland regions by occult deposition. *Nature*. **302**, 241–247.

Dröscher,F., Nickel,J., Mikisch, E., (1989) Measurement of dew and fogwater deposition in forest stands. In: H.-W.Georgii (ed), *Mechanisms and Effects of Pollutant-Transfer into Forests*, Kluwer Academic Publishers, pp, 205–212.

Duan,B. C.W.Fairall, D.W.Thomson, (1988),Eddy Correlation Measurements of the Dry Deposition of Particles in Wintertime *J. App. Met.* **27,** 642–52.

Duyzer,J.H, H.L.M Verhagen,J.H.Westrate,F.C.Bosveld (1992), Measurement of the dry deposition flux of NH_3 onto coniferous forest. *Environ. Pollut.* **75**, 3–13.Duyzer, J.H., Bosveld, F.C. (1988) Measurements of the dry deposition of fluxes of O_3, NO_x, SO_2, and particles over grass/heathland vegetation and the influence of surface inhomogeneity. MT-TNO Delft, report R 88/111.

Duyzer, J., Westrate, H., Beswick, K., Gallagher, M.W., (1994), Measurements of the dry deposition flux of sulfate and nitrate aerosols to the Speulderbos using Micrometeorological Methods. TNO Report TNO-MW-R 94/255 Oct.21,1994 Dutch Priority Programme on Acidification, TNO Institute of Environmental Sciences, P.O.Box 6011, NL-2600 JA, Delft Netherlands.

Duyzer, J.H., J.H. Westrate, H.S.M.A. Dierderen, A. Vermetten, P. Hofschreuder, P. Wyers, F.C. Bosveld, J.W. Erisman, (1994) The deposition of acidifying compounds and ozone to the Speulderbos derived from gradient measurements in 1989 and 1989. TNO Report R94/095.

Duyzer, J.H., Bowman, A.M.H., Diederen, H.M.S.A., von Aalst, R.M. (1987), Measurements of dry deposition of NH_3 and NH_4 over natural terrains. Research Report no. R87/273, Netherlands Organisation for Applied Scientific Research (TNO)

Duyzer, J., H. Westrate, K. Beswick, M. Gallagher, (1994). Measurements of the dry deposition flux of sulfate and nitrate aerosols to the Speulderbos using Micrometeorological Methods. TNO Report, TNO-MW-R 94/244, Oct. 21, 1994 No. 51628, Dutch Priority Programme on Acidification.

Duyzer ,J.H., Bosveld, F.C. (1988) Measurements of the dry deposition of fluxes of O_3, NO_x, SO_2, and particles over grass/heathland vegetation and the influence of surface inhomogeneity. MT-TNO Delft, report R 88/111.

Erisman, J.W. (1993) Acid deposition onto nature areas in the Netherlands. Part I: Throughfall measurements compared to deposition estimates. *Water Air and Soil Pollut.* **71**, 51–80.

Erisman, J.W., Mennen, M.G., Hogenkamp, J.E.M. Kemkers, E., Goedhart, D., van Pul, W.A.J., Boermans, G.M.F., Duyzer, J.H., Wyers, G.P. (1993). Evaluation of dry deposition measurements for monitoring applicatiom over the Speulder forest. RIVM, Bilthoven, report 722108002.

Fairall,C.W. (1984), Interpretation of Eddy Correlation Measurements of Particulate Deposition and Aerosol flux. *Atmos. Environ.* **18**, 1329–37.

Fitzgerald, J.W. (1975) Approximation formulas for the equilibrium size of an aerosol particle as a function of its dry size and composition and the ambient relative humidity. *J. App. Met.* **14**, 1044–1049.

Fontan, J., Lopez, A, Lamaud, E., Druilhet, A, (1995) Vertical Flux Measurements of the Sub-Micronic Aerosol Particles and Parameterization of the Dry Deposition Velocity, this volume.

Fowler, D., Duyzer, J.H., Baldocchi, D.D. (1991) Inputs of Trace Gases, Particles and Cloud Droplets to Terrestrial Surfaces. *Proc. Roy. Soc. Edinburgh* **97B**, 35-59.

Fowler,D., Gallagher, M.W., Lovett, G.M. (1992), Wet, Cloud Water and Fog Deposition", In: G.Lövland,J.W.Erisman, D.Fowler (eds), *Models and Methods for the Quantification of Atmospheric Input to Ecosystems,* Nordiske Seminar-og Arbejds-rapporter, pp. 51–74.

Fowler, D., Cape, J.N., Unsworth, M.H., (1989), Deposition of Atmospheric Pollutants on Forests, *Phil. Trans. Roy. Soc. London* **B324**, 247–265.

Fowler, D., Cape, J.N., Leith, I.D., Choularton, T.W., Gay, M.J., Jones, A. (1988). The influence of altitude on rainfall composition at Great Dun Fell. *Atmos.Environ.* **22**, 1355–1362.

Gallagher, M. (1995). Deposition of Aerosol to Forests: Measurements from the Speulderbos Experiment. EUROTRAC BIATEX workshop on Deposition and Biogenic Emission, Madrid, January 16–17, 1995.

Gallagher, M.W.,Choularton, T.W., Morse, A.P., Fowler, D. (1988) Measurements of the size dependence of cloud droplet deposition at a hill site. *Quart .J. Roy. Met. Soc.* **114**, 1291–1303.

Gallagher, M.W., Beswick, K.M., Choularton, T.W., Coe, H., Fowler, D., Hargreaves, K. (1992). Measurements and Modelling of Cloudwater Deposition to Moorland and Forests. *Environ. Pollut.* **75**, 97–107.

Gallagher, M.W., Beswick, K.M. and Choularton, T.W.(1992). Measurements and modelling of cloud water deposition to a snow covered forest. *Atmos. Environ.* **26A**, 2893–2904.

Gallagher, M.W., T.W. Choularton, A. Wicks, K.M. Beswick, H. Coe, M. Sutton, D. Fowler, J. Duyzer, P. Wyers (1992) Measurements of aerosol exchange to Heather Moor. in: P.M. Borrell, P. Borrell, T. Cvitaš, W. Seiler (eds), *Proc. EUROTRAC Symp. '92*, SPB Academic Publishing bv, The Hague 1993, pp. 694–698.

Garland,J.A., Cox,L.C., (1982) Deposition of small particles to grass, *Atmos. Environ.* **16**, 2699–2702.

Gould, T.R., Davidson, C.I. (1992) Variability and uncertainty in particle dry deposition modelling. In: S.E. Schwarz, W.G.N.Slinn (eds), *Precipitation scavenging and atmosphere-surface exchange processes.* Hemisphere pub., Washington DC pp. 1125–1142.

Grosch,S., Schmitt,G. (1988) Experimental Investigations on the Deposition of Trace Elements in Forest Area, in: Grefen, K., Löbel, L. (eds), *Environmental Meteorology*, Kluwer Academic Publishers, pp. 201–216.

Hatakayama, S., Izumi, K., Fukuyama, T., Akimoto, H. (1989) Reactions of ozone with α-pinene and β-pinene in air: yields of gaseous and particulate products. *J. Geophys. Res.* **94**, 13013–13024.

Hicks, B.B., Baldocchi, D.D., Meyers, A.J., T.P., Hosker, R.P., Matt, D.R. (1987). A preliminary multiple resistance routine for deriving dry deposition velocities from measured quantities. *Water Air and Soil Pollut.* **36**, 311–330.

Hicks, B.B. (1984). Deposition both Wet and Dry. in: Hicks (ed), *Acid Precipitation Series* Vol. 4. Butterworth.

Hinds,W.C., (1982) *Aerosol technology, properties, behaviour and measurement of airborne particles.* John Wiley and Sons.

Hoefken,K.D., Gravenhorst,G. (1983) Utersuchung ueber die Deposition Atmosphaerischer Spurenstoffe an Buchen-und Fichten Wald. In UBA - *Berichte 6/83*, Teil II, Schmidt-Verlag, Berlin.

Hori, T. (ed), (1953), *Studies on Fogs.* Tanne Trading Co. Hokaido, Sapporro.

Hummelshøj, P., Jensen, N.O., Larsen, S.E. (1992). Particle Dry Deposition to a Sea Surface. In: Schwarz, S.E., Slinn, W.G.N. (eds), *Precipitation Scavenging and Atmosphere-Surface Exchange,* Vol. 2. Hemisphere Pub.Corp. USA, pp. 892–840.

Hummelshøj, P. (1992) Dry Deposition of Particles and Gases, Ph.D.Thesis RISØ National Laboratory Denmark, 1992

Hummelshøj, P., Ruijgrok, W., Semb, A.,Westling, O., Wyers, P. (1992) Working Group report on dry depositon of particles. In: G. Lövland, J.W. Erisman, D. Fowler (eds), *Models and Methods for the Quantification of Atmospheric Input to Ecosystems*, Nordiske Seminar-og Arbejds-rapporter, 1993:573 pp. 33–35.

Ibrahim, M., Barrie, L.A., Fanaki, K., (1983) , An experimental and theoretical investigation of the dry deposition of particles to snow, pine trees and artificial collectors. *Atmos. Environ.* **17**, 781–788.

Izumi, K., Fukuyama, T. (1990) Photochemical Aerosol Formation from Armatic Hydrocarbons in the presence of Nox. *Atmos. Environ.* **24A**, 1433–1441.

Iversen, T., Halvorsen, N., Mylona, S., Sandnes, H. (1991). Calculated budgets for airborne acidifying components in Europe. *EMEP Meteorological Synthesizing Centre West*, Norwegian Meteorological Institute, Oslo.

IPCC (1992) Climate Change 1992, in: Houghton *et al.* (eds), *Supplement Report to IPCC Scientific Assessment*, Cambridge University Press.

Johnson, A.H., (1987), Deterioration of red spruce in the northern Appalachian Mountains. In: T.C. Hutchison, K.M.Meema (eds), *Effects of atmospheric pollutants on forests, wetlands and agricultural ecosystems*, NATO ASI series, vol. 916 Springer-Verlag, Heidelberg pp. 83–99.

Joutsenoja, T. (1992), Measurements of Aerosol Deposition to a Cereal Crop. In: Choularton, T.W. (ed), *Measurements and Modelling of Gases and Aerosols to Complex Terrain* - NERC Report 1992, GR3/7259 A.P.I. Appendix X1.2.

Katen, P.C., Hubbe, J.M., (1983), Size resolved measurements by eddy correlation of the dry deposition velocity of atmospheric aerosols. In: Pruppacher, Semonin, R.G., Slinn, W.G. (eds), *Precipitation Scavenging, Dry Deposition and Resuspension*, Elsevier Science Publishing Co. New York pp. 953–961.

Kowalski, A.S., Anttioni, P.M., Vong, R.J., Delany, A.C., Maclean, G.D. (1996) Deployment an evaluation of a system for ground-based measurements of cloud liquid water turbulent fluxes. To be submitted.

Krause, G.M.H., Arndt, U., Brandt, D.J., Bucher, J., Kenk, G., Matzner,E., (1986), Forest decline in Europe: development and possible causes. *Water Air Soil Pollut.* **31**, 647–688.

Kroll, G., Winkler, P. Trace substance input to coniferous forests via cloud interception. (1989) In: H.-W.Georgii (ed), *Mechanisms and Effects of Pollutant-Transfer into Forests*, Kluwer Academic Publishers pp. 205–212

Lamaud, E., Y. Brunet, A. Labatut, A. Lopez, J. Fontan, A. Druilhet, (1994a) The Landes Experiment: Biosphere-Atmosphere Exchanges of Ozone and Aerosol Particles Above a Pine Forest, *J. Geophys.Res.* **99D8**, 16511–16521.

Lamaud, E., A.Labatut, J. Fontan, A. Lopez, A. Druilhet, Y. Brunet, (1994b) Biosphere Atmosphere exchanges: Ozone and aerosol dry deposition velocities over a pine forest. Non CO_2 greenhouse gases. Why and how to control. Maastricht, *Environmental Monitoring Assessment* **31**, 175–181.

Lamaud, E., J. Fontan, A. Lopez, A. Druilhet, (1994c) Parameterization of the dry deposition velocity of sub-micron aerosol particles. *Air Pollution II, Vol 2. Pollution Control and Monitoring*. Computational Mechanics Publications, Southamption, Boston pp 433–440.

Lelieveld, J., Heintzenberg, (1992) Sulfate cooling effect on climate through in-cloud oxidation of anthropogenic SO_2. *Science*, **258**, 117–120

Little, P. (1977) Deposition of 2.75, 5.0 and 8.5 μm particles to plant and soil surfaces. *Environ. Pollut.* **12,** 293–305.

Lewellen, W.S. (1985). Modeling Turbulent Exchange in Forest Canopies. In : Hutchison, B.A., Hicks, B.B. (eds), *The Forest-Atmosphere Interaction*, pp.481–499.

Lindberg, S.E., Lovett,G.M. (1985) Field measurements of particle dry deposition rates to foliage and inert surfaces in a forest canopy, *Environ. Sci.. and Tech.* **19**, 238–244.

Lorenz, R., Murphy JR, C.E., (1989), Dry Deposition of Particles to a Pine Plantation. *Boundary-Layer Met.* **46**, 355–366

Lövblad, G. (1992). Mapping deposition over Sweden. In. Schwartz, S.E., Slinn, W.G.N. (eds), *Precipitation scavenging and atmosphere surface exchange processes.* Hemisphere. Publ. Washington DC pp. 1533–1542.

Lovett,G.M. (1981) Forest Structure and Atmospheric Interactions:Predictive Models for Subalpine Fir Forests. Ph.D.Thesis, Dartmouth College, Hanover,New Hampshire.

Lovett,G.M. (1984). Rates and mechanisms of cloud water deposition to a subalpine balsam fir forest. *Atmos. Environ.* **19**, 361–371.

Lovett,G.M. (1986). Canopy structure and cloud water deposition in a subalpine coniferous forest. *Tellus* **36**, 319–327.

Milne,R., Crossley, A., Unsworth, M.H. (1989), Physics of cloudwaterdeposition and evaporation at Castlelaw, S.E.Scotland. In: M.H. Unsworth, D. Fowler (eds), *Acid deposition processes at high elevation sites.* Klewer, Dordrecht pp.299–307.

Murphy,C.E. JR, Lorenz,R., (1985) Analysis of Forest Environmental Measurements to Estimate Parameters of Microclimate and Air Pollution Deposition Velocity Models. In: Hutchison, B.A., Hicks, B.B. (eds), *The Forest-Atmosphere Interaction*, pp. 383–393.

Neumann,H.H., Den Hartog,G. (1985) Eddy correlation measurements of Atmospheric fluxes of ozone, sulfur and particulates during the Champaign Intercomparison Study. *J. Geophys. Res.* **90D1**, 2097–2110.

Nicholson, K.W. (1988) The dry deposition of small particles: a review of experimental measurements. *Atmos. Environ.* **22,** 2653–2666.

Noone, K.J., Charlson, R.J., Covert, D.S, Ogren, J.A., Heintzenberg, J., (1988) Chemical composition differences in fog and cloud droplets of different sizes. *J. Geophys. Res.* **93**, 9477–9482.

Noone, K.J., Covert, D.S, Ogren, J.A., Heintzenberg, J., (1990) An examination of clouds at a mountain-top site in central Sweden: the distribution of solute within cloud droplets. *Atmos. Res.* **25**, 3–15.

Ogren, J.A., Charlson, R.J., (1992) Implications for models and measurements of chmeical inhomogeneities among cloud droplets, *Tellus* **44B**, 208–225,.

Ogren, J.A., Heintzenberg ,J., Zuber, A, Noone, K.J., Charlson, R.J., (1989) Measurements of the size dependence of solute concentrations in cloud droplets, *Tellus* **41B**, 24–31.

Ogren, J.A., Noone, K.J., Hallberg, J., Heintzenberg ,J.,Schell, D., Berner, A., Solly, I., Kruisz, C., Reischle, G., Arends, B.G., Wobrock, (1992) Measurements of the size dependence of the concentration of non-volatile material in fog droplets, *Tellus* **44B**, 570–580.

Peters, K., Eiden, R (1992) Modelling the dry deposition velocity of aerosol particles to a spruce forest. *Atmos. Environ.* **26A**, 2555–2564.

RGAR (1987) *Acid Deposition in the UK 1981-1985.* In The Second Report of the UK Review Group on Acid Rain. Stevenage: Warren Spring Laboratory 104 pp.

RGAR (1990) *Acid Deposition in the UK 1986-1988.* In The Third Report of the UK Review Group on Acid Rain. Stevenage: Warren Spring Laboratory 104 pp.

RGIND (1994) *UK Review Group on Impacts of Nitrogen Deposition on Terrestrial Ecosystems.* Pitcairn, C.E.R. (ed), UK Dept.of Environ.

Ruijgrok, W., H. Tieben, P. Eisinga, (1994). Dry Deposition of Acidiyinf and Alkaline Particles to Douglas Fir - A Comparison of Measurements and Model Results. KEMA Report Dutch Priority Programme on Acidification. 20159-KES/MLU 94-3216 No 83397.KC.17-94P02

Ruijgrok, W., Nicholson, K.W., Davidson, C.I. (1993) Dry Deposition of Particles, In: G. Lövland, J.W. Erisman, D. Fowler (eds), *Models and Methods for the Quantification of Atmospheric Input to Ecosystems*, Nordiske Seminar-og Arbejds-rapporter, 1993:573 Appendix 5, pp 145–162.

Ruijgrok, W., (1992) Uncertainty in models calculating the dry deposition of aerosols to forests. In: N.Fukuta and P.E. Wagner (eds) *Nucleation and atmospheric aerosols.* Deepak Publ., Hampton pp. 481–485.

Sehmel, G. (1973). Particle Eddy Diffusivities and Deposition Velocities for isothermal Flow and Smooth Surfaces. *Aerosol Sciences* **4**, 125–138.

Sehmel, G. (1980) Particle and Gas Dry Deposition: A Review. *Atmos. Environ.* **14**, 983–1011.

Sehmel, G.A., Hodgson, W.H. (1980) A Model for predicting dry deposition of particles and gases to environmental surfaces. In *AICHE Symposium Series* **76**, pp 218–230.

Seinfeld, J.H. (1986) *Atmospheric Chemistry and Physics of Air Pollution.* Wiley Interscience Publication, New York, USA.

Shuttleworth, W.J. (1977), The exchange of wind driven fog and mist between vegetation and the atmosphere. *Boundary-Layer Met.* **12**, 463–489.

Sievering, H. (1982) Profile measurements of Particle Dry Deposition Velocity at an Air-Land Interface. *Atmos. Environ.* **16**, 301–306.

Sievering, H. (1983) Eddy Flux and Profile measurements of Small particle dry-deposition velocity at the Boulder Atmospheric Observatory.In: Pruppacher, Semonin, R.G., Slinn, W.G. (eds), *Precipitation Scavenging, Dry Deposition and Resuspension,* pp. 963–977.

Sievering, H. (1987) Small particle dry deposition under high wind speed conditions: eddy flux measurements at the Boulder Atmospheric Observatory. *Atmos. Environ.* **21**, 2179–2185.

Sievering, H. (1988) Small-Particle Dry Deposition Measurements: A comparison of Gradient and Eddy-Flux Techniques over Agricultural Fields. In: *Annual Meeting of Air Pollution Central Association*, Dallas, Texas, June 19–24, 1988, 6, 88–101.

Slinn, W.G., L.Hasse, B.N. Hicks,A.W. Hogan,D. Lai, P.S. Liss, K.O. Munnich,G.A. Sehmel, O.Vittori. (1978) Some Aspects of the Transfer of Atmospheric Trace Constituents past the Air-Sea Interface. *Atmos. Environ.* **12**, 2055–2087.

Slinn, W.G.N. (1982) Predictions for Paricle Deposition to Vegetative Canopies. *Atmos. Environ.* **16**, 1785–1794.

Slinn, W.G.N. (1983) A Potpourri of Deposition and Resuspension Questions. In: Pruppacher, Semonin, R.G., Slinn, W.G.N. (eds), *Precipitation Scavenging, Dry Deposition and Resuspension*, Elsevier Science Publishing Co., New York pp. 1361–1416.

Seinfeld, J.H. (1986) *Atmospheric Chemistry and Physics of Air Pollution.* John Wiley and Sons Inc.

Spyros, N., Pandis, S.N. *et al.* (1991) Aerosol formation in the Photooxidation of isoprene and β-pinene. *Atmos. Environ.* **25A**, 997–1008.

Sutton, M. (1995). Generalised description of ammonia exchange. *EUROTRAC BIATEX workshop on Deposition and Biogenic Emission*, Madrid, January 16–17, 1995.

Twomey, S. (1977) *Atmospheric Aerosols*, Elsevier Scientific Publishing Co.

Unsworth, M.H., (1984) Evaporation from forests in cloud enhances the effects of acid deposition. *Nature* **312**, 262–264.

Unsworth, M.H., Crossley, A. (1987) Capture of wind-driven cloud by vegetation. In: P.J.Coughrey, M.H.Martin, M.H.Unsworth (eds), *Pollutant transfer and fate in ecosystems British Ecological Society special publication no. 6*, Blackwell Scientific, Oxford pp.125–137.

Vincent, J.H. (1989) *Aerosol Sampling: Science and Practice*. John Wiley Sons.

Vong, Richard J., Andrew S. Kowalski, (1995). Eddy Correlation Measurements of Size-Dependent Cloud Droplet Turbulent Fluxes to Forest, *Tellus*, accepted.

Vong, R.J., Bailey, B.H., Markus, M.J., Mohnen, V.A. (1990), Factors governing cloud water composition in the Appalachian mountains. *Tellus* **42B**, 435–453.

Vong,R.J., Mohnen, V., Kadlecek, J., Bailey, B., Mueller,S., Meagher, J., Baumgardner, R. (1990). Network Measurements of Droplet Deposition of Atmospheric Pollutants. In: D.Sisterson (ed), *NAPAP State of Science and Technology Report No. 6,* National Acid Precipitation Assessment Program Washington, DC, June 1990.

Vong, R. (1991) Cloud water deposition to Appalachian Forests. *Environ. Sci.Technol.* **25,6**, 1014–1021.

Wang *et al.* (1992a) Aerosol Formation and Growth in Atmospheric Organic NO_x Systems - I. Outdoor Smog Chamber Studies of C7 and C8 Hydrocarbons. *Atmos. Environ.* **26A**, 403–420.

Wang *et al.* (1992b) Aerosol Formation and Growth in Atmospheric Organic NO_x Systems - II. Aerosol Dynamics. *Atmos. Environ.* **26A**, 421–434.

Warraghai, A. and G. Gravenhorst.(1989), Dry Deposition of Atmospheric Particles to an Old Spruce Stand. In: H.-W.Georgii (ed), *Mechanisms and Effects of Pollutant-Transfer into Forests*, Kluwer Academic Publishers pp. 77–86

Wesely, M.L., Hicks, B.B., Dannevik, W.P.,Frisella, S., Husar, R.B. (1977) An eddy correlation measurement of particulate deposition from the atmosphere. *Atmos. Environ.* **11**, 561–563.

Wesely, M.L. Cook, D.R., Hart, R.L. (1983). Fluxes of Gases and Particles Above a Deciduous Forest in Wintertime. *Boundary Layer Met.* **27**, 237–255.

Wesely, M.L. Cook, D.R., Hart, R.L. ,Hicks, B.B., Durham, G.L., Speer, R.E., Stedman, D.H., Tropp, R.J. (1983). Eddy correlation measurements of dry deposition of particulate sulfur and sub-micron particles. In: Pruppacher, Semonin, R.G., Slinn, W.G.N. (eds), *Precipitation Scavenging, Dry Deposition and Resuspension*, Elsevier Science Publishing Co. New York p. 943.

Westrate, H., Duyzer, J., (1994) Evaluation of the gradient method for use in monitoring dry depositiion at Speuld based on measurements for ozone. *TNO - IMW Delft Report*, **R94–104**.

Whitby, K.T., Cantrell, B. (1976) Fine Particles, in: Int.Conf. on Environmental Sensing and Assessment, Las Vegas, NV, IEEE.

Williams, R.M., (1982) A model for the dry deposition of particles to natural water surfaces. *Atmos. Environ.* **17**, 1933–1938.

Wiman, B.L.B., Agren, G.I. (1985). Aerosol depletion and deposition in forests - a model analysis. *Atmos. Environ.* **19**, 335–362.

Wyers, P. (1995). Overview of Ammonia and Aerosols. EUROTRAC BIATEX Workshop on Deposition and Biogenic Emission, Madrid, January 16–17, 1995.

Wyers, G.P. Veltkamp, Vermeulen, A.T., Geusebroek, M., Wayers, A., Möls, J.J., (1994), Deposition of Aerosol to Coniferous Forest.ECN Report December -C-94-051 , Netherlands Energy Research Foundation.

Appendix

Summary of parameterisations of deposition velocity to real, natural surfaces based on field measurements

Equations 4, 5 and 6 in the tables below are recommended for general modelling applications with typical aerosol loadings and a mass median diameter of around 0.4 to 0.5 µm and $k_1 = 0.002$ for soil and short grass and 0.004 for tall vegetation.

The sub-micron aerosol surface deposition velocity, V_{ds} , as a function of particle size may be somewhat more complex, as discussed later, however, the parameterisations in equations 8 and 9 are suggested in the interim for size specific applications.

For turbulent deposition of cloud or fog droplets equation 12 is adequate, particularly for hill cloud applications with the addition of the appropriate settling velocity contribution which will become more important for large cloud droplets *e.g.* in fogs. It is certainly adequate enough to describe bulk pollutant fluxes by cloud deposition *e.g.* for sulfate, despite a significant additional resistance to transfer of droplets in the size range 1 to 6 µm size. Table C is a summary of the best parameterisations obtained to date, Ruijgrok *et al.* (1994), for the collection efficiencies of various important ionic species to a forest canopy, from the Speulderbos programmes, taking into account relative humidity, surface wetness and friction speed.

Table A: Aerosol dry deposition parameterisations from micro meteorological experiments

Experiment DDIE	*lush grass*	Wesely *et al.* (1985)
Experiment ANL	*pine forest*	Hicks *et al.* (1982)
	leafless deciduous	Wesely *et al.* (1983a)
	grazed pasture	Wesely *et al.* (1983b)

Stable Conditions

$$V_{ds} = 0.002\ u_* \qquad \text{(for } L > 0\text{)} \quad \text{Stable} \qquad (1)$$

Unstable Conditions

$$V_{ds} = 0.002\ u_*(1 + (-300/L)^{2/3}) \qquad \text{(for L} \leq 0\text{)} \quad \text{Unstable} \qquad (2)$$

$$V_{ds} = 0.002\ u_*\{1 + (-0.3\ z_i/L)^{2/3}\} \qquad (3)$$

where z_i = boundary layer height

Landes Experiment Landes forest: *Pinus pinaster Ait*

$h = 15$ m, LAI = 3, $d = 11$ m, $z_o = 1.4$ m

Stars I Experiment Sahel Niger, Africa: *bare clay soil*

$z_o = 0.20$ cm

Stars II Experiment Ebre valley: *bare soil, almond trees*

$z_o = 0.25–1.0–10$ cm Lamaud *et al.*(1994 a,b,c) Fontan *et al.* (1995)

Stable Conditions $L \geq 0$

$$V_{ds}/u_* = k_1 \qquad (k_1 = 0.004 \text{ Landes}) \qquad (4)$$

$$V_{ds}/u_* = k_1 \qquad (k_1 = 0.002 \text{ Stars I and II}) \qquad (5)$$

Unstable Conditions $L < 0$

$$V_{ds}/u_* = k_1 + k_2\, W*^2/u_*^2 = k_1 + k_2\,(\beta H z_i)^{2/3}/u_*^2 = k_1 + k_2\,(-z_i/L)^{2/3} \qquad (6)$$

where $k_2 = 0.0009$
H = sensible heat flux
β = buoyancy term
W_* = convective velocity scale

Speulderbos Experiments Speulderbos forest, Netherlands, *Douglas fir*

$h = 18–20$ m, $d = 11$ m, $z_o = 2.5$ m, LAI = 10–17

Duyzer *et al.* (1994),Weyers *et al.* 1994), Ruijgrok *et al.* (1994), Beswick *et al.* (1994), Gallagher *et al.* (1995)

Stable Conditions L ≥ 0

$$V_{ds}/u_* = 0.003–0.004 \quad (\text{MMD} = 0.4\text{-}0.5\ \mu\text{m}) \qquad (7)$$

Unstable Conditions $L<0$

$$V_{ds}/u_* = 0.0135\, d\ \{1 + (-300/L)^{2/3}\} \qquad (8)$$

$$V_{ds}/u_* = 0.0135\, d\ \{1 + (-0.3\, z_i/L)^{2/3}\} \qquad (9)$$

where $(-0.04 < L < 0)$† and $(0.1\mu\text{m} \leq d \leq 0.5\mu\text{m})$
d - particle size in microns (optical diameter from Mie Scattering)
† - stability range of experiment

$$V_{ds} = 0.0380\, u_* - 0.004 \ \text{m s}^{-1} \ (r^2 = 0.895) \qquad (10)$$

ammonium bisulfate particles MMD = 0.8 μm

$$V_d = 0.0380\, u_* \ \text{m s}^{-1} \ (r^2 = 0.903)† \qquad (11)$$

† recommended over equation 10

Cloud droplet deposition parameterisations

Cloud droplet parameterisations Dollard and Unsworth (1982)

Bulk cloud water fluxes Morse (1988), Gallagher *et al.* (1988) Fowler *et al.* (1989), Beswick *et al.* (1992), Gallagher *et al.* (1992), Wyers *et al.* (1994)

$$V_{ds} \approx V_m \quad (12)$$

V_m - momentum deposition velocity

$$V_T = 0.195\, u_*^2 \qquad (r = 0.83) \quad (13)$$

$$V_T = 0.5\, V_m \qquad (r = 0.75) \quad (14)$$

V_d as a function of Cloud Droplet Size Coe *et al.* (1991)

$$V_d(r) = V_m(-0.019\, r^2 + 0.353\, r - 0.443) \quad (15)$$

($1 < r < 10$ µm)
r - particle *radius* (microns)

$$V_{ds} \approx V_m \qquad (r > 5\ \mu\text{m}) \quad (16)$$

Table B: Collection efficiency (E) parameterisations for major ionic species for forest canopies from Ruijgrok *et al.* (1994)

	Wet surface conditions		Dry surface conditions	
Comp	rh ≤ 80	rh > 80	rh ≥ 80	rh > 80
NH_4	$0.066\, u_*^{0.41}$	$0.066\, u_*^{0.41}[1+0.37e^{(rh-80)/20}]$	$0.05\, u_*^{0.23}$	$0.05\, u_*^{0.23}[1+0.18e^{(rh-80)/20}]$
SO_4	$0.08\, u_*^{0.45}$	$0.08\, u_*^{0.45}[1+0.37e^{(rh-80)/20}]$	$0.05\, u_*^{0.28}$	$0.05\, u_*^{0.28}[1+0.18e^{(rh-80)/20}]$
NO_3	$0.10\, u_*^{0.43}$	$0.10\, u_*^{0.43}[1+0.37e^{(rh-80)/20}]$	$0.063\, u_*^{0.25}$	$0.063\, u_*^{0.25}[1+0.18e^{(rh-80)/20}]$
Na	$0.679\, u_*^{0.56}$	$0.10\, u_*^{0.43}[1+0.37e^{(rh-80)/20}]$	$0.14\, u_*^{0.12}$	$0.14\, u_*^{0.12}[1-0.09e^{(rh-80)/20}]$

Table C: Measurements of aerosol fluxes by micrometeorological and chemical surface techniques to mostly rough surfaces

$\overline{d_p}$ (µm)	V_{ds} (cm s^{-1})	Surface	z_0 (cm)	u_* (m s^{-1})	z (m)	u (m s^{-1})	z/L	Method	Reference
0.07	0.64	Grass	3				$-1 \rightarrow +1$	Grad	Wesely (1977)
0.15–3	0.38 ± 0.29	Agric./soil	1–20	< 0.30				Grad	Sievering (1982)[1]
0.03–1	0.7	Loblolly pine		0.2–0.7	z = 23	h = 17	d = 13 m	EC	Hicks *et al.* (1982)*[2]
0.47	0.37 ± 0.04	Agric./wheat	4 ± 2	0.49	5.5	6.4	+0.08	EC	Sievering (1988)
0.47	1.19 ± 0.18	Agric./wheat	4 ± 2	0.59	5.5	8.7	–0.13	EC	"
0.21	0.38	Agric./rye 1m	1–20				$-1 \rightarrow 0$	Grad	"(1982)
0.22	< 0.05	Grass 30–40 cm?					$-0.8 \rightarrow +0.7$	EC	Neumann (1985)†
0.21	0.034	Snow	0.1				$0 \rightarrow +1$	EC	Duan (1988)
0.71	0.021	Snow	0.1				$0 \rightarrow +1$	EC	Duan (1988)
0.099	0.16	Barley 5 cm	≤1.0				$0 \rightarrow +1$	Grad	Hummelshøj (1992)

Table C: continued

$\overline{d_p}$ (µm)	V_{ds} (cm s^{-1})	Surface	z_0 (cm)	u_* (m s^{-1})	z (m)	u (m s^{-1})	z/L	Method	Reference
0.24	0.20	Barley 5 cm	≤1.0				0 → +1	Grad	Hummelshøj (1992)
0.6	0.28	Barley 5 cm	≤1.0				0 → +1	Grad	Hummelshøj (1992)
0.10	0.14	Wheat	2	0.6	3.1		0 → +0.5	EC	Joutsenoja (1992)
0.10	0.08	Heather	1	0.3–0.4	5.0		+0.05	EC	Gallagher (1992)
0.62	0.15–0.30	Heather/grass	0.3–0.4	5.0			+0.05	EC	
0.56	0.18	"	"	0.5 ?				EC	Duyzer (1986)
2.0	0.3	"							"
		Tussock grass	1.1	0.56	1	6.2	Cloud	Grad	Gallagher (1988)
		Note Values are for V_T							

d	3.0	5.0	7.0	9.0	11.0	13.0	15.0	17.0	19.0	21.0	23.0	25.0	27.0	29.0	31.0
V_T	*	0.6	4.0	3.0	4.6	5.9	5.2	7.1	7.9	9.0	8.6	6.7	6.5	4.9	5.2

Table C: continued

$\overline{d_p}$ (μm)	V_d (cm s–1)	Surface	z_o (cm)	u_* (m s^{-1})	z (m)	u (m s^{-1})	z/L	Method	Reference
		Sitka spruce 4.2 m tall	30	0.37	1	1.9	Cloud	EC	Beswick (1990)¥
		d = 2.81							
		Measurement height, 9.5 m			Values are for V_T				

d	3.0	5.0	7.0	9.0	11.0	13.0	15.0	17.0	19.0	21.0	23.0	25.0	27.0	29.0	31.0
V_T	0.4f	0.7	0.9	1.0	1.0	1.1	1.3	1.6	1.7	2.7	3.1	5.1	5.7	13.3	12.6

$\overline{d_p}$ (μm)	V_d (cm s–1)	Surface	z_o (cm)	u_* (m s^{-1})	z (m)	u (m s^{-1})	z/L	Method	Reference
		Snow covered	15.7	0.7				EC	Gallagher (1992)
Sitka Spruce height 4.2 m		Measurement height 9.5 m			Values are for V_T				

d	3.0	5.0	7.0	9.0	11.0	13.0	15.0	17.0	19.0	21.0	23.0	25.0	27.0	29.0	31.0
V_T	–4.4*¥	–1.0*	2.0	2.7	3.2	5.2	6.1	5.8	6.2	7.2	6.4	16.8	10.9	50.9	27.1

Table C: continued

$\overline{d_p}$ (µm)	V_d (cm s^{-1})	Surface	Method	Reference
0.5–1.0	0.34 ± 0.35	Pine canopy *Pinus Taeda* 10 yrs old h = 9 m	Grad	Lorence (1989)
1.0–2.0	0.78 ± 0.64	LAI = 6 in winter, 12 at end growing season		
2.0–5.0	0.92 ± 0.76	reference height = 10 m OPC		
0.5–1.0	1.5 ± 0.34	Spruce forest	SEM	Waraghai (1989)•
1.0–2.0	1.1 ± 0.36	"		
2.0–5.0	0.86 ± 0.39	"		
5.0–8.0	1.1 ± 0.41	"		
8.0–10	0.93 ± 0.70	"		
> 10.0	1.64 ± 0.59	"		
0.4	1.1–2.6	Spruce forest	CTF	Grosch (1988)•
1.1	0.9–1.7	"		

Table C: continued

$\overline{d_p}$ (μm)	V_d (cm s^{-1})	Surface		Method	Reference
4.4	1.2–2.2	"			
0.7	0.9–1.8	Spruce forest		CTF	Brückman (1988)•
0.4	0.9–2.1	Spruce forest		CTF	Höfken (1983)•
0.8	0.8–1.6	"			
1.4	0.6–1.4	"			
3.4	1.3–2.5	"			
0.15*	0.8 ($z/L \leq 0$)	Maritime pine forest h = 15 m, LAI = 3, d = 11 m, z_0 = 1.4 m,		EC	Lamaud (1994a)
0.15*	0.1 ($z/L > 0$)	measuring height = 25 m, * MMD = 0.5 μm			
0.4 AMD	$V_{ds} = 0.73 \pm 0.49$ cm s^{-1}	Speulder forest	^{214}Pb particles	GS	Wyers *et al.* (1994)
0.8 MMD	$V_{ds} = 1.48$ cm s^{-1}	($u_* = 0.4$ m s^{-1})	$V_d = 0.04\ u_*$ m s^{-1} ammonium bisulphate	Grad	Wyers *et al.* (1994)

Table C: continued

$\overline{d_p}$ (μm)	V_d (cm s^{-1})	Surface	Method	Reference
0.5MMD	1.0 V_{ds}, ($z/L \leq 0$)	Douglas fir h = 18m, d = 15 m, z_0 = 2.2 m, LAI = 10–17	EC	Duyzer (1995)
	0.1	(z/L > 0)		See also Ruijgrok *et al.* (1994)
0.4-0.5MMD	$V_{ds} \approx 1.0$ cm s^{-1}	V_{ds}/u_* strong function of size for sub-0.4 micron particles	EC	Gallagher (1995)

Key:

EC	Eddy Correlation	Grad	Flux Gradient
SEM	Scanning Electron Microscopy	OPC	Optical Particle Counter
CTF	Chemical Throughfall	GS	Gamma Spectroscopy
†	Statistical Counting Problems ?	•	Estimated from Figures in Warraghai and Gravenhorst (1989)
*	Positive fluxes due to evaporation (see Vong *et al.* 1994)		
¥	V_d was found to be significantly smaller for snow covered canopy	*f*	For errors see reference.

1 Sievering (1982 b) suggests that this may be an underestimate due to anthropogenic influence.
Fluxes were confounded by inverse dependence of V_d found with wind speed. Further complications may have arisen from removal of fine particulate matter from senescent vegetation.

2 Eddy fluxes of very small particles were predominantly found to be out of the canopy with the surface acting as a sink during only part of the day- the authors suggest that this may be due to gas to particle conversion.

Chapter 4

Assessment of Dry Deposition and Total Acidifying Loads in Europe

Jan Willem Erisman and Addo van Pul

National Institute of Public Health and the Environment (RIVM) P.O. Box 1, 3720 BA Bilthoven, The Netherlands

Summary

In describing the effects of acidification on the level of ecosystems, acid loads should be available at least on the size of ecosystems. Until now no deposition maps on this resolution are available, hampering accurate estimation of exceedances of critical loads in Europe. Current methods and models for estimating deposition of acidifying components differ in the used horizontal resolution *i.e.* varying from 5 × 5 km up to 150 × 150 km. These include inferential models applied in different countries of Europe, where enough measurements of concentrations and meteorological parameters are available. For Europe as a whole long-range transport model results based on emission estimates (EMEP, TREND) have been available. Too few measurements are available in Europe to apply the inferential method for deposition mapping. Within the framework of BIATEX a so called merging method was developed, the EDACS model, combining long-range transport modelling and inferential methods to map deposition of acidifying components on a small scale of Europe. Maps of small scale fluxes of Europe with this method are presented in this review. The maps are produced in close co-operation with EMEP.

4.1 Introduction

Despite the large efforts which have been put in deposition research and measurements during the last decades, it is not possible to derive deposition maps entirely based on deposition monitoring results. This is caused by the stochastic nature of the dry deposition. On an annual basis, wet deposition can be treated as smoothed fields over regions. By installing a monitoring network with enough samplers to cover the spatial gradients, wet deposition can be monitored and wet deposition maps can be derived from interpolated annual averaged data. For several years now this is done on national levels (*e.g.* RIVM, 1994; UK RGAR, 1990; Lövblad *et al.*, 1993; see van Leeuwen *et al.*, 1995 for an overview). For

Europe as a whole it is currently possible to obtain a reasonable coverage with all available national and European monitoring networks. Van Leeuwen *et al.* (1995) compiled wet deposition maps of different components of Europe using monitoring results of about 800 sites.

Contrary to wet deposition, dry deposition maps cannot be obtained from interpolating long-term dry deposition measurements. Furthermore, there is a serious lack of monitoring data. Currently dry deposition of SO_2 and O_3 is monitored at two sites, in Portugal and Spain (EC SREMP project), SO_2 and NH_3 at the Oostvaardersplassen in the Netherlands (Mennen *et al.*, in prep.; Wyers *et al.*, in prep.) and SO_2, NH_3 and NO_x at three sites, in Scotland, Germany and the Netherlands (EC LIFE project). This makes only six monitoring sites in Europe, with only one forested area. There are three ways of deriving dry deposition maps, *i.e.* using measurements of concentrations, meteorological parameters and surface condition parameters from which the dry deposition flux can be inferred (inferential technique, Hicks *et al.*, 1987); using emission estimates and long-range transport model (LRTAP) calculations; or using a combination of the two (Erisman and Draaijers, 1995). This review will provide some background information on these three methods, but will mostly focus on the combination method, the merging technique. A special project, which aimed at generalisation of deposition in Europe using this method, was initiated in BIATEX and executed by RIVM (van Pul *et al.*, 1992; 1995). The main results will be highlighted here.

4.2 Long-range transport modelling

The way deposition is modelled depends on its purpose. It may be considered as a loss term in long-range transport modelling, or as an input term in determining the deposition to ecosystems. For the latter it is important to consider the scale on which the deposition is desired. If the scale is comparable to the scale considered in the long-range transport (LRTAP) model, the deposition estimates of this model can be used directly. But if the desirable scale is much smaller than that considered in the LTRAP model, sub-grid processes should be taken into account.

Modelling the dispersion of trace gases in the atmosphere, including their physical and chemical transformations, is an essential element of the general study of trace gas behaviour and in the determination of the functional relationships between emissions and concentration or deposition levels. Measurements and models are strongly intertwined. On the one hand, measurements are necessary for parametrisation and validation of models, while on the other model results may render support in the evaluation, generalisation or extrapolation of measurements. Dispersion models exist in all sizes and complexities. The choice for a model or model concept is determined by the spatial and temporal scale on which air quality and deposition estimates have to be made (van Dop, 1986). In general, the models can be divided into different categories according to their mathematical framework: Eulerian models with a fixed co-ordinate system, Lagrangian models, where the co-ordinate system moves with the representative parcel of air or hybrid models merging both concepts. Here we will describe two Lagrangian-type

LRTAP models that are used to determine the annual spatial distribution of pollution in Europe. These are the EMEP (Eliassen and Salbones 1983; Iversen *et al.*, 1991) and OPS (Operationeel Prioritaire Stoffen model, Van Jaarsveld, 1990; 1995) models. Eulerian-type models will not be addressed here because up to now they have not been used to map annual deposition in Europe. In the USA the RADM (Regional Acid Deposition Model) model is applied for deposition mapping (Chang *et al.*, 1987). The EURAD (European Acid Deposition Model) model, used and developed in EUMAC, was designed to simulate the transport, chemical transformation and deposition of atmospheric pollutants over Europe (Jakobs, 1994). Until now this model has not been applied for mapping annual deposition fluxes. Several studies have been done simulating episodes.

The deterministic Lagrangian EMEP model is a receptor-oriented one-layer trajectory model originally developed at the Norwegian Meteorological Institute (Eliassen and Saltbones, 1983) and operationally applied in the framework of the 'Co-operative programme for monitoring and evaluation of the long-range transmission of air pollutants in Europe'. The European Monitoring and Evaluation Programme (EMEP) under the UN-ECE Convention on Long Range Transboundary Air Pollution, consists of a monitoring programme and modelling of atmospheric transport, conversion and deposition. The aim of EMEP is to provide basic information about emissions, levels of concentrations and deposition. The model calculates 96-h trajectories in reverse time on the basis of the 850 hPa wind field from the arrival of an air parcel at a receptor. Receptor points are the grid points of a 150 × 150 km grid system covering Europe. The trajectories arrive at the receptor points four times a day. During the air parcel's travel along the trajectory, the gridded sulfur and nitrogen emissions are immediately mixed in the vertical up to the top of the constant mixed layer. The model is used every year to estimate sulfur and nitrogen pollution in Europe and to determine the country-to-country receptor matrices and country budgets (*e.g.* Tuovinen *et al*, 1994).

The statistical transport model OPS is able to describe both short- and long-distance transport, and average concentrations and depositions can be computed for time scales from 1 day to more than 10 years (van Jaarsveld, 1990; 1995). It can account for both point sources of various heights and area sources of various shapes and heights. The sources need not be distributed on a regular grid system. It yields realistic results both within area sources and near point sources, as well as at long distances from sources. The receptor system is therefore determined by the resolution of the emissions. The concentrations and depositions in Europe are described on a 50 × 50 km grid system, whereas those in the Netherlands are estimated on a 5 × 5 km grid system. Computations are made for a limited number of meteorological situations (classes) with a representative meteorology for each class. Among the discretisations, a total of 12 wind-direction sectors and 6 atmospheric stability classes are distinguished. The basis for the model is formed by the Gaussian plume formulation for a point source. It is assumed that the plume is reflected only once at the surface and at the top of the boundary layer. Moreover, it is assumed that at larger distances from the source the plume is vertically distributed homogeneously over the whole boundary layer, apart from an attenuation near the surface due to dry deposition (van Jaarsveld, 1994; Asman and van Jaarsveld, 1992).

4.3 Deposition modelling

There is a difference between deposition modelling and long-range transport modelling, *i.e.* the focus of the two is different: deposition modelling is aimed at estimating the input to ecosystems as accurately as possible; long-range transport modelling, on the other hand, is aimed at determining the distribution of pollutants (concentration and deposition) based on emission estimates and dispersion calculations. In long-range transport modelling, deposition is a loss term in which the accuracy needed is dependent on the modelled scale. When both the pollution distribution and the deposition needs to be known as accurately as possible, a compromise is required on the details. Some of the aspects will be modelled in order to take the main processes determining transport and deposition into account. Deposition modelling is split in methods merely based on measurements and on a merge between measurements and LRTAP model results

4.3.1 Inferential technique

As an alternative to the direct measuring methods, knowledge on dry deposition processes can be used to infer dry deposition fluxes from basic information on routinely measured air concentrations and meteorological parameters. The flux is inferred as the product of ambient concentration of the chemical of interest and its dry deposition velocity. The dry deposition velocity is derived using a multiple-resistance transfer model (Hicks *et al.,* 1987). The resistance model provides a framework for coupling individual processes, some being surface dependent or pollutant dependent (Wesely, 1989. Erisman *et al.*, 1994; Erisman and Draaijers, 1995). The three major resistances are the aerodynamic resistance R_a, a near surface boundary resistance R_b, and the surface resistance R_c.

Flux measurements on a routine basis using the inferential technique are made in two networks in the USA (Hicks *et al.*, 1991; Clarke *et al.*, 1992). Weekly mean measurements of HNO_3, SO_2, NO_3^-, SO_4^{2-} and NH_4^+ concentrations in these networks are combined with meteorological parameters to obtain weekly average fluxes (Hicks *et al.*, 1985). This technique has also been applied in studies done in the UK (UK Review Group on Acid Rain, 1990), in Sweden (Lövblad *et al.,* 1991) and in the Netherlands (Erisman *et al.,* 1989; Erisman, 1992; 1993).

The quality of results obtained by this method depends on the availability and quality of data, and on the description of the resistances. R_a and R_b can be estimated from measurements of wind speed and radiation and information on surface roughness, together with parametrisations given by Garland (1977) and Wesely and Hicks (1977). R_c values can be obtained from flux measurements and from chamber methods. The most important parameters determining R_c should be included on a relevant time and spatial scale (see also section 4.4).

Eder and Dennis (1990) developed an inference technique to allow estimation of the annual and monthly dry deposition of Ca^{2+}, Mg^{2+}, Na^+ and K^+. The technique is based on the strong correlation between concentrations within precipitation and the surface-level air at 23 stations in Ontario, Canada. This method has been applied for the Netherlands by Erisman *et al.* (1994a), and for Europe by Erisman and Draaijers (1995) and Draaijers *et al.* (1995).

4.3.2 Merging technique

In applying abatement strategies on a national or European scale, it is desirable to know both the deposition and the distribution of pollution as accurately as possible. When emission abatement strategies are based on the critical load concept, the deposition has to be determined on the ecosystem level. Desired emission reductions based on the target levels should be known per country and/or per individual source or activity. Therefore, long-range transport modelling and ecosystem-scale deposition modelling should be linked. The deposition model which was developed within the framework of BIATEX is the EDACS model (Estimation of Deposition of Acidifying Components in Europe, Van Pul *et al.*, 1995). Wet deposition in this model is estimated using interpolation of wet deposition measurements (Van Leeuwen *et al.*, 1995). Dry deposition is determined using the inferential technique (Hicks *et al*, 1987). The dry deposition flux is inferred from LRTAP concentrations and land-use-specific dry deposition velocities. The resistance framework used in the inferential and merging technique will be explained in the next section.

4.3.3 Framework for the description of atmosphere - surface exchange

A common framework has been provided to describe the exchange of a range of gases and particles with very different chemical and physical properties. The most widely used is that of the resistance analogy (Chamberlain, 1966). In this framework, three stages in dry deposition to the surface are distinguished. The flux of a trace gas is given as:

$$F = V_d(z)\, c(z) \tag{1}$$

where $c(z)$ is the concentration at height z and V_d is the dry deposition velocity; z is the reference height above the surface. If the surface is covered with vegetation, a zero-plane displacement is included: $z = z - d$. d is usually taken as 0.6–0.8 times the vegetation height (Thom, 1973). The absorbing surface is often assumed to have zero surface concentration and the flux is therefore viewed as being linearly dependent on atmospheric concentration. This holds only for depositing gases and not for gases that might also be emitted, such as NH_3 and NO. For these gases a non-zero surface concentration, a compensation point c_p, might exist, which can be higher than the ambient concentration, in which case the gas is emitted. For these gases the flux is estimated as:

$$F = V_d(z)\,[c(z) - c_p] \tag{2}$$

V_d is widely used to parametrise gas uptake at the ground surface. V_d is the inverse of three resistances:

$$V_d(z) = \frac{1}{R_a(z-d) + R_b + R_c} \tag{3}$$

The three resistances represent the three stages of transport, the aerodynamic resistance R_a for the turbulent layer, the laminar layer resistance R_b for the quasi-laminar layer enveloping receptor surfaces and the surface or canopy resistance R_c for the receptor itself. The atmospheric resistance to transport of gases across the constant flux layer is assumed to be similar to that of heat (*e.g.* Hicks *et al.*, 1989). R_a is approximated following the procedures used by Garland (1978), while the quasi-laminar layer resistance R_b can be approximated by the procedure presented by Hicks *et al.* (1987).

4.3.3.1 Surface resistances

The surface or canopy resistance, $R_{c,}$ is the most difficult of the three resistances to describe. R_c values can be obtained from theoretical considerations based, for instance, on solubility and equilibrium calculations in combination with simulations of vegetation specific processes, such as accumulation, transfer processes through stomata, mesophyll and cuticles, absorption, *etc.* (Baldocchi *et al.*, 1987; Wesely, 1989). R_c values presented in the literature are primarily based on field measurements of V_d and on chamber studies. By determining R_a and R_b from the meteorological measurements, R_c can be calculated from V_d measurements as the residual resistance using Equation (3). Values of R_c can then be related to surface conditions, time of day, *etc.*, yielding paramtrisations. For use in a generalisation framework R_c parametrisations must be based on routinely available data, *i.e.* ambient temperature, relative humidity, global radiation, wind speed, land use class, type of soil and vegetation. In the next section the derivation of R_c values from experimental and modelling work, among others conducted within EUROTRAC, is explained. The parametrisation is extensively described in Erisman *et al.* (1994). The basic assumptions are only briefly discussed here, and some example values resulting from application of the parametrisation under different conditions are given.

The surface resistance of gases consists of a series of resistances reflecting different pathways of uptake. They are either determined by the actual state of the receptor, or by a memory effect. R_c is a function of the canopy stomatal resistance R_{stom} and mesophyll resistance R_m; the canopy cuticle or external leaf resistance R_{ext}; the soil resistance R_{soil} and in-canopy resistance R_{inc}, and the resistance to surface waters or moorland pools R_{wat} (Equation 4). In turn, these resistances are affected by leaf area, stomatal physiology, soil and external leaf surface pH, and presence and chemistry of liquid drops and films. The stomatal resistance, external leaf surface resistance and soil resistance act in parallel:

$$
\begin{aligned}
&\textit{vegetative surface:} && R_c = [\frac{1}{R_{stom} + R_m} + \frac{1}{R_{inc} + R_{soil}} + \frac{1}{R_{ext}}]^{-1} \\
&\textit{water surfaces:} && R_c = R_{wat} \\
&\textit{bare soil:} && R_c = R_{soil} \\
&\textit{snow cover:} && R_c = R_{snow}
\end{aligned} \tag{4}
$$

In Table 4.1 values of surface resistances obtained with the parametrisations presented in Erisman *et al.* (1994) are given for three different pollution climates as examples.

Table 4.1: Examples of averaged surface resistances (s m^{-1}) for different gases during daytime under different pollution climates, with the stomatal resistance (R_{stom}) also given

	Moist (rain or *rh* > 90 %), moderate temperature		Winter, snow covered surface		Dry, warm summer (*rh*<60%),well watered soil	
	Coniferous forest	Grassland	Coniferous forest	Grassland	Coniferous forest	Grassland
R_{stom}	200	60	400	–	150	50
SO_2	0#	0#	500	500	300	100
NH_3	0#	0#	500	500	500*	500*
NO_2	320	100	640	1000	240	80
HNO_2	0# (?)	0# (?)	500	500	300	100
HNO_3	0#	0#	10	50	0#	0#
HCl	0#	0#	10	50	0#	0#
PAN	500	500	1000	1000	500	500
O_3	320	100	640	1000	240	80

* Under these conditions NH_3 is expected to be emitted from the vegetation. The upward flux is estimated with this R_c value in Equation (3).

\# R_c values will be set to 1 s m^{-1} to avoid unrealistically high V_d at low values of R_a and R_b.

SO_2

The deposition of SO_2 on vegetation has been shown to be regulated mainly by stomatal resistance and the presence of surface water on the foliage. The stomatal uptake can be modelled satisfactorily. However, an understanding of the variations of surface resistance due to surface wetness chemistry is limited. It is assumed here that when the surface is wet, deposition of SO_2 is regulated by external leaf uptake. At low SO_2 concentrations, the buffering capacities of vegetation leaves are probably sufficient to maintain low R_c values. At high SO_2 concentrations other neutralising components, such as NH_3, might become important. The surface wetness may be provided by increased relative humidity, dew, guttation, rain or fog, and cloud deposition. Deposition of SO_2 to soil is dependent on the soil conditions (snow cover, frozen), relative humidity and on the soil pH.

NH_3

NH_3 is emitted from fertilised soils and from pasture during grazing. In addition, emission takes place from crops, ungrased pasture and semi-natural vegetation under dry, warm conditions during daytime at low ambient concentrations. In other cases NH_3 is mainly deposited. Deposition in these cases is regulated by stomatal uptake and by deposition to external leaf surfaces. R_c values for semi-natural vegetation and forests are low (in the range of 0–50 s m^{-1}), and depend on surface wetness.

NO_x

NO is not taken into account here as a depositing pollutant. Deposition of NO_2 is mainly regulated by stomata. The deposition to external leaf surfaces, water surfaces and soils is one order of magnitude less than stomatal uptake.

PAN, HNO_2, HNO_3, HCl

PAN deposition is slow with an overall surface resistance of 500–1000 s m^{-1}. HNO_2 is treated as SO_2 until more information becomes available. Thus uptake of HNO_2 is regulated by stomata and external leaf surfaces when the surface becomes moist. HNO_3 and HCl deposit efficiently to all surfaces with negligible R_c values. Only at low temperatures with a surface covered with snow does R_c increase up to 50–100 s m^{-1}

O_3

The uptake of ozone by vegetation is mainly via the stomata. The resistance of the remaining plant parts (the cuticle) to ozone is found to be larger than 1000 s m^{-1}. Ozone is destroyed slowly at a wetted surface or free water due to its low water solubility, and may depend on the chemical content. The surface resistance to ozone deduced from measurements above water and snow are about 1000–2000 s m^{-1}. The soil resistance to ozone is largely dependent on the type of soil and the soil water content. The soil resistance is typically 100 s m^{-1} under dry soil conditions. This value increases typically up to 500 s m^{-1} at a waterlogged stage of the soil (20–30 % water content).

4.3.3.2 Particles

The resistance analogy is not used for particles. Transport from the free atmosphere to the receptor surface is more or less similar to gas transport. However, transport processes through the quasi-laminar layer differ considerably between gases and particles. Whereas gases are transported primarily through molecular diffusion, particle transport and deposition basically take place through sedimentation, interception, impaction and/or Brownian diffusion. The currently available particle deposition models cannot be used for generalisation because most of the input data are not available for different kinds of surfaces. Ruigrok *et al.* (1994) gives an overview of available particle deposition models and ranges in V_d resulting from these models when applied to a specific situation. This range is very large. They simplified the Slinn (1982) model to include all relevant routinely available parameters. Slinn (1982) proposes use of a simplification of his model by splitting V_d for *forests* in a turbulent contribution and a surface deposition velocity, the latter being dependent on the collection efficiency of the surface and thus surface properties:

$$V_d = [V_{dt}^{-1} + V_{ds}^{-1}]^{-1}, \tag{5}$$

where the surface deposition velocity is given as $V_{ds} = E\, u_*^2/u_h$, with u_h the wind speed at canopy height, E is the collection efficiency and V_{dt} is the turbulent contribution, represented by the inverse of the aerodynamic resistance. E is different for different size classes as different processes become important. E is determined by parametrisation of modelled values based on data such as u_*, z_0, d, u_{50}, rh, surface wetness, etc. (Ruigrok *et al.*, 1994).
Current models for generalisation, such as DEADM (Erisman, 1993a), use a parametrisation of V_d for *low vegetation* based on relations of Wesely *et al.* (1985):

$$V_d = \frac{u_*}{300} \quad \textit{for stable conditions and} \tag{6}$$

$$V_d = \frac{u*}{300}\,[1 + (\frac{-300}{L})^{2/3}] \quad \textit{for unstable and neutral conditions}$$

Routinely available data for generalisation include meteorological parameters such as wind speed, radiation, temperature, relative humidity, u_* and L, and land use statistics and z_0 maps.

4.3.3.3 Evaluation of the SO_2 *atmosphere - surface exchange parametrisation*

The surface resistance parametrisation for surface uptake of SO_2 was tested by Erisman (1994) using eddy correlation measurements over a deciduous forest in Canada and vertical gradient measurements over a coniferous forest, a grassland and a heathland in the Netherlands. The surface resistance parametrisation is described in section 4.3.2. It was concluded that the modelled V_d compared reasonably well with the measurements, yielding no systematic differences for the coniferous forest and the heathland (more than 40 % of variance accounted for). Comparison of modelled and measured values is shown in Fig. 4.1. There is much less agreement with the measurements for the deciduous forest and grassland, showing systematic differences. However, it was concluded that the errors were not due to the surface resistance parametrisation, but rather to possible random errors in the (low) concentration measurements over the deciduous forest and to underestimating the time the surface was assumed wet (for the grassland measurements). The parametrisation was tested for four different classes (dry and wet conditions for day and night). No systematic error could be detected from this comparison other than what is described above.

The surface resistance parametrisation developed by Wesely (1989) was also tested using the same data set. Results showed a systematic underestimation of V_d values when compared to measured values. The underestimate was mainly the result of the parametrisation of R_c during and after precipitation. Wesely assumes an increase in R_c, whereas measurements showed a decrease to values close to zero. This led to an underestimation of V_d (SO_2) by a factor of 2 for low vegetation to a factor of 8 for forests.

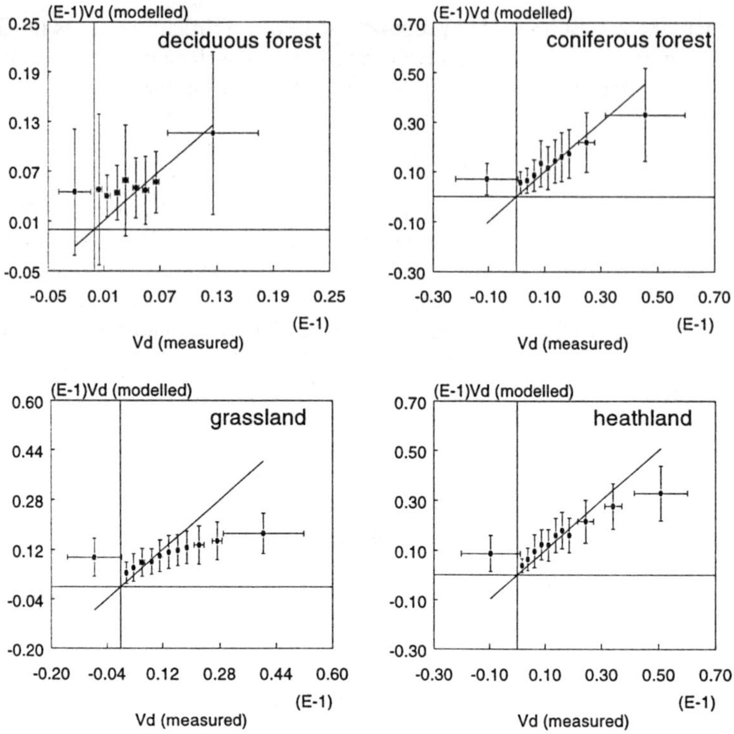

Fig. 4.1: Comparison of modelled V_d (m s^{-1}) with those obtained from measurements: A. deciduous forest; B. coniferous forest; C. grassland and D. heathland. Solid dots represent average modelled values for class averaged measured values. The line represents the 1:1 ratio. Small overbars represent measured class averages ± standard deviation; large overbars represent modelled averages ± standard deviation.

The applicability of the parametrisation for SO_2 has been demonstrated for the Netherlands and may be representative for larger areas in Europe. Other types of vegetation can be modelled by adjusting the stomatal resistance. The parametrisation might not be representative for those areas which have a completely different pollution climate from that occurring during the measurements. This might be the case for very dry areas or areas where ammonia concentrations are negligibly small compared to those for SO_2. Unfortunately, there is a lack of long-term measurements for such areas. It is therefore recommended to collect such measurements over these areas to evaluate and/or improve surface resistance parametrisations.

The surface resistance parametrisation was sent to every principal investigator of BIATEX, representing most of the research groups involved in deposition measurements in Europe. These PIs were asked to test the parametrisation and if necessary extend or improve it. The response was very limited because there is lack of suitable measurements. However, most reactions confirmed that the proposed schemes represent the state of knowledge on deposition processes.

4.4 EDACS

The outline of the EDACS model to estimate local and regional scale deposition fluxes is presented in Fig. 4.2. The basis for the two estimates is formed by calculations with the EMEP long-range transport model. With this model dry, wet and total deposition is estimated on a 150 × 150 km grid over Europe using emission maps for SO_2, NO_x and NH_3 (Eliassen and Saltbones, 1983; Iversen *et al.*, 1991; Tuovinen *et al.*, 1994). The model results are used for estimating country-to-country budgets, as a basis of sulfur and nitrogen protocols, and for assessments. The local-scale approach used by RIVM depends strongly on LRTAP model results. Calculated ambient concentrations of the acidifying components (daily averages) and concentrations in precipitation (monthly averages) are used along with a detailed land-use map and meteorological observations to estimate small-scale dry deposition fluxes (Fig. 4.2). By using calculated concentration maps, the relationship between emissions and deposition

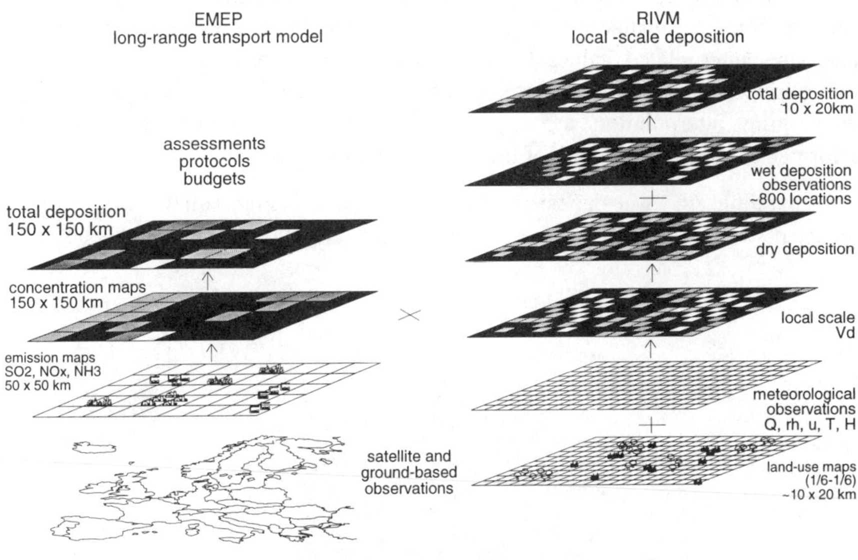

Fig. 4.2: Outline of method to estimate local scale deposition fluxes.

is maintained and scenario studies, budget studies and assessments can be carried out on different scales. Wet deposition is added to the dry deposition to estimate total local scale deposition in Europe. Wet deposition can either be obtained directly from the EMEP model, or from measurements made in Europe. The latter method is used here and will be described in the next section. The method for estimating dry deposition will be explained in section 4.4.2.

4.4.1 Wet deposition

Up to now, wet deposition maps on a European scale are based on long-range transport model results (*e.g.* the EMEP model; Iversen *et al.*, 1991), whereas for most components wet deposition maps based on measurements are only available on national scales. Van Leeuwen *et al.* (1995) recently used measurements in Europe to estimate the wet deposition in Europe. Acidifying components and basic cations, based on results of field measurements made at approximately 750 locations (the number of locations differs per component), were mapped on a 50 × 50 km scale over Europe for 1989. Information on concentrations of ions in precipitation in 1989 was obtained from the EMEP database (European Monitoring and Evaluation Programme) and from organisations responsible for wet deposition monitoring in their countries. Concentrations measured with bulk samplers were corrected for the contribution of dry deposition onto the funnels of these samplers. Sulfate concentrations were corrected for the contribution of sea salt, because only non-marine sulfate contributes to the acidification process. By assuming that the ratio of sodium to sulfate in sea-spray is the same as in bulk sea water, and that all the sodium in a sample is of marine origin, it is possible to calculate non-marine sulfate. At about 200 sites only information about the sulfate concentration was available and the sodium concentration was not measured. At those sites interpolated values from the sodium map were used for correction. Point observations were interpolated to a field covering the whole of Europe using the Kriging interpolation technique. An extensive uncertainty analysis was performed to assess the quality of the maps (see section 4.4).

As an example, the wet deposition flux map of non-marine sulfate (Fig. 4.3) is presented. The map clearly resemble European emission and climate patterns.

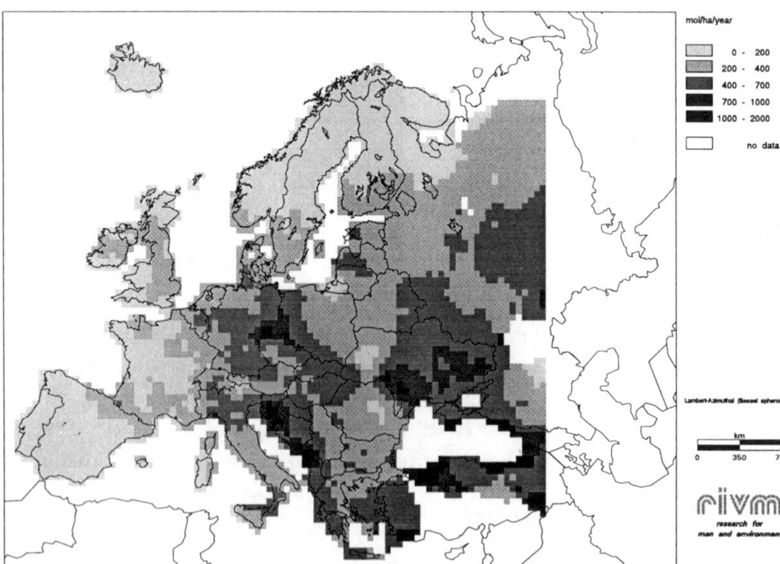

Fig. 4.3: Wet deposition of SO_4^{2-} in Europe on a 50 × 50 km basis in 1989 in mol ha^{-1} a^{-1}(van Leeuwen *et al.*, 1995)

Large emission sources can be recognised in the concentration maps (not shown here), whereas in the flux maps climate patterns and orographic effects are also observed. In mountainous areas large fluxes due to orographic rains can be seen. Especially in former Yugoslavia the large long-term mean precipitation cause large fluxes in all flux maps. Besides originating from large amounts of rainfall, large fluxes can also arise from high concentrations in precipitation.

Large fluxes (400–1000 mol ha^{-1} a^{-1}) can be observed in eastern and central Europe (Germany, Poland, Czech and Slovak Republics, Austria, Hungary, northern Italy, the Ukraine and former Yugoslavia). Large sulfate concentrations are mainly caused by SO_2 emissions from industry and power stations. Concentrations and fluxes obtained in Turkey should be interpreted with care, as they are not based on measurements in Turkey itself but are solely the result of interpolation (in this case extrapolation) from surrounding countries.

4.4.2 Dry deposition

The method for dry deposition has been developed by van Pul *et al.* (1992; 1994; 1995). Details can be found in these references. It is based on that used for the Netherlands (Erisman, 1992; 1993a). Dry deposition in EDACS is inferred from the combination of long-range transport model concentrations provided by EMEP and parametrised dry deposition velocities (Hicks *et al.*, 1987; Van Pul *et al.*, 1995). Concentrations at 50 m above the surface (blending height) are used. At this height it is assumed that concentrations and meteorological parameters are not influenced by surface properties to a large extent. Dry deposition velocities of gases and particles at this height are calculated on a small scale using a land-use map, routinely available meteorological data and the inferential technique (Erisman *et al.*, 1994; Van Pul *et al.*, 1995) Resistances are modelled using observations for meteorological parameters in Europe and parametrisation of surface exchange processes. Parametrisations of the surface resistance used for different receptor surfaces and pollution climates in Europe are described in section A.3. Meteorological parameters are obtained from the ODS (Observational Data Set: Potma, 1994) every six hours and interpolated over Europe on a $1/6^0 \times 1/6^0$ grid. Meteorological parameters consist of wind speed, friction velocity, radiation, temperature, humidity and precipitation.

The land-use map for Europe, based on a $1/6^0 \times 1/6^0$ (*ca.* 10×20 km), is constructed by RIVM from ground-based and satellite observations (van Velde *et al.*, 1994). Fig. 4.4 shows roughness length maps for the summer season as derived from the land-use map and z_0 classifications according to Erisman (1992). The roughness length is used for estimating atmospheric transport to the surface. For each land-use class, the coverage within a grid cell and the deposition velocity are calculated. The average deposition velocity for a grid cell is then calculated by weighting the land-use specific deposition velocities with the surface area within that specific grid cell. Site-specific dry deposition might be calculated likewise, provided that detailed land-use and surface-roughness data of the site is available. Here aggregated estimates are presented.

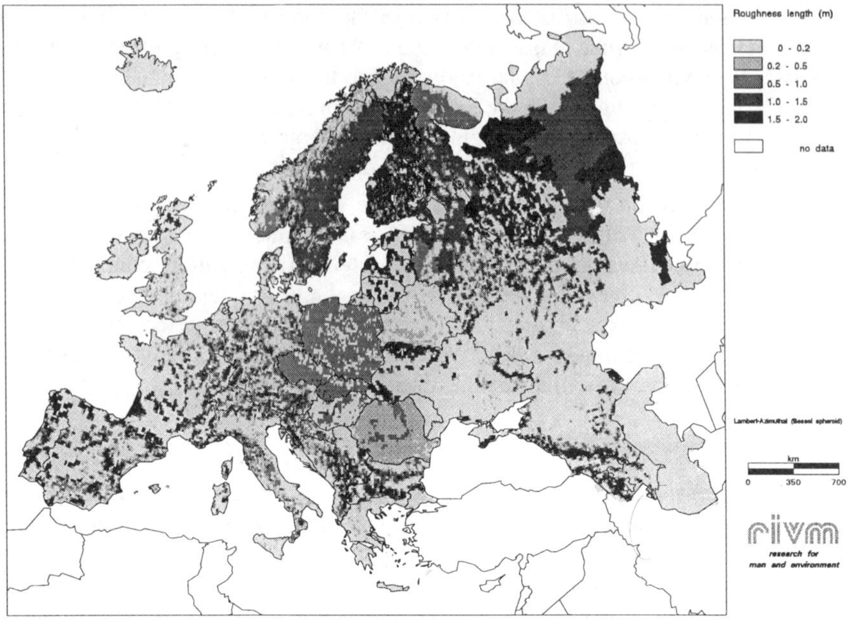

Fig. 4.4: Summer roughness length map of Europe on a 1/6 ^{0} × 1/6 ^{0} grid (in cm) (after van Pul *et al.*, 1995).

Figs. 4.5 to 4.7 show maps of the annual average dry deposition on a $1/6^0 \times 1/6^0$ scale of NO_y, NH_x and So_x deposition in Europe. The effect of land use (roughness) and the difference in V_d is clearly shown. In areas with forested terrain the dry deposition is increased relative to the EMEP values, and, for example, in dry areas, the dry deposition is decreased as a result of a difference in V_d estimates. There is a clear sub-grid effect. The variation in dry deposition within EMEP grids is determined by variation in land use. This is the result of a variation in the resistance due to atmospheric transport due to differences in roughness (see Fig. 4.4), or of a variation of the surface resistance due to differences in vegetation and surface conditions.

4.4.3 Total deposition of **potential** *acid*

Total deposition of potential acid, the sum of these components expressed in mol H^+ ha^{-1} a^{-1} is given in Fig. 4.8. Potential acid can be used to determine the maximum acid input to ecosystems. The actual acid load depends on the input of alkaline components and the extent that NH_3 is nitrified in the soil (*e.g.* Heij and Schneider, 1991, Erisman, 1993a). Highest deposition of potential acid occurs in the 'Black Triangle'. Relatively high fluxes are also found in parts of the Ukraine and former Yugoslavia, the Czech and Slovak Republics, Poland, northern France, Belgium, the Netherlands and Germany. Lower depositions are found over southwestern Europe, Ireland and northern Scandinavia.

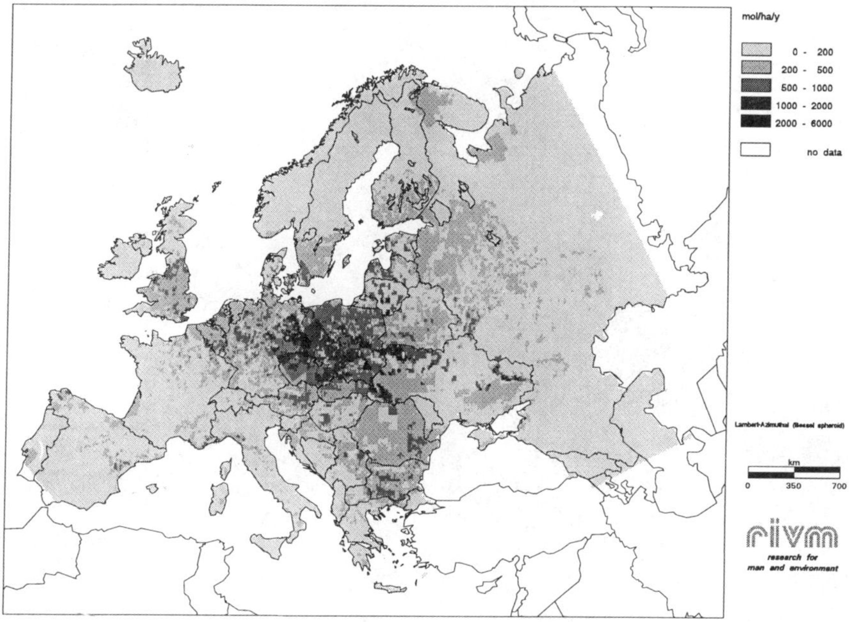

Fig. 4.5: Dry deposition of SO_x in Europe on a $1/6^0 \times 1/6^0$ scale in mol ha^{-1} a^{-1} (van Pul *et al.*, 1995)

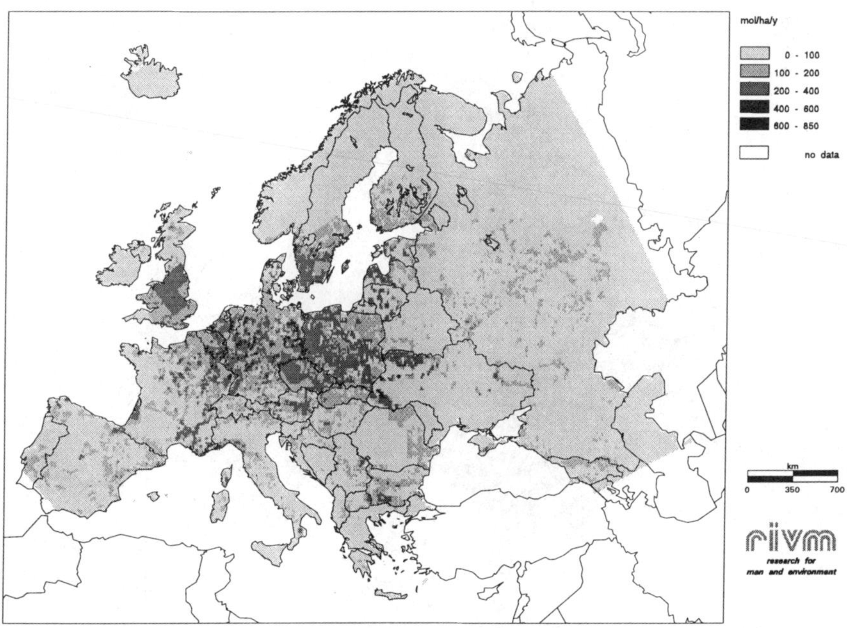

Fig. 4.6: Dry deposition of NO_y in Europe on a $1/6^0 \times 1/6^0$ scale in mol $ha^{-1}a^{-1}$ (van Pul *et al.*, 1995)

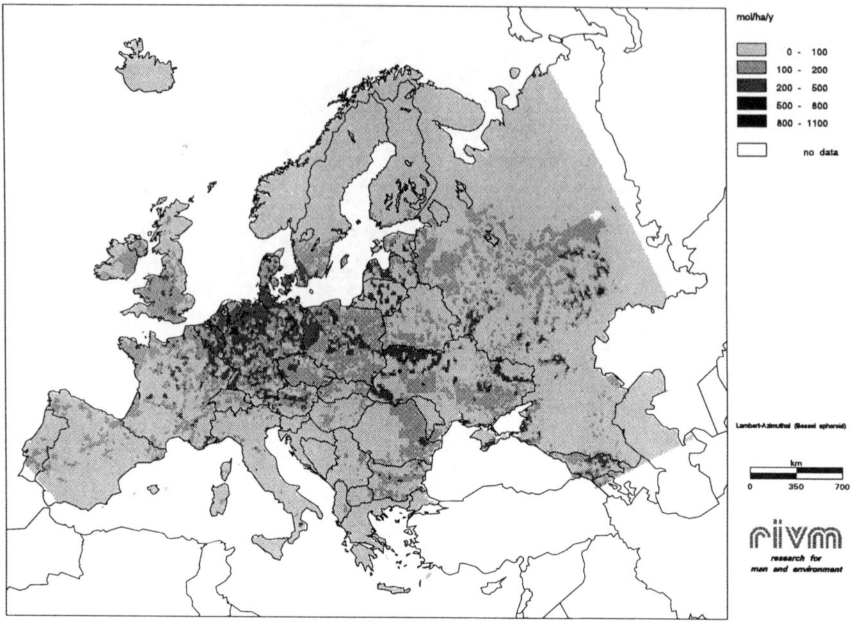

Fig. 4.7: Dry deposition of NH_x in Europe on a $1/6^0 \times 1/6^0$ scale in mol ha^{-1} a^{-1} (van Pul *et al.*, 1995)

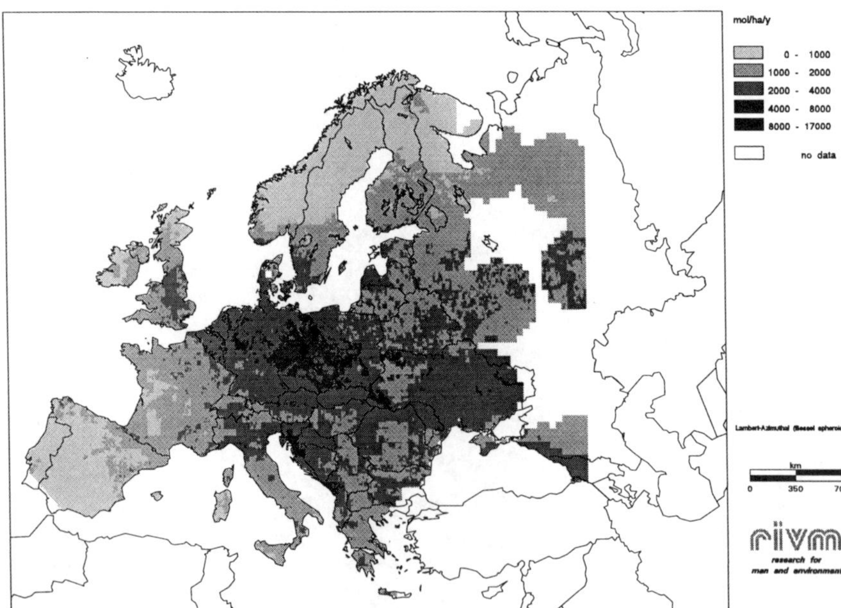

Fig. 4.8: Total potential acid deposition in Europe on a $1/6^0 \times 1/6^0$ scale in mol ha^{-1} a^{-1} (van Pul *et al.*, 1995)

4.4.4 Uncertainty in the EDACS results

4.4.4.1 Wet deposition

The wet deposition maps are subject to several sources of uncertainty. These are divided into three main categories:

(1) uncertainty associated with the measurements,
(2) uncertainty associated with assumptions and simplifications in the methods used, and
(3) uncertainty caused by the interpolation procedure.

Van Leeuwen *et al.* (1995) give an extensive description of these uncertainties. Furthermore, they quantify the contribution of all individual uncertainty sources to the total uncertainty. To assess the total uncertainty in the wet deposition maps obtained in this study is very difficult and therefore the estimates themselves are uncertain as well. The numbers presented here should therefore be regarded as a best guess based on expert judgement. The main source of uncertainty appears to be different for different regions of Europe. Therefore, a division into three categories is made:

(a) west, northwestern and central Europe as 'good quality areas' where data quality is assumed to be good and sufficient (representative) data is present,

(b) areas on edges of the maps, *i.e.* east, southeastern and southwestern Europe as 'poor quality areas', where less data is available of which the representativeness and quality can be questionable, and

(c) mountainous areas (*e.g.* the Alps) and upland areas (*e.g.* United Kingdom and Scandinavia), even though located in areas with much data, as 'complex terrain areas with additional uncertainties'.

Total uncertainty in concentrations and in rainfall amounts have been estimated separately. For both variables the most dominating uncertainty (which could be quantified either in this study or by literature) was taken to represent total uncertainty, as all uncertainties are distributed around the same mean. The mathematical procedure followed for the error propagation is that of the error propagation. In the case of calculation of the wet deposition fluxes the correlation coefficient is unknown. Therefore, two cases are considered: first a propagation of errors without correlation (conservative), and second, a propagation with full positive correlation (worst case). The numbers of total uncertainty are also presented as percentages, which show the total relative deviation from the estimated value in terms of one standard deviation. As the estimates are rather crude, no distinction between different components was made. In the conservative case (1), the overall uncertainty for an average 50×50 km grid cell is estimated to be 50 % in the west, northwest and central Europe, whereas in the worst case (2) the overall uncertainty is estimated to be 70 % here. In areas on the edges of the maps, *i.e.* east, southeastern and southwestern Europe, (1) is estimated to be 65 %

and (2) 90 %, whereas in complex terrain areas (1) is estimated to be 85 % and *(2)* 120 %.

4.4.4.2 Dry deposition

Probably the highest uncertainty contribution is the result of using the simple resistance formulation for such a complex process as dry deposition, and, more specifically, the surface resistance parametrisations used to determine the surface exchange of gases and particles. The resistance model represents a simple approach to a highly variable process. It assumes a constant flux layer, *i.e.* there are no surface inhomogeneities, edge effects or chemical reactions. How much these simplifications contribute to the total uncertainty in the annual average deposition fluxes has not been quantified. The uncertainty in the surface resistance parametrisation is the largest uncertainty in this simplified scheme. Therefore more and more accurate parametrisations are needed for various vegetation species and surface types. Moreover, there is a lack of measurements which can be used to test these parametrisations, especially for southern and eastern European climates and surfaces. The vegetation and land use classes used in EDACS should be expanded when more parametrisations become available. Surface wetness is found to be one of the major factors influencing the deposition process of soluble gases. In the present version of EDACS only rain and indication of dew are used. Because of the great influence of surface wetness, processes leading to surface wetness should be taken into account in greater detail. The overall uncertainty in the surface resistance due to these factors is different for each component and surface type. This uncertainty varies on an annual basis, between 20 percent or more to more than 100 %.

In the current version of EDACS, the EMEP-LRTAP modelled concentrations on a 150 × 150 km grid are used. The uncertainty in the concentrations are estimated to amount 40–70 % using a statistical analysis with the EMEP measurements (Krüger, 1993). These concentrations represent the background situation in Europe. In source areas the uncertainty can be even higher. It is assumed that the concentration distribution within a grid is homogeneous. This is not the case in a grid which contains industrialised areas or many scattered sources such as with NH_3 and NO_x. For such conditions, sub-grid concentration variations are present and will lead to underestimates of the deposition in that grid. To obtain an indication of the errors, a small-scale, short range model can be useful here to resolve sub-grid concentration gradients for dense source areas. The uncertainty in the deposition in an EDACS grid cell due to these gradients is estimated at 25 % (Berg and Schaug, 1994).

The deposition in EDACS is based on the EMEP-LRTAP concentrations which in turn are dependent on EMEP deposition parametrisations. The deposition in the EMEP model and in EDACS are parametrised in different ways. By using other dry deposition velocities in EDACS, a mass inconsistency between EMEP calculated deposition and the small-scale maps of EDACS might be introduced.

Erisman *et al.* (1994) demonstrated that the mass inconsistency between EMEP-calculated deposition and the small-scale maps of EDACS does not lead to large differences in country average depositions. As the result of a close cooperation between EMEP and RIVM, EMEP is currently using the same V_d parametrisations and land use data as those used in EDACS. Therefore, in the near future there will be no problem of mass inconsistencies between the two models.

The accuracy of the presented results depends on the availability and quality of the input data, such as the land-use map and the meteorological observations. In the gridded version of the RIVM land-use database, forest is not subdivided into deciduous and coniferous forest. All forest is classified as coniferous forest. This will probably lead to overestimates of the deposition velocity to deciduous forests for all components during winter. However, the stomatal resistance in winter will generally be large due to the low temperatures. This overestimate will therefore not be very large.

The dry deposition is calculated on a daily basis. However, due to the daily averaging of the concentration and deposition velocity, a loss in temporal interpolation is introduced. This error is component-specific and estimated to be smaller than 20 % (van Pul *et al*, 1993).

4.4.4.3 Total deposition

An intercomparison between budgets for a range of acidifying components for the Netherlands using different models was presented in Derwent *et al.* (1989). They compared results from four different long-range transport models, amongst which were EMEP and OPS model, with deposition inferred from airborne and precipitation concentrations (Erisman *et al.*, 1989). Erisman *et al.* (1993) compared budgets and regional distribution of deposition in the UK, Sweden and the Netherlands using seven model approaches, including EMEP and OPS, the national models and a preliminary version of the EDACS model. The results of these two studies showed that the country budgets calculated with different models were reasonably in agreement. However, on a regional scale large differences between estimated total deposition fluxes were identified.

The uncertainty in regional scale deposition estimates strongly depends on the pollution climate and on landscape complexity of the area under study. The uncertainty is determined by the uncertainty in wet, dry or cloud and fog deposition. Furthermore, deposition estimates yield higher uncertainty in areas built up by complex terrain and with strong horizontal concentration gradients. The influence of roughness transition zones and forest edges to the deposition in the Netherlands is quantified by Draaijers (1993). He shows that there is a systematic underestimation of the dry deposition in the order of 5–10 % as the result of these factors. Estimates for other types of complex terrain, such as mountainous regions, are not available.

Table 4.2: Uncertainty of key factors influencing deposition estimates of S and oxidised and reduced N in different pollution regions in Europe

Key Factors	NO_y							NH_x							SO_x						
Regions[a]:	1	2	3	4	5	6	7	1	2	3	4	5	6	7	1	2	3	4	5	6	7
emission/type of	–	++	++	+	++	+	+	–	++	+	+	+	+	–	–	+	++	+	+	+	–
source	2	2	2	2	2	2	2	2	3	3	3	3	3	2	1	1	3	1	3	1	1
concentration	++	++	++	++	++	++	++	++	++	++	++	++	++	++	++	++	++	++	++	++	++
	2	2	3	3	3	3	2	1	3	3	3	3	3	2	1	2	3	3	3	2	2
wind speed	–	+	+	–	–	+	–	+	++	++	+	++	++	–	+	++	++	+	++	++	–
	1	1	1	1	1	1	1	1	1	1	1	1	1	1	1	1	1	1	1	1	1
roughness length	–	+	+	+	+	+	–	+	+	+	+	+	+	+	+	+	+	+	+	+	+
	1	1	2	2	2	2	2	1	1	2	2	2	2	2	1	1	2	2	2	2	2
surface wetness	–	–	–	–	–	–	–	+	++	++	+	++	+	+	+	++	++	+	++	++	+
	1	1	1	1	1	1	1	3	2	3	2	3	3	3	3	2	3	2	3	3	3
orography	–	–	–	–	–	++	–	++	–	–	+	+	++	–	+	–	+	+	+	++	–
	1	1	1	1	1	1	1	2	1	1	2	2	2	1	2	2	2	2	2	2	1
co-deposition	–	–	–	–	–	–	–	–	++	++	+	+	+	–	–	++	++	+	+	+	+
	1	1	1	1	1	1	1	3	2	3	3	3	3	3	3	2	3	3	3	3	3
surface resistance	++	++	++	++	++	++	++	++	++	++	+	++	+	+	++	++	++	+	++	+	–
	2	2	2	2	2	2	2	3	3	3	3	3	3	3	3	3	3	3	3	3	3
dry deposition	–	+	+	+	+	+	–	–	++	++	+	++	+	–	–	++	++	+	++	+	–
	3	3	3	3	3	3	3	3	3	3	3	3	3	3	3	3	3	3	3	3	3
wet deposition	++	+	+	+	+	++	+	++	+	+	+	+	++	+	++	+	+	+	+	++	+
	1	1	1	1	2	2	2	1	1	1	1	2	2	2	1	1	1	1	2	2	2
cloud and fog	+	++	–	–	+	++	–	+	++	–	–	+	++	–	+	++	–	–	+	++	–
	3	3	3	3	3	3	3	3	3	3	3	3	3	3	3	3	3	3	3	3	3

legend:	importance of key factors:	uncertainty:
	++ dominating	1 low < 30 %
	+ important	2 median 30–70 %
	– neutral	3 high > 70 %
	— unimportant	[a]the distribution of these regions is given in Fig. 4.9

Table 4.2 shows the uncertainties associated with major key factors in deposition for seven (pollution) regions in Europe. The seven regions are shown on a European map in Figure 4.9. These regions are chosen because there is a marked difference in dominating deposition process: region 1: main input is wet

deposition: region 6: main input is fog or cloud deposition; or there is a marked difference in climate: region 2 exhibits a sea climate, whereas region 7 exhibits a Mediterranean sea climate and region 4 mainly an inland climate; or there is a marked difference in industrial activity: region 1 is remote, region 3 is mainly dominated by large old industrial complexes and region 2 exhibits high reduced nitrogen emission densities. This is only a crude classification. Furthermore, Table 4.2 shows only a crude estimate of uncertainty. Because of the strong local variations in dry deposition and the associated uncertainty local scales should be considered.

Fig. 4.9: Distribution of the pollution regions in Europe in Table 4.2 (see text for explanation).

Acknowledgements

Anton Eliassen, Erik Berge and Helge Styve of EMEP MSC-W are thanked for their cooperation in providing the EMEP model concentration data. ECMWF anf KNMI are acknowledged for the WHO synops data.

References

Asman, W.A.H., Jaarsveld, J.A. van (1992) A variable-resolution transport model applied for NHx for Europe. *Atmos. Environ.* **26A**, 445–464.

Baldocchi, D.D., Hicks, B.B., Camara, P. (1987) A canopy stomatal resistance model for gaseous deposition to vegetated surfaces. *Atmos. Environ.* **21**, :91–101.

Berg, T. and Schaug, J. (1994) *Proc. EMEP workshop on the accuracy of measurements*, with sessions on determining the representativeness of measured parameters in a given grid square as compared to model calculations, 22–26 November 1993, Passau (Germany).

Chamberlain, A.C. (1966) Transport of gases from grass and grass-like surfaces. *Proc. Roy. Soc. Lond.* **A290,** 236–265.

Chang, J.S., Brost, R.A., Isaksen, I.S.A., Madronich, S., Middloton, P., Stockwell, W.R., Walcek, C.J. (1987) *J. Geophys. Res.* **92**, 14681.

Clarke, J.F., Edgerton, E.S. and Boksleitner, R.P. (1992) Routine estimation and reporting of dry deposition for the USA Dry Deposition Network. In: S.E. Schwartz and W.G.N. Slinn (eds), *Proc. 5th IPSASEP conference, Richland, June 1991.* Hemisphere Publishing Corp., Washington.

Davidson, C.I. and Wu, Y.L. (1990) Dry deposition of particles and vapors. In: S.E. Lindberg, A.L. Page, S.A. Norton (eds), *Acidic Precipitation*, Vol. **3**. Springer Verlag, New York.

Dop, H. van (1986) Models for deposition processes. In: S. Sandroni (ed), *Regional and Long-range transport of air pollution.* Esevier Science Publishers, Amsterdam.

Draaijers, G.P.J. (1993) The variability of atmospheric deposition to forests. Ph.D.thesis, University of Utrecht, The Netherlands.

Draaijers, G.P.J., *et al.*, in preparation

Eliassen, A., Saltbones, J. (1983) Modelling of long-range transport of sulfur over Europe: a two year model run and some model experiments. *Atmos. Environ.* **17**, 1457–1473.

Erisman J.W., Leeuw F.A.A.M.de, and Aalst R.M.van (1989) Deposition of the most acidifying compounds in the Netherlands during 1980–1986. *Atmos. Environ.* **23**, 1051–1062

Erisman, J.W. (1992) Atmospheric deposition of acidifying compounds in the Netherlands. Ph.D. Thesis, Utrecht University, The Netherlands.

Erisman, J.W. (1993) Acid deposition onto nature areas in the Netherlands; Part I. Methods and results. *Water Soil Air Pollut.* **71**, 51–80.

Erisman, J.W., Jaarsveld, J.H. van, Pul, A. van, Fowler, D., Smith, R. and Lövblad, G. (1993) Comparison between small scale and long-range transport modelling. In: G. Lövblad, J.W. Erisman, D. Fowler (eds), Models and methods for the quantification of atmospheric input to ecosystems, *Report No. 1993:573,* Nordic Council of Ministers, Copenhagen, Denmark.

Erisman, J.W. (1994) Evaluation of a surface resistance parameterization of SO_2. *Atmos. Environ.* **28**, 2583–2594.

Erisman, J.W., Draaijers, G.J.P., Duyzer, J.H., Hofschreuder, P., Leeuwen, N. van, Römer, F.G., Ruijgrok, W., Wyers, G.P. (1994a) Contribution of aerosol deposition to atmospheric deposition and soil loads onto forests. *Report No. 722108005*, National Institute of Public Health and Environmental Protection, Bilthoven, The Netherlands.

Erisman, J.W., Pul, A. van, Wyers, P. (1994) Parameterization of dry deposition mechanisms for the quantification of atmospheric input to ecosystems. *Atmos. Environ.* **28**, 2595–2607.

Erisman, J.W., Draaijers, G.P.J. (1995) *Atmospheric deposition in relation to acidification and eutrophication. Studies in Environmental Research*, Elsevier, the Netherlands.

Garland, J.A. (1977) The dry deposition of sulfur dioxide to land and water surfaces. *Proc. Roy. Soc. Lond.* **A354,** 245–268.

Garland, J.A. (1978) Dry and wet removal of sulfur from the atmosphere. *Atmos. Environ.* **12**, 349.

Heij, G.J., Schneider, T. (1991) Final report, Dutch Priority Programme on Acidification, Second Phase. *Report No. 200-09*, National Institute of Public Health and Environmental Protection, Bilthoven, The Netherlands.

Hicks, B.B., Baldocchi, D.D., Meyers, T.P., Hosker Jr., R.P., Matt, D.R. (1987) A preliminary multiple resistance routine for deriving dry deposition velocities from measured quantities. *Water Air Soil Pollut.* **36**, 311–330.

Hicks, B.B., Draxler, R.R., Albritton, D.L., Fehsenfeld, F.C., Hales, J.M., Meyers, T.P., Vong, R.L., Dodge, M., Schwartz, S.E., Tanner, R.L., Davidson, C.I., Lindberg, S.E., Wesely, M.L. (1989a) *Atmospheric processes research and procecess model development. State of Science/Technology,* Report No. 2. National Acid Precipitation Assessment Program.

Hicks, B.B., Hosker, R.P., Meyers, T.P. and Womack, J.D. (1991) Dry deposition inferential measurement techniques - I. Design and tests of a prototype meteorological and chemical system for determining dry deposition. *Atmos. Environ.* **25A**, 2345–2359.

Iversen, T., Halvorsen, N., Mylona, S., Sandnes, H. (1991) *Calculated budgets for airborne acidifying components in Europe, 1985, 1987, 1988 and 1990.* MSC-West, Norwegian Meteorological Institute, Oslo.

Jaarsveld, H.J.A.van (1990) A quantative model analysis of year to year changes in concentration and deposition. *Presented at the NATO CCMS meeting*, Vancouver, Canada.

Jaarsveld, H.J.A.van (1995), modelling the long-term atmospheric behaviour of pollutants on various spatial scales. Ph.D. thesis, University of Utrecht, the Netherlands.

Jakobs, H.J. (1994) Use of nested models for air pollution studies.in: P.M. Borrell, P. Borrell, T. Cvitaš, W. Seiler (eds), *Proc. EUROTRAC Symp. '94*, SPB Academic Publishing bv, The Hague 1994, pp. 799–806.

Krüger, O. (1993) The applicability of the EMEP code to estimate budgets for airborne acidifying components in Europe. In: *Proc. CEC/BIATEX Workshop 4–7 May 1993*, Aveiro, Portugal.

Leeuwen, E.P. van, Potma, C., Draaijers, G.P.J., Erisman, J.W., Pul, W.A.J. van (1995) European wet deposition maps based on measurements. *RIVM Report 722108006*, National Institute for Public Health and Environmental Protection, Bilthoven, The Netherlands.

Lövblad, G., Erisman, J.W., Fowler, D. (1993) Models and methods for the quantification of atmospheric input to ecosystems. *Report No. Nord 1993:573* Göteborg, Sweden, 3–7 November 1992. Nordic Council of Ministers, Copenhagen, Denmark.

Pul, W.A.J. van, Erisman, J.W., Jaarsveld, J.A. van and Leeuw, F.A.A.M. de (1992) High resolution assessment of acid deposition fluxes. In: T. Scneider (ed), *Acidification research: evaluation and policy application. Studies in Environmental Science.* Elsevier, Amsterdam.

Pul, W.A.J., Potma, C., Leeuwen, E.P. van, Draaijers, G.P.J. and Erisman, J.W. (1995) EDACS: European Deposition maps of Acidifying Compounds on a Small scale. Model description and results. *RIVM Report 722401005.*

RIVM (1994) Jaarrapport 1993. *Report No. 222101006*, National Institute of Public Health and Environmental Protection, Bilthoven, The Netherlands.

Ruijgrok, W., Tieben, H., Eisinga, P. (1994) The dry deposition of acidifying and alkaline particles on Douglas fir. *Report No. 20159-KES/MLU 94-3216*, KEMA, Arnhem, The Netherlands.

Slinn, W.G.N. (1982) Predictions for particle deposition to vegetative surfaces. *Atmos. Environ.* **16**, 1785–1794.

Tuovinen, J.P., Barrett, K., Styve, H. (1994) Transboundary acidifying pollution in Europe: Calculated fields and budgets 1985–1993. *EMEP/MSC-W, Report 1/94*, Norwegian Meteorological Institute, Oslo.

UK Review group on acid rain (1990) *Acid deposition in the United Kingdom 1986–1988.* Warren Spring Laboratory, UK.

Velde, R.J. van, Faber, W.S., Katwijk, V.F. van, Kuylenstierna, J.C.I., Scholten, H.J., Thewessen, T.J.M., Verspuij, M., Zevenbergen, M. (1995) The preparation of a European land use database. *Report no. 712401001*, National Institute of Public Health and the Environment, Bilthoven, the Netherlands.

Wesely, M.L., Hicks, B.B. (1977) Some factors that affect the deposition rates of Sulfur Dioxide and similar gases on vegetation. *J. Air Pollut. Contr. Assoc.* **27**, 1110–1116.

Wesely, M.L., Cook, D.R., Hart, R.L. (1985) Measurements and parameterization of particulate sulfur dry deposition over grass. *J. Geophys. Res.* **90**, 2131–2143.

Wesely, M.L. (1989) Parametrization of surface resistances to gaseous dry deposition in regional scale numerical models. *Atmos. Environ.* **23**, 1293–1304.

Chapter 5

Biological Mechanisms involved in the Exchange of Trace Gases

J. Kesselmeier[1], K. Bode[1], J.K. Schjørring[2] and R. Conrad[3]

[1] Max Planck Institute for Chemistry, Biogeochemistry Department, Johann-Joachim Becher Weg 29, D-55128 Mainz, Germany
[2]Royal Veterinary and Agricultural University, Centre of Ecology and Environment & Plant Nutrition Laboratory, Thorvaldsenvej 40, DK-1871 Frederiksberg C, Denmark
[3]Max Planck Institute for terrestrial Microbiology, Karl-von-Frisch-Straße, D-35043 Marburg, Germany

5.1 Introduction

The chemistry of the atmosphere is significantly influenced by exchange processes between the biosphere and the atmosphere (Warneck 1988; Schlesinger 1991; Sharkey *et al.* 1991; Butcher *et al.* 1992; Graedel and Crutzen 1993, 1994). A number of trace gases, which can dramatically affect atmospheric chemistry, are not only released from but also adsorbed and consumed by the biosphere. Within this context we examine plant physiological and soil microbial processes, potentially influenced by environmental factors such as temperature, humidity, light and trace gas composition of the surrounding air. In contrast to adsorption/desorption processes of trace gases produced by soils and only mediated by the plant tissue (*i.e.* methane) it is obvious that volatile components have to be produced prior to the emission. Thus we need to understand why these compounds are produced: interactions, special functions, unimportant by-products or accumulations due to blocking metabolic pathways. In all cases the status of the soil and plant or the soil-plant system and external factors have a significant influence on emissions. These are the crucial points in addition to stress, senescence and seasonal effects. Furthermore, the measurement techniques used for plant or micro-organism studies influence the exchange behaviour sometimes dramatically. Another important factor without direct involvement of plant metabolism is the influence of leaf surface micro-organisms, the epiphytes. All plant surfaces are more or less colonised by bacteria, fungi, yeasts or algae which form an ecosystem of their own, called the phyllosphere (Blakeman 1981; Fokkema and van den Heuvel 1986; Andrews and Hirano 1991). Information about their influence on trace gas exchange is lacking. However, as the epiphytes have their own metabolism, including the exchange of metabolites with their environment, an interaction with the surrounding air has to be assumed.

5.2 Plant anatomical and morphological factors

All substances produced by plant cells have to pass a number of borderlines, all with their own resistances, before being released to the atmosphere. To understand the mechanisms controlling emissions, a rough overview about the structure and chemical properties of the leaf compartments is given below.

5.2.1 Membranes

All eucaryotic cells and cell compartments are enclosed by lipid bilayers (plasmalemma, tonoplast, plastid envelope *etc.*) mainly consisting of lipids and proteins. Assuming that the substance considered is not actively excreted by active transport mechanisms, the resistance provided by the actual membrane depends on the molecular size and the lipophilic character of the specific compound. The latter is estimated by using the octanol/water partition coefficient ($K_{o/w}$). Thus, for a $K_{o/w}$ ranging between 10^2 to 10^6 a range of resistances of 5×10^5 to 50 s m^{-1} can be calculated (Schönherr and Riederer 1989).

5.2.2 Cell wall

Additionally, all cells are surrounded by a solid cell wall, consisting mainly of cellulose, acidic carbohydrates and some proteins. The acidic carbohydrates contain weak acidic groups which are cross-linked by calcium or magnesium ions. Due to the high amount of mainly negative charges the cell wall is a cation exchanger. According to Grignon and Sentenac (1991) this structure contains micropores with diameters in the range of 3–8 nm, thus allowing the transport of water and hydrated inorganic ions. For the purpose of estimating its resistance the cell wall is considered an unstirred aqueous solution. Assuming a thickness around 5 μm the resistance will be *ca.* 50,000 s m^{-1} (Schönherr and Riederer 1989).

5.2.3 Apoplast

The extracellular aqueous phase, including the continuum of the cell walls, is called the apoplast. It occupies between 5 and 20% of leaves, see Table 5.1 (Cosgrove and Cleland 1983; Speer and Kaiser 1991). Expressed relative to the total leaf volume, both the volume of apoplastic water and the of air-filled intercellular cavities may decrease with leaf age.

Whereas the cytosolic pH for most plants is assumed to be fairly constant at *ca.* pH 7 (Martin *et al.* 1982; Grignon and Sentenac 1991), the conditions in the apoplast may be much more variable, depending on species, their specific tissues, state of development and nutritional status (Hoffmann *et al.* 1992; Kurkdjian and Guern 1989). The majority of values measured, range between pH 5 and 6.5 thus resulting in a pH gradient across the cell membrane, which is important for nutrient uptake by cells (Grignon and Sentenac 1991). Values below pH 5.5 are

Table 5.1: Apoplastic pH, ammonium and estimated ammonia compensation points in leaves at four different positions on *Brassica napus* plants (means ± SE, n= 4-6). Data from Husted and Schjørring (1995).

Leaf position*	Air volume (%)	Water volume (%)	NH_4^+ concentration (mM)	pH
7	27.2	14.4 ± 1.1	1.10 ± 0.24	5.77 ± 0.19
9	24.5	11.3 ± 0.5	1.06 ± 0.16	5.73 ± 0.16
14	18.3	12.2 ± 1.8	1.03 ± 0.14	5.83 ± 0.08
18	15.0	8.3 ± 1.0	1.30 ± 0.08	5.78 ± 0.07

* acropetal direction

measured by use of fluorescent dyes, a technique which may lead to underestimation of apoplastic pH due to auto-quenching of the dye following interaction with cell wall components. Values and changes of the actual pH affect the transfer properties of the apoplast.

The significance of the actual state of the apoplast for the exchange of trace gases is discussed for ammonia as an example for trace gases with acid-base properties. Apoplastic ammonium concentrations measured in leaves of *Brassica napus* plants with an intermediate N-status were around 1 mM (Table 5.1). Together with an apoplastic pH around 5.7 this corresponds to an ammonia concentration in the air within the substomatal cavities of about 5 nmol per mol air. The concentration of NH_4^+ in the leaf apoplast will depend on the concentration of NH_4^+ in the transpiration stream (xylem sap). Xylem sap NH_4^+ concentrations in barley plants may increase to well above 500 μM and may be higher in NH_4^+ than in NO_3^- fed plants (Mattsson and Schjørring 1995). Similar results have been reported for wheat and maize by Cramer and Lewis (1993). Thus plants supplied via the roots with NH_4^+ show a larger NH_3 emission than plants fed with NO_3^-, in agreement with the different NH_4^+ concentrations in the xylem sap (Mattsson and Schjørring 1995).

Because of lower pH values in the apoplastic system than in the cytoplasm (pH *ca.* 7) or in the chloroplasts (pH *ca.* 8) , the intra-cellular concentration of NH_4^+ may be 10 to 100 times lower than the apoplastic NH_4^+ concentration and still maintain an intra-cellular concentration of dissolved NH_3 that is higher than that in the apoplast. Taking for example a chloroplastic NH_4^+ concentration of 20 μM, which is close to the K_m for GS (12-22 μM NH_4^+ ; O'Neal and Joy 1974), then the concentration of dissolved ammonia in the chloroplasts at 20 °C would be *ca.* 0.5 μM. In the apoplast a concentration of 1 mM NH_4^+ at pH 5.7 would support a concentration of dissolved NH_3 of only 0.19 μM. Considering the rapid diffusion of NH_3 across biomembranes, a concentration gradient of dissolved NH_3 may result in a significant outward diffusion of dissolved NH_3 across the plasmalemma.

5.2.4 Cuticle

The last barrier between the plant and the atmosphere is the cuticle, a lipophilic and strongly water resistant layer. The cuticle consists of extractable waxes (soluble lipids) attached to a polymer matrix. The latter is a lipid polyester primarily constructed from C16 and C18-hydroxy and epoxy fatty acids, which are linked together mainly by ester bridges (Holloway 1982). The wax fraction consists of a heterogeneous mixture of mostly unsaturated long chain aliphatic components between C12 and C72, *i.e.* alkanes, alkane-1-oles, esters, fatty acids, aldehydes and ketones as well as secondary plant products like sterols, flavonoids and terpenes (Baker 1982). These compounds sometimes may be organised in a crystalline form constructing different structures like filaments, plates or rods on the leaf surfaces (Baker 1982). For most of the trace gases the cuticle is a barrier. However, it is not as effective in limiting trace gas uptake as for the water loss from the leaf (Schönherr 1982; Lendzian 1987). Permeability for gases is by 2–3 orders of magnitude higher than for H_2O. As for water, the main resistance for gases is caused by the soluble cuticular lipids. The matrix without these lipids shows a higher permeability. Thus, it is obvious that the cuticle represents a second way for gases to enter the plant in the case of closed stomata, though the stomata have to be regarded as the controlling site for water and gas exchange.
Due to its chemical properties the cuticle is an excellent barrier for lipophobic compounds. For these substances resistance is reported to range between 10^6 and 10^{10} s m^{-1} (Schönherr and Riederer 1989). In case of incision or injuries due to mechanical or chemical treatment or simply due to seasonal (senescence) effects this barrier will change. In this context, erosion and microbial decomposition of the cuticular components may be important (Kolattukudy 1980). For lipophilic organic compounds the cuticle serves as an absorber (Sablicj *et al.* 1990). Thus, exchange processes due to adsorption/desorption processes may occur without involving plant metabolism. In addition surface properties change under environmental influences such as fog, cloud, rain or dew, and guttation (Fowler and Unsworth 1979; Bieder and Schatz 1993). SO_2 dry deposition for example is enhanced over wet surfaces (Garland and Branson 1977; Fowler and Unsworth 1979; Fowler 1985; Vermetten *et al.* 1992; Erisman *et al.* 1993a; Erisman and Wyers 1993) and wetted cuticles lead to a rapid absorption of ammonia (Sutton *et al.* 1993a, b; Duyzer *et al.* 1994). Due to the presence of bound microscale water in or on the leaf cuticle ammonia may even be adsorbed to the cuticle at lower air humidities. The adsorption capacity of the cuticle for ammonia increases strongly with atmospheric humidity between 45-95% (van Hove *et al.* 1989).

5.2.5 Stomata

To meet the needs for the exchange of carbon dioxide all higher plants posses stomata which can adapt to different environmental conditions. The total stomatal area, a product of pore density and maximum pore width, ranges for most of the plants between 0.5 and 1.2 % of the leaf surface (Larcher 1984; 1994). Concerning stomatal movement, one has to be aware of the dilemma to which plants are subjected. On the one hand they need atmospheric CO_2. On the other hand, in the processes of gaining CO_2, they will lose water. Thus, they have to control this exchange very efficiently and balance between hunger and thirst if there is not enough water available. In this context, the calculation of stomatal conductance may be essential for the evaluation of exchange data because it gives a relative impression of the actual size of stomatal opening (Ball 1987; Tenhunen *et al.* 1987; Nobel 1991). A comparison of boundary layer resistances and stomatal resistances is given by Monteith and Unsworth (1990).

5.3 Regulation of emission from plants

We may distinguish principally between two groups of compounds being emitted: emission dependent from the actual production and emission from storage pools. The following scheme gives a survey of the different emission protocols:

A) Production dependent case:

Primary metabolism → secondary metabolism → actual production → emission

Triggers: light (energy), temperature → production rates
temperature (light) → vapour pressure
light, CO_2-level, water potential → stomatal opening

B) Storage (pre-production) dependent case:

- *External localisation:*

Storage pools (hairs, glands on external leaf surface) → emission

Triggers: temperature (light) → vapour pressure

- *Internal localisation:*

Storage pools (resin ducts, hairs, glands inside leaf tissue) → emission

Triggers: temperature (light) → vapour pressure
light, CO_2-level, water potential → stomatal opening

The trigger in both cases is light and/or temperature, providing energy for production pathways as well as influencing the vapour pressure of trace gases. The production may be directly or indirectly coupled to CO_2 assimilation as discussed by Kesselmeier *et al.* (1995) for monoterpenes emitted from holm oak (*Quercus ilex L.*). After production the compound has to leave the plant. Because most of the gases do not penetrate the cuticle in significant quantities, they have to pass the stomata to reach the atmosphere. If a direct relation between emission and transpiration or stomatal conductance is found, we can assume that the compound is coming from the interior of the leaf. However, this is not a proof for an actual production. If there are pools localised inside the leaf, *e.g.* resin ducts, inner hairs and glands, and if cuticular transport can be excluded, the emission from these pools is under stomatal control and depends on factors triggering stomatal aperture such as leaf water potential, internal CO_2 level and light. In contrast, if the compounds are stored outside on plant surfaces their release is not under stomatal control and their volatilisation may for example underlie only the leaf temperature. Similar principles rule of course the deposition to the vegetation.

5.4 Interaction of plants with environmental conditions

5.4.1 Developmental stage

One of the most important factors for the exchange behaviour is the developmental stage of the vegetation during the season. Development of new leaves on deciduous and coniferous trees, growing and flowering of crops and trees, seed ripening and senescence all influence the emission and deposition pattern of trace gases. Changes in the demand for trace elements by the vegetation and plant surface conditions also affect exchange. Starting with bud breaking in spring the size of the leaves, the thickness of the cuticle, protein content, physiological state and other properties change, and these factors influence metabolic processes which might be involved in the production of volatile compounds. Seasonal as well as diurnal cycles control plant metabolism via climatic factors like temperature, radiation, air humidity and precipitation. Variations in ammonia exchange with developmental stage of herbaceous plants for example are strongly dependent on age-dependent nitrogen transformation processes within the plant and on the availability of nitrogen in the root medium. It is well known that senescing tissue may have relatively high NH_3 emission (O'Deen and Porter 1986; Parton *et al.* 1988; Schjørring 1991). However, in cases with a very high external N supply seedlings may also lose NH_3 to the atmosphere (Mattsson and Schjørring 1995), although NH_3 emission from young vegetatively growing plants normally seem to be lower than those from plants in growth stages between anthesis and full senescence (Morgan and Parton 1989). Probably due to differences in soil N availability, agricultural crops seem in most cases to be a source of atmospheric NH_3 pollution (Harper *et al.* 1987; Sutton 1990; Schjørring *et al.* 1993a; b; Schjørring 1995), while forests and moorlands generally act as efficient NH_3 sinks (Duyzer *et al.* 1987; Sutton *et al.* 1992; Wyers, Vermeulen and Slanina 1992). Plant N deficiencies during vegetative growth may result in NH_3 absorption instead of NH_3 emission (Harper *et al.* 1987).

5.4.2 Stress and defence mechanisms

In addition to the daily and seasonal conditions and environmental changes we have to consider other effects such as chilling, drought, wounding or air pollution. In response to these stresses plants show various alterations or adaptations of their physiology through changes in enzymatic activity and the hormonal state. Variations of respiration and photosynthesis will have effects on the energy status of the cells, generally defined by the ATP/ADP ratio, and thus will influence metabolism including exchange processes. Additional synthesis and/or accumulation of secondary plant compounds like polyphenols or anthocyans can be seen and increasing amounts of antioxidative substances indicate the involvement of oxygen radicals (Elstner 1982; Larcher 1994). Micro-organisms or insects may trigger a series of defence reactions, initialising the production of different compounds like jasmonic or salicylic acid, coumarin, different terpenes or ethene acting as a signal or a chemical defence components (Ennos and Swales 1988; Kauss 1991; Boland *et al.* 1992; Raskin 1992; Hopke *et al.* 1994; Kauss *et al.* 1994). Extreme climatic conditions, attack by pathogens or just disturbing or wounding by an experimentator may change plant metabolism and alter production and/or emission of different volatiles. Hence, knowing the function of the volatile compound and recognising the developmental stage of the plant are important factors to understand regulation processes and to interpret exchange measurements. In this context diel patterns of environmental conditions such as temperature and water availability are also important. For example, increasing temperature increases the partial pressure of NH_3 gas in equilibrium with an NH_3 solution. In accordance, plant-atmosphere NH_3 fluxes exhibit temperature sensitivity: increasing temperature results in increased NH_3 emission (Farquhar *et al.* 1980) or decreased NH_3 absorption (Husted and Schjørring, unpublished results). Mild water deficiency may also give rise to a relatively rapid reduction in NH_3 exchange, probably due to a reduction of stomatal opening. Another important factor affecting plant-atmosphere ammonia exchange is atmospheric pollution. Thus, there is increasing evidence showing that plant communities in natural terrestrial ecosystems may respond to long term nitrogen deposition by temporarily emitting ammonia (Langford and Fehsenfeld 1992; Sutton *et al.* 1992). Additionally, increased concentrations of SO_2, and thus of acidic SO_4^{2-} aerosol, may deplete gaseous NH_3 and lead to NH_3 emission (Sutton *et al.* 1995; see also section 5.2, compensation points).

5.5 Experimental conditions and exchange

5.5.1 Adaptation to cuvette conditions

The techniques provided for chamber measurements have been significantly improved during the last decade. But even the most sophisticated cuvette changes the environment. To prevent "stress" effects, the conditions within the chamber should be carefully controlled to reflect the natural environment as closely as possible. Some plants need time to adapt to the altered conditions of the cuvettes. Adaptation may be reflected by a shift in the exchange rates during the experiment even over a long time period. Therefore, exchange studies should be carried out over several days. As observed by Bode (1994) a shift in the emission of organic acids has been observed with spruces. The release of formic acid by the test plant declined steadily during the course of incubation and reached constant rates not earlier than after 60 hours. Whether such adaptations are due to stress is not easy to verify. Some plants react already after rough handling without obvious wounding, showing altered emission, as was also observed for enhanced terpene emissions or the production and release of substances like hexenol (Juuti *et al.* 1990, Arey *et al.* 1991). Wounding has to be considered as a special kind of stress and could also cause the reactions mentioned above. Additionally, transport resistances may decrease due to some damage as was observed for terpenes stored within resin channels (Schürmann 1993).

5.5.2 Purified air and the compensation point

The decision whether to use purified air or environmental air causes a major dilemma. The necessity for example to prevent oxidation processes when measuring reactive hydrocarbons may lead to the use of purified and thus oxidant free air. Additionally, the use of purified air may improve the accuracy and sensitivity of the measurements. On the other hand this procedure causes some disadvantages, as the chamber conditions may be far from those of the natural environment including the background concentration of the trace gas investigated. Exchange processes between plants or soils and the atmosphere are under the control of the compensation point for many trace gases (Kesselmeier *et al.* 1993; Conrad 1994; Sutton *et al.* 1995; see also Kesselmeier *et al.* this volume). A compensation point reflects the balance between production and consumption processes. The simultaneous production and consumption implies a condition at which the rates of both reactions are equal. This may be achieved by temperature or light effects as well as by the trace gas concentration. For example, if trace gas concentrations are too low there will be no uptake. But at higher values the uptake will start and increase with increasing concentration. Thus the same plant (or soil) may even switch between emission and uptake caused by changes of ambient gas concentrations. The example understood best is the exchange of CO_2 governed by photosynthesis, respiration and photorepiration and is described in all fundamental textbooks (Mohr and Schopfer 1985; Strasburger 1991; Larcher 1984, 1994).

This knowledge described above has to be transferred in principal to all trace gases. Recently Kesselmeier and Merk (1993) showed the occurrence of a compensation point for carbonylsulfide (COS) in higher plants. Also microbiological exchange is under the rule of compensation points. Many trace gases (maybe all if properly investigated) are simultaneously produced and consumed in soils. Thus, the existence of compensation concentrations in soil is shown for H_2, CO, N_2O, NO and COS (Conrad 1994; 1995a). The magnitude of the compensation concentration depends on the relative rates of production and consumption reactions and changes with the conditions in soil. In the case of NO, for example, the compensation concentrations change with the supply of O_2 (Krämer and Conrad 1991), with soil temperature (Saad and Conrad 1993), and are dependent on soil type and season (Baumgärtner and Conrad 1992). The magnitude of the compensation concentration is a critical variable for the trace gas flux between soil and atmosphere (Conrad 1994). The soil acts as a source, if the compensation concentration is higher than the atmospheric concentration, and acts as a sink if the compensation concentration is lower than the atmospheric concentration. This is of special interest as many studies on the COS exchange between soil or plants and the atmosphere in the past have been done with purified air (*c.f.* Chin and Davis 1993). But we know today that purified air may trigger an emission though under ambient conditions there would have been deposition. Consequently the assumption that soils act as a sink for COS may be wrong. As recently shown by Castro and Galloway (1991) the exchange of reduced sulfur gases between soils and the atmosphere is dependent on the enclosure techniques used. After flushing their cuvettes with sulfur-free air, they found an emission of COS and CS_2, whereas the use of ambient air resulted in the deposition of both trace gases. Similar observations were made by Demello and Hines (1994) for the exchange of COS between peatlands and the atmosphere. These findings underline the existence of compensation points for the exchange of COS not only for higher plants but also for the soil. Especially for the soils the question arises which process is responsible for the consumption of COS. Cyanobacteria (Miller *et al.* 1989; Badger and Price 1990) as well as the fresh water algae *Chlamydomonas reinhardtii* (Kesselmeier *et al.* 1992; Protoschill-Krebs *et al.* in prep.) utilise this trace gas due to the activity of carbonic anhydrase, an enzyme which has also been made responsible for the COS consumption in higher plants (Protoschill-Krebs and Kesselmeier 1992). As carbonic anhydrase is an ubiquitous enzyme in all organisms, it may be assumed that soil micro-organisms may also use it for the consumption of COS.

Similar observations are reported for the exchange of ammonia with plants. The existence of an NH_3 compensation point in plants, *i.e.* the NH_3 molar fraction at which NH_3 absorption balances NH_3 loss and net NH_3 exchange is zero, was established by Farquhar *et al.* (1980). For several plant species they determined compensation points ranging between 1 and 4 μg NH_3 m^{-3}. Micrometeorological measurements indicate NH_3 compensation points in the range 1–7 μg NH_3 m^{-3} for canopies of grass clover (Denmead *et al.* 1976), ungrazed agricultural grassland (Sutton *et al.* 1993a), soybean (Lemon and van Houtte 1980, Harper *et al.* 1989), and barley (Schjørring *et al.* 1993a, b). Compensation points below 1 μg m^{-3} were measured in canopies of pine, spruce fir, and aspen (Breiding and Gravenhorst 1990; Langford and Fehsenfeld 1992; Kesselmeier *et al.* 1993). Much higher NH_3 compensation points were observed in spring wheat by Morgan and Parton (1989): during early grain filling the compensation point was about 13 μg NH_3 m^{-3}. It increased to 23 μg NH_3 m^{-3} during late grain filling and was higher than 25 μg NH_3 m^{-3} in the end of the growing period. Measurements of NH_3 concentration gradients in a winter wheat crop suggested a NH_3 compensation point around 15 μg NH_3 m^{-3} (Harper *et al.* 1987). However, in both the studies of Harper *et al.* (1987) and Morgan and Parton (1989) background levels of NH_3 were very high, giving differences between background and compensation point levels similar to those obtained in other studies. Thus, the NH_3 compensation point seems to be fairly close to natural ambient levels and explains why NH_3 may either be adsorbed or emitted by plants. Being aware of compensation points for the exchange of trace gases between plants and the atmosphere we can understand and discuss questions of emission and deposition under environmental conditions in a reliable way.

5.6 Biological and chemical processes in soils

5.6.1 Microbial metabolism

Production and/or consumption of a trace gas are often due to metabolism by soil micro-organisms. Metabolism implies that the reaction is useful for the micro-organism, *e.g.* generation of energy and/or biosynthetic reactions. A typical example is the production and consumption of NO by denitrifying bacteria which are active within anoxic microniches of the soil. Nitric oxide is an obligatory intermediate in the sequential reduction of nitrate to N_2:

$$NO_3^- \rightarrow NO_2^- \rightarrow NO \rightarrow N_2O \rightarrow N_2$$

This reduction pathway is used by denitrifiers as the energy-generating process under anaerobic conditions. Each step in the sequential reduction pathway is catalysed by a specific enzyme. These enzymes are only synthesised, if the O_2 partial pressure drops below a particular level that itself is a function of the bacterial species and of soil conditions, such as pH, temperature, availability of NO_3^- and organic substrates. All these variables regulate the extent of NO production and consumption in a micro-organism, a microbial community and a soil (Tiedje 1988; Zumft 1993; Conrad 1995b). However, NO is in addition produced and consumed by other microbial processes and by other mechanisms.

5.6.2 Enzyme reactions

An example of an enzymatic consumption reaction also involves NO consumption in soils. In some soils NO consumption cannot be explained by the activity of denitrifiers, since the kinetic characteristics of the NO consumption reaction are completely different from those of soils dominated by denitrification (Baumgärtner 1991; Schuster and Conrad 1992; Rudolph 1995; Conrad 1995c). A heterotrophic soil bacterium was isolated which contains enzymes that can oxidise NO to NO_3^- without any obvious profit (Koschorreck and Conrad, in prep.). Enzymes may persist in soil even after the death of soil micro-organisms. Atmospheric H_2, for example, is obviously oxidised by such "soil enzymes" (Conrad 1995a).

5.6.3 Chemical reactions

Production of atmospheric trace gases can also be caused by chemical reactions in soils. Production of CO in soil is obviously due to chemical decomposition of humic material (Conrad and Seiler 1985). Chemical reactions are at least of partial importance for COS formation in soil, as COS production is not completely abolished by sterilisation (unpublished results). The kind of chemistry involved is unknown. The mechanisms of chemical NO production are better known. They all depend on the availability of nitrite (Chalk and Smith 1983; McKenney *et al.* 1990), *e.g.*

$$3\,NO_2^- + 2\,H^+ \rightarrow 2\,NO + NO_3^- + H_2O$$

$$NO_2^- + Fe^{2+} + 2H^+ \rightarrow NO + Fe^{3+} + H_2O$$

In most soils, nitrite is only present in small amounts, so that chemical NO production would stop after a while, if nitrite would not be produced by microbial nitrification or denitrification. Hence, NO production is under microbial control, even if some of the NO originates from chemical decomposition of nitrite.

5.6.4 Interactions between different reactions

All kinds of interactions between metabolic, enzymatic and chemical reactions are feasible, although they are not always easy to prove for a particular case. Davidson (1991; 1992), for example, proposed an interaction between metabolic nitrite production and chemical decomposition of nitrite to NO in soil microniches.

There, nitrifiers may produce both NO_2^- and H^+ which may then be concentrated during freezing or drying of water films thus creating conditions that are ideal for chemical decomposition of nitrite to NO.

Another example is the microbial reduction of NO_3^- to NO_2^- combined with a microbial reduction of Fe^{3+} to Fe^{2+}. These metabolic reactions would trigger the chemical NO production. Such a reaction sequence has been demonstrated in *Escherichia coli* (Brons *et al.* 1991).

The two most important metabolic reactions of NO and N_2O production in soil are nitrification and denitrification. The importance of denitrification for the production of NO and N_2O is often assessed by specific inhibition of nitrification using acetylene or nitrapyrin (Klemedtsson *et al.* 1988; Tortoso and Hutchinson 1990; Schuster and Conrad 1992; Davidson 1992; Skiba *et al.* 1993). However, in soils and sediments denitrifiers may be limited by the supply of nitrate through nitrifiers (Rysgaard *et al.* 1993; Nielsen and Revsbech 1994). The two processes may be so tightly coupled, that inhibition of nitrification will automatically also result in inhibition of denitrification, so that the relative role of the two processes for NO production cannot be assessed.

References

Andrews, J.H., S.S. Hirano (eds), (1991). *Microbial Ecology of Leaves*. Springer Verlag, New York.

Arey, J., Winer, A.M., Atkinson, R., Aschmann, S.M., Long, W.D., Morrison, C.L. (1991) The emission of (Z)-3-hexen-1-ol, (Z)-3-Hexenylacetate and other oxygenated hydrocarbons from agricultural plant species, *Atmos. Environ.* **25**, 1063–1075.

Badger, M.R., Price, G.D. (1990) Carbon oxysulfide is an inhibitor of both CO_2 and HCO_3 uptake in the cyanobacterium Synechococcus PCC7942. *Plant Physiol.* **94**, 35–39.

Baker, E.A., (1982). Chemistry and Morphology of plant epicuticular waxes. in: Cutler, D.F., K.L. Alvin and C.E: Price (eds), *The Plant Cuticle.* Academic Press, 139–165.

Ball, J.T. (1987) Calculations related to gas exchange. In: Zeiger, E., Farquhar, G.D., Cowan, I.R. (eds), *Stomatal function.* Stanford University Press, Stanford California, pp. 445–467

Bieder, U., Schatz, A. (1993). The deposition of air pollutants to wet vegetation. in: P.M. Borrell, P. Borrell, T. Cvitaš, W. Seiler (eds), *Proc. EUROTRAC Symp. '92*, SPB Academic Publishing bv, The Hague 1993, pp. 714–717

Blakeman, J. (ed.), (1981). *Microbial Ecology*. Academic Press, London.

Baumgärtner, M., (1991). Umsetzung von Stickoxiden (NO_x) in Böden, auf Gebäudeoberflächen und in Mikroorganismen. Konstanzer Dissertationen Nr. 327. Hartung-Gorre Verlag, Konstanz.

Baumgärtner, M. and R. Conrad, (1992). Effects of soil variables and season on the production and consumption of nitric oxide in oxic soils. *Biol. Fertil. Soils,* **14**, 166–174.

Bode, K. (1994) Die Abgabe von organischen Säuren an die Atmosphäre durch Pflanzen verschiedener Entwicklungsstufen. Dissertation, Fachbereich Biologie, Universität Mainz.

Boland, W., Z. Feng, J. Donath, A. Gäbler, (1992). Are Acyclic C_{11} and C_{16} Homoterpenes Plant Volatiles Indicating Herbivory? *Naturwissenschaften,* **79**, 368–371.

Breiding, H., Gravenhorst, G. (1990) Der NH_3-Transfer zwischen Atmosphäre und Nadelbäumen. In: *Ammoniak in der Umwelt* (Kuratorium für Technik und Bauwesen in der Landwirtschaft e.V. (KTBL) und Verein Deutscher Ingenieure (VDI), eds) KTBL–Schriften Vertrieb im Landwirtschaftsverlag GmbH, Münster, Germany).

Brons, H. J., W. R. Hagen, A. J. B. Zehnder, (1991). Ferrous iron dependent nitric oxide production in nitrate reducing cultures of Escherichia coli. *Arch. Microbiol.*, **155**, 341–347.

Butcher, S.S., Charlson, R.J., Orians,G.H., Wolfe, G.V. (1992) *Global Biogeochemical Cycles*. Academic Press, Inc., San Diego, USA.

Castro, M..S., Galloway, J.N. (1991) A comparison of sulfur-free and ambient air enclosure techniques for measuring the exchange of reduced sulfur gases between soils and the atmosphere. *J. Geophys. Res.* **96**, 15,247–15,437.

Chalk, P. M. and Smith, C.J. (1983). Chemodenitrification. *Dev. Plant Soil Sci.,* **9**, 65–89.

Chin, M., Davis, D.D. (1993) Global sources and sinks of OCS and CS_2 and their distributions. *Global Biogeochemical Cycles* **7**, 321–337

Conrad, R., (1994). Compensation concentration as critical variable for regulating the flux of trace gases between soil and atmosphere. *Biogeochem.*, **27**, 155–170.

Conrad, R., (1995a). Soil microbial processes and the cycling of atmospheric trace gases. *Phil. Trans. R. Soc. London A*, **351,** 219–230.

Conrad, R., (1995b). Metabolism of nitric oxide in soil and soil microorganisms and regulation of flux into the atmosphere. in: J.C. Murrell and D.P. Kelly (eds), *The Microbiology of Atmospheric Trace Gases; Sources, Sinks and Global Change Processes.* NATO ASI Series. Springer, in press.

Conrad, R., (1995c). Soil microbial processes involved in production and consumption of atmospheric trace gases. *Adv. Microb. Ecol.*, in press.

Conrad, R., W. Seiler, (1985). Characteristics of abiological CO formation from soil organic matter, humic acids, and phenolic compounds. *Environ. Sci. Technol.*, **19**, 1165–1169.

Cosgrove, D.J., Cleland, R.E. (1983). Solutes in the free space of growing tissues. *Plant Physiol.*, **72**, 326–331.

Cramer, M.D. and Lewis, O.A.M. (1993). The influence of nitrate and ammonium nutrition on the growth of wheat (Triticum aestivum) and maize (Zea mays) plants. *Ann. Bot.*, **72**, 359–365.

Davidson, E. A., (1991). Fluxes of nitrous oxide and nitric oxide from terrestrial ecosystems. in: J. E. Rogers, W. B. Whitman (eds), *Microbial Production and Consumption of Greenhouse Gases: Methane, Nitrogen Oxides, and Halomethanes,* American Society for Microbiology, Washington D.C., pp. 219–235.

Davidson, E. A., (1992) Sources of nitric oxide and nitrous oxide following wetting of dry soil. *Soil Sci. Soc. Am. J.* **56**, 95–102.

Demello, W.Z., Hines, M.E. (1994) Application of static and dynamic enclosures for determining Dimethyl Sulfide and Carbonyl Sulfide exchange in Sphagnum peatlands - Implications for the magnitude and direction of flux. *J. Geophys. Res.* **99**, 14601–14607.

Denmead, O.T., Freney, J.R., Simpson, J.R. (1976) A closed ammonia cycle within a plant canopy. *Soil Biol. Biochem.* **8**, 161–164.

Duyzer, J.H., Bouman, A.M.H., Diederen, H.S.M.A., Aalst, R.M. van (1987) Measurement of dry deposition velocities of NH_3 and NH_4^+ over natural terrains. *Report R87/273, MT-TNO* Delft, The Netherlands.

Duyzer, J.H., Verhagen, H.L.M., Westtrate, J.H., Bosveld, F.C., Vermetten, A.W.M. (1994). The dry deposition of ammonia onto a douglas fir forest in the Netherlands. *Atmos. Environ.* **A28**, 1241–1253.

Elstner, E.F. (1982). Oxygen activation and oxygen toxicity. *Ann. Rev. Plant Physiol.,* **33**, 73–96

Ennos, R.A., K.W. Swales, (1988). Genetic variation in tolerance of host monoterpenes in a population of the ascomycete canker pathogen Crumenolopsis sororia. *Plant Pathol.* **37**, 407–416.

Erisman, J.W., Wyers, G.P. (1993). Continuous measurements of surface exchange of SO_2 and NH_3; implications for their possible interaction in the deposition process. *Atmos. Environ.* **27A**:1937–1949.

Erisman, J.W., Versluis, A.H., Verplanke, T.A.J.W., de Haan, D., Anink, D., van Elzakker, B.G., Mennen, M.G., Aalst, R.M. van (1993a). Monitoring the dry deposition of SO_2 in the Netherlands. *Atmos. Environ.* **27**,1153–1161.

Farquhar, G.D., Firth, P.M., Wetselaar, R., Weir, B. (1980) On the gaseous exchange of ammonia between leaves and the environment: Determination of ammonia compensation point. *Plant Physiol.* **66**, 710–714.

Fokkema, N. J., J. van den Heuvel (eds), (1986). *Micobiology of the Phyllosphere.* Cambrige University Press, Cambridge.

Fowler, D. (1985. Dry deposition of SO_2 onto plant canopies. in: Winner, W.E., Mooney H.A., Goldstein R.A. (eds), *Sulphur dioxide and vegetation*, Stanford University Press, California, pp. 75–95.

Fowler, D., Unsworth, M.H. (1979). Turbulent transfer of sulfur dioxide to a wheat crop. *Quart. J. Roy. Met. Soc.* **105**,767–783.

Garland, J.A., Branson, J.R. (1977). The deposition of sulphur dioxide to pine forest assessed by a radioactive tracer method. *Tellus* **29**,445–454.

Graedel, T.E., Crutzen, P.J. (1993) *Atmospheric Change: An Earth System Perspective.* W.H. Freemann and Co., New York, USA.

Graedel, T.E., Crutzen, P.J. (1994) *Chemie der Atmoshäre: Bedeutung für Klima und Umwelt.* Spektrum Akademischer Verlag GmbH, Heidelberg, Germany.

Grignon, C., Sentenac, H. (1991) pH and Ionic Conditions in the Apoplast. *Annu. Rev. Plant Physiol. Plant. Mol. Biol.*, **42**, 103–128.

Harper, L. A., Giddens, J. E., Langdale, G. W. (1989). Environmental effects on nitrogen dynamics in soybean under conservation and clean tillage systems. *Agronomy. J.* **81**, 623–631.

Harper, L.A., Sharpe, R.R., Langdale, G.W. and Giddens, J.E. (1987) Nitrogen cycling in a wheat crop: Soil, plant and aerial nitrogen transport. *Agronomy. J.* **79**, 965–973.

Hoffmann, B., Plönker, R., Mengel, K. (1992). Measurement of pH in the apoplast of sunflower leaves by means of fluorescence. *Physiologia Plantarum*, **84**, 146–153.

Holloway, P.J. (1982). The chemical constitution of plant cutins in: Cutler, D.F., Alvin, K.L. and Price, C.E. (eds), *The Plant Cuticle*. Academic Press, 45–85

Hopke, J., J. Donath, S. Blechert, W. Boland, (1994). Herbivore-induced volatiles: the emission of acyclic homoterpenes from leaves of *Phaseolus lunatus* and *Zea mays* can be triggered by a-glucosidase and jasmonic acid. *FEBS Letters*, **352**, 146–150.

Hove, L.W.A. van, Adema, E.H., Vredenberg, W.J., Pieters, G.A. (1989) A study of the adsorption of NH_3 and SO_2 on leaf surfaces. *Atmos. Environ.* **A, 23**, 1479–1486.

Husted, S., Schjørring, J.K. (1995) Apoplastic pH and ammonium concentration in leaves of Brassica napus L., *Plant Physiol.* submitted.

Juuti, S., J.Arey, R. Atkinson, (1990). Monoterpene emission rate measuremts from a Monterey pine, *J. Geophys. Res.* **95 D**, 7515–7519.

Kauss, H., (1991), Phosphorprotein-controlled changes in ion transport are common events in signal transduction for callose and phytoalexin induction. in: Hennecke, H. and D.P.S. Verma (eds), *Advances in Molecular Genetics of Plant-Microbe Interations*, **1**, 428–31.

Kauss, H., W. Jeblick, J. Ziegler,W. Krabler, (1994). Pretreatment of Parsley (*Petrosilenum crispum* L.) Suspension Cultures with Methyl Jasmonate Enhances Elicitation of Activated Oxygen Species. *Plant Physiol.*, **105**, 89–104.

Kesselmeier,J., Hofmann,U., Protoschill-Krebs,G., Hofmann,R., Merk,L., Bartell,U., Velmecke,F., Andreae.M.O. (1992) Exchange of reduced sulfur compounds between atmosphere and biosphere. *EUROTRAC Annual Report 1991, Part 4, BIATEX,* EUROTRAC ISS, Garmisch-Partenkirchen pp. 202–208.

Kesselmeier, J.,Merk, L. (1993) Exchange of Carbonyl Sulfide (COS) between Agricultural Plants and the Atmosphere: Studies on the Deposition of COS to Peas, Corn and Rapeseed. *Biogeochemistry* **23**, 47–59.

Kesselmeier, J., Merk, L., Bliefernicht, M.,Helas,G. (1993). Trace gas exchange between terrestrial plants and atmosphere: carbon dioxide, carbonyl sulfide and ammonia under the rule of compensation points. In: Slanina, J., Angeletti, G. and Beilke, S. (eds) *General Assessment of Biogenic emissions and Deposition of Nitrogen Compounds, sulphur compounds and oxidants in Europe.* CEC Air Pollution Research Report **47**, E. Guyot SA, Brussels, pp. 71–80

Kesselmeier,J. Schäfer, L., Ciccioli, P., Brancaleoni, E., Cecinato, A., Frattoni, M., Foster, P., Jacob, V., Denis, J., Fugit, J.L., Dutaur, L.,Torres, L. (1996) Emission of monoterpenes and isoprene from a Mediterranean oak species *Quercus ilex* L. measured within the BEMA (Biogenic Emissions in the Mediterranean Area) project. *Atmos. Environ.* **30**, 1841–1850

Klemedtsson, L., B.H. Svensson,T. Rosswall, (1988). A method of selective inhibition to distinguish between nitrification and denitrification as sources of nitrous oxide in soil. *Biol. Fertil. Soils*, **6**, 112–119.

Kolattukudy, P.E., (1980). Cutin, suberin and waxes. In: Stumpf, P.K. (ed) *The Biochemistry of Plants.* Vol.4, Academic Press, New York pp. 571–645.

Krämer, M.,R. Conrad, (1991). Influence of oxygen on production and consumption of nitric oxide in soil. *Biol. Fertil. Soils*, **11**, 38–42.

Kurkdjian, A.,Guern, J. (1989). Intracellular pH: measurement and importance in cell activity. Ann. Rev. Plant Physiol. *Plant Mol. Biol.,* 40, 271–303.

Langford, A.O. and Fehsenfeld, F.C. (1992) Natural vegetation as a source or sink for atmospheric ammonia: A case study. *Science,* **255**, 581–583.

Larcher, W. (1984) *Ökologie der Pflanzen,* 4. Auflage. Eugen Ulmer Verlag, Stuttgart, Germany.

Larcher, W. (1994) *Ökophysiologie der Pflanzen,* 5. Auflage. Eugen Ulmer Verlag, Stuttgart, Germany.

Lemon, E., Houtte, R. (1980). Ammonia exchange at the land surface. *Agronomy J.* **72**, 876–883.

Lendzian, K.J. (1987) Aufnahme und zellphysiologische Wirkungen von Luftschadstoffen. *Naturwissenschaften* **74**, 282–288.

Martin, J.-B., Bligny, R., Rebeille, F., Douce, R., Leguay, J. J., Mathiey, Y., Guern, J. (1982). A 31P nuclear magnetic resonance study of intracellular pH of plant cells cultivated in liquid medium. *Plant Physiology,* **70**, 1156–1161.

Mattsson, M.,Schjørring, J.K. (1995) Ammonia emission from the leaves of barley grown with controlled nitrogen availability. *J. Exp. Bot.* submitted.

McKenney, D.J., C. Lazar,W.J. Findlay, (1990). Kinetics of the nitrite to nitric oxide reaction in peat. *Soil Sci. Soc. Am. J.*, **54**, 106–112.

Miller, A.G., Espie, G.S., Canvin, D.T. (1989) Use of carbon oxysulphide, a structural analog of CO_2, to study active CO_2 transport in the cyanobacterium Synechococcus UTEX 625. *Plant Physiol.* **90**, 1221–1231.

Mohr, H., Schopfer, P. (1985) *Lehrbuch der Pflanzenphysiologie*, 3.Auflage. Springer Verlag, Berlin, Germany

Monteith, J.L., Unsworth, M.H. (1990) *Principles of Environmental Physics.* Edwal Arnold, London.

Morgan, J.A, Parton, W.J. (1989) Characteristics of ammonia volatilization from spring wheat. *Crop Sci.* **29**, 726–731.

Nielsen, T. H., N. P. Revsbech, (1994). Diffusion chamber for nitrogen-15 determination of coupled nitrification-denitrification around soil manure interfaces. *Soil Sci. Soc. Am. J.,* **58**, 795–800.

Nobel, P.S., (1991). *Physicochemical and Environmental Plant Physiology.* Academic Press. San Diego, USA.

O'Deen, W.A., Porter, L.K. (1986) Continous flow system for collecting volatile ammonia and amines from senescing wheat. *Agronmy. J.* **78**, 746–749.

O'Neal, D., Joy, K. W. (1974). Glutamine synthetase of pea leaves: Divalent cation effects, substrate specificity, and other properties. *Plant Physiology*, **54**, 773–779.

Parton, W.J., Morgan, J.A., Altenhofen, J.M., Harper, L.A. (1988) Ammonia Volatilization from spring wheat plants. *Agronomy J.* **80**, 419–425.

Protoschill-Krebs, G., J. Kesselmeier, (1992). Enzymatic pathways for the consumption of carbonyl sulphide (COS) by higher plants. *Bot. Acta* **105**, 85–96.

Raskin, I. (1992), Role of Salycylic Acid in Plants, *Ann. Rev. Plant Physiol. Plant. Mol. Biol.*, **43**, 439–463.

Rudolph, J., (1995). *Umsetzung und Transport von Stickstoffmonoxid (NO) in Böden.* Verlag Shaker, Aachen.

Rysgaard, S., N. Risgaard-Petersen, L. P. Nielsen and N. P. Revsbech, (1993). Nitrification and denitrification in lake and estuarine sediments measured by the 15N dilution technique and isotope pairing. *Appl. Environ. Microbiol.*, **59**, 2093–2098.

Saad, O.A.L.O., R. Conrad, (1993). Temperature dependence of nitrification, denitrification, and turnover of nitric oxide in different soils. *Biol. Fertil. Soils,* **15**, 21–27.

Sablijc, A., H. Güsten, J. Schönherr, J. and M. Riederer, (1990). Modeling Plant Uptake of Airborne Organic Chemicals. 1. Plant Cuticle/ Water Partitioning and Molecular Connectivity. *Environ. Sci. Techno.*, **24**, 9, 1321–1328.

Schjørring, J.K. (1991) Ammonia emission from the foliage of growing plants. In: T.D. Sharkey, E.A. Holland, H.A. Mooney (eds), *Trace Gas Emissions by Plants*, Academic Press, San Diego pp. 267–292.

Schjørring, J.K. (1995) Long-term quantification of ammonia exchange between agricultural cropland and the atmosphere. I. Evaluation of a new method based on passive flux samplers in gradient configuration. *Atmos. Environ.* **29**, 885–893.

Schjørring, J.K., Kyllingsbäk, A., Mortensen, J.V., Byskov-Nielsen, S. (1993a) Field investigations of ammonia exchange between barley plants and the atmosphere. I. Concentration profiles and flux densities of ammonia. *Plant, Cell and Environment* **16**, 161–167.

Schjørring, J.K., Kyllingsbäk, A., Mortensen, J.V., Byskov-Nielsen, S. (1993b) Field investigations of ammonia exchange between barley plants and the atmosphere. II. Nitrogen remobilization, free ammonium content and activities of ammonium assimilating enzymes in different leaves. *Plant, Cell and Environment* **16,** 169–178.

Schlesinger, W.H. (1991) *Biogeochemistry, An Analysis of Global Change*. Academic Press, San Diego, USA..

Schönherr, J. (1982) Resistance of plant surfaces to water loss: Transport properties of cutin, suberin and associated lipids. In: Pirson, A. and Zimmermann, M.H. (eds), *Encyclopedia of Plant Physiology* Vol. 12 B, Springer Verlag Berlin p. 154.

Schönherr ,J., M. Riederer, (1989), Foliar penetration and Accumulation of Organic Chemicals in Plant Cuticles. *Reviews of Environmental Contamination and Toxicology*, **108,** 1–69.

Schürmann, W. (1993). Emission von Monoterpenen aus Nadeln von *Picea abies* (L.) Karst. sowie deren Verhalten in der Atmosphäre. Dissertation, Technische Universität München

Schuster, M., R. Conrad, (1992). Metabolism of nitric oxide and nitrous oxide during nitrification and denitrification in soil at different incubation conditions. *FEMS Microbiol. Ecol.*, **101**, 133–143.

Sharkey, T.D., Holland, E.A., Mooney, H.A. (1991) *Trace Gas Emissions from plants*. Academic Press, San Diego, USA.

Skiba, U., K. A. Smith, D. Fowler, (1993). Nitrification and denitrification as sources of nitric oxide and nitrous oxide in a sandy loam soil. *Soil Biol. Biochem.*, **25**, 1527–1536.

Speer, M., Kaiser, W.M. (1991). Ion relations of symplastic and apoplastic space in leaves from Spinacia oleracea L. and Pisum sativum L. under salinity. *Plant Physiol.* **97**, 990–997.

Strasburger, E. (1991) *Lehrbuch der Botanik* (Sitte, P., Ziegler, H., Ehrendorfer, F., Bresinsky, A., (eds) Fischer Verlag, Stuttgart, Germany.

Sutton, M.A. (1990) The surface/atmosphere exchange of ammonia. Ph.D. Thesis, University of Edinburgh, Edinburgh, UK.

Sutton, M.A., Moncrieff, J.B., Fowler, D. (1992) Deposition of atmospheric ammonia to moorlands. *Environ.l Pollut.* **75**, 15–24.

Sutton, M.A., Fowler, D., Moncrieff, J.B., Storeton-West, R.L. (1993a). Exchange of atmospheric ammonia with vegetated surfaces. I.Unfertilized vegetation. *Quart. J. R.oy. Metr. Soc.*, **119**, 1023–1045.

Sutton, M.A., Pitcairn, C.E.R., Fowler, D. (1993b) The exchange of ammonia between the atmosphere and plant communities. *Adv. Ecol. Res.* **24**, 301–393.

Sutton, M., Schjoerring, J.K., Wyers, P. (1995) Plant-atmosphere exchange of ammonia. *Phil. Trans. Roy. Soc. London* **351**, 261–278.

Tenhunen, J.D., Pearcy, R.W., Lange, O.L. (1987) Diurnal variations in leaf conductance and gas exchange in natural environments. In: Zeiger, E., Farquhar, G.D., Cowan I.R. (eds), *Stomatal function.* Stanford University Press, Stanford California, pp. 323–351

Tiedje, J. M., (1988). Ecology of denitrification and dissimilatory nitrate reduction to ammonia. in: A. J. B. Zehnder (ed), *Biology of Anaerobic Microorganisms*, Wiley, New York, pp. 179–244.

Tortoso, A. C., G. L. Hutchinson, (1990). Contributions of autotrophic and heterotrophic nitrifiers to soil NO and N_2O emissions. *Appl. Environ. Microbiol.*, **56**, 1799–1805.

Vermetten, A.W.M., Hofschreuder, P., Duyzer, J.H., Diederen, H.S.M.A., Bosveld, F.C., Bouten, W. (1992). Dry deposition of SO_2 onto a stand of Douglas fir: the influence of canopy wetness. In: Schwartz S.E., Slinn W.G.N. (eds), *Proc. 5th IPSASEP conference,* Richland, 15–19, June 1991, USA., Hemisphere publishing corporation, Washington.

Warneck, P. (1988) *Chemistry of the Natural Atmosphere,* Academic Press, San Diego, USA. ISBN 0-12-735630-4.

Wyers, G.P., Vermeulen, A.T., Slanina, J. (1992) Measurement of dry deposition of ammonia on a forest. *Environ. Pollut.* **75**, 25–28

Zumft, W. G., (1993). The biological role of nitric oxide in bacteria. *Arch. Microbiol.*, **160**, 253–264.

Chapter 6

Atmosphere - Surface Exchange of Nitrogen Oxides and Ozone

D. Fowler[1], F.X. Meixner[2], J.H. Duyzer[3], G. Kramm[4] and L. Granat[5]

[1]Institute of Terrestrial Ecology, Bush Estate, Penicuik, Midlothian, EH26 0QB
[2]Max-Plank-Institut-für Chemie, Abteilung Biogeochemie, Postfach 3060, D-55020 Mainz, Germany
[3]Institute of Environmental Sciences TNO, PO Box 6011, 2600 JA Delft, The Netherlands
[4]Fraunhofer Institut für Atmosphärisches Umweltforschung, Kreuzeckbahnstrasse 19, D-82467 Garmisch-Partenkirchen, Germany
[5]Department of Meteorology, Stockholm University, S-106 91 Stockholm, Sweden

Introduction

During the development of the EUROTRAC research programme on atmosphere-surface exchange it was recognised that while for SO_2 much valuable progress has been made, there were few published studies of NO_2 deposition, a very weak understanding of the processes which regulate NO_2 exchange and many problems both in measurement and interpretation. For emissions of nitric oxide from soil, the available literature was also very limited and yet projected global emissions using the literature available suggested a major contribution from soils and that rates of emission are strongly influenced by land-use and anthropogenic influences.

The measurement of NO_2 and NO exchange therefore became a primary focus of the BIATEX programme.

Recognising the practical difficulties in measurement and the need to make progress in understanding processes as well as obtaining flux measurements for field studies, the programme included the following parallel activities.

- Laboratory studies of NO emissions from soil.
- Laboratory measurements of NO_2 uptake by plants.
- Field measurements of NO emission using chamber methods.
- Field measurements of NO, NO_2 and O_3 using micrometeorological methods (flux-gradient and eddy co-variance).
- Development of models to analyse and interpret micrometeorological measurements and to extrapolate fluxes spatially over regions.

- Provision of annual deposition and emission fluxes of nitrogen at 5×5 km, 20×20 km and at the country scale.

While the primary focus of the research programme of BIATEX on NO and NO_x was clearly to address straightforward scientific questions, much of the work was focused on the specific questions required to address the policy issues of individual governments and of the EU. For example, what are the annual fluxes of NO and NO_2 at the landscape and at the country scale for Europe? There were therefore clear scientific goals and a strategy to achieve them.

During the last 3 years of the programme, the development of critical loads approaches throughout Europe as a means of quantifying environmental effects and developing policies to maximise the environmental gain for the investment in emission reduction, led to further specific applications of the BIATEX research. To some extent, the demands of critical loads mapping focused the efforts of the BIATEX community during these recent years.

This brief review of NO_2 and NO exchange between the atmosphere and terrestrial surfaces summarises the work within the EUROTRAC BIATEX programme and the scientific achievements within a context of other contemporary literature. The applications of the work are also considered but as they have been described within the Application Project in some detail, this review provides a broader summary of the science, only some of which has been applied to date. The work includes studies of NO_y species other than NO and NO_2 including HONO and HNO_3 and at some stage of the work, measurements of emission fluxes of N_2O were made to complete the inventory of trace gas emission fluxes. These extensions of the remit were within the original description of the work but are considered here only briefly.

Emissions of nitric oxide from soil

Emissions of nitric oxide (NO) represent an important source of NO_x in the lower atmosphere (Galbally *et al.*, 1980). Measurements made of the emissions from pasture soil and from arable land and forest prior to the EUROTRAC project showed fluxes in the range 0.1 ng NO–N m^{-2} s^{-1} to 70 ng NO–N m^{-2} s^{-1} (Galbally and Roy, 1978; Johansson and Granat, 1984; Kim, 1973).

The measurements, in addition to identifying a wide range of fluxes between different soils and land uses also demonstrated considerable spatial and temporal variability.

During the period 1986 to 1992, the BIATEX programme of NO flux measurement included field and laboratory studies of processes and fluxes.

No attempt was made to systematically study NO fluxes from a large number of individual soils to provide a statistically based sample of the population of European soils for the extrapolation. The approach adopted was to study NO emission (and in some cases deposition) processes in soils which are representative of major land uses throughout Europe, to provide a mechanistic basis for the extrapolation based on an understanding of the process. These soils were in many cases taken from sites of extensive field measurements of NO_x exchange using micrometeorological and enclosure methods. In this way the relationships between soil-atmosphere exchange of NO and the major variable soil temperature, moisture, NO_3^- concentration, and the fluxes of other NO_x species became the focus for study.

The soils and study sites extended over a broad range of European soils with agricultural and semi-natural landuses:

Landuse	**Country**	
Forest	Sweden	areas of high and low atmospheric N inputs
UK	Dunslair Heights	
Germany	Högelwald measurement sites	
Netherlands	Speulder measurement sites	
Arable	UK	East Lothian measurement sites
(wheat)	Germany	Manndorf measurement sites
Grassland	UK	Halvergate and Stirling measurement sites
Heathland	Netherlands	Leende Heide measurement sites
Moorland	UK	Dunslair Heights measurement sites

Enclosure methods

These methods, also known as box, cuvette or chamber methods are the main source of NO flux measurements. The advantages of the method are that very small fluxes may be detected, effects of ozone may be excluded or corrected and the approach may be adopted for a wide range of surfaces.

The enclosure technique has many forms, and may be based on either the static method, in which the rate of increase in concentration with time within a closed system is used to estimate the flux, or more commonly, a dynamic method in which the concentration and mass flow into and out of an enclosure are used to estimate the net exchange within the chamber (Woodward and Sheehy, 1983; Remde *et al.*, 1993). A summary of the measurements of NO_x emissions made by dynamic chamber methods in Europe is shown in Table 6.1.

NO, NO_2 and NO_x concentrations are measured by chemiluminenscence.

Chamber systems may include stirring fans to reduce the atmospheric resistance (*i.e.* $r_a + r_b$) to gas transport within the chamber. Differences in chamber design and operation have been widely discussed in the literature and may have profound effects on the ability to study processes. In particular, in a dynamic enclosure

Table 6.1: Field measurements of NO_x emissions by dynamic chamber methods in Europe.

Location and Date	Soil type	Land use	NO flux ng N m^{-2} s^{-1}	Reference
Halvergate Marshes, UK Sept 1989	Drained marsh	Grazed grassland	7.2–14.6	Remde *et al.* 1993
Nyon, CH Sept 1991	Cambisol	Green manure incorporation fallow	2–41 1.5–25	Vos *et al.*, 1994
Mainz, Germany July–Sept 1983	Sandy loam	Mowed grass/clover/ herbs bare soil/no N fertiliser loss rates %% applied N	0.19 ± 0.5 –1.4–0.86 0.003%–0.154%	Slemr & Seiler, 1991
Manndorf, Germany Jul–Aug 1990	Silt, loam, luvisol	Wheat before harvest After harvest +40 kg N/ha	0.14 0.4–2.2 0.8–4.4	Ludwig,1994 Meixner 1994,
Höglwald, Germany	Acid Forest soil	80 year old spruce	0.14–6.7	Papen *et al.* 1993
Hurley, UK July 1985	Loam	Ryegrass + urine control	0–360.32	Colbourn *et al.*, 1987
SE Scotland, UK May–Oct 1991	Sandy clay loam	Cut lawn Alder Southern birch Birch Sitka spruce Sitka + acid mist (85 kg N ha^{-1} y^{-1})	0–0.2 0.26–2.74 0.001–1.7 –0.33 0.29–0.96 3.32–6.23	Skiba *et al.*, 1992, 1994
Stirling, UK April 1992	Clay	Ryegrass	1.4–2	Smith *et al.*, 1994
Central Sweden	clay loam	arable	0.1–62	Johansson 1984
Central Sweden	grey brown podzol	forest fertilized	1–60	Johansson 1984
Central Sweden	iron podzol	forest unfertilized	0.1–0.8	Johansson 1984

[a]DC = dynamic chamber

system, the flux measured is of course dependent on the difference between inlet and outlet, and if as is the case for NO emission, the net flux varies with ambient NO concentration, chamber systems which introduce NO free air as the inlet gas will tend to yield larger net emissions than chambers which use ambient NO as the inlet gas (Meixner, 1994).

The chamber system used by Johansson *et al.* 1988 and by Rondon *et al.*, 1994 is shown in Fig. 6.1.

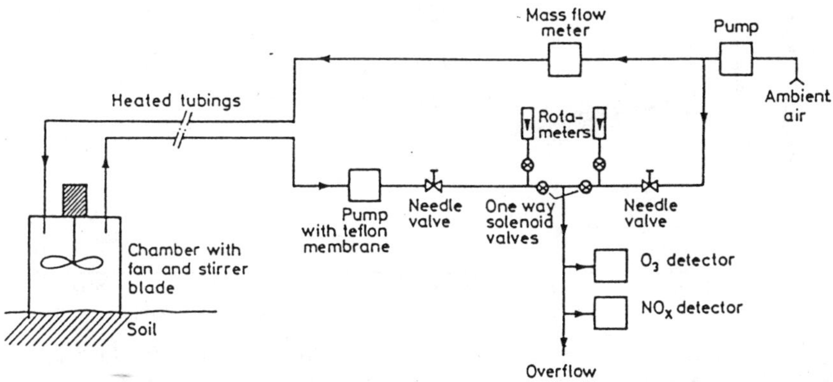

Fig. 6.1: Schematic diagram of chamber sampling system. (From Johanssen 1984).

To summarise the NO flux measurements it is helpful to show the observed relationships between NO emission flux and the major variables which regulate them.

Soil temperature

The emission of NO has been shown to increase with temperature by Williams and Fehsenfeld (1991). Similarly, laboratory measurements performed on wheet/soil samples taken from the Mandorf feild site (Ludwig *et al.*, 1992) using a dynamic chamber technique show a clear variation with temperature (Fig. 6.2). Skiba *et al.* (1992) also report a marked temperature response of NO emission from arable soil.

An analysis of the data to provide activation energies (Fig. 6.3) yielded values in the range 44 to 103 k J mol^{-1} similar to those reported by Johansson and Granat (1984). However, the positive temperature response to NO emission is not always observed. In particular, long term field measurements of NO emission from an arable field at Manndorf during a very dry summer showed only a weak response of NO flux to temperature (Ludwig, 1994). Similar very small responses of NO emission to temperature for dry soils have also been reported by Cardenas *et al.* (1994) for tropical soils. Clearly, the evidence from these and other recent work shows that the temperature response of NO is positive within an intermediate range

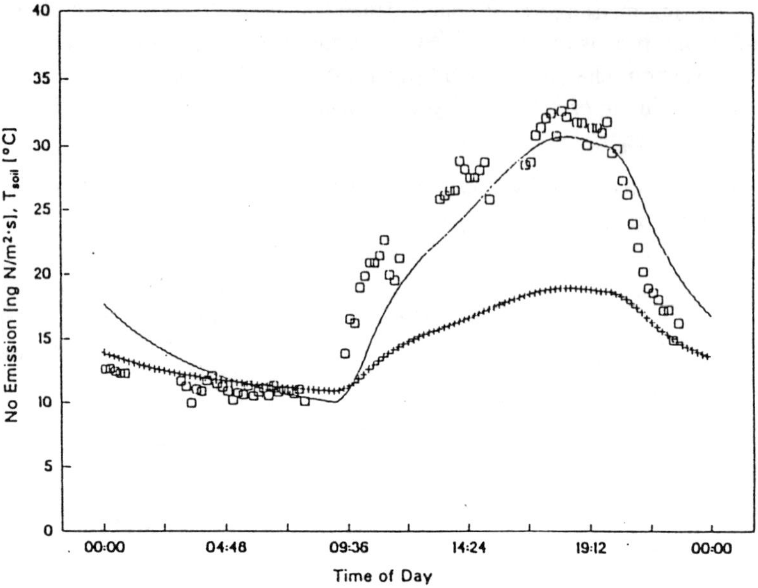

Fig. 6.2: Variation of NO emission (open squares) during 27 June 1991 (Ludwig *et al.* 1992). Plants from the ecosystem ' wheatfield' (Mandorf,) were fumigated in a growth chamber with NO < 0.5 ppb. Also shown : mean soil temperature over depth of 1–4 cm ($T_{soil,}$ crosses), and the calculated NO emission flux (F(NO), straight line) assuming a linear dependence on soil temperature (F(NO)= $-18.2 + 2.58 \times T_{soil}$). Measurements were performed by a dynamic enclosure technique (see Remde *et al.* 1993). (From Meixner *et al.* 1994)

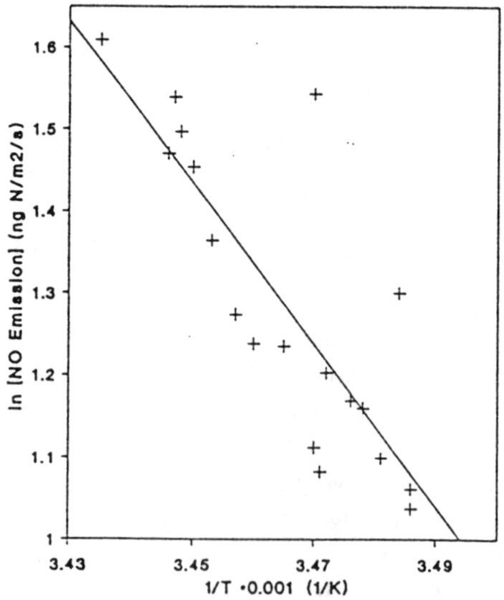

Fig. 6.3: Temperature dependence of the NO emission from a fertilized clay loam cropped with ryegrass at Glencorse Mains, 14–15 June 1990. (From Skiba *et al.* 1992)

of soil moisture contents, but that other processes may exert the primary control over both the rate and direction of exchange for very dry or for very wet soils.

The temperature responses reported refer to soil temperature at different depths or in some cases, air temperature. Since the bulk of the evidence of the zone of maximum production of NO show that it lies very close to the surface (in the 1 to 3 cm surface layer, Galbally and Johansson, 1989; Remde *et al.*., 1993) the surface temperatures or that within the upper 2 or 3 cm, generally provide the most appropriate reference depth for temperature measurements. In some cases the NO deposition which has been reported by Skiba *et al.* (1992) is also temperature dependent as shown in Fig. 6.4, who interpret the small NO deposition flux as uptake by plants within the chamber, as has been reported by Slemr and Seiler (1984), Johansson and Granat (1984) and Ludwig (1994). The mechanistic basis of NO exchange between vegetation and the atmosphere is unclear but clearly complicates the procedures for extrapolating NO fluxes over the rural landscape. The deposition of NO however is small, the majority of NO exchange studies show net NO emission.

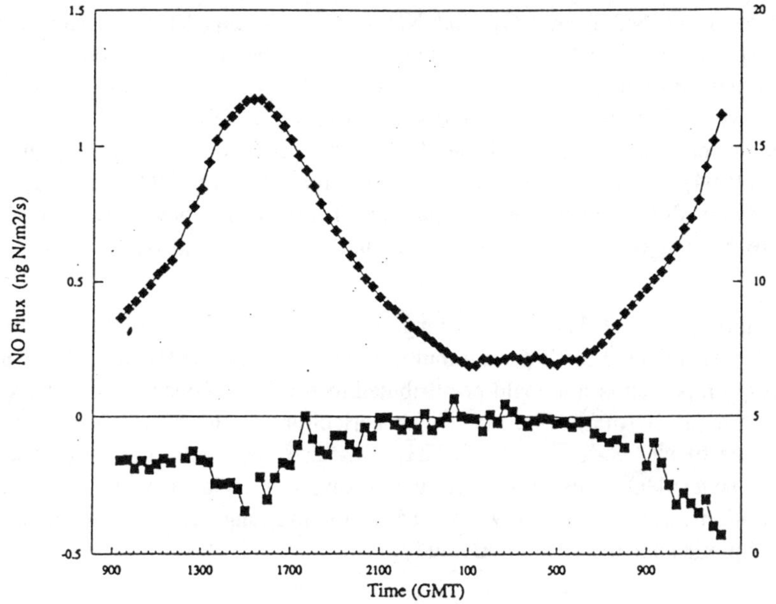

Fig. 6.4: Diurnal variations in the NO deposition rate from an unfertilized ryegrass plot on clay loam soil at Glencorse mains, 22–23 April 1990). ■ - NO flux, ◆ - soil temperature ($n = 5$ for every data point). (From Skiba *et al.* 1992).

Soil moisture

Effects of soil moisture on microbiological processes have been widely documented and these, as has been described by Meixner (1994) and Ludwig (1994) from the measurements at the Manndorf field experiment, may override the temperature effects in very dry soils. The effects of soil moisture have also been observed in examining the response of NO fluxes to rainfall events Skiba *et al.* (1992) who found a stronger link between rainfall amount and NO flux than between NO flux and the average soil moisture content of the top 10 cm of soil. Those measurements were interpreted as further evidence that the site of control over NO emission lies close to the soil surface as the precipitation events only influenced the top 3 cm of the soil. Some of the field measurements show a decrease in NO emission following heavy rain (Johansson and Granat, 1984), which is consistent with a more abrupt change within the soil profile to conditions with restricted aeration and an environment less favourable to nitrification. Davidson (1991) suggests that effects of soil water on NO emission via nitrification and denitrification are best characterised by the water-filled pore space (WFPS) and a value of 60 % WFPS appears to be the threshold between water limited and aeration limited microbial processes in a range of soils.

Effects of soil NO_3^- and NH_4^+

The emission of NO from NO_3^- and NH_4^+ fertilised soils has been shown by several studies within BIATEX (Ludwig, 1994; Skiba, 1992) as well as many other studies (Williams and Fehsenfeld, 1991; Hutchinson *et al.*, 1993). Fig. 6.5 from the work of Ludwig (1994), shows clear effects of N fertiliser application and subsequent increases in soil available NH_4^+ + NO_3^- concentrations on NO fluxes. Ludwig (1994) was able to demonstrate the different effects of NO_3^- and NH_4^+ on NO emission. Both the substrate and product of nitrification have been shown to stimulate NO emission by Williams (1991) and Fehsenfeld and Hutchinson *et al.* (1993).

By examining the NO fluxes over a range of agricultural and semi-natural soils, Skiba (1994) was able to show in a multiple regression analysis that 60 % of the variability on NO emission could be attributed to soil NO_3^- concentration and soil temperature (at 30 mm) (Fig. 6.6). For agricultural soils, a significant linear dependence of NO emission on soil NH_4^+ was also observed. For semi-natural soils, however NO emissions were not strongly associated with soil NH_4^+ concentrations, as a result of very low nitrification rates and NH_4^+ accumulation in the acid upland soils included in this study.

Up scaling annual NO emission fluxes to regions

The measurements of NO emission within BIATEX and elsewhere are mainly restricted to campaigns or a series of campaigns at a single field site over a growing season. However, the long term (annual) net fluxes over regional, countrywide and continental spatial scales has remained a key objective of the work. In principle the detailed microbiological studies of NO production and the influence of the key soil variables of temperature, water and NO_3^- as well as some of the physical properties of soils were designed to provide the underlying scientific understanding of processes. The extrapolation from enclosures to annual

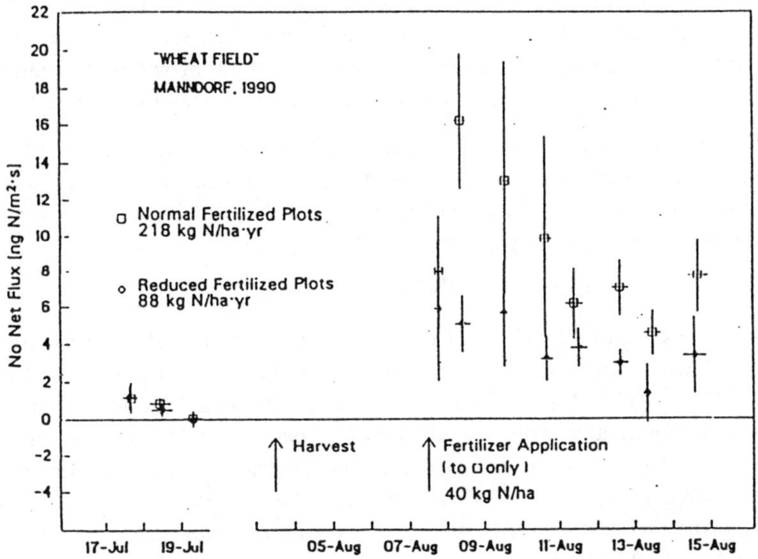

Fig. 6.5: Results of measurements on net NO flux (emission) from a wheatfield ecosystem (Manndorf, 48.71 °N, 12.96 °E) before and after harvest and fertilization (40 kg N/ha to 'normal' fertilized plots only). Error bars (± 95% confidence interval) indicate the temporal and spatial variation of the corresponding daily mean value obtained by a set of 4 dynamic chambers on 'N' as well as on 'R' plots (adapted from Ludwig 1994). (From Meixner 1994).

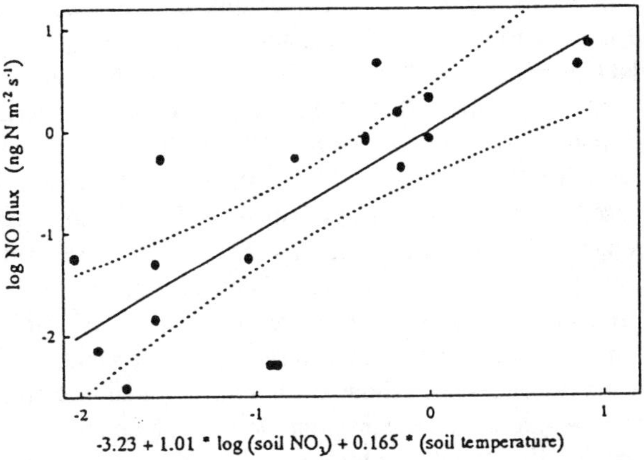

Fig. 6.6: Dependence of NO flux on soil NO_3^- concentration and soil temperature (30 mm), The solid line was fitted by linear regression, $r^2 = 0.61$, df =18. The dotted line shows the 95% confidence interval. (From Skiba *et al.* 1994).

fluxes for Europe was regarded as a task for models which at the landscape scale could be validated in the short term using micrometeorological studies, and are described in later sections of the report. The modelling of annual NO emission fluxes for regions is at a very rudimentary stage, and is essentially empirical.

The first of these attempts to provide annual fluxes was that of Williams *et al.* (1992), who estimated the soil emissions of NO for the United States of America. The approach by Williams *et al.* (1992) assumed that the flux of NO from soils (F_{NO} could be estimated according to

$$F_{(NO)} = A.\exp(B.T_{soil})$$

in which F is the flux of NO (ng N m^{-2} s^{-1}). A is a factor to characterise the soil properties with units ng N m^{-2} s^{-1} and B is a temperature coefficient (0.071 ± 0.007) $°C^{-1}$. Values for A were obtained from land use maps and B was determined empirically from the relationship between NO emission and soil temperature. This approach indicated that soils contributed approximately 6 % of the anthropogenic source. A similar exercise by Simpson (1993) for Europe suggested that NO emissions for individual countries within Europe contributed between 5 and 10 % of anthropogenic emissions. However, the simplicity of the approach fails to satisfactorily predict NO fluxes for many specific cases such as the wheat field data from the Manndorf experiment (Meixner, 1994). In his analysis of the approach, Meixner points out that the simple factor (A in the above equation) is inadequate to describe the temporal and geographical variability in NO emissions.

An alternative but equally simplistic approach to that of Williams *et al.* was adopted by Skiba *et al.* (1994) to estimate the soil emissions of NO in the UK. The relationship between soil NO_3^-, temperature and NO emission (Fig. 6.6) formed the basis of the method. The equation defining the data from Fig. 6.6, yielded 1 kt NO–N per year. The relatively large soil NO emission following fertiliser application (assuming a 0.15 to 0.75 % loss of the NH_4^+ fertiliser applied) contributes a further emission of 1.4 to 6.8 kt NO–N. The total annual emission of NO from soil was therefore estimated to be of the order of 2 to 8 kt NO–N. This estimate was based on many assumptions, which need to be more clearly outlined. More measurements are needed to verify and narrow the range of the percentage of N fertiliser released as NO. The established dependence of soil NO emission on the soil NO_3^- concentration does not provide an easy way of estimating the soil NO emission and produce emission inventories. Measurements of soil NO_3^- are not always available and these show even more spatial variability than NO measurements. It would be more appropriate to base NO emission estimates of N inputs, which are readily available at high resolution from agricultural surveys and atmospheric deposition maps and models. More recent estimates of the UK soil NO emission (Skiba, unpublished results) suggest that emissions are more likely to be of the order of 20 kt NO–N y^{-1}. This estimate was based on published NO emission measurements for temperate climates, which were categorised into 4 land classes (Table 6.2). Assumptions were made that emissions will only occur for 274 days y^{-1} and that for agricultural fertilised soils the fertiliser effect of two nitrogen fertiliser applications will only result in high emission rates for 50 days y^{-1} after which the lower rate in Table 6.2 will apply. The drawback of this approach is that considerable emphasis is placed on very few published data and the effects of temperature variations were not taken into account. However, the estimate agrees quite well with that of Simpson (1993), who adopted the Williams *et al.* (1992) approach, which does include temperature variations to calculate soil N emissions for European countries. To produce more accurate NO emission estimates, the effects of changes of NO emission rates with soil temperature, rainfall, N input and soil type and land use need to be incorporated, and for this our understanding of the underlying process must be improved considerably. Such approaches are well suited to the GIS methodology and modelling, which should yield much better estimates of annual NO fluxes and their spatial variability. This is an important goal for future work.

Table 6.2: Typical soil NO emission rates for different land use classes in temperate climates. Emission rates were calculated from published measurements made in Germany, Switzerland, UK and Central Tennessee.

Land use	NO emission (ng N M^{-2} s^{-1}) mean ± standard deviation	Number of observations
Recently fertilised agricultural fields (tilled land)	25 ± 24	16
Not recently fertilised agricultural land	4.7 ± 4.2	11
Grassland	8 ± 9	14
Forests	1.1 ± 1.6	7

The exchange of NO_2 between vegetation and the atmosphere

The laboratory and field measurements of NO_2 exchange (Johannson, 1987; Thoene *et al.*, 1991) show that NO_2 uptake is primarily through stomata and that leaf cuticular surfaces are a negligible sink for NO_2. These conclusions are central to our current understanding of NO_2 dry deposition and the modelling of annual fluxes across the landscape. A brief description of the work from which the current understanding has been developed is therefore necessary. Measurements by Hanson *et al.* (1989) and Skarby *et al.* (1981) show for a range of deciduous and conifer forest trees, that stomatal uptake by vegetation is the primary NO_2 deposition pathway to terrestrial surfaces. These species show negligible uptake on cuticular surfaces or on bark or leaf litter. The complications in the development of understanding of NO_2 exchange with vegetation in controlled conditions are that NO_2 might be emitted from foliage as a consequence of metabolic processes or that the sub-stomatal cavity may not be a perfect sink for NO_2. In these circumstances, which may of course be related through the presence of a compensation point, both the magnitude and direction of NO_2 exchange with vegetation would be regulated by ambient NO_2 concentration and a range of physiological conditions in the vegetation.

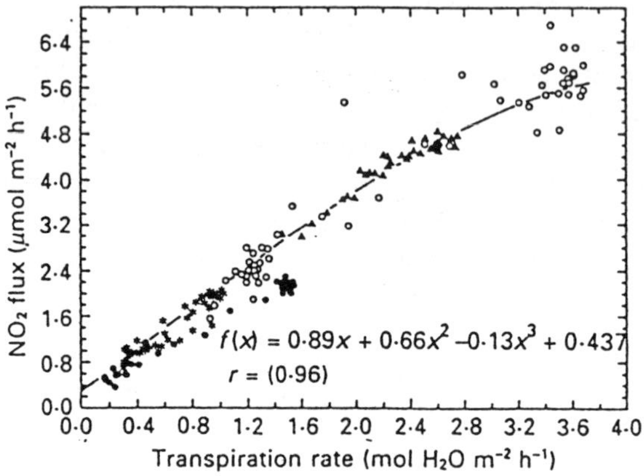

Fig. 6.7: NO_2 fluxes to spruce needles as dependent on transpiration rate at various photon flux rates ($\mu E\ m^{-2}\ s^{-1}$). Δ , 1000; O,600; ●,300; *, 150. Needles were exposed during four day/night rhythyms to 49.5 nl $NO_2\ l^{-1}$. NO_2 fluxes to shoots were expressed as $\mu mol\ NO_2\ m^{-2}\ h^{-1}$. Data represent means of continuous readings in 30 min intervals during day and night. (From Thoene *et al.*1991).

The work by Johannson (1987) suggested the existence of a compensation point for NO_2 uptake by Scots pine (*Pinus sylvestris*) in using enclosure methods with no net uptake at concentrations below about 1 ppb. A similar study, again using enclosure methods by Thoene *et al.* (1991) showed that Norway spruce (*Picea abies*) uptake of NO_2 increased linearly with the transpiration H_2O flux at NO_2 concentrations in excess of 5 ppb NO_2 (Fig. 6.7). At smaller concentrations than 2.6 ppb NO_2, no uptake was observed and the stomatal flux of NO_2 at higher concentrations was consistently smaller than that predicted from the stomatal conductance for water vapour.

These data supported those of Johannson (1987) indicating the presence of internal resistances, but with no evidence of NO_2 emission. The presence of an internal resistance to NO_2 uptake at low ambient NO_2 concentrations was not detected in the studies of NO_2 uptake by Norway spruce by Rondon *et al.* (1993) (Fig. 6.8). Similarly the studies of NO_2 uptake by Red spruce of Hanson *et al.* (1989) found no evidence of internal resistance. Vegetation is capable of producing and emitting NO_2 (Klepper, 1979, 1990) but the degree to which such mechanisms may be involved in field conditions, in the absence of herbicides is in doubt. A limited uptake of NO_2 by cuticular surfaces and bark observed by Hanson *et al.* (1989) and by Thoene *et al.* (1991), but this generally represents much less than

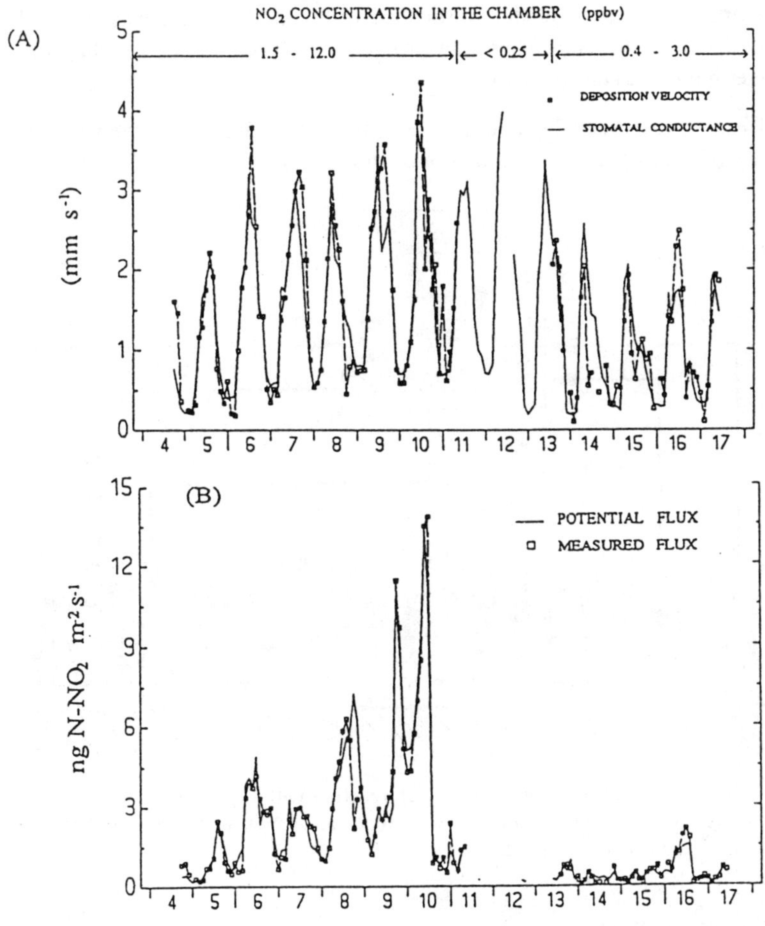

Fig. 6.8: (A) Diurnal variation of the stomatal conductance and deposition velocity for NO_2 per prjected needle area at different ranges of NO_2 concentrations in the pine stand in Simlångsdalen. (B) Diurnal variation of the potential and measured flux rates of NO_2 per projected needle area. (From Rondon *et al.* 1993).

10% of the surface flux and may also be a consequence of an experimental artefact.

Field measurements of NO_x exchange

The micrometeorological methods used widely to measure field-scale SO_2 fluxes (Garland, 1977) were used to measure NO_x exchange by Duyzer *et al.* (1983) and Delany and Davies (1983). In practice, the methods are subject to large errors arising from non-stationarity, including storage and advection effects as well as effects of air chemistry in the surface layer. With many complicating processes contributing to determine the vertical profile in NO_2 concentration above vegetation it is hardly surprising that progress towards an understanding of NO_x exchange in the field has been slow.

The different processes which contribute to the land-atmosphere exchange and air chemistry of NO_x compounds is illustrated in Fig. 6.9.

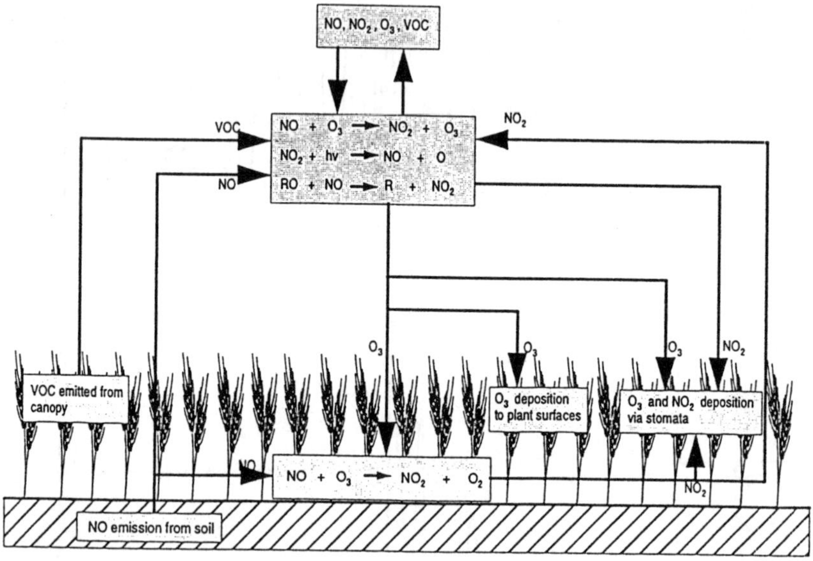

Fig. 6.9: Interactions of soil emissions of NO, reaction with O_3 in plant canopies and NO_2 uptake by vegetation in determining the net exchange of NO_x between soil-plant and the atmosphere. (From Fowler *et al.* 1996).

Advection and storage

Vehicles represent the major source of ground level NO_x and as they are so numerous in the countryside and widely distributed it is difficult to find appropriate field sites where substantial horizontal gradients in NO_2 are not present. Such horizontal gradients NO_2 result in the flux inferred from vertical concentration profiles differing from the actual surface flux. In practice, gradients as small as 1 $\mu g\ NO_2\ m^{-3}\ km^{-1}$ may introduce errors as large as the actual surface flux. Similarly, changes in concentration with time during the measurement period may introduce errors of a similar magnitude with temporal changes in ambient concentration of 6 $\mu g\ NO_2\ m^{-3}\ hour^{-1}$. Storage and advection errors represent a divergence in the flux between the reference height for measurement and the absorbing surface. Small changes in wind direction over the measurement period may introduce a different combination of upwind sources to the site of measurement and the resulting ambient NO_2 at the site may change. The

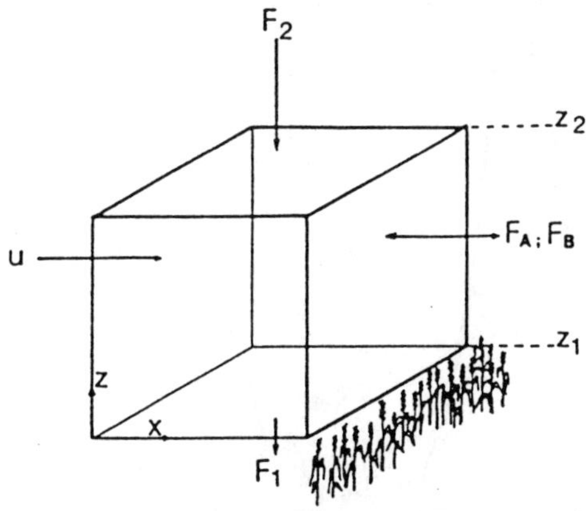

F_1 ACTUAL FLUX TO THE SURFACE
F_2 MEASURED FLUX
F_A NET HORIZONTAL FLUX
F_B STORED FLUX
z VERTICAL AXIS
x HORIZONTAL AXIS
u HORIZONTAL WINDSPEED m s^{-1}

Advection

Height range (z_1 to z_2) m	Horizontal conc. gradient $\mu g\ m^{-3}\ km^{-1}$	Actual flux (F_1) $\mu g\ m^{-2}\ s^{-1}$	Advection error (F_A) $\mu g\ m^{-2}\ s^{-1}$	$F_A/F_1 \times 100$ %
1	1	0.1	0.002	2
1	10	0.05	0.02	40
5	5	0.05	0.05	100

Storage

Height range (z_1 to z_2) m	Rate of change in concentration $\mu g\ m^{-3}\ h^{-1}$	Actual flux F_1 $\mu g\ m^{-2}\ s^{-1}$	Storage F_s $\mu g\ m^{-2}\ s^{-1}$	$F_S/F_1 \times 100$ %
1	6	0.1	0.0017	2
1	60	0.05	0.017	40
5	60	0.1	0.085	85

Fig. 6.10: Notation for discussion of advection (F_A) and (F_B) errors in measurements of vertical fluxes (F_1). (From Fowler & Duyzer 1989).

magnitude of the storage error may be surprisingly large for modest changes in NO_2 concentration. The advection and storage errors are illustrated in Fig. 6.10 in which the magnitude of error for typical daytime conditions is also given. The advection and storage errors may be minimised by careful selection of the field site and by rejecting data from periods during which non-stationarity

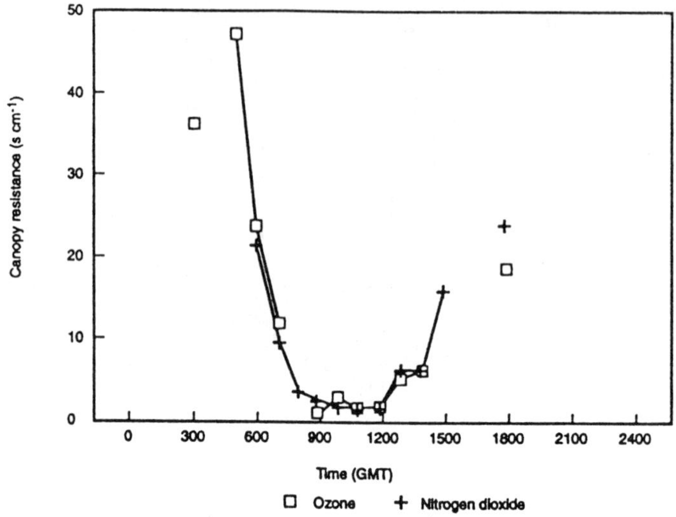

Fig. 6.11: Canopy resistance to the deposition of NO_2 and O_3, Halvergate, 15 September 1989. (From Hargreaves *et al.* 1992).

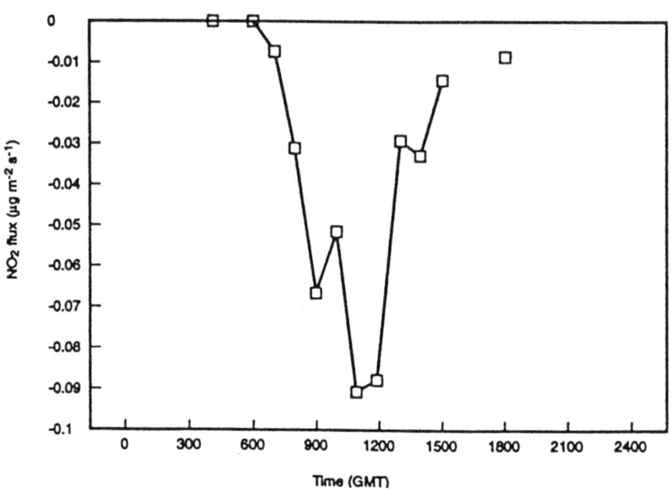

Fig. 6.12: NO_2 flux, Halvergate, 15 September 1989. (From Hargreaves *et al.* 1992),

effects may be detected within the data. This is a much easier task for storage than for advection errors, since in the latter case, the concentration at the site of measurements at a specific reference height could be constant throughout the measurement period, yet be subject to a large advection error.

The measured vertical NO_2 gradients, in conditions with minimal error due to advection and storage may also fail to quantify the flux at the surface as a consequence of atmospheric chemistry in the surface layer (Fig. 6.9).

In conditions in which fluxes are measured over a small height range (< 4m) and concentrations of NO are very small (>1 ppb) and rates of NO_2 uptake are large. Corrections to the NO_2 flux are likely to be small (Duyzer *et al.*, 1995).

At the BIATEX field experiment at the Halvergate Marshes in 1989, NO and NO_2 fluxes were measured using a range of micrometeorological methods. Correction procedures using a one-dimensional surface layer model to solve the differential equations describing turbulent transport and reactions between NO, NO_2 and O_3 were used with the field measurements of eddy diffusivity K_m, radiation and measured profiles of NO, NO_2 and O_3 concentrations. The corrected profiles show that errors in the flux measurements in these conditions were of the order of 30 %. The measurements however provided strong evidence of a link between NO_2 flux and stomatal conductance. The data from one of the best days (*i.e.* with smallest errors due to non-stationarity effects) on the 15th September show a clear diurnal cycle in NO_2 and O_3 deposition to the grassland surfaces (Figs. 6.11 and 6.12). The minimum canopy resistance of 80 s m^{-1} during the day and nocturnal value in excess of 500 s m^{-1} strongly support the hypothesis that cuticular surface uptake of NO_2 is negligible and that canopy resistance is not greatly enhanced by internal resistances for such surfaces.

Micrometeorological techniques and modelling NO_x exchange processes

Micrometeorological modelling techniques have been developed to determine the dry deposition of atmospheric trace constituents within BIATEX and these range from modified Bowen ratio and aerodynamic profile techniques to inferential methods and more complex turbulence models. Bowen ratio and aerodynamic profile techniques are based on the constant flux approximation. These techniques are widely used to estimate dry deposition fluxes of gaseous and particulate matter on the basis of vertical profile data of concentration, wind speed, temperature, and humidity measured at two or more heights above vegetation-soil and natural water systems (*e.g.*, Fowler and Duyzer, 1989; Hargreaves *et al.*, 1992; Müller *et al.*, 1993).

Assuming horizontal homogeneity (advection and horizontal divergence terms are neglected) the budget equation for the concentration c_i (expressed in ng m^{-3} or ppb) of a trace constituent, transported through the atmospheric surface layer, such as O_3, NO, NO_2 etc., may be expressed by (*e.g.*, Businger, 1986)

$$\frac{\delta c_i}{\delta t} = -\frac{\delta F_i}{\delta z\, f_i}(c_1,...,c_p) \quad (1)$$

Here, F_i is the vertical flux, and $f_i(c_1, ..., c_p)$ is the chemical production or depletion rate. By analogy with the mixing laws of micrometeorology, F_i may be described by (*e.g.*, Businger, 1986)

$$F_i = -(D_i + K_i)\frac{\delta c_i}{\delta z} - v_T\, c_i \quad (2)$$

where D_i is the molecular diffusivity, K_i is the eddy diffusivity, and v_T is the terminal settling velocity. In the case of trace gases as considered here the gravitational effects are negligible. Furthermore, the quantity D_i can be ignored in the fully turbulent region of the atmospheric surface layer.

$$V_{d,i} = -\frac{F_i}{c_i} \tag{3}$$

Dry deposition may be described in terms of c_i, F_i and the so-called deposition velocity $V_{d,i}$ defined Chamberlain, (1961) wherein all three quantities generally vary with height and time (see Eq. (1)). If the deposition velocity determined for a certain level z_R (*e.g.*, $z_R = 1$ m) is nearly independent of the surface properties for different atmospheric conditions, the dry deposition flux can be estimated by $F_i = - V_{d,i}(z_R)\ c_i(z_R)$. The flux can then be considered as a function of the mean concentration at this reference height z_R, and, hence, the atmospheric input of pollutants by dry deposition can be estimated for larger areas and different periods on the basis of the horizontal concentration distributions. However, as shown previously by Sehmel (1980) as well as Hanson and Lindberg (1991) for several trace species and confirmed by the results of BIATEX, the range of deposition velocity varies strongly for different surface properties, sometimes of more than one order of magnitude, so that such simple flux parameterisation may be quite misleading.

Aerodynamic profile techniques

A formulation for fluxes based on the aerodynamic profile method is given by (*e.g.*, Arya, 1988;

$$F_i = -\frac{k^2 (U_R - U_\delta)(c_{R,i} - c_{\delta,i})}{(\ln\frac{z_R - d}{\delta - d} - \Psi_i(\varsigma_R, \varsigma_\delta))(\ln\frac{z_R - d}{\delta - d} - \Psi_m(\varsigma_R, \varsigma_\delta))} \tag{4}$$

Kramm *et al.*, 1995a) where $\boldsymbol{k} = 0.4$ is the von Kármán constant, U_δ and U_R are the mean wind speeds at the heights δ and z_R, the lower and upper boundaries of the fully turbulent atmospheric surface layer, $c_{\delta,i}$ and $c_{R,i}$ are the concentrations at the same heights, $d < \delta$ is the zero-plane displacement, ψ_m $(\varsigma_R, \varsigma_\delta)$ and ψ_i $(\varsigma_R, \varsigma_\delta)$ are the integral stability functions for momentum and mass (Paulson, 1970; Kramm, 1989) which are based on the Monin-Obukhov similarity theory (Monin and Obukhov, 1954), and $\varsigma = (z - d)/L$ (where L is the Monin-Obukhov stability length) is a dimensionless height. Equation (4) is based on the assumption that F_i is independent of z and $c_i(z)$ (Kramm, 1991a; Kramm *et al.*, 1992). The height z_R has to fulfil the condition $z_R \leq z_{max} \sim 0.3\cdot$ (fetch)$^{1/2}$ (Raabe, 1991), where the fetch is the distance between the change of the surface properties in the upwind region of a measuring tower and the tower itself, and z_{max} is the height of the undisturbed new equilibrium layer (Rao *et al.*, 1974).

The flux-gradient relationship for this turbulent region of the atmospheric surface layer may also be expressed in the sense of the Monin-Obukhov similarity theory by (*e.g.*, Hargreaves *et al.*,

$$F_i = -\frac{k^2}{\phi_i(\varsigma)\,\phi_m(\varsigma)} \frac{\partial U}{\partial ln\,(z-d)} \frac{\partial c_i}{\partial ln\,(z-d)} \tag{5}$$

1994) where U is the mean wind speed, ϕ_m(ζ and (I(() are the dimensionless local stability functions for momentum and matter. Equation (5) can be approximated by finite differences

$$F_i = -k^2 \frac{(U_R - U_\delta)(c_{R,i} - c_{\delta,i})}{<\Phi_i(\zeta)><\Phi_m(\zeta)>} \left(\ln\frac{z_R - d}{\delta - d}\right)^2 = const \tag{6}$$

(Kramm et al., 1995a), where the brackets (< >) denote the quantities representative for the layer $\delta \le z \le z_R$. Equations (4) and (6) agree well for inversion conditions. Whereas for lapse conditions the finite difference approximation (Eq. (6)) has to be considered as crude.

Energy balance (Bowen ratio) method

$$\begin{Bmatrix} F_i \\ H \\ E \end{Bmatrix} = -\frac{1}{R_T} \begin{Bmatrix} \Delta c_i \\ (\rho c_p)\Delta\Theta \\ (\varepsilon\lambda\rho / p)\Delta e \end{Bmatrix} = \text{constant} \tag{7}$$

For (quasi) constant flux conditions Eq. (2) may be integrated to relate the flux F_i to a finite difference _c_i as described in Eq. (4). Expressing the meteorological fluxes of sensible heat, H, and latent heat, E, as relations analogous to Eq. (2) with K_h and K_q being the respective eddy diffusivities, and assuming $K_i = K_q = K_h$, the integrated relations for the various fluxes may by written as (*e.g.*, Müller *et al.*, 1993) : with the turbulent bulk resistance r_t given by (Hicks, 1976; Kramm, 1989)

$$t_t = \int_\delta^{z_R} K_h^{-1} dz = \frac{1}{u.K}(\ln\frac{z_R - d}{\delta - d} - \psi_h(\zeta_R, \zeta\delta)) \tag{8}$$

where Ψ_i (ζ_R, ζ_δ) is the integral stability function for sensible heat. Equations (7) express the concept of Bowen ratio similarity, and one immediately derives the basic relations of the so-called modified Bowen ratio method (*e.g.* Biscoe *et al.*, 1975; Businger, 1986):

$$F_i = (\rho\ c_p)^{-1}\ H\ \frac{\Delta c_i}{\Delta\Theta} = \frac{p}{\varepsilon\lambda\rho}\ E\ \frac{\Delta c_i}{\Delta e} \tag{9}$$

The inferential method

The inferential method provides estimates of dry deposition in the absence of direct measurements. It is based on the 'big leaf' multiple resistance approach given by (Wesely and Hicks, 1977; Hicks *et al.*, 1987)

$$F_i = - \frac{c_{R,i}}{r_t + r_{mt,i} + r_{s,i}} \tag{10}$$

where $r_{mt,i}$ is the corresponding bulk resistance of the underlying thin molecular-turbulent sublayer formulated in terms of the roughness Reynolds number and the Schmidt number (*e.g.*, Kramm *et al.*, 1992); and $r_{s,i}$ is the bulk resistance of the canopy which characterises the biochemical and physical processes taking place in the soil and within the plants (*e.g.*, Baldocchi *et al.*, 1987).

The canopy resistance summarises the substrate resistances $r_{st,i}$ (stomatal resistance), $H_i\ r_{int,i}$ ('effective' internal resistance), $r_{cu,i}$ (cuticular resistance), $r_{sl,i}$ (resistance of nearly dry soil), $H_i\ r_{w,f,i}$ and $H_i\ r_{w,sl,i}$ ('effective' resistances of leaf wetness and soil water), as well as the molecular-turbulent resistance $r_{mt,fg,i}$ to a total 'substrate' resistance using Kirchhoff's laws of electrostatics to characterise series and parallel pathways of diffusive mass transfer. The stomatal resistance can be estimated with a sufficient degree of accuracy using Jarvis's (1976) expression. The manipulation of additional surface resistance terms to quantify internal (physiological) resistances within plants or for the effects of leaf surface water on NO_2 uptake on cuticles is not justified by current understanding of these processes. The lack of detail in calculating these effects requires laboratory and field measurements to demonstrate the controlling processes which should be a focus of future work in this area.

Modelling procedures to estimate the surface fluxes of NO, NO_2 and O_3 from boundary layer measurements

Nitrogen dioxide occurs predominantly in the troposphere as a result of the oxidation of NO which is mainly released in combustion processes. Its photo dissociation by sunlight ($\lambda < 430$ nm) leads to the formation of ozone. Ozone rapidly reacts with NO to regenerate NO_2 so that in the absence of transfer processes and other chemical reactions a photostationary state between NO, NO_2 and O_3 can be achieved. This simple reaction scheme can be expressed by (reaction rate constants, k_1 and k_2, in ppm-min units; all reaction rate coefficients mentioned here are taken from Baulch *et al.*, 1984, and Atkinson *et al.*, 1989)

$$NO_2 + h\nu \xrightarrow{O_2} NO + O_3 \qquad k_1 = \text{variable} \tag{R1}$$

$$NO + O_3 \rightarrow NO_2 + O_2 \qquad k_2 = 7.82 \cdot 10^2\ p/T \exp(-1{,}370/T) \tag{R2}$$

where p_L is the air pressure in hPa, and T is the air temperature in K. Results from recent field experiments over agricultural crops and unmanaged grassland at different sites (Delany *et al.*, 1986; Hicks and Matt, 1988; Nestlen *et al.*, 1993) indicate that in the case of the highly reactive trace gases NO and NO_2 the eddy fluxes measured directly and the mean concentrations determined concurrently are not in agreement with the flux-resistance relationship expressed by Eq. (14). As presented by these authors, some disagreement could be caused by emission of NO and NO_2 from the soil.

Profile methods for determining the fluxes of NO, NO_2, and O_3

Recent results from numerical studies with first-order and higher-order closure models performed by a number of authors outside and within BIATEX (Lenschow, 1982; Fitzjarrald and Lenschow, 1983; Kramm, 1989; Duyzer *et al.*, 1995; Kramm *et al.*, 1991; Gao *et al.*, 1991, Duyzer, 1992) suggest that the fluxes of NO and NO_2 vary strongly with height owing to the flux divergence, for instance, for NO :

$$-\partial F_{NO_2}/\partial z = \partial F_{O_3}/\partial z = \partial F_{NO}/\partial z$$

$$\frac{\partial F_{NO}}{\partial z} = k_1\,[NO_2] - k_2\,[NO]\,[O_3] \tag{11}$$

caused by chemical reactions among these nitrogen species and O_3. The calculated vertical profiles of concentrations and fluxes which are based on the first-order closure (see Eq. (2)) only differ slightly from those derived with higher-order closure principles. However, as shown in Figs. 6.13 and 6.14, the flux profiles derived with chemical reactions differ considerably from those based on the constant flux approximation. Hence, some disagreement concerning the flux-resistance relationship (see Eq. (10)) could also be attributed to chemical reactions. The differences between the fitted concentration profiles are so small that an evaluation of chemistry transport models only on the basis of calculated and observed concentration distributions (necessary condition) appears doubtful. A final evaluation of such models can only be made by a comparison of the calculated fluxes with fluxes concurrently measured by eddy correlation techniques at two or more heights. In doing so, a first attempt was performed during the joint BIATEX/SANA field experiment at Eisdorf near Leipzig in eastern Germany (Nestlen *et al.*, 1993).

The turbulent resistance, r_t, is only used to determine the fluxes

$$F_j = \begin{cases} F_{NO} + F_{NO_2} & \text{for } j = 1 \\ F_{NO_2} + F_{O_3} & \text{for } j = 2 \\ F_{NO} - F_{O_3} & \text{for } j = 3 \end{cases} \tag{12}$$

of the composed chemically conservative trace gases with 'group' concentrations

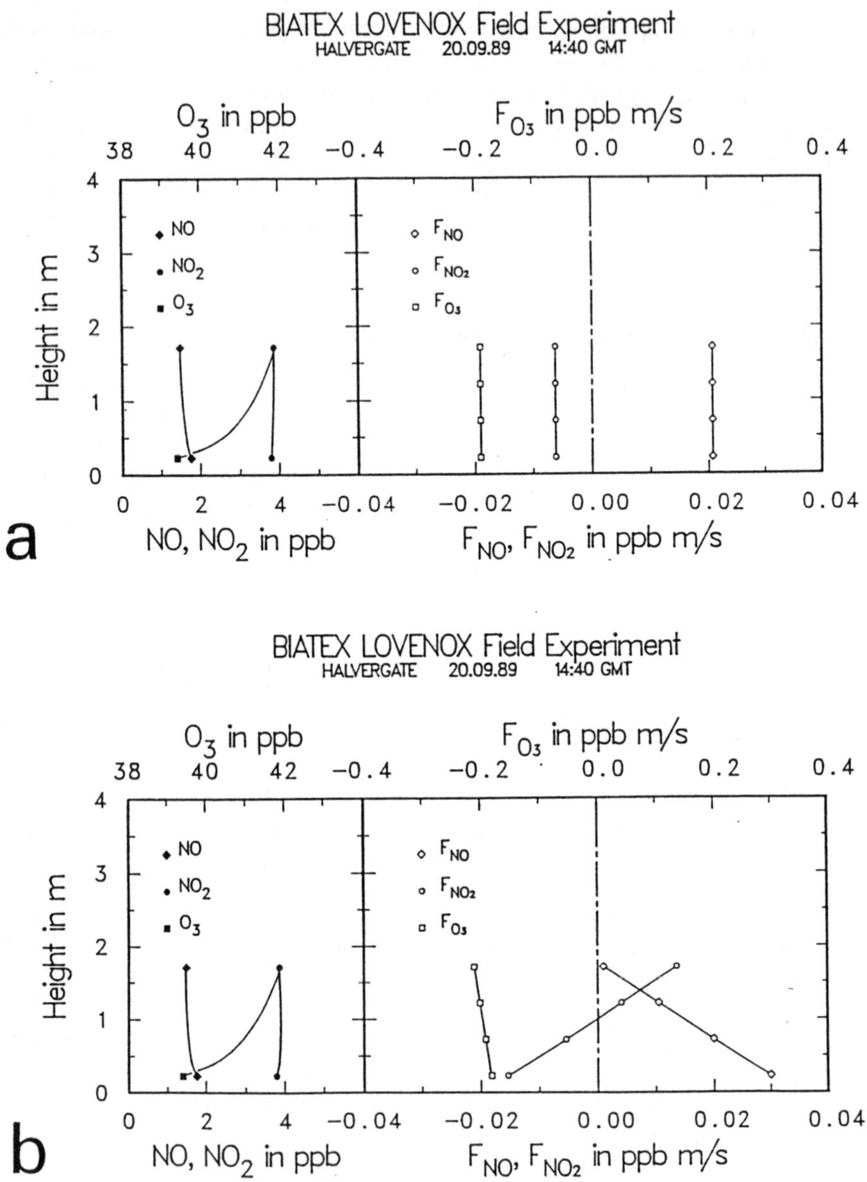

Fig. 6.13: Vertical profiles of concentrations and turbulent fluxes of NO, NO_2 and O_3. Solid symbols represent the measured concentrations (taken from Kramm *et al.* 1991), and the lines the calculated vertical profiles. Calculations were performed (a) with the constant flux approximation, and (b) with chemical reactions (Eqs. (2) and (11) to (13).

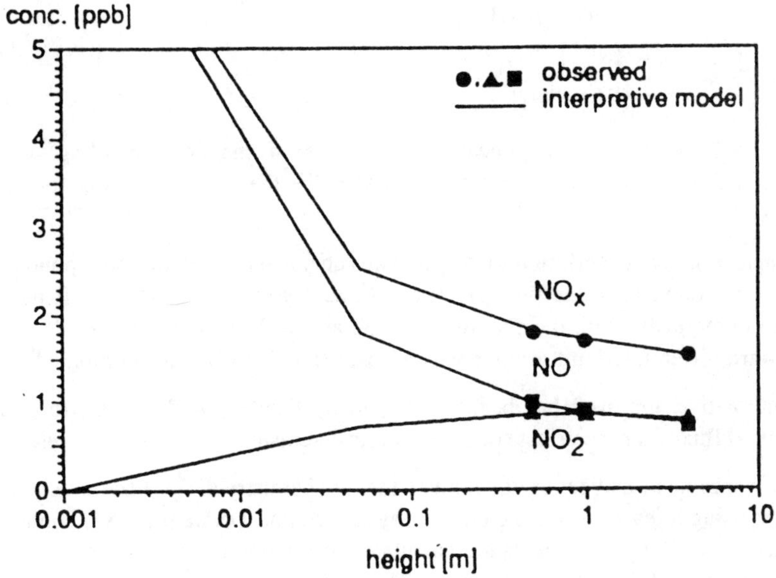

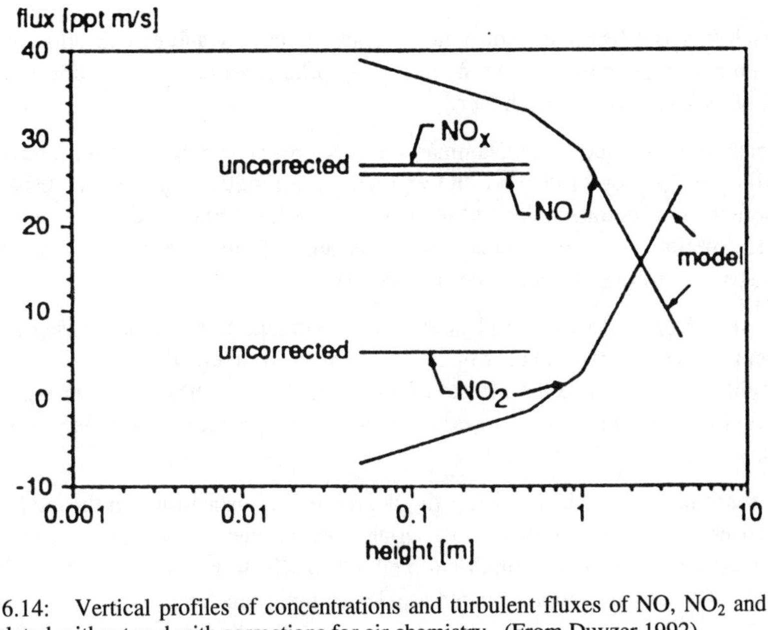

Fig. 6.14: Vertical profiles of concentrations and turbulent fluxes of NO, NO_2 and NO_x calculated without and with corrections for air chemistry. (From Duyzer 1992).

$$c_j = \begin{cases} [NO] + [NO_2] & \text{for } j = 1 \\ [NO_2] + [O_3] & \text{for } j = 2 \\ [NO] + [O_3] & \text{for } j = 3 \end{cases} \qquad (13)$$

where both F_j and c_j are linear combinations of the corresponding $N = 3$ 'individual' fluxes and concentrations with $M = 2$ is the rang of the transformation matrix.

The mathematical description of this transfer chemistry problem corresponds to a two-point boundary value problem (Lenschow, 1982; Kramm, 1989). It may be solved numerically with the shooting method and with reference to Gear's (1971) backward differentiation formulae where Eqs. (2), (11) to (13) are combined.

An alternative approach has been developed by Duyzer *et al.* (1995) to correct measured fluxes for the effects of atmospheric chemistry in the surface layer.

In their approach, Duyzer *et al.* simulate the air chemistry of 43 different chemical species using a reaction scheme devised by Hough and Derwent (1987) containing 85 reactions. Rate coefficients are calculated as a function of air temperature and zenith angles of incident radiation. The turbulent exchange between the 20 logarithmically spaced layers of the model is calculated using a *k* theory approach in which the turbulent exchange for height z of the layer is given by

$$k_{(z)} = \frac{k\ U_* z}{\phi_c(z/L)} \qquad (14)$$

in which ***k*** is von Karman's constant, U_* friction velocity and ϕ_c s the flux/profile function for gases derived from ϕ_h the flux/profile function for heat, and L is the Monin-Obukou stability parameter.

The canopy resistance r_c and laminar sub-layer resistance r_b are both calculated according to Wesley (1989) ad NO emission is estimated using a soil resistance and soil NO concentration derived in the test procedures from direct measurements at the experimental sites. This method is able to quantify the effects of air chemistry on profiles of concentration and fluxes.

The comparison of detailed models with field data tend to be compromised by the inherent noise of field data which are subject to many different sources of variability in addition to the effects of air chemistry. In these circumstances the detection of deviations of 10 to 40 % in measured profiles from those expected becomes difficult.

With these difficulties in matching the degree of sophistication within models of the process with our ability to appropriate measurements, Duyzer *et al.* (1995) developed a much simpler 'empirical method' in which the flux divergence $\partial F/\partial$ for NO, NO_2 and O_3 is approximated as a logarithmic function.

Taking the simple equation of Lenshaw and Delany (1987), for a height range restricted to the lowest 4 metres the flux divergence is described by

$$\left(\frac{\partial f}{\partial z}\right)_z = a\ \mathrm{Ln}(z) + b \tag{15}$$

integrating the functions for NO, NO_2 and O_3 separately provides expressions for the flux with height which may then be used to derive concentration profiles for the reactive gases. A 'correction factor' approach is also described by Duyzer *et al.* In which the requirement for low noise measurement data is relaxed and the approach has been shown to provide satisfactory results in a range of different conditions.

To compare the different procedures Duyzer *et al.* used concentration profiles for NO, NO_2 and O_3 calculated using a simulation model as input for the different correction procedures. In the case of NO_2 uptake by vegetation, it is probable that

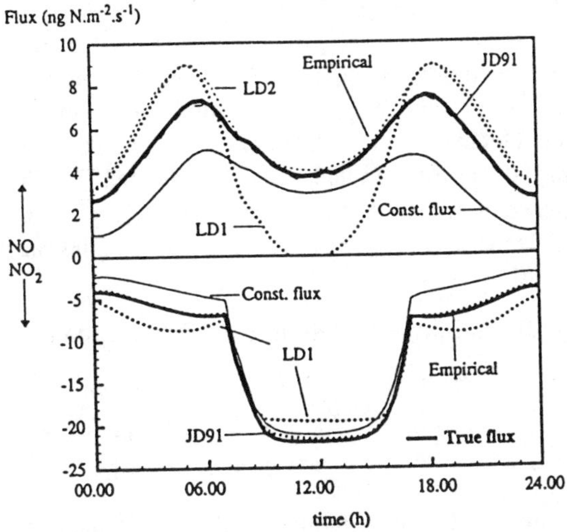

Fig. 6.15: The true surface flux of NO and NO_2 from the base run and the surface fluxes calculated from the concentration at 0.5, 1, 2 and 4 m using the classical gradient approach. (From Duyzer 1995).

surface deposition fluxes may be estimated with greater accuracy as the product of bulk canopy stomatal conductance for NO_2 (derived from energy balance or other micrometeorological methods) and the ambient NO_2 concentration at canopy height than directly using flux profile relationships and any current model. In order to estimate the net exchange of NO_x between soil/vegetation and the atmosphere it is of course necessary, in addition to NO_2 uptake by the plants, to know the soil NO emission flux.

The result illustrated in Fig. 6.15 shows large discrepancies in NO fluxes estimated using the constant flux and Lenshaw and Delany (1987) methods. Smaller but significant errors are also present in the NO_2 fluxes, especially at night.

An inferential method for NO, NO_2, and O_3

Since the model results mentioned before indicate that the flux profiles can be affected by chemical reactions, a modified inferential method has been developed within the framework of BIATEX to evaluate the vertical transfer of NO, NO_2, and O_3 according to Eqs. (2), (11) to (13) where emission of NO and NO_2 emissions: from the canopy are considered, too (Kramm *et al.*, 1995b). This method is based on (1) an improved Deardorff-type soil - vegetation - atmosphere transfer scheme which provides, among other things, the stomatal resistances for water vapour and trace gases using Jarvis's (1976) approach as well as the atmospheric fluxes of momentum, sensible and latent heat and the soil fluxes of sensible heat and moisture (Deardorff, 1978; Kramm *et al.*, 1995b), and (2) the flux-resistance relationship for the canopy layer and the interfacial sublayer given by (Kramm, 1991b)

$$F_i = - \frac{c_{\delta,i}}{r_{mt,i} + r_{s,i}} (1 - \mu_i) \tag{16}$$

where $c_{\delta,i}$ is the concentration at height δ, the outer edge of the interfacial sublayer; the index *I* stands for NO, NO_2, or O_3. Similar equations have already been derived, for example, by O'Dell *et al.* (1977) and Farquhar *et al.* (1983). As shown in Kramm *et al.* (1991), Kramm and Dlugi (1994), and Kramm *et al.* (1995a) the quantity μ_I depends on the substrate concentrations $C_{int,i}$, $C_{cu,i}$, $C_{sl,i}$, $C_{w,f,i}$ and $C_{w,sl,i}$, the substrate resistances $r_{st,i}$, $r_{int,i}$, $r_{cu,i}$, $r_{sl,i}$, $r_{w,f,i}$ and $r_{w,sl}$, the canopy resistance, $r_{s,i}$, and the concentration $c\delta_{,i}$.

It is obvious that the quantity μI determines the flux direction, *i.e.*

$$F_i = \begin{cases} < 0, & \text{if } \mu_i < 1 \quad \text{(deposition)} \\ = 0, & \text{if } \mu_i = 1 \quad \text{(compensation)} \\ > 0, & \text{if } \mu_i > 1 \quad \text{(emission)} \end{cases} \tag{17}$$

The modified inferential method requires only routine data for concentrations of NO, NO_2, and O_3 as well as of wind speed, temperature, humidity and air pressure, which are easily provided by monitoring stations in a network. In addition, it requires the soil temperature and the soil moisture measured at a depth of a few centimetres (root zone), and the typical station parameters, such as roughness length, zero plane displacement, type of vegetation, leaf area index etc. The radiation input may be calculated on the basis of simple or more sophisticated radiative transfer models, but measured radiation quantities can also be used.

The routine data mentioned above serve as upper boundary conditions at z_R. The lower boundary conditions, *i.e.*, the concentrations of the 'individual' trace species at the height (and the 'group' concentrations $[NO_x]$, $[O_x]$, and $[NO_x] - [O_x]$ (see Eq. (13)) at the same height have to be calculated simultaneously (Kramm and Dlugi, 1994; Kramm *et al.*, 1995b). Here, the

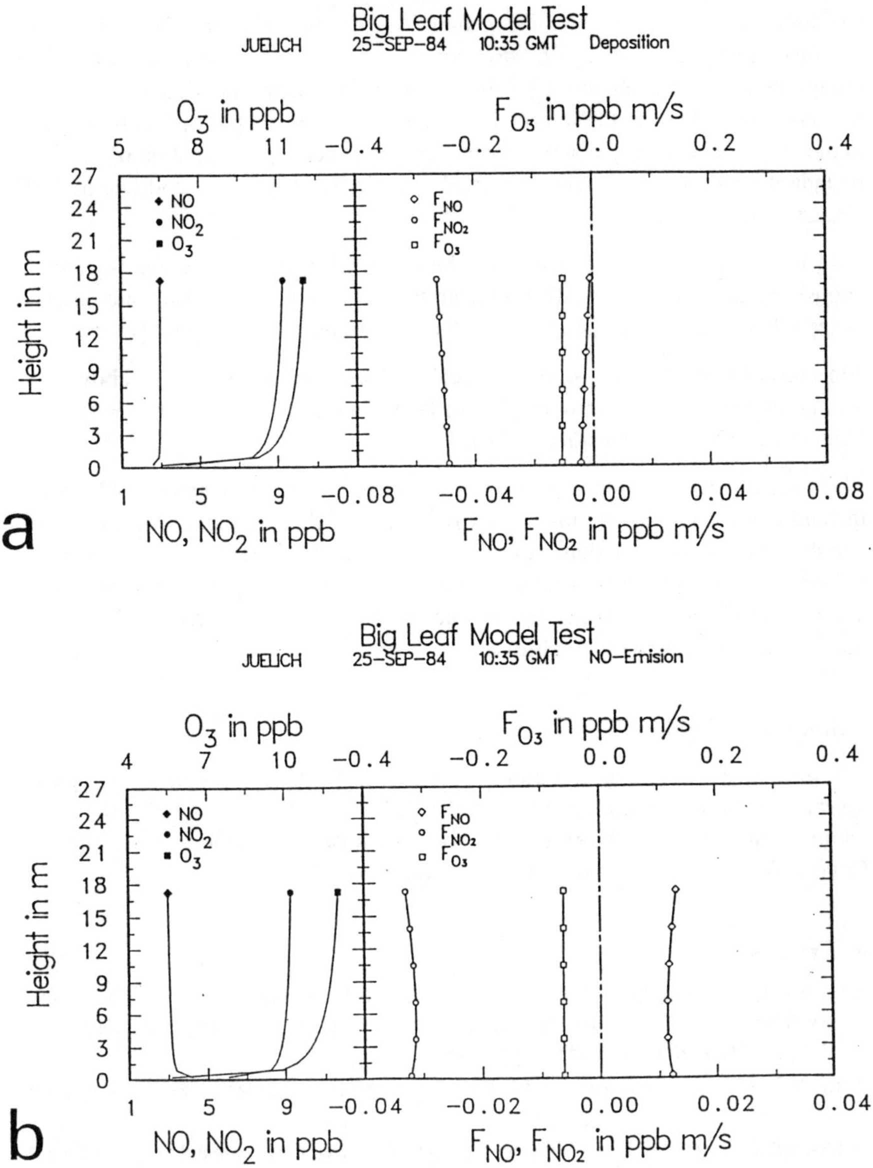

Figure 6.16: Vertical profiles of concentrations and turbulent fluxes of NO, NO_2 amd O_3 for (a) $c_{sl,NO} = c_{sl,\ NO2} = 0$ and (b) $c_{sl,NO} = 22$ ppb, $c_{sl,NO2} = 12$ ppb. The solid symbols represent the measured concentrations (Höfken *et al.*, 1986), and the lines represent the calculated vertical profiles (from Kramm *et al.*, 1995b).

calculations of the lower boundary conditions are based on the requirement that the flux divergences ($\partial F_{NO}/\partial z$, $\partial F_{NO2}/\partial z$, and $\partial F_{O3}/\partial z$) vanish at height ∂. This requirement is a consequence of the constant flux approximation on which the resistance approach for the interfacial sublayer and the canopy layer is based, even in the case of chemically reactive trace species. Hence, the calculation of the lower boundary conditions can only be carried out iteratively (Kramm and Dlugi, 1994; Kramm *et al.*, 1995b).

Test results of sensitivity studies with the modified inferential method referring to deposition and emission of NO are illustrated in Fig. 6.16. The results are based on observed values of wind speed, dry-bulb temperature, specific humidity, and

concentrations of NO, NO_2, and O_3 collected at a height of 17.2 m above level terrain covered with sugar beet (25 cm tall), near Jülich in western Germany in September 1984 (Höfken *et al.*, 1986).

The wide divergence in the approach to simulate the net exchange of NO_y species in field conditions may be taken as a reflection of the complexity of the process and our limited understanding of it. Thus, the complex models contain most of the known processes but have a requirement for a large number of parameters which are not readily available and the simple models are only correct within a very narrow range of boundary conditions.

Acknowledgements

The authors all acknowledge funding from the European Union which greatly assisted in the collaboration between the different national research groups. In addition, funding from the respective National Research Councils, Ministries and Environment Departments is gratefully acknowledged.

References

Atkinson, R., Baulch, D.L., Cox, R.A., Hampson, R.F. Jr., Kerr, J.A., Troe, J. (1989). Evaluated kinetic and photochemical data for atmospheric chemistry: Supplement III. *J. Phys. Chem. Ref. Data* **18** (2), 881–1097.

Ayra, S.P. (1988). *Introduction to Micrometeorology*. Academic Press, San Diego, 303 pp.

Baldocchi, D.D., Hicks, B.B., Meyers, T.P. (1987). A canopy stomatal resistance model for gaseous deposition to vegetated surfaces. *Atmos. Environ.* **21**, 91–101.

Baulch, D.L., Cox, R.A., Hampson, R.F., Kerr, J.A., Troe, J., Watson, R.T. (1984). Evaluated kinetic and photochemical data for atmospheric chemistry: Supplement II. *J. Phys. Chem. Ref. Data* **13** (4), 1259–1380.

Biscoe, P.V., Clark, J.A., Gregson, K., McGowan, M., Monteith, J.L., Scott, R.K. (1975). Barley and its environment. *J. Appl. Ecol.* 12, 227–257.

Businger, J.A. (1986). Evaluation of the accuracy with which dry deposition can be measured with current micrometeorological techniques. *J. Appl. Meteor.* **25**, 1100–1124.

Cardenas, L., Rondón, A., Johansson, C., Sanhueza, E. (1993). Effects of soil moisture, temperature and inorganic nitrogen on nitric oxide emissions from acidic tropical savannah soils. *J. Geophys. Res.* **98(D8)**, 14783–14790.

Chamberlain, A.C. (1961). Aspects of travel and deposition of aerosol and vapour clouds. A.E.R.E. Report HP/R **1261** (RP[14]). Harwell, Berkshire.

Colbourn , P., Ryden, J. C., Dollard, G.J. (1987). Emission of NO_x from urine-treated pasture. *Environ Pollut.* **46**, 253–261.

Delany, A.C., Fitzjarrald, D.R., Lenschow, D.H., Pearson, R.Jr., Wendel, G.J., Woodruff, B. (1986). Direct measurements of nitrogen oxides and ozone fluxes over grassland. *J. Atmos. Chem.* **4**, 429–444.

Delany, A.C., Davies, T.D. (1983). Dry deposition of NO_x to grass in rural east Anglia. *Atmos. Environ.* **17**, 1391–1394.

Duyzer, J.H., Meijer, G.M., Van Aalst, R.M. (1983). Measurement of dry deposition velocities of NO, NO_2 and O_3 and the influence of chemical reactions. *Atmos. Environ.* **17**, 2117–2120.

Duyzer, J.H. (1992). The influence of chemical reactions on surface exchange of NO, NO_2 and O_3: Results of experiment and model calculations. In: S.E. Schwartz and W.G.N. Slinn (eds) *Precipitation Scavenging and Atmosphere-Surface Exchange Processes.* Vol. 2 - The Semonin Vol. Hemisphere Publ., New York, pp. 1105–1114.

Duyzer, J., Weststrate, H. (1995). The use of the gradient method to monitor trace gas fluxes over forest: flux-profile functions for ozone and heat. In: G.J. Heij and J.W. Erisman (eds), *Acid Rain Research: Do We Have Enough Answers?*, Elsevier, Amsterdam.

Fowler, D., Duyzer, J.H. (1989). Micrometeorological techniques for the measurement of trace gas exchange. In: M.O. Andreae and D.S. Shimel (eds) *Exchange of Trace Gases between Terrestrial Ecosystems and the Atmosphere*, J. Wiley and Sons, Chichester/New York/Brisbane/Toronto/Singapore, 189–207.

Fowler,D., Sutton M.A., Skiba, U., Hargreaves, K.J. (1996) . Agriculural nitrogen and emissions to the environement. *Issues in Environ. Sci. Tech.* in press.

Galbally, I.E., Roy, C.R. (1978). Loss of fixed nitrogen from soils by nitric oxide exhalation. *Nature* **275**, 734–735.

Galbally, I.E., Roy, C.R. (1981). Ozone and nitrogen oxides in the southern hemisphere troposphere. In: *Proceedings of the Quadrennial International Ozone Symposium*, Boulder, Colorado, August 1980, pp. 431–438.

Galbally, I.E., Johansson, C., (1989). A model relating laboratory experimments of rates of nitric oxide production and field measurements of nitric oxide emissions from soils. *J. Geophys. Res.* **94**, 6473.

Gao, W., Wesely, M.L., Lee, I.Y. (1991). A numerical study of the effects of air chemistry on fluxes of NO, NO_2 and O_3 near the surface. *J. Geophys. Res.* **96** (D10), 18761–18769.

Garland, J.A. (1977). The dry deposition of sulphur dioxide to land and water surfaces. *Proc. Roy. Soc. Lond. A.* **354**, 245–268.

Gear, C.W. (1971). *Numerical Initial Value Problems in Ordinary Differential Equations.* Prentice-Hall, Englewood Cliffs, New Jersey, pp. 253.

Hanson, P.J., Rott, K., Taylor, G.E., Gunderson, C.A., Lindberg, S.E., Ros-Todd, B.M. (1989). O_2 deposition to elements representative of a forest landscape. *Atmos Environ.* **23**, 1783–1794.

Hanson, P.J., Lindberg, S.E. (1991). Dry deposition of reactive nitrogen compounds: A review of leaf, canopy and non-foliar measurements. *Atmos. Environ.* **25A**, 1615–1634.

Hargreaves, K.J., Skiba, U., Fowler, D., Arah, J., Wienhold, F.G., Klemedtsson, L., Galle, B. (1994). Measurement of nitrous oxide emission from fertilized grassland using micrometeorological techniques. *J. Geophys. Res.* **99**, 16569–16574.

Hicks, B.B. (1976). Some micrometeorological aspects of pollutant deposition rates near the surface. In: R.J. Engelmann and G.A. Sehmel (eds) *Atmosphere-Surface Exchange of Particulate and Gaseous Pollutants (1974).* Techn. Inf. Center, Office of Public Affairs, ERDA, Oak Ridge, TN, pp. 434–449.

Hicks, B.B., Baldocchi, D.D., Meyers, T.P., Hosker, R.P.Jr.,, Matt, D.R. 1987. A preliminary multiple resistance routine for deriving dry deposition velocities from measured quantities. *Water, Air and Soil Pollution* **36,** 311–330.

Hicks, B.B., Matt, D.R. (1988). Combining biology, chemistry and meteorology in modelling and measuring dry deposition. *J. Atmos. Chem.* **6**, 117–131.

Höfken, K.D., Meixner, F., Müller, K.P., Ehhalt, D.H. (1986). Untersuchungen zur trockenen Deposition und Emission von atmosphärischem NO, NO_2 und HNO_3 an natürlichen Oberflächen. *Berichte der Kfa Jülich*, No **2054**, Jülich, F.R.G.

Hough, A.M., Derwent, R.G. (1987). Studies of the distribution of photochemical ozone production between different hydrocarbons. *Atmos. Environ.* **21**, 2015–2033.

Hutchinson, G.L., Livingston, G.P., Brams, E.A. (1993). Nitric and nitrous oxide evolution from managed subtropical grassland. In: *Biogeochem of Global Change: Radiatively active trace gases.* Chapman and Hall, New York.

Jarvis, P.G. (1976). The interpretation of the variations in leaf water potential and stomatal conductance found in canopies in the field. *Phil. Trans. R. Soc. London*, B **273,** 593–610.

Johansson, C., Granat, L. (1984). Emission of nitric oxide from arable land. *Tellus* **36B**, 25–37.

Johansson, C., Rhode, H., Sanhueza, E. (1988). Emission of NO in a tropical savanna and a cloud forest during the dry season. *J. Geophys. Res.* **93**, 7180–7192.

Johansson, C. (1984). Field measurements of emission of nitric oxide from fertilized and unfertilized forest soils in Sweden. *J. Atmos. Chem.* **1**, 429–442.

Johansson, C. (1987). Pine forest: a negligible sink for atmospheric NO_x in rural Sweden. *Tellus* **39B**, 426–438.

Kim, C.M. (1973). Influence of vegetation types on the intensity of ammonia and nitrogen dioxide liberation from soil. *Soil Biol. Biochem.* **5**, 163–166.

Klepper, L. (1979). Nitric oxide (NO) and nitrogen dioxide (NO_2) emissions from herbicide-treated soybean plants. *Atmos. Environ.* **13**, 537–542.

Klepper, L. (1990). Comparison between NO_x evolution mechanisms of wild-type and nr_1 mutant soybean leaves. *Plant Physiology* **93**, 26–32.

Kramm, G. (1989). A numerical method for determining the dry deposition of atmospheric trace gases. *Boundary-layer Meteorol.* **48**, 157–176.

Kramm, G. (1991a). Modellierung des Vertikaltransports von Aerosolpartikeln in der bodennahen Luftschicht. *Meteorol. Rdsch.* **43**, 65–80.

Kramm, G., Neier, N., Foken, T., Müller, H., Schröder, P., Seiler, W. (1995b). A soil/vegetation/atmosphere transfer scheme for determining the dry deposition fluxes of NO, NO_2 and O_3 - part I: Model description and test results, *Meteorol. Atmos. Phys.* submitted.

Kramm, G., Dlugi, R., Dollard, G.J., Foken, T., Mölders, N., Möller, H., Seiler, W., Sievering, H. (1995a). On the dry deposition of ozone and reactive nitrogen species. *Atmos. Environ.* (in press).

Kramm, G., Dlugi, R. (1994). Modelling of the vertical fluxes of nitric acid, ammonia and ammonium nitrate in the atmospheric surface layer. *J. Atmos. Chem.* **18**, 319–357.

Kramm, G., Beheng, K.-D., Müller, H. (1992). Vertical transport of polydispersed aerosol particles in the atmospheric surface layer. In: S.E. Schwartz and W.G.N. Slinn (eds) *Precipitation Scavenging and Atmosphere-Surface Exchange Processes.* Vol. 2 - The Semonin Vol. Hemisphere Publ., New York, 1125–1141.

Kramm, G., Müller, H., Fowler, D., Höfken, K.D., Meixner, F.X. and Schaller, E. (1991). A modified profile method for determining the vertical fluxes of NO, NO_2, O_3 and HNO_3 in the atmospheric surface layer. *J. Atmos. Chem.* **13**, 265–288.

Lenschow, D.H., Hicks, B.B. (Eds.) (1989). *Global tropospheric Chemistry - Chemical Fluxes in the Global Atmosphere.* National Center for Atmospheric Research, Boulder Co., USA.

Lenschow, D.H. (1982). Reactive trace species in the boundary layer from a micrometeorological perspective. *J. Meteor. Soc. Japan* **60**, 472–480.

Lenschow, D.H., Delany, A.C. (1987). An analytic formulation for NO and no flux profiles in the atmospheric Ltr. Surface layer. *J. Atmos. Chem.* **5**, 301–309.

Louis, J.-F. (1979). A parametric model of vertical eddy fluxes in the atmosphere. *Boundary-Layer Meteorol.* **17**, 187–202.

Ludwig, J., Weber, P., Miexner, F.X., Rennenberg, H., Surface fluxes of NO and NO_2 by a dynamic chamber technique - Laboratory studies on wheat. in: Angeletti, G., Beilke, S., Slanina, S. (eds) *Air Pollution Research Report* **39**, CEC, Brussels 1992, pp. 257–265.

Ludwig, J. (1994). Untersuchungen zum Austausch von Stickoxiden zwischen Biosphäre und Atmosphäre. Ph. D. Thesis. University of Bayreuth.

Ludwig, J., Meixner, F.X. (1994). Surface exchange of nitric oxide (NO) over three European ecosystems. Proc. 6th European Symp. on Physico-Chemical Behaviour of Atmospheric Pollutants. October 18–22, Varese, Italy. Brussels (B): CEC, Directorate General for Science, Research and Development.

Meixner, F.X. (1994), Surface exchange of odd nitrogen oxides. *Nova Acta Leopoldina* **70 Nr 288**, 299–348.

Monin, A.S., Obukhov, A.M. (1954). Osnovnye zakonomernosti turbulentnoho peremesivanija v prizemnom slow atmosfery. *Trudy geofiz. Inst. AN SSSR.* **24** (151), 163–187 (in Russian).

Müller, H., Kramm, G., Meixner, F., Dollard, G.J., Fowler, D., Possanzini, M. (1993). Determination of HNO_3 dry deposition by modified Bowen ration and aerodynamic profile techniques. *Tellus* **45B**, 346–367.

Nestlen, M., Müller, H., Meixner, F. (1993). Measurements of vertical fluxes of CO_2, H_2O, NO, NO_2 and O_3 by eddy-correlation: First results from the BIATEX/SANA experiment 'Eisdorf', March 1991. in: P.M. Borrell, P. Borrell, T. Cvitaš, W. Seiler (eds), *Proc. EUROTRAC Symp. '92*, SPB Academic Publishing bv, The Hague 1993, pp. 727–731.

Papen ,H., Hellmann, H., Papke, H., Rennenberg, H. (1993). Emissions of N-oxides from acid irrigated and limed soils of a coniferous forest in Bavaria. In:, Chapman, New York, p.245

Raabe, A. (1991). Die Höhe der internen Grenzschicht. *Z. Meteorol.* **41**, 251–261.

Rao, K.S., Wyngaard, J.C., Coté, O.R. (1974). The structure of a two-dimensional internal boundary layer over a sudden change of surface roughness. *J. Atmos. Sci.* **31**, 738–746.

Remde, A., Ludwig, J., Meixner, F.X., Conrad, R. (1993). A study to explain the emission of nitric oxide from marsh soil. *J. Atmos. Chem.* **17**, 249–275.

Rondón, A. (1993). Photo-induced deposition of ozone on the plant leaf cuticle. *Science,* submitted.

Rondón, A., Johansson, C., Granat, L. (1993). Dry deposition of nitrogen dioxide and ozone to coniferous forests. *J. Geophys. Res.* **98,** 5159–5172.

Rondón, A., Granat, L. (1994). Studies on the dry deposition of NO_2 to coniferous species at low NO_2 concentrations. *Tellus* **46B**, 339–352.

Sehmel, G.A. (1980). Particle and gas dry deposition: A review. *Atmos. Environ.* **14**, 983–1011.

Simpson, D. (1993). Biogenic VOC emissions in Europe: modelling the implications for ozone control strategies. AWMA Conference, San Diego, California.

Skärby, L., Bengtson, C., Boström, C-A., Grennfelt, P., Troeng, E. (1981). Uptake of NO_x in Scots pine. *Silvae Fennica* **15**, 396–398.

Skiba, U., Hargreaves, K.J., Smith, K.A., Fowler, D. (1992). Fluxes of Nitric and Nitrous Oxides from Agricultural Soils in a Coll Temperate Climate, *Atmos. Environ.* **26A**, 2477–2488.

Skiba ,U., Fowler, D., Smith, K. 1994. Emissions of NO and N_2O from soils. *Environ. Mon. Assess.* **31**, 153–158.

Slemr, F., Seiler, W. (1984). Field measurements of NO and NO_2 emissions from fertilized and unfertilized soils. *J. Atmos. Chem.* **2**, 1–24.

Smith, K.A. *et al.* (1994). Micro meteorological and chamber methods for measurement of nitrous oxide fluxes between soils and the atmosphere: Overview and conclusion *J. Geophys. Res.*, **99**, 16541

Thoene, B., Schröder, P., Papen, H., Egger, A., Rennenberg, H. (1991). Absorption of atmospheric NO_2 by spruce (*Picea abies* L. Karts.) Trees. 1. NO_2 influx and its correlation with nitrate reduction. *New Phytol.* **117**, 575–585.

Vos, G.J.M., Bergevoet, I.M.J., Vedy, J.C., Neyroud, J.A. (1994). The fate of spring applied fertilizer N during the autumn-winter period: Comparison between winter-fallow and green manure cropped soil. *Plant and Soil*, **160**, 201–214.

Wesely, M.L., Hicks, B.B. (1977). Some factors that affect the deposition rates of sulfur dioxide and similar gases on vegetation. *J. Air Poll. Control Assoc.* **27,** 1110–1116.

Wesely, M.L. (1989). Parameterization of surface resistances to gaseous dry deposition in regional-scale numerical models. *Atmos. Environ.* 23, 1293–1304.

Williams, E.J., Hutchinson, G.L., Fehsenfeld, F.C. (1992). NO_x and N_2O emissions from soil. *Global Biogeochem. Cycles* **6**, 351–388.

Williams, E.J., Guenther, A., Fehsenfeld, F.C. (1991). An inventory of nitric oxide emission from soils in the United States. *J. Geophys. Res.* **97** D7, 7511–7519.

Woodward, F.I., Sheehy, J.E. (1983). *Principles and Measurements in Environment Biology*. Butterworths, London.

Chapter 7

Exchange of Sulfur Gases between the Biosphere and the Atmosphere

J. Kesselmeier[1], P. Schröder[2] and J.W. Erisman[3]

[1]Max Planck Institute for Chemistry, Biogeochemistry Department, P.O. Box 3060, D-55020 Mainz, Germany
[2]GSF-Forschungszentrum für Umwelt und Gesundheit GmbH, Institut für Biochemische Pflanzenpathologie, D-85758 Oberschleißheim, Germany
[3]National Institute of Public Health and the Environment (RIVM), Laboratory for Air Research (LLO), P.O. Box 1, NL-3720 BA Bilthoven, the Netherlands

Sulfur is one of the essential elements. It plays a dominant role in atmospheric chemistry as well as in biological processes. Recent estimates of total natural sulfur emissions into the atmosphere, including the contribution of both plants and soils (4–15 Tg (S) a^{-1} range around 65 (± 25) Tg S a^{-1} (Andreae and Jaeschke, 1992), which is equivalent to estimates of man-made sulfur emissions of 93 ± 15 Tg S a^{-1} (Cullis and Hirschler, 1980). Organisms use sulfur for the synthesis of a large number of substances, for example amino acids and proteins. In addition resistance, defence and detoxification mechanisms including sulfur containing compounds are discussed (see Ernst 1990; Haines *et al.* 1989; Rennenberg 1991). Thus sulfur compounds are found in primary as well as secondary metabolism. Volatile sulfur compounds are exchanged between biosphere, atmosphere and hydrosphere. The original assumption that microbial activities were the main source for biogenic H_2S in terrestrial and aquatic ecosystems (Hill 1973, Granat *et al.* 1976) changed following reports on biogenic emissions from vegetation (see reviews: Aneja and Cooper, 1989; Rennenberg, 1991). Today it is well accepted that higher plants are involved in the atmospheric sulfur cycle. Both, uptake and emission of volatile reduced sulfur compounds by plants have been reported (Kluczewski *et al.*, 1985; Taylor *et al.*, 1983, De Kok *et al.*, 1989, 1991; Brown *et al.*, 1986; Goldan *et al.*, 1988, Fall *et al.*, 1988; Rennenberg *et al.*, 1990, Rennenberg 1991; Kesselmeier 1991, Schröder 1993).

One must discern between volatile reduced and oxidised sulfur compounds such as mainly anthropogenically produced sulfur dioxide (SO_2) and reduced sulfur gases. Due to their different atmospheric residence times (Warneck 1988), two groups of reduced sulfur compounds exist: (i) the more labile gases such as hydrogen sulfide (H_2S), methylmercaptan (CH_3SH), dimethyl sulfide (CH_3SCH_3, DMS) and CS_2, and (ii) the most stabile compound COS with a residence time of one year and above. With the exception of CS_2, which is partly oxidised to COS (by 50 %), the reduced sulfur gases are removed from the atmosphere by photolysis (Crutzen 1976) or after reactions with hydroxyl radicals (Ravishankara *et al.* 1980; Adewuyi 1989). Both groups of reduced S compounds affect climate.

The reduced, naturally emitted compounds H_2S, CH_2SH, DMS and partly CS_2 are oxidised to give SO_2/SO_4^{2-} (Fig. 7.1) thus, principally have the same effect on aerosol and cloud production as is known for SO_2 emission from combustion processes and are therefore relevant to the earths's energy budget. SO_2 and SO_4^{2-} are scavenged from the atmosphere by dry and wet deposition, which are addressed in one of the following chapters. COS, the most abundant sulfur gas in the atmosphere is relatively inert in the troposphere and is transported into the stratosphere where it's photodissociation and photo-oxidation are believed to be the major source of stratospheric sulfur aerosol, the "Junge layer" (Crutzen 1976; Turco *et al.* 1980; Hofmann 1990). Therefore, the quantification of COS sources and sinks is of considerable importance. Although COS may be a sulfur gas emitted from vegetation (Lamb *et al.*, 1987, Guenther *et al.*, 1989), other studies show that it is taken up by vegetation (Taylor *et al.*, 1983, Kluczewski *et al.*, 1985, Brown *et al.*, 1986, Brown and Bell, 1986, Fall *et al.*, 1988, Goldan *et al.*, 1988). Several field studies support the deposition of COS to plants (Goldan *et al.*, 1987, Mihalopoulos *et al.*, 1989, Hofmann *et al.* 1992a) but another found no evidence of COS uptake (Berresheim and Vulcan, 1992). The main problems in interpreting experimental studies are the gaps in our knowledge on the mechanisms of production and consumption of the sulfur gases by vegetation. The use of purified air as an artificial atmospheric environment for the investigation of sulfur fluxes may result in emission instead of deposition (Kesselmeier *et al.* 1993b). Such data may partly be responsible for the still considerable uncertainties reported for the exchange of sulfur compounds (Andreae and Jaeschke 1992; Chin and Davis 1993). Additionally, algae, mosses, fungi and symbionts like lichens may play significant roles in some environments. The role of oceanic phytoplankton is quite well understood (see Andreae and Jaeschke, 1992) however lower plants and terrestrial microorganisms or symbionts are poorly investigated.

In addition to its aerosol and cloud generating potential, sulfur dioxide was one of the first air pollutants recognised as harmful to humans and ecosystems. Therefore, it is one of the most studied pollutants, particularly in the early seventies. In the eighties, interest in Europe moved from sulfur compounds to nitrogen compounds because abatement strategies for sulfur emissions during the last decade have been successful in western Europe, Canada and the USA. In addition nitrogen compounds are now recognised as being harmful. SO_2 emission reductions on the order of 50 % have been realised. On regional scales both concentration and deposition of sulfur compounds decreased to the same extent. Despite the large research effort put into the study of dry deposition of SO_2, no unequivocal process description is available. Factors influencing deposition are numerous and result in

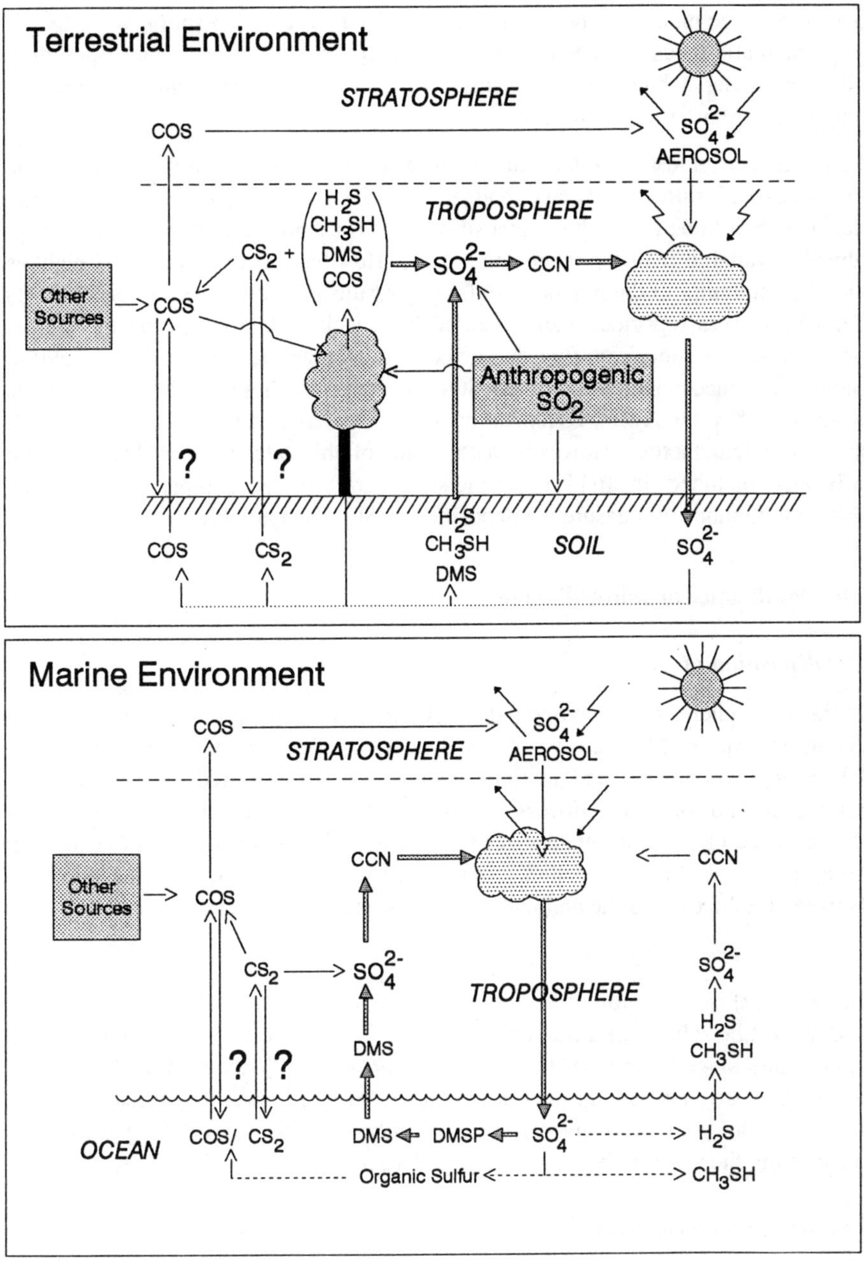

Fig. 7.1: Sulfur cycles over the terrestrial and marine environment.

part from large natural variation in surface resistances and deposition velocities. It is problematic to use earlier research results for generalisation because pollution climates under which measurements were made (surface wetness, humidity, ammonia) are often not available.

In addition, we must consider reduced sulfur compounds as well as sulfur dioxide to understand sulfur loads and budgets. At the beginning of BIATEX it was decided to exclude research on deposition of SO_2 because more data for SO_2 were already available than for other compounds. However, at the end of the eighties and the beginning of the nineties, conflicting results regarding the influence of SO_2 deposition were reported, mainly caused by a lack of knowledge on the factors determining the deposition flux. In view of this problem, it was decided in 1990 to incorporate mechanistic studies on SO_2 deposition. These studies included the influence of water layers, codeposition with other compounds such as ammonia, and biological factors. However, as a result of this late decision SO_2 was not officially included in field experiments. Therefore, no comparisons between different methods to measure the deposition of SO_2 are available.

The significance of sulfur dioxide

Dry deposition of SO_2

In the atmosphere, sulfur compounds are ultimately removed through dry, wet and occult deposition. Mechanisms of dry deposition of SO_2 have been examined in laboratory experiments and in field studies. Once transported to vegetation by turbulence and molecular diffusion, SO_2 is absorbed readily through stomata. In the presence of moisture uptake of SO_2 by external leaf surfaces and the soil can be a major sink. Most field studies report deposition velocities (V_d), *i.e.* the ratio between the flux F and the ambient concentration c(z):

$$V_d(z) = F/c(z) \qquad (1)$$

Because c(z) is a function of height z above the ground, V_d is also a function of z, and should be related to a reference height. F is assumed to be constant over the appropriate range of heights. The use of the concept of V_d is valid for SO_2 because its flux is nearly always uni-directional (downwards) because the surface is a sink for SO_2. The goal of many early studies was to support modelling efforts by quantifying the magnitude and variability of V_d over a variety of surfaces.

Dry deposition measurements

SO_2 dry deposition measurements have been made by micrometeorological, surface analysis and chamber methods (for a review see Erisman *et al.*, 1994b). Micrometeorological methods include the eddy correlation, gradient, variance, conditional sampling and eddy accumulation methods. The first two are most used. These methods are mainly used to quantify the short-term flux and to determine the regulating processes frequently related to meteorology, surface conditions, and vegetation behaviour).

There are several important limitations to all of these methods (see Businger, 1986; Baldocchi *et al.*, 1988; Fowler and Duyzer, 1990). Surface analysis methods include analysis of snow, leaf washing techniques and the throughfall technique. The throughfall technique measures the total sulfate flux beneath the canopy. Several assumptions have to be made to estimate SO_2 dry deposition from these measurements though they are well suited for deposition measurements where micrometeorological methods cannot be applied (complex terrain, forest edges). A comparison between results of throughfall measurements and other methods is presented elsewhere (Erisman *et al.*, 1994b; Erisman and Draaijers, 1995).

Chamber methods in which gas uptake by the depositing surface is measured are used in the laboratory or in the field. In open-top chambers or in closed chambers the characteristics of the vegetation is studied by controlling the environment. Factors thought to affect the plants are carefully controlled and measured. The response of different species of plants or different plant components can be studied in semi-isolation. The deposition to the surfaces (soil or vegetation) in the chambers can be estimated by comparing the in- and outlet of the chamber over time (*e.g.* Milne *et al.*, 1979; Taylor and Tingey, 1983; Taylor *et al.*, 1983; Granat and Johansson, 1983). In addition, cuvette and mini cuvette studies with branches or single leaves, provide information about fundamental physical, chemical and biological processes related to trace gas exchange. However, there are several important limitations to these methods (see Murphy and Sigmon, 1989).

Deposition velocity and resistance to transports:transport

In the literature V_d ranges are reported for areas of 0.1 to 2 cm s^{-1} (conifer forests [1]; deciduous forests [2]; heathland [3]; agricultural [4] grasslands [5]; bodies of water [6] and bare soil [7]). Daytime values over a variety of plant surfaces are in the order of 0.8 to 1.2 cm s^{-1} [8]. Higher values (exceeding 2 cm s^{-1}) are observed over water surfaces [9] and forests [10] and smaller values (< 0.13 cm s^{-1}) are measured over snow [11] and bare soil [12].

[1] Galbally (1979); Garland (1977); Hicks *et al.* (1982); Fowler and Cape (1984); Lorenz and Murphy Jr. (1985); McMillen *et al.* (1987); Erisman *et al.* (1993b);

[2] Petit *et al.* (1976); Matt *et al.* (1987); Meyers and Baldocchi (1988);

[3] Erisman *et al.* (1994c);

[4] Fowler (1978); Fowler and Unsworth (1979); Hicks *et al.* (1989a,b);

[5] Garland (1977); Platt (1978); Van Dop *et al.* (1983); Hicks *et al.* (1983); Neumann and den Hartog (1985); Duyzer and Bosveld (1988); Davies and Mitchell (1983); Davies and Wright (1985); Erisman *et al.*(1993a);

[6] Welphdale and Shaw (1974); Garland (1977);

[7] Garland (1977);

[8] Garland (1977); Wesely and Hicks (1977); Nicholson and Davies (1988); Erisman *et al.* (1993a; 1994c);

[9] Whelpdale and Shaw (1974);

[10] Vermetten *et al.* (1992); Erisman *et al.* (1993b); Erisman and Draaijers (1995);

[11] Barrie and Warmsley (1978), Welphdale and Shaw (1974), Cadle *et al.* (1985), Granat and Johansson (1983), Valdez *et al.* (1987); Davidson and Wu (1990);

[12] Payrissat and Beilke (1974); Garland (1977)

V_d depends on many variable parameters, such as wind speed, temperature, radiation, etc. and surface conditions. The transport of pollutants from the atmosphere to the ground comprises three stages: first, the materials must be transported through the atmosphere to the receptor surface (turbulent layer); second, there is a transport through the quasi-laminar layer at the surface; last, the gas (or particle) must be captured by the surface. To describe the surface exchange processes, a general framework in close analogy to resistances is provided by separating three stages of dry deposition (Thom, 1975; Garland, 1977; Wesely and Hicks, 1977; Fowler, 1978): the aerodynamic resistance R_a, the quasi-laminar resistances: resistance R_b, and the canopy or surface resistance R_c. The inverse sum of the resistances gives the deposition velocity.

$$V_d(z) = 1/[R_a(z) + R_b + R_c] \qquad (2)$$

The aerodynamic resistance determines the rate of transport of gases between a given level in the atmosphere and the height of the effective surface sink. Knowing the roughness length, R_a can be estimated from wind speed, temperature and radiation data. R_b is associated with the transfer through the quasi-laminar layer in contact with the surface; it quantifies the way in which pollutant transfer differs from momentum transfer in the immediate vicinity of the surface. R_b can be estimated from the friction velocity u_* and from component specific constants. R_c values found in the literature are usually based on measurements of V_d, and are commonly determined as a residual resistance. By determining R_a and R_b from the meteorological parameters, R_c can be calculated as the residual resistance using equation 2. R_c has been shown to be very variable. Its value is often quite uncertain since errors in the measurements are accentuated in this term.However, in ideal conditions values of R_c can be related to surface conditions, time of day, season, *etc.*, yielding parameterisations.

The resistance analogy has been formalised in a 1-D model for estimating dry deposition rates (Hicks *et al.*, 1987). In this 'big-leaf' model, the most important processes that are known to control dry deposition are parameterised. The flux is estimated from the concentration, wind speed, temperature, radiation, roughness length and R_c. R_c values can be estimated from theoretical considerations based on solubility and equilibrium calculations in combination with simulations of vegetation specific processes, such as accumulation, transfer processes through stomata, mesophyll and cuticles, absorption, *etc.* R_c values thus obtained are, however, hard to validate by measurements because of the complexity of the processes involved.

Surface resistance

Numerous field studies show that the surface resistance R_c is the major controller of SO_2 deposition. R_c is the most difficult to describe of the three resistances. R_c is a function of the canopy stomatal resistance R_{stom}, the canopy cuticle resistance $R_{cuticle}$, and the soil resistance R_{soil}. In turn, these resistances are affected by leaf area, stomatal physiology, soil pH, and the presence and chemistry of liquid drops and films. The stomatal, leaf surface (cuticle) and soil resistances act in parallel:

$$R_c = 1/[R_{stom} + 1/R_{soil} + 1/R_{cuticle}] \quad (3)$$

Physiological control

Most SO_2 enters plants through the stomata though SO_2 may also penetrate the cuticle with a similar permeability coefficient as O_2 (Lendzian, 1987). As gas molecules enter the leaf, deposition occurs as molecules react with the moist cells in the sub-stomatal chamber and the mesophyll. Several conclusions can be drawn about environmental effects and inter specific differences on SO_2 deposition based on known stomatal physiology. Thus, factors that increase stomatal resistance should increase the SO_2 $R_{stom.}$. Stomatal resistance decreases hyperbolically with increasing light and increases linearly with increasing vapour pressure deficits (Jarvis, 1976). Soil water deficits cause stomata closure after some threshold deficit level is exceeded. Low and high temperatures cause stomatal closure and moderate temperatures promotes stomatal opening. Leaf age, nutrition and adaptation are other factors affecting stomatal resistance. Moderate SO_2 mixing ratios (15 ppb) cause stomatal opening whereas elevated exposure (450 ppb) to SO_2 causes stomata to close (Larcher, 1994, p. 328). Stomatal resistance is different for different types of vegetation. Values of R_{stom} under ideal conditions during daytime, range between 30 and 300 s m^{-1} for a range of herbaceous annuals and woody perennials (Fowler, 1995; Baldocchi *et al.*, 1987; Matt *et al.*, 1987).

The canopy cuticle resistance exceeds the canopy stomatal resistance by far; $R_{cuticle}$ ranges between 3000 and 40000 s m^{-1} (van Hove, 1989). However, it is observed that the surface resistance decreases as relative humidity increases (van Hove, 1989; Garsed, 1985; Erisman *et al.*, 1993a; Erisman and Wyers, 1993; Draaijers and Erisman, 1993). This might indicate that the cuticle resistance is lowered at increasing humidity (van Hove, 1989) or that the surface becomes moist at increasing humidity, *e.g.* through deliquesce of salts at the external leave surface. When the cuticle is wetted or affected by pre-deposited particles or gases, SO_2 deposition to the surface also occurs via absorption and via chemical oxidation reactions with the surface.

Physical-chemical control

SO_2 dry deposition is enhanced over wet surfaces (Garland and Branson, 1977; Fowler and Unsworth, 1979; Fowler, 1985; Vermetten *et al.*, 1992; Erisman *et al.*,

1993a; Erisman and Wyers, 1993); the absorption is associated with a series of reactions (Seinfeld, 1986):

$$SO_2(g) + H_2O \rightarrow SO_2{*}H_2O \quad (4a)$$

$$SO_2{*}H_2O \;<\!\!-\!\!>\; H^+ + HSO_3^- \quad (4b)$$

$$HSO_3^- \;<\!\!-\!\!>\; H^+ + SO_3^{2-} \quad (4c)$$

The total dissolved S(IV) in chemically clean solutions is related to the partial pressure of SO_2 over the solution (pSO_2) and the pH of the solution:

$$[S(IV)] = K_h \times pH_2O\ (1 + K_1/[H^+] + K_1 \times K_2/[H^+]2) \quad (5)$$

K_h is Henry's coefficient for SO_2 (5.4 mol L^{-1} atm^{-1} at 15 °C) and K_1 and K_2 are the rate coefficients for reactions 1 and 2, respectively. From Eq. 5 it can be concluded that an increase in acidity will reduce the amount of dissolved SO_2. The oxidation of S(IV) to S(VI) by O_2, O_3 and H_2O_2 provides another source of protons. The oxidation of aqueous SO_2 by O_3 and H_2O_2 is fast, oxidation by O_2 is slow unless there is a metal catalyst (iron or manganese). An increase in acidity due to solution and oxidation of SO_2 in water films will reduce SO_2 dry deposition to wetted surfaces. The presence of acid neutralising species either from the atmosphere (base cations, NH_3) or from the inner parts of the leaves, will reduce the acidity of the water film, abolishing the SO_2 dry deposition reduction. At present too little information is available to quantify these processes in terms of the influence on the magnitude of the surface resistance.

The pH of a wetted surface is the deposition rate limiting factor. Thus, the origin of surface wetness is important: by fog, cloud, rain or dew, guttation (Fowler and Unsworth, 1979). The pH of rain, cloud and also fog is less than that of pure water. The chemistry of wetted surfaces is affected by uptake of acidic precursors and oxidants, particles on the leaf surface, chemical lifetime, droplet size, leaching of nutrients from leaf and evaporation rate of wetness (Brimblecombe, 1978; Chameides, 1987; Wesely *et al.*, 1990)

The deposition of NH_3 neutralises an acid solution. Thus simultaneous deposition of SO_2 and NH_3 should experience higher SO_2 (and NH_3) dry deposition rates to water surfaces than would be found by SO_2 (or NH_3) alone (Wesely *et al.*, 1990; Erisman and Wyers, 1993; Erisman *et al.*, 1994c). Van Hove (1989), Fowler *et al.* (1991) and Erisman and Wyers (1993) present evidence, showing that the surface SO_2 uptake at wet surfaces is enhanced in the presence of NH_3. This process is generally referred to as co-deposition. One site of reaction is in the water-filled pores of the cuticle at relative humidities above about 70 % and the other site is on the wetted leaf surface.

Deposition to soil and litter

Deposition to canopies involves deposition to both vegetation and soil. Early studies assumed that deposition to soils under vegetation was relatively small

(5–10 % of the total flux; Fowler, 1978). Recent work shows that a substantial amount of material can be deposited to the soil below vegetation. Meyers and Baldocchi (1988) report 20 to 30 % of SO_2 deposited within a deciduous forest occurs at the forest floor, because large scale intermittent eddies are able to penetrate through the vegetation to the soil. Differentiation of deposition to the canopy and to the soil is difficult to measure. Furthermore, it may be variable both in time and for different canopies.

Deposition to soil decreases at soil pH below 4 and increases with relative humidity (Garland, 1977; Payrissat and Beilke, 1974). When surface temperatures fall below zero or the surface is covered with snow, Rc values increase up to values of 200–500 s m^{-1} (Voldner *et al.*, 1986; Erisman and Wyers, 1993). The deposition of SO_2 to snow covered surfaces depends on pH, snow temperature and probably the amount of SO_2 already scavenged by the snow pack (Caddle *et al.*, 1985; Hicks *et al.*, 1989a,b).

Significance of reduced volatile sulfur compounds

Techniques developed and used within BIATEX

Sampling and analysis of sulfur gases

Cryogenic sampling is a reliable technique for the quantification of reduced sulfur compounds in air samples. However, cryogenic pre-concentration of trace gases leads to a co-trapping of water and thus ice may limit sample volumes as it blocks the sample flow in the trap. Additionally, reactive gases may be dissolved in water and water could be injected onto columns.

Within the BIATEX work, the Max Planck Institute Mainz (MPI) combined the cryogenic trapping of reduced sulfur compounds with a commercially available Nafion drier (Perma-Pure) and tested the technique under laboratory and field conditions (Hofmann *et al.* 1992 b). It was suitable for measurements of reduced sulfur compounds such as H_2S, COS, CH_3SH, CH_3SCH_3 (DMS), and CS_2 in dry and humid air in the ppt range. For the scavenging of oxidants this Nafion drier technique was combined with a cotton oxidant scrubber. Results obtained under laboratory conditions with humidified samples and in the course of a small field experiment confirmed the oxidant scrubber qualities of cotton. In fact the cotton oxidant scrubber in combination with a Nafion drier is suitable for measurements of nearly all reduced sulfur compounds except CH_3SH. The results clearly showed that labile compounds such as H_2S and especially DMS may be protected from being destroyed by oxidants during sampling and transfer. COS and CS_2 are hardly affected.

The Fraunhofer Institute in Garmisch-Partenkirchen (IFU) used a stepwise desorption at different temperatures to desorb sulfur compounds from cryogenically trapped air samples without the use of Nafion driers. Desorption was performed in a three step procedure. In a first step, O_2 was allowed to escape from the sample loop by rinsing it with carrier gas N_2 at –186 °C. In a second step, CO_2, H_2S, COS, SO_2 and CH_3SH were expelled from the sample unit by immersing it into a low temperature bath consisting of EtOH and dry ice at –78 °C. The sample was focused in a second but smaller focusing loop connected to the carrier gas stream and immersed in liquid argon. Water vapour, which would lead to significant losses of H_2S and SO_2 was retained in the sample loops. The sample was swept onto the column by heating the focusing loop to 20 °C and analyzed with the sample loop still in the low temperature bath. After termination of the analysis the focusing loop was cooled to –186 °C and the sample loop was removed from the bath and allowed to reach ambient temperature. CS_2 and DMS were now vaporised and transferred to the focusing loop with the carrier gas. After focusing the sample, it was injected onto the column by heating the focusing loop again. Water vapour that had entered the column during this second analysis was removed from the system by evacuating the tubings and heating the column. The results of both analyses were tabulated and yielded the sulfur gas content of the air sample.

Calibration of analysers and intercomparison

For the comparison of data obtained by different groups it is necessary to intercompare the sampling and analysis techniques. Therefore, IFU and MPI intercompared their techniques described above. For the calibration of the GC standardized sample loops were used instead of syringes. The loops were loaded

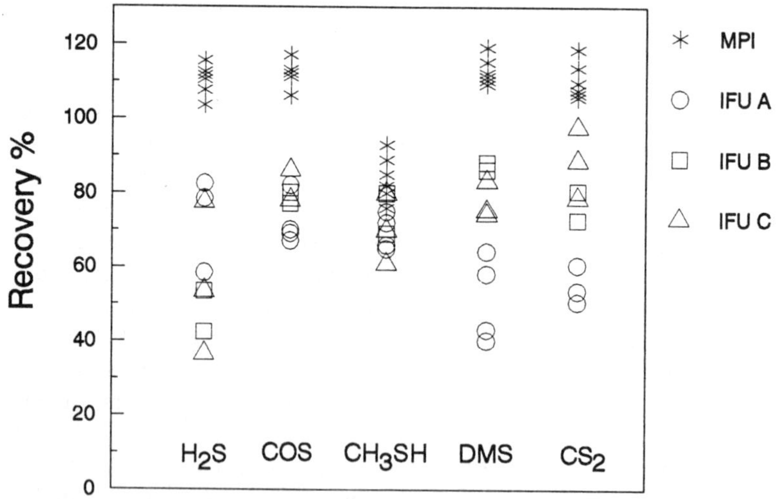

Fig. 7.2: Recovery of reduced sulfur compounds during an intercomparison exercise of the Max Planck Institute for Chemistry (Mainz) and the Fraunhofer Institute for Atmospheric Research (Garmisch-Partenkirchen). MPI: Cryogenic trapping using a Nafion drier and direct transfer to the GC (see text). IFU: A: two step desorption and two step analysis; IFU: B: two steps desorption one step analysis; IFU C: same as IFU B but with an additional Nafion drier during sampling.

with the respective sulfur gases from a device containing permeation tubes. After loading, the loops were sealed and connected to the inlet valve of the GC. Injection was via the valve. By this procedure calibrations could be repeatedly performed with the high amount of accuracy which is necessary when handling reactive gases. Within an intercomparison between the MPI Mainz and the IFU Garmisch the Nafion technique could be compared with a "stepwise adsorption/desorption" technique (IFU). The results are presented in Fig. 7.2. The data obtained within the intercomparison showed a quite homogenous recovery of reduced sulfur compounds by the MPI procedure, however a slight overestimation in the range of 10 to 20 %. The IFU data show some scattering and an underestimation between 10 and 50 %, depending on the final modification of sampling, desorption and analysis. Methylmercaptan is underestimated by both techniques.

Studies on natural emissions of reduced sulfur compounds

In addition to the knowledge on oceanic emissions, one of the most important discoveries of recent years was the observation that plants are emitting sulfur compounds throughout the day even without excess sulfur taken up from the environment (Goldan *et al.* 1987, Lamb *et al.* 1987, Fall *et al.* 1988, Rennenberg *et al.* 1990). The emission patterns are modulated on the one hand by plant age and developmental stage, on the other hand by environmental factors, *e.g.* light intensity, temperature, humidity and photosynthetic CO_2 fixation. These emissions of reduced sulfur compounds are presently explained as regulatory steps in order to balance the sulfur pools in the plant. This maintainance of constant sulfur pools is important because the plant sulfur status and requirements are subject to changes during ontogenic development. Whereas the proteins of vegetative parts of the plants are generally found to be sulfur-rich, a significant shift to sulfur poor amino acids occurs in the fruits during ripening (see also Rennenberg 1991).

Emission budgets in coastal areas

Because high emissions of DMS from oceans are significant for the sulfur budget (for review see Andreae and Jaeschke, 1992) comparing natural emissions with those of anthropogenic origin is of interest. Within BIATEX, Arends *et al.* (1995) report studies within the close vicinity of the North Sea coast and estimate the significance for the local budget. Within their studies they found a contribution of DMS and DMS reaction products to the Dutch sulfur concentration in air to be about 3 % in summer and 1–2 % over the whole year. They concluded that the local Wadden Sea was not an important source for DMS. However, a day with north winds showed high DMS concentrations, indicating that the northern part of the North Sea is a source region. Similar observations with very high concentrations of DMS have been reported for the northeast of Scotland by Davison and Hewitt (1992). However, from summer average values it was concluded that the contribution of biogenic and reduced compounds to the sulfur deposition in the Netherlands is less than 4 %. This value fits well into the frame calculated by Tarrason *et al.* (1994). Based on their model, they showed that the biogenic contribution to total sulfur depostion in Europe ranges between 0 and 5 %, depending on whether a coastal area with higher values or a continental area with a lower biogenic contribution is considered.

Exchange studies with crop plants

The large uncertainty for biogenic emissions is due to the lack of "reliable" data, that results from the technical difficulty in measuring atmospheric sulfur and from insufficient information on vegetation type and developmental status. Most crucial is the lack of knowledge on diurnal, seasonal, and annual variations. Therefore, within BIATEX a large field experiment was conducted measuring the emission and deposition of the reduced sulfur compounds COS, CS_2, H_2S, and DMS over a wheat field at Manndorf in the Danube valley during the three vegetation periods: germination (April 1990), growth (May 1990) and seed ripening (grainfilling, June 1990). All field experiments included the measurements of photosynthetically active radiation, relative humidity and temperature. The cooperation between the MPI-Mainz and the IFU-Garmisch allowed a comparison of results obtained by the gradient-technique with those obtained by cuvette measurements. Exchange measurements of H_2S, CS_2 and DMS made by the gradient method were reported by Hofmann *et al.* (1992a) and Hofmann (1993). Measurements performed with the dynamic gas exchange cuvettes are summarized by Huber (1994).

The gradient technique revealed that H_2S gradients were mainly positive (deposition) in spring whereas CS_2 gradients pointed to an emission from the soil/plant system to the atmosphere. The cuvette technique showed opposite results: CS_2 was deposited to the plants during the day, whereas H_2S showed clear emission in correlation with photosynthesis and transpiration of the plants. No DMS fluxes were observed in the cuvettes. In May and June positive gradients for H_2S and CS_2 over the day (deposition) were observed with the gradient technique, whereas the H_2S fluxes measured by the cuvettes indicated an emission of the gas in correlation to the rate of photosynthesis and transpiration. During the night, the H_2S gradient increased in both systems and the CS_2 gradients decreased. The DMS pattern indicated soil emission during dawn and night with the gradient system, the enclosure technique showed additional clear emissions from the plants during midday. Daytime DMS data obtained with the gradient technique are not available as the oxidant scavenger was not used during these experiments and thus were affected by oxidants present under sunlight conditions. In summary, the wheat plant/soil system in May and June was responsible for a deposition of atmospheric CS_2 under daylight conditions. At the early stage of development in April H_2S was deposited to the soil plant system except during midday when emission occurred, whereas CS_2 was emitted by the soil/plant system and obviously driven by temperature. Exchange of COS between the wheat field and the atmosphere (gradient method) as reported by Hofmann *et al.*(1992a) and Hofmann (1993) showed a variation of COS concentration gradients and fluxes over the season in principal agreement with the cuvette studies done by IFU. Thus, the developmental stage of plants is of great importance for sulfur flux measurements. Furthermore, the results demonstrated that the deposition of COS to plants is closely coupled to the photosynthetic CO_2 assimilation. Field data obtained by flux measurements over a meadow (Bartell *et al.* 1993) also show a deposition of COS to the soil/plant system in close correlation to the CO_2 uptake and evapotranspiration.

Another important finding was the detection of a compensation point for the exchange of COS between vegetation and the atmosphere. As shown by Rennenberg *et al.* (1991), Schröder (1993) and Huber (1994) the compensation point changed with the developmental stage of the wheat plants. These findings are in close correlation with the results obtained by cuvette studies under laboratory

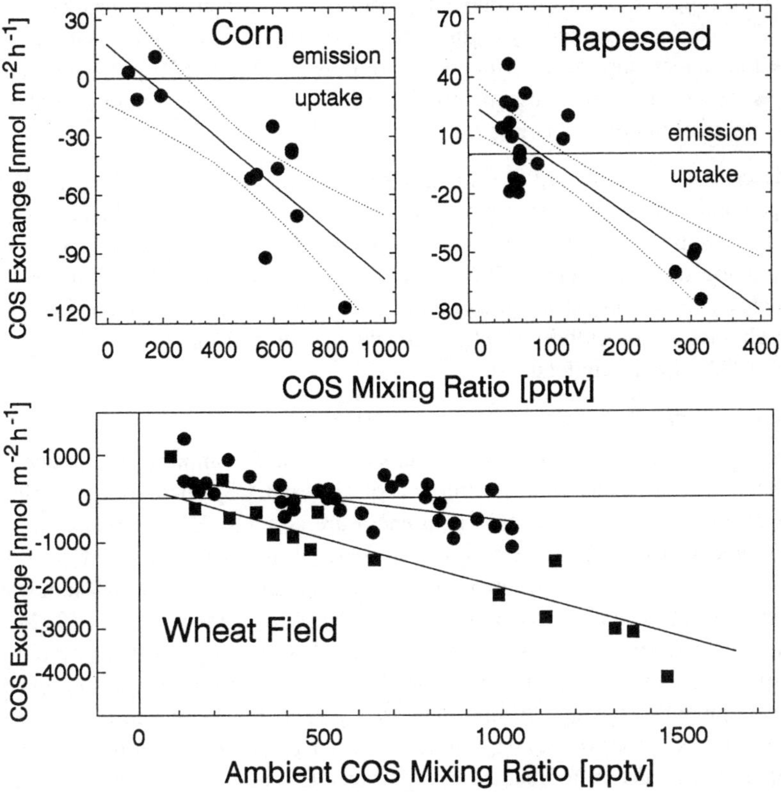

Fig. 7.3: Exchange of carbonyl sulfide (COS) between corn and rapeseed leaves when exposed to different COS mixing ratios in flushed ambient air. Solid lines show linear regression (r2 = 0.67 for corn; r2 = 0.66 for rapeseed); dotted lines express 95 % confidence level. Compensation points calculated were 89 (57–135) ppt and 144 (0–328) ppt for rapeseed and corn, respectively (Kesselmeier and Merk, 1993). Exchange of COS between wheat plants and the atmosphere during spring and summer measured by a cuvette enclosing wheat plants under field conditions and flushed with ambient air (Schröder, 1993).

conditions (Kesselmeier *et al.* 1992a; Kesselmeier and Merk 1993; Kesselmeier *et al.* 1993b; see Fig. 7.3 and below).

Compensation points for the exchange of a trace gas depend on numerous environmental factors like ambient concentration of the gas, light or temperature as well as development stage of the plants or the field. For young plants of corn and rape, COS compensation points were found around 144 ppt and 90 ppt, respectively (Kesselmeier *et al.* 1992a; Kesselmeier and Merk 1993). These values as well as a compensation point around 160 ppt found for wheat plants in springtime (Fig. 7.3; Rennenberg *et al.* 1991; Schröder 1993;Huber 1994) are consistent with the findings of Goldan *et al.* (1988) who found the uptake resistance of soybean increases when ambient COS concentrations are below 200 ppt. This suggests that COS uptake approaches zero at concentrations below 200 ppt. Though accompanied by significant errors, these COS compensation points show that under normal atmospheric conditions ([COS] > 550 ppt) COS should always be taken up by these plants. Furthermore, we may conclude that other environmental factors are negligible.

To describe and understand a compensation point (*i.e.* emission = deposition) two independent and counteracting processes need to be investigated. The pathways of COS consumption by plants are known to be principally the same as those for the assimilation of CO_2 (Protoschill-Krebs and Kesselmeier 1992; Kesselmeier 1992, see below). However, COS emissions under ambient air, as sometimes observed (Kesselmeier *et al.* 1992b; 1993a; Berresheim and Vulcan 1992; *c.f.* Chin and Davis 1993), are not understood.

Exchange studies with trees

During an experiment at the Schachtenau stand in the Bavarian forest, spruce trees exhibited deposition as well as emission of H_2S and CS_2 but only deposition of COS when fluxes were measured with enclosure systems (Huber 1994). Gradient measurements over the spruce forest in May at the same site did not show significant COS gradients, whereas very small gradients found for H_2S indicated a depostion of this sulfur gas to the forest area (Kesselmeier *et al.* 1992a). In parameterization experiments with young cloned spruce trees, it became obvious that spruce trees modulate their emission/deposition pattern of S gases in response to developmental status, seasonal conditions and ambient concentrations (Huber 1994). Highest emissions of H_2S were recorded prior to and during bud break. This peroid was also accompanied by significant CS_2 and COS emissions, whereas these gases were deposited during the rest of the year.

The importance of investigating plants at different developmental stages was also shown by a field expedition into a rain forest. Concentration measurements as well as emission rate measurements of reduced sulfur compounds were performed over and within an equatorial African rain forest in Cameroon (Kesselmeier *et al.* 1993a). Results obtained from a dirigible hot air balloon and a tethered balloon system in the early morning hours reflect a multilayered structure of the atmospheric boundary layer with gradients of COS and CS_2 indicating respectively an uptake or production of these trace gases by the soil and vegetation system. For additional studies of the emission and deposition rates of volatile reduced sulfur compounds from tropical tree species cuvettes were operated with young trees in a forest clearing near ground level as well as at a mature tree on top of the forest canopy, employing a specially designed tree platform ("Treetop raft III, dirigible version") placed on the canopy crown. The results show qualitative and quantitative disparities between different tree species as well as between individuals of the same species near ground level (young) and at the top of the canopy (mature). Here some correlations between photosynthetic CO_2 assimilation and emission of sulfur compounds were also found. Comparison between measurements at the ground and at the canopy top show that these tree species adapt their photosynthetic CO_2 assimilation in response to the microclimate and micrometeorology at the canopy top. This is accompanied by a quantitative change in trace gas emission especially by an extremely high emission of DMS. The increase of DMS emission at the canopy top may have resulted from a demand of nitrogen in the foliage as this would cause an accumulation of sulfur.

The dominant sulfur species emitted from all trees species investigated was DMS. H_2S emission was of minor importance. This conclusion may be applied to some mid-latitude trees (Kesselmeier *et al.* 1993c). Spruces, beech, ash and oak trees all showed a DMS emission and a COS deposition within cuvette experiments under laboratory conditions. Within these investigations, the oak species *Quercus petrea*, one of the typical European oak species howed a significant correlation between DMS emission and photosynthetic CO_2 assimilation (Fig. 7.4), as was found with tropical tree species (Kesselmeier *et al.* 1993a).

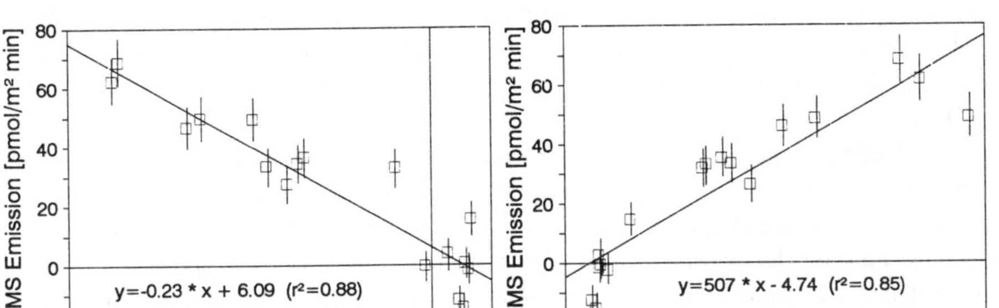

Fig. 7.4: Emission of DMS by an oak species (*Quercus petrea* L.) in close correlation to the CO_2 assimilation and stomatal conductance. Measurements were performed within a dynamic minicuvette flushed with ambient air as described for COS studies by Kesselmeier and Merk, 1993. Error bars give the total errors as obtained by parallel sampling form a permeation device and transferred to the calculation of the differences between the in- and outlet of the cuvette.

COS uptake/emission depended on the physiological status (assimilation or respiration) of the plant. Under active photosynthetic CO_2 assimilation an uptake of COS was always observed. In addition, H_2S emission was never found during these experiments with young trees if sampling started after an adaptation time of at least 1 hour, but was seen if sampling started too early or the plant was mechanicaly injured (even outside the cuvette). The minor significance of H_2S emissions are supported by experiments comparing young and old crop plants. As shown in the course of these latter studies (Hofmann 1994) H_2S emissions were mainly found with very young plants. Additionally, the question is open, how H_2S may escape from the assumed alkaline site of production, the chloroplast (for review see Rennenberg *et al.* 1991). These findings raise the question whether H_2S emission may at least be partly triggered by "stress" effects. However, this interpretation is in conflict with reports on the possible regulatory function of H_2S emissions in plants during early developmental stages. Additionally, H_2S emission occurs in response to excess sulfur in any form, including S-containing amino acids (Sekiya *et al.* 1982a,b). Furthermore, the production and release of H_2S has been shown to be light dependent and to follow a diurnal pattern (Rennenberg *et al.* 1990). In young spruce trees under low light, H_2S emissions proceed during the whole daylight period and is under stomatal control. However, the emissions are low and can be stimulated when branches are wounded or detached from the trees or exposed to SO_2 in moderate concentrations. These effects also raise the question of whether the dynamic chamber environment may represent a latent stress to the plants leading to H_2S emissions, even if "normal" stress responses are not observed. We think these questions have to be addressed by future research to clarify meaning of S-compound emissions.

Physiological conditions for the emission of sulfur gases

Single plant emission studies reflect a complex relationship between the ontogenetic status of the whole plant, the plant part under consideration, nutritional effects, the amount of sulfur applied and plant physiological as well as meteorological parameters. The influence of these factors on the emission of reduced sulfur compounds has been reviewed by Rennenberg (1989; 1991) and Anderson (1990). Although some data are available on the emission of methyl mercaptan after amino acid uptake (Schmidt *et al.* 1985), H_2S is the most abundant species emitted by various plant species in response to excess sulfur (see compilation by Rennenberg 1991). Because all plants seem to be able to emit H_2S, the formation of this substance has been subject to investigations in numerous studies.

Emission as reaction to excess sulfur

Plants may be exposed to excess inorganic sulfur by fertilization or by growing in soils with high organic sulfur content because of incorporated decaying organic matter. Under these circumstances, the roots absorb various sulfur compounds such as sulfite, sulfide, sulfate, but also sulfur containing amino acids like methionine or cysteine (Spaleny 1977; Filner *et al.* 1984; Rennenberg 1989). Hydrogen sulfide emissions in response to high sulfate loads occur in correlation to the amount of sulfur in the fertilizer, but are also highly dependent on the developmental stage of the plant (Rennenberg and Filner 1983). Sulfate/sulfite uptake into the plant is regulated by the roots (Rennenberg and Thoene 1987; Rennenberg 1991) due to inhibitory concentrations of glutathione. This regulatory mechanism may prevent the accumulation of high burdens of inorganic sulfur in the tissue under circumstances when H_2S emission does not occur. Detatched spruce branches, for example, emit large amounts of H_2S during winter time after sulfite/sulfate additions to the transpiration water. The whole tree, however, does not emit any H_2S, even if the soil concentrations of sulfate are increased to 30 mM. After cutting the roots, normal H_2S emission rates are found. This experiment demonstrates either an immense barrier potential of the roots (Huber, pers. comm.) or an intensive stress effect triggering the release of H_2S (see above). The present data set is, however, limited and needs augmenting with accurate uptake and emission studies under controlled conditions. The emission of sulfur compounds under the influence of high concentrations of sulfur containing amino acids has only been studied with leaf discs or suspension culture cells. Additions of *L*- as well as *D*-cysteine to cell cultures resulted in the emission of H_2S (Sekiya *et al.* 1982b), and surplus amounts of methionine led to the volatilization of CH_3SH from the cells. The production of these gases from amino acids was always light dependent and under developmental control (Rennenberg and Filner 1983; Rennenberg 1984). Although it has been postulated in the past decades, *L*-cysteine seems not to be an intermediate in the H_2S formation from sulfate (Rennenberg 1989). It is unclear, whether the emission of H_2S under the influence of *L*- and *D*-cysteine is mediated by cysteine-desulfhydrase, because the enzymes activity was not coupled to the emission of the gas (Schuetz *et al.* 1991). In its reverse reaction, however, cysteine desulfhydrase has been shown to participate in the assimilation of H_2S from atmospheric sources (Schütz *et al.* 1991). Presently there is no clear cut evidence available whether the hydrogen sulfide precursor *D*-cysteine is formed from *L*-cysteine by a racemase in higher plants (Rennenberg *et al.* 1992).

Emission after fumigation

The effects of plant fumigation with SO_2 have attracted interest because of a possible connection to pollution damage and forest dieback. In ornamental plants, the deposition of SO_2 was proportional to the atmospheric concentration but decreased with time (Elkiey and Ormrod 1981). Growth reductions under the influence of SO_2 of about 20 to 50 %, depending on the species under consideration, are reported in trees by Freer-Smith (1985) and Whitmore and Freer-Smith (1982). The latter authors did not observe re-emission of sulfur in any form. This is contrary to earlier results obtained by Materna (1966) and DeCormis (1968) which indicated a positive correlation between SO_2 fumigation and H_2S emission in several plants as a detoxification reaction. Forest canopies have been identified as sinks as well as sources of volatile sulfur by Hicks *et al.* (1982). However, only a low amount of the sulfur taken up by the trees seemed to be re-emitted. Sekiya and co-workers reported an emission of 15 % of the SO_2-sulfur in reduced form, some 60 % was further oxidised and stored in the vacuole (Sekiya *et al.* 1982a). Increasing amounts of S-containing amino acids in treated cells indicate that plants are able to utilize atmospheric SO_2 or H_2S in low concentrations to fill or maintain their sulfur pool. In this context the role of glutathione (*p*-glutamyl-cysteinyl-glycine) as a storage peptide and its formation (Brunold 1990; Bergmann and Hell 1990) or degradation (Anderson 1990) has been reviewed extensively (Rennenberg and Lamoureux 1990, Rennenberg 1991). COS is deposited into plants very effectively (Fall *et al.* 1988; Hofmann *et al.* 1992c), and is clearly being consumed in close analogy to CO_2 (Protoschill-Krebs and Kesselmeier 1992; see below).

Biological background

Enzymatic background for the synthesis and emission of sulfur gases

The synthesis of H_2S in higher plants has been investigated in numerous studies. Whereas the enzymatic degradation of cysteine by cysteine desulfhydrase action was favoured as source for H_2S from plant tissue, recent studies (Rennenberg 1983; Rennenberg *et al.* 1982, 1983, 1987) Rennenberg (1984) do not support this hypothesis. Instead, a reduction of sulfate or sulfite under the influence of light is favoured. This would include the removal of the thiol group from cysteine to yield *O*-acetylserine and sulfide, the latter of which will be transformed to sulfite, sulfate and sulfide, and finally emitted as hydrogen sulfide. Critical pools are represented by sulfate and cysteine. Whereas sulfate uptake by the roots can be regulated by glutathione, the cysteine pool size is maintained by the cysteine desulfhydrase. As already mentioned above, it is presently not clear whether *L*-cysteine could be transformed to *D*-cysteine by racemase action and degraded by a *D*-cysteine specific cysteine desulfhydrase. In higher plants methyl mercaptan and DMS are derived from the breakdown of methionine and *S*-methylmethionine, as has been shown by Schwenn *et al.* (1982) with *Catharanthus roseus* cells and by Fall *et al.* (1988) with *Triticum aestivum*. No enzymatic pathways are presently known for the evolution of COS and CS_2 from plant tissues. Recent investigations of Kesselmeier and co-workers indicate a possible correlation between formation of COS and the photosynthetic and respiratory carbon cycle (Hofmann *et al* 1990, Protoschill-Krebs and Kesselmeier 1992, see below). The sources of CS_2 remain obscure and may be related more to microbial metabolism than to higher plants. Giovanelli (1987) and Rennenberg (1991) suggested that COS and CS_2 are volatile intermediates in the metabolism of nonprotein sulfur amino acids and unusual sulfur compounds. However, the pathways of biosynthesis and degradation of such compounds are still poorly understood.

Enzymatic background for the uptake of COS

The close relation of the uptake of COS with CO_2 assimilation found in many studies was the reason to investigate which enzyme system is responsible for the consumption of COS by vegetation. Investigations with rat hepatocytes and erythrocytes (Chengelis and Neal 1979, 1980) and with cyanobacteria (Miller *et al.* 1989) implied that COS is directly consumed by carbonic anhydrase (CA). However, at the beginning of BIATEX there was no experimental evidence allowing the transfer of this model to the plant system. Results of studies reported by Protoschill-Krebs (1991), Protoschill-Krebs and Kesselmeier (1992) and Kesselmeier (1992) showed that the carboxylating enzymes ribulose-1,5-bisphosphate carboxylase (Rubisco) and phosphoenolpyruvate carboxylase (PEP-Co) may use COS as a substrate. The latter enzyme was already identified as an enzyme metabolizing COS (Lorimer and Pierce, 1989). However contrary to CO_2, the COS was not an activator of ribulose-1,5-bisohosphate-carboxylase, which might limit its importance for photosynthesis (Lorimer and Pierce 1989). Although these early investigations claimed a direct uptake of COS by Rubisco, recent investigations (Protoschill-Krebs 1991; Kesselmeier *et al.* 1991; Protoschill-Krebs and Kesselmeier 1992; Protoschill-Krebs *et al.* 1996) showed that COS is first metabolized by the action of a carbonic anhydrase whose COS affinity is 1000 times higher than its CO_2 affinity. In enzyme assays the sulfur

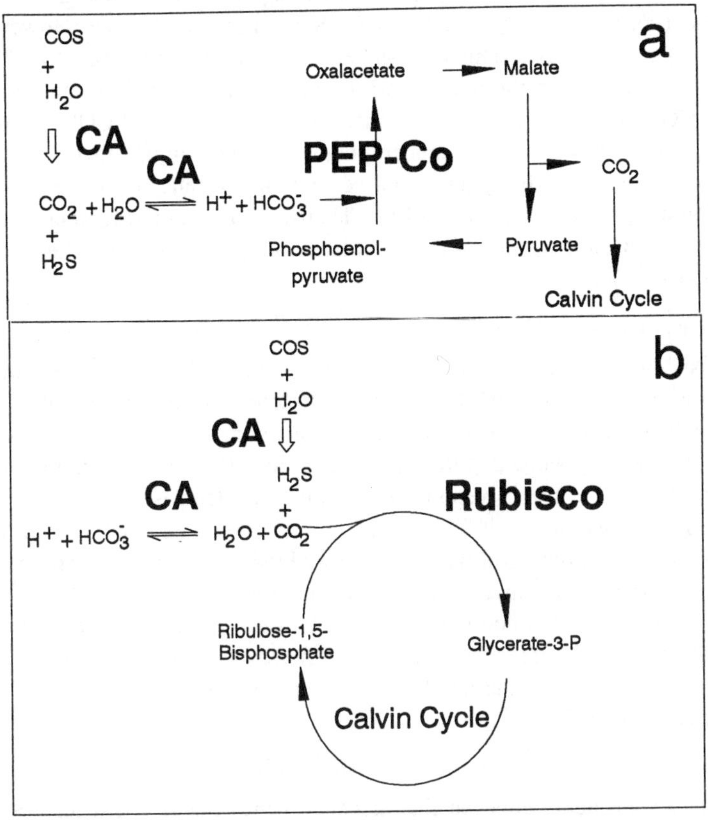

Fig. 7.5: Proposed scheme for the uptake and metabolism of COS in higher plants according to Protoschill-Krebs and Kesselmeier (1992). (a) Mechanism for C_4 plants; (b) Mechanism for C_3 plants. PEP-Co: Phosphoenolpyruvate carboxylase; CA: Carbonic anhydrase; Rubisco: Ribulosebisphosphate carboxylase.

group is rapidly cleaved, reduced and set free as H_2S. About 60 % of the S taken up as COS is re-emitted as H_2S rapidly . However, this observation may not correspond to *in vivo* conditions with intact plants. Though a correlation between COS uptake and H_2S emission is expected, Bartell *et al.* (1993) report about only 1 % of the sulfur released in comparison to the COS-sulfur taken up. Only with intact algal cells (*Chlamydomonas reinhardtii*) a higher release of 60 % can be reported (Protoschill-Krebs 1991; Kesselmeier at al. 1991; Protoschill-Krebs *et al.*, 1995) possibly due to an extracellular localized carbonic anhydrase. Recently, evidence for the consumption of COS by marine phytoplankton species has also been reported (Kesselmeier *et al.* 1994). Thus we accept the model for the uptake of COS as given in Fig. 7.5.

New important sinks for COS *recently identified*

As described above, we now have a basic understanding of the exchange of COS between vegetation and the atmosphere, of compensation points, and about the enzymatic processes for the uptake of COS by plants and algae. However, the global budget for COS is still not balanced, as recently reviewed by Chin and Davis (1993). Based on the net primary production as a basis for the correlation between CO_2 and COS uptake we can make a rough estimate of COS deposited to the vegetation. However, the number obtained is not higher than 0.86-1.0 Tg (COS) consumed globally by terrestrial vegetation and thus not higher than taken into account by Chin and Davis (1993). Therefore, it is not sufficient to balance the sources. Within this context, it is of special interest to identify either new sinks or to unmask sources.

Soil: source or sink for COS *?*

On a basis of very variable data (gradients in the early morning) during field measurements in Cameroon, the soil/plant system may act as a sink for COS during night (Kesselmeier *et al.* 1993a). Measurements of the vegetation at the same time of the day show that trees may be excluded as a sink during night. Thus we speculate that the soil is responsible for this COS decrease. Studies on the role of the soil in the exchange of COS have been conducted by Conrad *et al.* (1994). They investigated several soil types and showed that soils may act a sink for COS. According to their results two processes may be involved: abiotic delivery of COS and COS uptake due to microbial activities. It is too early to use these studies to recalculate soil sink/source strengths, but in accordance with Castro and Galloway (1991), we can state that the numbers used for soils as a global source for COS may be overestimated.

Leaf litter involved in the exchange of volatile sulfur

Additional experiments done on the exchange behaviour of plant leaf litter also brought some interesting results. As shown by Kesselmeier *et al.* (1993c), leaf litter may switch between emission (mainly DMS) and deposition (only COS) of S-compounds, depending on the moisture content and the microbial activity in the dead leaf material. However, in terms of global meaning, neither DMS emission nor COS deposition from/to leaf litter seems to be of significance.

Lichens involved in the exchange of reduced sulfur compounds

The northern latitude boreal and arctic regions are potentially important in terms of COS exchange. Johnson *et al.* (1993) report about a 19 % COS depletion in air masses originating from boreal or arctic regions of central and northern Canada, travelling for at least 2000 km across tundra and transitional forest in central and northern Canada along the west coast of Hudson Bay. The sudden decrease of COS concentrations in these air masses was explained by a large unknown sink for COS in this region. Studies by Fried *et al.* (1993) with bog microcosms showed that peat, mosses and vascular plants within this bog system can not account for such a large sink. We know that these regions are lichen dominated communities (Kershaw, 1985) and for this reason study of COS exchange between lichens and the atmosphere has started (Gries *et al.* 1994). Under laboratory conditions, 10 green algal lichens and four cyanolichens, as well as an isolated cyanobacterium all emitted significant quantitities of H_2S and DMS, but were excellent sinks for COS. Although some interspecific variation occurred, exchange rates were relatively uniform and were not influenced by irradiance conditions. In contrast to DMS and H_2S emission, COS uptake was strongly influenced by thallus hydration. In the meantime field investigations (May/June 94 and November/December 1994) were made at a site in California with a high lichen biomass load and they confirm the laboratory investigations (Kuhn, personal communication). Because lichen dominated systems cover extensive terrestrial habitats, their COS uptake may be potentially important in the world's sulfur budget.

Concluding remarks

In retrospect we have a somewhat strange situation. At the beginning of BIATEX research was focused on reduced sulfur compounds and it was decided not to include investigations on deposition of SO_2. Later on (1990) SO_2 was incorporated but the financial support for reduced sulfur compounds (at least in Germany) was nearly terminated. One of the major goals of BIATEX was to establish a sound data-base of sulfur fluxes between biosphere and atmosphere for the development of large scale models of deposition and emission over Europe. Therefore, it was essential to quantify the fluxes of reduced sulfur from important European plant species and to elucidate which factors govern the exchange. However, it was not possible to continue the promising research of the first three years within BIATEX. Nevertheless, the outcome of the sulfur research contributes significantly to our knowledge.

It was the work conducted within BIATEX that showed the important influence of water layers on the deposition of SO_2 measured by gradient experiments. The influence of water layers had been discovered already in the eighties, however, it was not included in models describing the surface exchange of SO_2, because it was thought to be negligible. Up to that time a limited number of measurements had been made under conditions where surface water chemistry played a limiting role in the surface exchange (*e.g.* high SO_2 concentrations to reduce measuring errors, saturated water layers). Our understanding of SO_2 dry deposition mechanisms has increased through the work conducted in BIATEX (*e.g.* Erismann and Baldocchi, 1994; Erisman *et al.*, 1994a; Erisman and Draaijers, 1995). However, it is still not possible to quantify the different pathways of deposition into the various parts of (eco)systems. Current measuring techniques and interpretation can be used for monitoring of the dry deposition of SO_2. As the result of this, several groups from the BIATEX community started to monitor deposition. Currently dry deposition of SO_2 (and O_3) is monitored at two sites, in Portugal and Spain (EC SREMP project), SO_2 (and NH_3) at the Oostvaardersplassen in the Netherlands and SO_2 (NH_3 and NO_x) at three sites, in Scotland, Germany and the Netherlands (EC LIFE project). This makes six monitoring sites in Europe, of which one is forested.

The emission of reduced sulfur compounds is regarded as a significant detoxification step under the influence of excess sulfur or atmospheric SO_2. It may also have regulatory function during plant development under moderate sulfur supply. The amounts of sulfur gases produced by plants depend on the internal concentrations of the metabolic precursors and are governed by the developmental stage of the plant and by environmental factors such as temperature, light intensity and air humidity. Recent studies also indicate a correlation between the emission of certain sulfur species and their atmospheric concentration. The BIATEX work on reduced sulfur compounds showed that emissions are not restricted to certain species or ecosystems, but appear ubiquitous. DMS and H_2S are the most important gases released by higher plants, followed by COS and methyl mercaptan. In addition to the observed but not understood emission of COS by higher plants, we have to face the open question, whether H_2S emission may at least be partly triggered by "stress" effects. We think these questions have to be addressed by future research to clarify meaning of S-compound emissions. The emission of CS_2 has only been observed in two wetland species and might therefore be restricted to plants in this ecosystem. Because COS is significantly involved in the climatic regulation of our planet, its sinks and sources are still of high interest, especially as there is uncertainty on whether the soils are a source or a sink for this trace gas (see above). The explanation of the enzymatic consumption of COS by the ubiquitous enzyme carbonic anhydrase may help to understand several sink processes, including those recently identified in lichens.

In addition to their role in acidification sulfur compounds have considerable impact on the earth's climate. For example DMS is a molecule involved in cloud condenstaion nuclei formation and in rapid reactions with radicals. The stable compounds COS and CS_2 may furthermore contribute significantly to the stratospheric sulfate budget and be involved in climate stability (Crutzen 1976). Though of high importance, the quantification of sulfur fluxes within ecosystems is still not possible. To overcome this lack as well as the uncertainties of global estimates of sulfur fluxes, further measurements are needed to characterize properly plant physiological, ecological and meteorological prerequisites for the emission of these compounds. This is of special importance because the anthropogenic emission of sulfur compounds is being reduced and as a consequence natural emissions will regain greater scientific interest. In this context also monitoring of reduced sulfur compounds, especially of COS, would improve our knowledge.

Acknowledgements

We would like to thank our wifes Andrea, Elsbeth and Maria as well as our children Christian, Jelske, Marja, Miriam and Peter for their patience and their support in the course of our field-, laboratory and paper-work within BIATEX.

References

Adewuyi, Y.G., (1980), Oxidation of biogenic sulfur compounds in aqueous media: Kinetics and environmetal implications. in: Saltzman,E.S., Cooper,W.J. (eds), *Biogenic Sulfur in the Environment*, ACS Symposium Series, Washington pp. 529–559.

Anderson J.W., (1990), Sulfur metabolism in Plants. in: Lea P.J., Miflin B.J. (eds), *The Biochemistry of Plants* **16**, 327–381.

Andreae, M.O., Jaeschke, W.A., (1992), Exchange of sulphur between biosphere and atmosphere over temperate and tropical regions, in: Howarth, R.W., Stewart, J.W.B., Ivanow, M.V. (eds), *Sulphur Cycling on the Continents* pp. 27–61.

Aneja,V.P., Cooper,W.J., (1995), Biogenic sulfur emissions. A review. in: Saltzman, E.S., Cooper,W.J.(eds), *Biogenic Sulfur in the Environment*, ACS Symposium Series, Washington 1989, pp. 2–13.

Arends, B.G., Mas, F., Geusebroek, M., Wyers, G.P., Veltkamp, A.C., Biogenic and reduced atmospheric sulphur compounds in the Netherlands. *ECN Report* ECN-C-95-014.

Baldocchi, D.D., Hicks, B.B., Camara, P., (1987), A canopy stomatal resistance model for gaseous deposition vegetated surfaces. *Atmos. Environ.* **21,** 91–101.

Bartell, U., Hofmann, U., Hofmann, R., Kreuzburg, B., Andreae, M.O., Kesselmeier, J., (1993), COS and H_2S fluxes over a wet meadow in relation to photosynthetic activity: An analysis of measurements made on 6 September 1990. *Atmos. Environ.* **27A,** 1851–1864.

Barrie, L.A., Warmsley, J.L., (1978), A study of sulphur dioxide deposition velocities to snow in northern Canada. *Atmos. Environ.* **12,** 2321–2322.

Bergmann, L., Hell, R., (1990), Determination and characterization of -glutamylcy-steine synthetase from higher plants. in: Rennenberg, H., Brunold, C., De Kok, L.J., Stulen, I. (eds), *Sulfur nutrition and sulfur assimilation in higher plants,* SPB Scientific Publishers, The Hague pp. 167–172.

Berresheim, H., V.D. Vulcan, (1992), Vertical distribution of COS, CS_2, DMS and other sulfur compounds in a loblolly pine forest. *Atmos. Environ.* **26A,** 2031–2036.

Brimblecombe, P., (1978), "Dew" as a sink for sulphur dioxide. *Tellus* **30,** 151–157.

Brown, K.A., Bell, J.N.B., (1986), Vegetation - The missing sink in the global cycle of COS. *Atmos. Environ.* **20,** 537–540.

Brown, K.A., Kluczewski, S.M., Bell, J.N.B., (1986), Metabolism of [35S]-carbonyl sulphide in perennial egrass (*Lolium Perenne L.*) and radish (*Raphanus Sativus* L.) *Environ. Experimental Bo.* **26,** 355–364.

Brunold, C., (1990), Reduction of sulfate to sulfide. in: Rennenberg, H., Brunold, Ch., De Kok, L.J., Stulen, I. (eds), *Sulfur Nutrition and Sulfur Assimilation in Higher Plants.* SPB Acad. Publ., The Hague pp. 13–31.

Businger, J.A., (1986), Evaluation of the accuracy with which dry deposition can be measured with current nometeorological techniques. *J. Climate Appl. Meteor.* **25,** 1100–1124.

Caddle, S.H., Dasch, J.M., Mulawa, P.A., (1985), Atmospheric concentrations and the deposition velocity to snow of nitric acid, sulfur dioxide and various particulate species. *Atmos. Environ.* **19,** 1819–1827.

Castro, M.S., Galloway, J.N., (1991), A comparison of sulfur-free and ambient air enclosure techniques for measuring the exchange of reduced sulfur gases between soils and the atmosphere. *J. Geophys. Res.* **96,** 15247–15437.

Chameides, W.L., (1987), Acid dew and the role of chemistry in the dry deposition of reactive gases to wetted surfaces. *J. Geophys. Res.* **92,** 11895–11908.

Chengelis, C.P., Neal, R.A., (1979), Hepatic carbonyl sulphide metabolism. *Biochem. Biophys. Res. Comm.* **90,** 993–999.

Chengelis, C.P., Neal, R.A., (1980), Studies of carbonyl sulphide toxicity: Metabolism by carbonic anhydrase.*Toxicol. App. Pharmaco*l. **55,** 198–202.

Chin, M., Davis, D.D., (1993), Global sources and sinks of OCS and CS_2 and their distributions. *Global Biogeochem. Cycles* **7,** 321–337

Conrad, R., Bollmann, A., Lehmann, S., Rudolph, J., (1994), Mechanisms of release of NO and OCS from soil.in: *EUROTRAC Annual Reports, Part 6,* EUROTRAC ISS Garmisch-Partenkirchen pp. 58–63.

Crutzen, P.J., (1976), The possible importance of COS for the sulfate layer of the stratosphere. *Geophys. Res.Lett.* **3,** 73–76.

Cullis, C.F., Hirschler, M.M., (1980), Atmospheric sulfur: Natural and man-made sources. *Atmos. Environ.* **14,** 1263–1278

Davidson, C.I., Wu, Y.L., (1990), Dry deposition of particles and vapors, in: Lindberg S.E., Page A.L., Norton S.A. (eds), *Acidic precipitation, Volume 3* Springer-Verlag, Berlin pp, 108–209.

Davies, T.D., Mitchell, J.R., (1983), Dry deposition of sulphur dioxide onto grass in rural Eastern England with some comparison with other forms of sulphur deposition, in: Pruppacher *et al.* (eds) *Precipitation Scavenging, Dry deposition and Resuspension,* Elsevier Science Publishing Co., Inc.

Davies, T.D., Wright, R.G., (1985), Sulphur dioxide deposition velocity by a concentration gradient measurement system. *J. Geophys. Res.* **90,** 2091–2095.

Davison, B., Hewitt, C.N., (1992), Natural sulphur species from the North Atlantic and theit contribution to the United Kingdom sulphur budget. *J. Geophys. Res.* **97,** 2475–2488.

De Kok, L., Stahl, K., Rennenberg, H., (1989), Fluxes of atmospheric hydrogen sulphide to plant shoots. *New Phytol.* **112,** 533–542

De Kok, L.J., H. Rennenberg, P.J.C. Kuiper, (1991) , The internal resistance in spinach leaves to atmospheric H_2S deposition is determined by metabolic processes. *Plant Physiol. Biochem.* **29,** 463–470.

De Cormis, L., (1968), Release of hydrogen sulfide into an atmosphere containing sulfur dioxide. *C.R. Acad. Sci.* **266,** 683–685.

Draaijers, G.P.J., Erisman, J.W., (1993), Atmospheric sulphur deposition onto forest stands: throughfall estimates compared to estimates from inference, *Atmos. Environ.* **27A,** 43–55.

Duyzer, J.H., Bosveld, F.C., (1988), Measurements of dry deposition fluxes of O_3, NO_x, SO_2 and particles over grass/heathland vegetation and the influence of surface inhomogeneity. *Report no. R 88/111,* TNO, Delft, the Netherlands.

Elkiey, T., Ormrod, D.P., (1981), Absorption of ozone, sulphur dioxide, and nitrogen dioxide by petunia plants. *Environ. Exp. Bot.* **21,** 63–70.

Erisman, J.W., (1993a), Acid deposition onto nature areas in the Netherlands; Part I. Methods and results. *Water Soil Air Pollut.* **71,** 51–80.

Erisman, J.W., (1993b), Acid deposition onto nature areas in the Netherlands; Part II. Throughfall measurements compared to deposition estimates. *Water Soil Air Pollut.* **71,** 81–99.

Erisman, J.W., Wyers, G.P., (1993), Continuous measurements of surface exchange of SO_2 and NH_3; implications for their possible interaction in the deposition process. *Atmos. Environ.* **27A**, 1937–1949.

Erisman, J.W., Versluis, A.H., Verplanke, T.A.J.W., de Haan, D., Anink, D., van Elzakker, B.G., Mennen, M.G., Aalst, R.M. van, (1993a), Monitoring the dry deposition of SO_2 in the Netherlands. *Atmos. Environ.* **27,** 1153–1161.

Erisman, J.W., Mennen, M., Hogenkamp, J., Goedhart, D., Pul, A. van, Boermans, J., (1993b), Dry deposition over the Speulder forest. in: J. Slanina, G.Angeletti, S. Beilke (eds), *Air Pollution Report 47*, CEC, Brussels.

Erisman, J.W., Baldocchi, D.D., (1994), Modelling dry deposition of SO_2. *Tellus* **46B,** 159–171.

Erisman, J.W., Pul, A. van, Wyers, P., (1994a), Parameterization of dry deposition mechanisms for the quantification of atmospheric input to ecosystems. *Atmos. Environ.* **28,** 2595–2607.

Erisman, J.W., Beier, C., Draaijers, G., Lindberg, S., (1994b), Review of deposition monitoring methods. *Tellus* **46B,** 79–93

Erisman, J.W., Elzakker, B.G. van, Mennen, M. G., Hogenkamp, J., Zwart, E., Beld L. van den, Römer, F.G., Bobbink, R., Heil, G., Raessen, M., Duyzer, J.H., Verhage, H., Wyers, G.P., Otjes, R.P., Möls, J.J., (1994c), The Elspeetsche Veld experiment on surface exchange of trace gases: summary of results. *Atmos. Environ.* **28,** 487–496.

Erisman, J.W., Draaijers, G.P.J., (1995), Atmospheric deposition in relation to acidification and eutrophication. *Studies in Environmental Research 63*, Elsevier, the Netherlands.

Ernst,W.H.O., (1990), Ecological aspects of sulfur metabolism. in: Rennenberg *et al.* (eds), *Sulfur Nutrition and Sulfur Assimilation in Higher Plants*, Academic Publishing bv, The Hague pp. 131–144.

Fall, R., Albritton D.L., Fehsenfeld, F.C., Kuster, W.C., Goldan, P.D., (1988) , Laboratory studies of some environmental variables controlling sulfur emissions from plants. *J. Atmos. Chem.* **6,** 341–362

Filner, P., Rennenberg, H., Sekiya, J., Bressan, R.A., Wilson, L.G., Le Cureux, L., Shimei, T., (1984), Biosynthesis and emission of hydrogen sulfide by higher plants. in: Koziol, M.J., Whatey, F.R. (eds), *Gaseous air pollutants and plant metabolism*, Butterworth, London. pp. 291–312.

Fowler, D., (1978), Dry deposition of SO_2 on agricultural crops. *Atmos. Environ.* **12,** 369–373.

Fowler, D., (1985), Dry deposition of SO_2 onto plant canopies. in: Winner, W.E., Mooney H.A., Goldstein R.A. (eds), *Sulphur dioxide and vegetation*, Stanford University Press, California pp. 75–95.

Fowler, D., Unsworth, M.H., (1979), Turbulent transfer of sulphur dioxide to a wheat crop. *Quart. J. R. Met. Soc.* **105,** 767–783.

Fowler, D., Cape, J.N. (1984), The contamination of rain samples by dry deposition on rain collectors. *Atmos. Environ.* **18,** 183–189.

Fowler, D., Duyzer, J.H., (1990), Micrometeorological techniques for the measurement of trace gas exchange. in: Andrae, M.O., Schimel, D.S. (eds),.*Exchange of trace gases between terrestrial ecosystems and the atmosphere*, John Wiley and Sons pp. 189–207.

Fowler, D., Duyzer, J.H., Baldocchi, D.D., (1991), Inputs of trace gases, particles and cloud droplets to terrestrial surfaces. *Proc. Roy. Soc. Edinburgh* **97B,** 35–59.

Freer-Smith, P., (1985), The influence of SO_2 and NO_2 on the growth, development and gas exchange of Betula pendula Roth. *New Phytol.* **99,** 417–430.

Fried, A., Klinger, L.F., Erickson III, D.J., (1993), Atmospheric carbonyl sulfide exchange in bog microcosms. *Geophys. Res. Lett.* **20,** 129–132.

Galbally, I., (1979), Sulphur uptake from the atmosphere by forest and farmland. *Nature* **280,** 49–50.

Garland, J.A., (1977), The dry deposition of sulphur dioxide to land and water surfaces. *Proc. Roy. Soc. Lond.* **A354,** 245–268.

Garland, J.A., Branson, J.R, (1977), The deposition of sulphur dioxide to pine forest assessed by a radioactive tracer method. *Tellus* **29,** 445–454.

Garsed, S.G., (1985), SO_2 uptake and transport. in: Winner W.E., Mooney H.A., Goldstein R.A. (eds), *Sulphur dioxide and vegetation*, Stanford University Press, California pp. 75–95.

Giovanelli, J., (1987), Sulfur amino acids of plants: an overview. in: Jacoby, W.B., Griffith, O.W. (eds), *Methods in Enzymology* Vol 143,. Academic Press, New York pp. 419–426.

Goldan, P.D., Kuster, W. C. Albritton, D.L., Fehsenfeld, F.C., (1987), The measurement of natural sulfur emissions from soil and vegetation: Three sites in the eastern United States revisited. *J. Atmos. Chem.* **5,** 439–467.

Goldan, P.D., Fall, R., Kuster, W. C., Fehsenfeld, F.C. (1988), The uptake of growing vegetation. A major tropospheric sink. *J. Geophys. Res.* **93,** 14186–14192

Granat, L., Rohde, H., Halberg, R.O., (1976), The global sulfur cycle. *Ecol. Bull.* **22,** 89–134.

Granat, L., Johansson, C., (1983), Dry deposition of SO_2 and NO_x in winter. *Atmos. Environ.* **17,** 191–192.

Gries, C., Nash III, T.H., Kesselmeier, J., (1994), Exchange of reduced sulfur gases between lichens and the atmosphere. *Biogeochem.* **26,** 25–39.

Guenther, A., Lamb, B., Westberg, H., (1989), U.S. national biogenic sulfur emissions inventory. in: Saltzman, E.S., Cooper, W.J. (eds), *Biogenic Sulfur in the Environment*, American Chemical Society, Washington D.C. pp. 15–30.

Haines, B., Black, M., Bayer, C., (1989), Sulfur emission from roots of the rain forest tree Stryphnodendron excelsum. in: Saltzman, E.S., Cooper, W.J. (eds), *Biogenic sulfur in the environment*, ACS Symp. Ser. Washington pp. 58–69.

Hicks, B.B., Wesely, M.L., Durham, J.L., Brown, M.A., (1983), Some direct measurements of atmospheric sulphur fluxes over a pine plantation. *Atmos. Environ.* **16**, 2899–2903.

Hicks, B.B., Wesely, M.L., Coulter, R.L., Hart, R.L., Durham, J.L., Speer, R.E., Stedman, D.H., (1983), An experimental study of sulphur and NO_x fluxes over grassland. *Boundary-Layer Meteor.* **34,** 103–121.

Hicks, B.B., Baldocchi, D.D., Meyers, T.P., Hosker, Jr. R.P., Matt, D.R., (1987), A preliminary multiple resistance routine for deriving dry deposition velocities from measured quantities. *Water, Air and Soil Pollut.* **36,** 311–330.

Hicks, B.B., Matt, D.R., McMillen, R.T., (1989), A micrometeorological investigation of surface exchange of O_3, SO_2 and NO_2 : a case study. *Boundary-Layer Meteor.* **47,** 321–336.

Hicks, B.B., Draxler, R.R., Albritton, D.L., Fehsenfeld, F.C., Hales, J.M., Meyers, T.P., Vong, R.L., Dodge, M., Schwartz, S.E., Tanner, R.L., Davidson, C.I., Lindberg, S.E., Wesely M.L., (1989), Atmospheric processes research and procecess model development. *State of Science/Technology report no. 2.* National Acid Precipitation Assessment Program, NAPAP office, Washington.

Hill, F.B., in: G.M. Woodwell, E.V. Pecan (eds), (1993), Carbon and the Biosphere. *Tech. Inform. Ctr.*, U.S. Atomic Energy Commission pp 159–181.

Hofmann, D.J., (1990), Increase in the stratospheric background sulfuric acid aerosol mass in the past 10 years. *Science* **248,** 996–1000.

Hofmann, U., Merk, L., Bingemer, H., Kesselmeier, J., Andreae, M.O., (1994), Mesurements of reduced sulphur compounds with a newly developed cryo trap in: *EUROTRAC Annual Reports, Part 4*, EUROTRAC ISS Garmisch-Partenkirchen pp 120–125.

Hofmann, U., (1994), Vergleichende Untersuchungen zum Austausch von reduzierten Schwefelverbindungen zwischen Vegetation und Atmosphäre im Feldversuch und unter kontrollierten Bedingungen. Dissertation, Fachbereich Biologie, Universität Mainz.

Hofmann, U., Hofmann, R., Kesselmeier, J., (1992a), Field measurements of reduced sulfur compounds over wheat during a growing season. in: S.E.Schwartz, W.G.N.Slinn (eds), *Precipitation Scavenging and Atmosphere-Surface Exchange. Volume 2-The Semonin Volume:* Atmosphere-Surface Exchange Processes, Hemisphere, Washington DC pp. 967–977.

Hofmann,U., Hofmann,R., Kesselmeier,J., (1992b), Cryogenic trapping of reduced sulfur compounds under the influence of a NAFION Dryer and cotton wadding as an oxidant scavenger. *Atmos. Environ.* **26,** 2445–2449.

Hofmann, U., Protoschill-Krebs, G., Merk, L., Bingemer, H., Kesselmeier, J. Andreae, M.O., (1991), Field measurements of reduced sulphur compounds and first physiological results on the uptake of COS by higher plants. in: P. Borrell, P.M. Borrell, W. Seiler (eds), *Proc. EUROTRAC Symp. '90*, SPB Academic Publishing bv, The Hague pp. 149–154.

Huber, B., (1994), Austausch flüchtiger Schwefelverbindungen in land- und forstwirtschaftlichen ökosystemen. Phd Thesis, Technische Universität München, Schriftenreihe des Fraunhofer-Institutes Band 26, Wissenschaftsverlag Maraun, Frankfurt.

Jarvis, P.G., (1976), The interpretation of the variations in leaf water potential and stomatal conductance found in canopies in the field. *Phil. Trans. Roy. Soc. Lond.* **B273,** 593–610.

Johnson, J.E., Bandy, A.R., Thornton, D.C., Bates, T.S., (1993) , Measurements of atmospheric carbonyl sulfide during the NASA CITE-3 project: Implications for the global COS budget. *J. Geophys. Res.* ,23443–23448.

Kershaw, K.A., (1985), *Physiological Ecology of Lichens,* Cambridge University Press, Cambridge, U.K.

Kesselmeier, J., (1991), Emission of sulfur compounds from vegetation and global-scale extrapolation, in: Th.D.Sharkey, E.A.Holland, H.A.Mooney (eds), *Trace Gas Emissions from plants,* Academic Press, San Diego pp. 261–265.

Kesselmeier, J., Hofmann, U., Protoschill-Krebs, G., Hofmann, R., Merk, L., Bartell, U., Andreae, M.O., (1991), Exchange of reduced sulfur compounds between atmosphere and biosphere, in: *EUROTRAC Annual Reports, Part 4,* EUROTRAC ISS Garmisch-Partenkirchen pp. 162–167.

Kesselmeier, J., (1992), Plant physiology and the exchange of trace gases between vegetation and the atmosphere, in: S.E.Schwartz, W.G.N.Slinn (eds), *Precipitation Scavenging and Atmosphere-Surface Exchange. Volume 2-The Semonin Volume: Atmosphere-Surface Exchange Processes Hemisphere*, Washington DC pp. 949–966.

Kesselmeier, J., Hofmann, U., Protoschill-Krebs, G., Hofmann, R., Merk, L., Bartell, U., Velmecke, F., Andreae. M.O., (1992a), Exchange of reduced sulfur compounds between atmosphere and biosphere. in: *EUROTRAC Annual Reports, Part 4,* EUROTRAC ISS Garmisch-Partenkirchen pp. 202–208.

Kesselmeier, J., Meixner, F.X., Hofmann, U., Ajavon, A., Leimbach, St., Andreae, M.O., (1992b), Distribution of ozone over a tropical rain forest and exchange of reduced sulfur compounds between tropical tree species and the atmosphere: Investigations from the atmospheric boundary layer, at ground level and from the top of the canopy, in: Halle, F., Pascal, O. (eds), *Biologie d'une Canopee de Foret Equatoriale - II. Rapport de mission: Radeau des Cimes Octobre / Novembre 1991*, Reserve de Campo, Cameroun pp. 247–263.

Kesselmeier, J., Meixner, F.X., Hofmann, U., Ajavon, A., Leimbach, St., Andreae, M.O., (1993a), Reduced sulfur compound exchange between the atmosphere and tropical tree species in southern Cameroon, *Biogeochem.* **23,** 23–45.

Kesselmeier, J., Merk, L., Bliefernicht, M., Helas,G., (1993b), Trace gas exchange between terrestrial plants and atmosphere: carbon dioxide, carbonyl sulfide and ammonia under the rule of compensation points, in: Slanina, J., Angeletti,G., Beilke, S. (eds), *General assessment of biogenic emissions and deposition of nitrogen compounds, sulphur compounds and oxidants in Europe*, CEC Air Pollution Research Report 47, E. Guyot SA, Brussels pp 71–80.

Kesselmeier, J., Hubert, A., Velmeke, F., Blezinger, S., Hofmann, U., Andreae, M.O., (1993c), Exchange of reduced sulfur compounds between leaf litter, living leaves and the atmosphere, in: *EUROTRAC Annual Reports, Part 4,* EUROTRAC ISS Garmisch-Partenkirchen pp. 154–159.

Kesselmeier, J., Merk, L., (1993), Exchange of Carbonyl Sulfide (COS) between Agricultural Plants and the Atmosphere: Studies on the Deposition of COS to Peas, Corn and Rapeseed. *Biogeochem.* **23,** 47–59.

Kesselmeier, J., Blezinger, S., Wilhelm, C., Gries, C., Nash III, T.H., Andreae, M.O., (1994), Exchange of reduced sulfur compounds between the biosphere and the atmosphere: marine phytoplankton and lichen species, in: *EUROTRAC Annual Reports, Part 4,* EUROTRAC ISS Garmisch-Partenkirchen pp. 200–206.

Kluczewski, S.M., Brown, K.A., Bell, J.N.B., (1985), Deposition of [35S]-carbonyl sulphide to vegetable crops, *Radiat. Protect. Dosim.* **11,** 173–177.

Lamb, B., Westberg, H., Allwine, G., Bamesberger, L., Guenther, A., (1987), Meaurement of biogenic sulfur emissions from soils and vegetation: application of dynamic enclosure methods with Natusch filter and GC/FPD analysis, *J. Atmos. Chem.* **5,** 469–491

Larcher, W., (1994), *Ökophysiologie der Pflanzen,* 5. Auflage. Eugen Ulmer Verlag, Stuttgart

Lendzian, K.J., (1987), Aufnahne und zellphysiologische Wirkunen von Luftschadstoffen, *Naturwissenschaften* **74,** 282–288.

Lorenz, R., Murphy, C.H. Jr., (1985), Sulphur dioxide, carbon dioxide, and water vapout fluxmeasurements utilizing a microprocessor-controlled data acquisition system in a pine plantation, in: Hutchison, B.A., Hicks, B.B. (eds), *The Forest-Atmosphere Interaction,* Kluwer, Dordrecht.

Lorimer, G.H., Pierce, J., (1989), Carbonyl sulphide: An alternate substrate for but not an activator of ribulose- 1,5-bisphosphate carboxylase, *J. Biol. Chem.* **264/5,** 2764–2772

Materna J., (1966), Die Ausscheidung des durch die Fichtennadeln absorbierten Schwefeldioxids, *Archiv für Forstwesen* **15,** 691–692.

Matt, D.R., McMillen, R.T., Womack, J.D., Hicks B.B., (1987), A comparison of estimated and measured SO_2 deposition velocities, *Water Air and Soil Pollut.* **36,** 331–347.

McMillen, R.T., Matt, D.R., Hicks, B.B., Womack, J.D., (1987), Dry deposition measurements of Sulphur dioxide to a Spruce-Fir forest in the Black forest: *A data report. NOAA Technical Memorandum* ERL ARL-152.

Meyers, T.P., Baldocchi, D.D., (1988), A comparison of models for deriving dry deposition fluxes of O_3 and SO_2 to a forest canopy, *Tellus* **40B,** 270–284.

Mihalopoulos, N., Bonsang, B., Nguyen, B.C., Kanakidou, M., Belviso, S., (1989), Field observations of carbonyl sulfide deficit near the ground: Possible implication of vegetation, *Atmos. Environ.* **23,** 2159–2166.

Miller, A.G., Espie, G.S. and Canvin, D.T., (1989), Use of carbon oxysulphide, a structural analog of CO_2, to study active CO_2 transport in the cyanobacterium Synechococcus UTEX 625, *Plant Physiol.* **90,** 1221–1231.

Milne, J.W., Roberts, D.B., Williams, D.J., (1979), *Atmos. Environ.* **13,** 373–379.

Murphy, C.E., Sigmon, J.T., (1989), Dry deposition of sulphur and nitrogen oxide gases to forest vegetation, in: Lindberg S.E., Page A.L., Norton S.A. (eds), *Acidic precipitation, Volume 3* Springer- Verlag, Berlin.

Neumann, H.H., den Hartog, C.D., (1985), Eddy correlation measurements of atmospheric fluxes of ozone, sulphur, and particles during the campaign intercomparison study. *J. Geophys. Res.* **90,** 2097–2110.

Nicholson K.W., Davies T.D., (1988), Letter to the editors: The dry deposition of sulphur dioxide at a rural site, *Atmos. Environ..* **22,** 2885–2889.

Payrissat, M., Beilke, S., (1974), Laboratory measurements of the uptake of sulphur dioxide by different European soils, *Atmos. Environ.* **9,** 211–217.

Petit, C., Ledoux, M., Trinite, M., (1976), Transfer resistance to SO_2 capture by some plane surfaces, water and leaves, *Atmos. Environ.* **11,** 1123–1126.

Platt, U., (1978), Dry deposition of SO_2, *Atmos. Environ.* **12,** 363–367.

Protoschill-Krebs, G., (1991), Untersuchungen zur Aufnahme und zum Metabolismus von Carbonylsulfid in Pflanzen. Ph.D. Thesis, Johannes-Gutenberg-University of Mainz.

Protoschill-Krebs, G., Kesselmeier, J., (1992), Enzymatic pathways for the metabolization of carbonyl sulphide (COS) by higher plants, *Botanica Acta* **105,** 206–212.

Protoschill-Krebs, G., Wilhelm, C., Kesselmeier, J., (1995), The consumption of carbonyl sulphide by carbonic anhydrase (CA) of *Chlamydomonas reinhardtii* grown under different CO_2 regimes. *Botanica Acta* **108**, 445–448.

Protoschill-Krebs, G., Wilhelm, C., Kesselmeier, J., (1996), Consumption of carbonyl sulfide by carbonic anhydrase (CA) polated from *Pisum sativum. Atmos. Environ.* 30, 3151–3156.

Ravishankara, A.R., Kreutter, N.M., Shah, R.C., Wine, P.H., (1980), Rate reaction of OH with COS,. *Geophys. Res. Lett.* **7,** 861–864.

Rennenberg, H., Sekijy, J., Wilson, L.G., Filner, P., (1982), Evidence for an intracellular sulfur cycle in cucumber leaves, *Planta* **154,** 516–524.

Rennenberg, H., (1983), Cysteine desulhydrase activity in cucurbit plants: stimulation by preincubation with L- or D-cysteine, *Phytochem.* **22,** 1557–1560.

Rennenberg, H., Filner, P., (1983), Developmental changes in the potential for H_2S emission in cucurbit plants, *Plant Physiol.* **71,** 269–275.

Rennenberg, H., Reski, G., Polle, A., (1983), Hydrogen sulfide emission by cultured tobacco cells, *Z. Pflanzenphysiol.* **111** 189–202.

Rennenberg, H., (1984), The fate of excess sulfur in higher plants, *Ann. Rev. Plant Physiol.* **35,** 121–153.

Rennenberg, H., Arabatzis, N., Grundel, I., (1987), Cysteine desulfhydrase activity in higher plants: Evidence for the action of L- and D-cysteine specific enzymes, *Phytochem.* **26,** 1583–1589.

Renneberg, H., Thoene, B., (1987), The mobility of sulphur in higher plants,. in: *Proc. Int. Symp. Elemental Sulfur in Agriculture*, Vol.2, Nice pp. 701–707.

Rennenberg, H., (1989), Synthesis and emission of hydrogen sulfide by higher plants, in: E.S. Saltzman, W.J. Cooper (eds), *Biogenic sulfur in the environment.*, ACS Symposium series pp. 44–57.

Rennenberg, H., Lamoureux, G.L, (1990), Physiological processes that modulate the concentration of glutathione in plant cells, in: Rennenberg, H., Brunold, C., De Kok, L.J., Stulen, I. (eds), *Sulfur nutrition and sulfur assimilation in higher plants*, SPB Scientific Publishers, The Hague pp. 53–65.

Rennenberg, H., Huber, B., Schröder, P, Stahl, K., Haunold, W., Georgii, H.-W., Slovik, S., Pfanz, H., (1990), Emission of volatile sulfur compounds from spruce trees, *Plant Physiol.* **92,** 968–971

Rennenberg, H., (1991), The significance of higher plants in the emission of sulfur compounds from terrestrial ecosystems, in: Sharkey,Th.D., Holland, E.A., Mooney,H.A. (eds), *Trace Gas Emissions from Plants,* Academic Press, San Diego pp. 217–260.

Rennenberg, H., Schröder, P., Huber, B., (1991), Emission of reduced sulfur compounds from agricultural and forest ecosystems, in: *EUROTRAC Annual Reports, Part 4*, ISS Garmisch-Partenkirchen pp. 156–162.

Rennenberg, H., Schröder, P., Huber, B., (1992), Emission of sulfur compounds from spruce trees, in: Schwartz, S.E., Slinn W.G.N. (eds), *Precipitation scavenging and atmosphere surface exchange processes.* Hemisphere Publ. Washington pp. 979–990.

Schmidt, A., Rennenberg, H., Wilson, L.G., Filner, P., (1985), Formation of methanethiol from methionine by leaf tissue, *Phytochem.* **24,** 1181–1185.

Schröder, P., (1993), Plants as sources of atmospheric sulfur, in: L.J. DeKok *et al.* (eds), *Sulfur Nutrition and Assimilation in Higher Plants*, SPB Academic Publishing bv, The Hague pp. 253–270.

Schütz, B., DeKok, L.J., Rennenberg, H., (1991), Thiol accumulation and cysteine desulfhydrase activity in H_2S fumigated leaves and leaf homogenates of cucurbit plants, *Plant Cell Physiol.* **32/5,** 733–736.

Schwenn, J.D., Schriek, U., Kilz, H.H., (1982), Dissimilation of methionine in cell suspension cultures from *Catharanthus roseus* L., *Planta* **158,** 540–549.

Seinfeld, J.H., (1986), *Atmospheric chemistry and physics of air pollution.* John Wiley and Sons, New York.

Sekiya, J., Schmidt, A., Rennenberg, H., Wilson, L.G., Filner, P., (1982), Hydrogen sulfide emission by cucumber leaves in response to sulfate in light and dark, *Phytochem.* **21,** 2173–2178.

Sekiya, J., Schmidt, A., Wilson, L.G., Filner, P., (1982), Emission of hydrogen sulfide by leaf tissue in response to L-cysteine, *Plant Physiol.* **70,** 430–436.

Spaleny, S., (1977), Sulphate transformation to hydrogen sulphide in spruce seedlings. *Plant and Soil* **48,** 557–563.

Tarrason, L., Turner, S., Floisand, I., (1995), An estimation of seasonal DMS fluxes over the North Atlantic Ocean and their contribution of European pollution levels. *J. Geophys. Res.* in press.

Taylor, G.E., Tingley, D.T., (1983),Sulphur dioxide flux into leaves of *Geranium carolinianum* L., *Plant Physiology* **72**, 237–244.

Taylor, G.E., McLaughlin, S.B., Shriner, D.S., Selvidge, W.J., (1983), The flux of sulfur-containing gases to vegetation, *Atmos. Environ.* **17,** 789–796.

Thom, A.S., (1975), Momentum, mass and heat exchange of plant communities, in: Monteith J.L. (ed), *Vegetation and Atmosphere*, Academic Press, London pp. 58–109.

Turco, R.P., Whitten, R.C., Toon, O.B., Pollack, J.B., Hamill, P., (1980), OCS, stratospheric aerosols and climate, *Nature* **283,** 283–286.

Valdez, M.P., Bales, R.C., Stanley, D.A., Dawson, G.A., (1987), Gaseous deposition to snow 1. Experimental study of SO_2 and NO_2 deposition, *J. Geophys. Res.* **92,** 9779–9787.

van Dop, H., (1983), Terrain classification and derived meteorological parameters for inter-regional transport models, *Atmos. Environ.* **17,** 1099–1105.

van Hove, L.W.A., (1989), The mechanism of NH_3 and SO_2 uptake by leaves and its physological effects. Thesis, University of Wageningen, the Netherlands.

Vermetten, A.W.M., Hofschreuder, P., Duyzer, J.H., Diederen, H.S.M.A., Bosveld, F.C., Bouten, W., (1992), Dry deposition of SO_2 onto a stand of Douglas fir: The influence of canopy wetness, in: S.E.Schwartz, W.G.N.Slinn (eds), *Precipitation Scavenging and Atmosphere-Surface Exchange. Volume 3- The Summers Volume: Applications and Appraisals,* Hemisphere, Washington DC pp. 1403–1414.

Voldner, E.C., Barrie, L.A., Sirois, A., (1986), A literature review of dry deposition of oxides of sulfur and nitrogen with emphasis on long-range transport modelling in North America, *Atmos. Environ.* **20,** 2101–2123.

Warneck, P., (1988), *Chemistry of the Natural Atmosphere.* International Geophysics Series Vol. 41 Dmowska,R., Holton,J.R. (eds), Academic Press, San Diego, USA.

Welphdale, D.M., Shaw, R.W., (1974), Sulphur dioxide removal by turbulent transfer over grass, snow, and water surfaces, *Tellus* **26,** 196–204.

Wesely, M.L., Hicks B.B., (1977), Some factors that affect the deposition rates of sulfur dioxide and similar gases on vegetation, *J. Air Poll. Contr. Ass.*, **27,** 1110–1116.

Wesely, M.L., Sisteron, D.L, Jastrow, J.D., (1990), Observations of the chemical properties of dew on vegetation that affect the dry deposition of SO_2, *J. Geophys. Res.* **95,** 7501–7514.

Whitmore, M.E., Freer-Smith, P., (1982), Growth effects of SO_2 an/or NO_2 on woody plants and grasses during spring and summer, *Nature* **300,** 55–56.

Individual Reports from BIATEX Contributors

Chapter 8

Measurement Techniques

8.1 On–line Measurements of Ozone Surface Fluxes: Instrumentation and Methodology

Hans Güsten, Günther Heinrich, Erbo Mönnich, Detlev Sprung and Joseph Weppner

Institut für Meteorologie und Klimaforschung,Forschungszentrum Karlsruhe, Universität Karlsruhe, D-76021 Karlsruhe, Federal Republic of Germany.

Summary

A novel fast-response ozone sensor was developed for various environmental applications. The low weight and low power consumption open up applications which are not possible with conventional ozone analysers. The ozone sensor was successfully used in various environmental applications such as the on-line measurement of the vertical turbulent ozone flux onto various ecosystems (vegetation, soil, desert, sea water), high-resolution measurements of vertical ozone profiles and other airborne ozone measurements.

Aims of the research

The principal interest of our work in recent years has concentrated on the development of the instrumentation and methodology of eddy flux measurements of ozone in the atmospheric surface layer, in particular using the eddy correlation or covariance technique [1–3]. Ozone destruction at the earth's surface is the most important sink. The rate of ozone destruction differs widely with the type of surface and its vegetation. Thus, in order to assess the rate of direct ozone deposition onto various ecosystems, the technology for long term measurements of the dry deposition fluxes had to be developed.

Principal scientific results

A novel fast-response ozone sensor

In order to meet the requirements of the eddy correlation technique, the ozone sensor must have a very fast response time with high sensitivity to respond to the small and rapid fluctuations of ozone concentration in the surface layer [4, 5]. The measuring principle of the novel fast-response ozone sensor is the chemiluminescence of an organic dye adsorbed on dry silica gel in the reaction with ozone. Its basic features are a cheap, blue-sensitive side-window photomultiplier, a high-performance miniature fan, and a thin chemiluminescent disc. Details about the prototype ozone sensor are given in [5–7].

In Table 8.1.1 the basic characteristics of the novel ozone sensor are compared with a conventional ozone analyser. The robust and inexpensive ozone sensor was used in continuous measurements of the eddy flux at unattended sites for periods of up to several months and longer [8–10]. Generally, it is very difficult to prove that a fast-response trace-gas sensor is capable of measuring the vertical turbulent flux of the trace gas in question.

Table 8.1.1: The basic characteristics of a conventional ozone analyser and a novel ozone sensor.

	Conventional	Novel
Detection limit	1 ppb	< 0.1 ppb
Response time	20 seconds	< 0.1 second
Weight	~ 15 kg	< 1 kg
Energy consumption	~ 100 Watt	< 10 Watt

A second independent and fast-response ozone sensor was built in order to demonstrate that the recently developed novel ozone sensor for direct eddy flux measurements [6, 7] yields true vertical ozone eddy fluxes in field experiments. Unlike the novel ozone sensor based on the dry chemiluminescence of ozone with an adsorbed organic dye, the second ozone sensor utilises the well known chemiluminescence reaction of ozone with ethylene in the gas phase. A commercial ozone analyser (Bendix model 8002), using this reaction, was modified to feature higher sensitivity (~ 500 ppt) and faster response time (~ 3 Hz). The size of the modified sensor was reduced by more than 60 %. In Fig. 8.1.1 the diurnal variation of the vertical turbulent eddy flux of ozone is depicted for both ozone sensors. The simultaneous measurements were performed

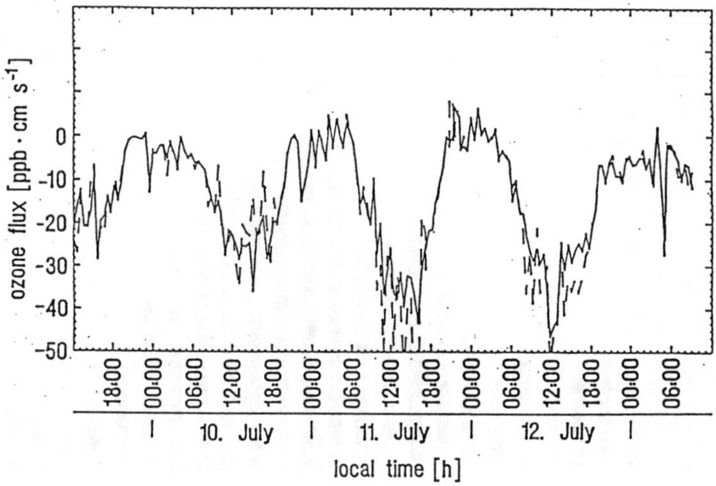

Fig. 8.1.1: Diurnal variation of the turbulent eddy flux of ozone over a wheat field in July 1991, measured simultaneously with two fast-response ozone sensors.

over a wheat field, with the two fast-response ozone sensors in cross correlation with one ultrasonic anemometer during four days.

Excellent agreement between the two ozone sensors is obtained which demonstrates that both fast-response ozone sensors measure directly true eddy fluxes.

Corrections of ozone flux measurements for density effects

Several researchers have shown that the measurements of the vertical turbulent flux of an atmospheric trace gas by micrometeorological techniques are affected by micro-turbulent density fluctuations caused by the fluxes of heat and/or water vapour [11–13]. In the atmospheric surface layer upward and downward fluxes of heat and water vapour cause expansion of the air and thus affect the density of the trace gas on its way to the surface. For appropriate corrections of the variations in density knowledge is required of the energy which is responsible for this effect. Thus, the sensible and latent heat fluxes during the eddy flux measurements of the trace gas in question have to be known.

In Fig. 8.1.2 the corrections for density fluctuations due to heat and water vapour transfer are shown for the mean diurnal turbulent ozone flux onto a barley field in July 1992. In accordance with the theory [11–13], corrections on the order of 10 % of the turbulent ozone flux during the day are needed.The contribution of the sensible heat flux is larger than that of the water vapour flux.

It is interesting to note that the often cited Webb correction [11] for density effects depends on sharing of the available energy between sensible heat and latent heat

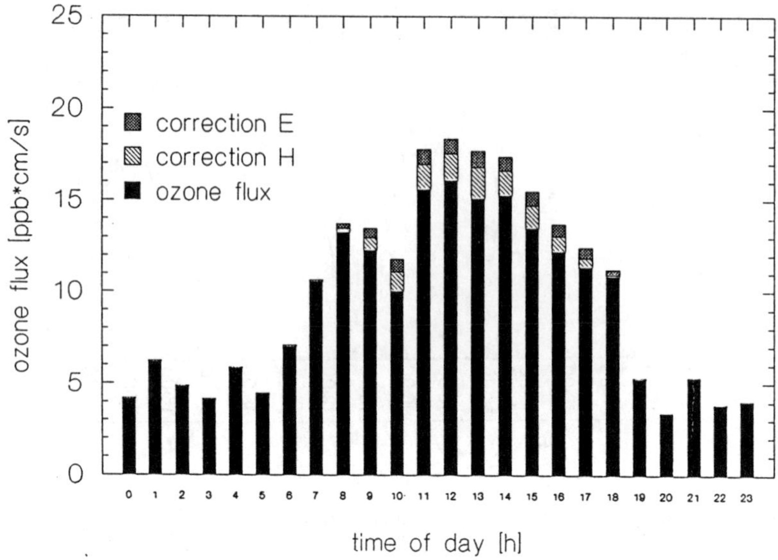

Fig. 8.1.2: Correction (Webb correction) of the mean vertical turbulent ozone flux onto a barley field in July 1992.

[13]. This is particular important at sites with high solar irradiation as in a desert, where a flux correction of up to 40% has to be made for the vertical turbulent ozone flux [10].

Long-term measurements of the ozone surface flux

The robust and light-weight ozone sensor was used successfully in continuous measurements of the eddy flux at our baseline measuring site near Bellheim (Palatinate) in southwest Germany. The measuring site is located at 49°10'33" N, 8 °16'35" E on a fairly flat agricultural terrain (elevation 130 m) of approximately 3×4 km^2 surface area in the Upper Rhine Valley providing a considerably homogeneous fetch of more than one kilometer upwind of the measuring tower. The slope of the entire field is less than 1 %. For long-term measurements the eddy sensors were mounted vertically on a single pole on top of a scaffold to minimise aerodynamic interferences. Fig. 8.1.3 shows the arrangement of the fast-response sensors – ultrasonic anemometer, ozone sensor and humidity sensor – mounted on a lean telescope tower.

For continuous measurement of the vertical turbulent deposition flux of ozone a real-time processing unit as well as a suitable software for on-line analysis of the data obtained were developed. Since at higher ozone concentrations in rural areas ozone leads to severe crop damage, it is necessary to assess the critical load of ozone onto various crops. The eddy correlation system described was used to measure the eddy flux of ozone onto various vegetations such as sunflowers [14],

Fig. 8.1.3: Eddy correlation system comprising a sonic anemometer, a fast-response ozone sensor, and a fast-response hygrometer.

wheat [9], barley [15], desert sand [10] or sea water. To prove the stability of the novel ozone sensor, the seasonal variations of the turbulent ozone flux was measured continuously over a barley field from April until August 1992 and over the flat harvested field from the beginning of August until the end of September.

A total of 2658 half hourly mean values of the turbulent ozone flux were obtained using the eddy correlation technique. In Fig. 8.1.4 the monthly averaged turbulent ozone fluxes are shown for the whole measuring period.

A statistical evaluation of the ozone fluxes divided into weekly mean periods provided the following information:

- The maximum weekly ozone fluxes onto barley vary by a factor of 3 (23–69 ppb cm s^{-1}). It is interesting to note that the lowest and the highest weekly maximum values of the ozone fluxes were measured in two consecutive weeks in July which indicates the strong influence of the weather on the ozone deposition flux (see Fig. 8.1.4).
- The weekly mean deposition velocities of ozone over the entire measurement period vary from 0.16 to 0.35 cm s^{-1}.

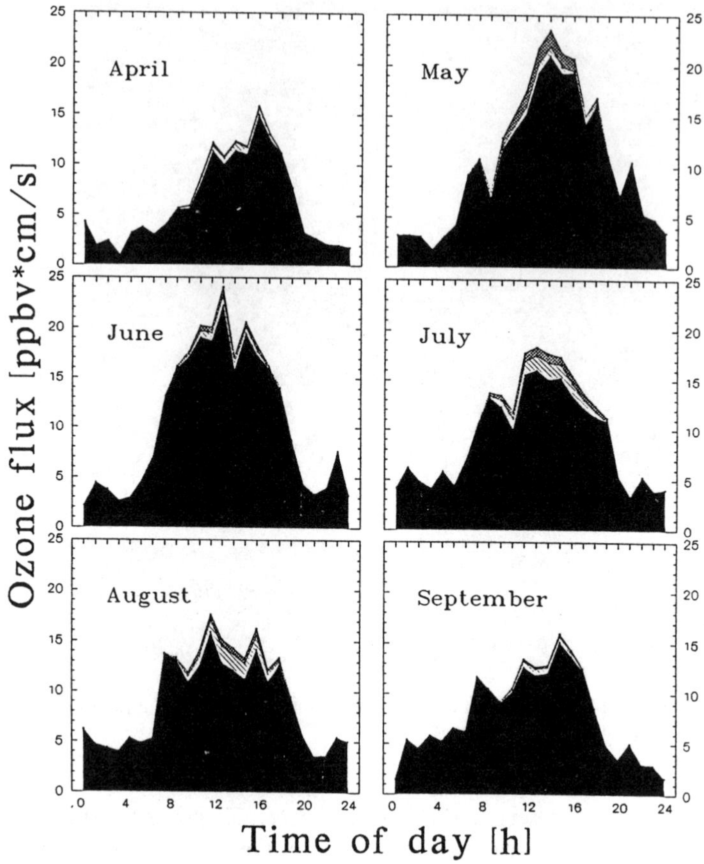

Fig. 8.1.4: Mean monthly averaged turbulent ozone fluxes onto a barley field and onto the harvested field, from April until October 1992. The ozone fluxes are corrected for density fluctuations (Webb correction).

- The largest values of the mean weekly ozone fluxes over the entire measurement campaign were obtained in June during the period of strong increase in plant growth (30 to 85 cm) as well as that of the leaf area index (1.5 to 7).

Vertical profiles of ozone in the Upper Rhine Valley during TRACT '92

During the TRACT field measurement campaign in September 1992, 42 vertical profiles of ozone and meteorological variables (temperature, pressure, humidity, wind velocity and wind direction) were measured in the so-called nested area using a prototype of the novel light-weight ozone sensor [16, 17]. The tethered balloon allowed vertical profiles to be measured up to 1000 m altitude. The vertical ozone profiles show the interesting phenomena of nocturnal surface inversion, its disappearance during the morning, and re-establishment in the evening. On sunny days in the Upper Rhine Valley, the nocturnal inversion layers due to radiative cooling at the surface are preferably formed in the evenings. The nocturnal inversion layers have often shallow mixing heights of less than 100 metres. Below the nocturnal inversion layer ozone is rapidly depleted in the evening hours. Downward diffusion of ozone rich air masses from aloft the inversion layer is inhibited during the evening by the atmospheric stability. In order to discriminate between two possible mechanisms for the nocturnal ozone depletion below the inversion layer – dry deposition onto the ground and/or chemical reactions - a simple model is developed (Fig. 8.1.5).

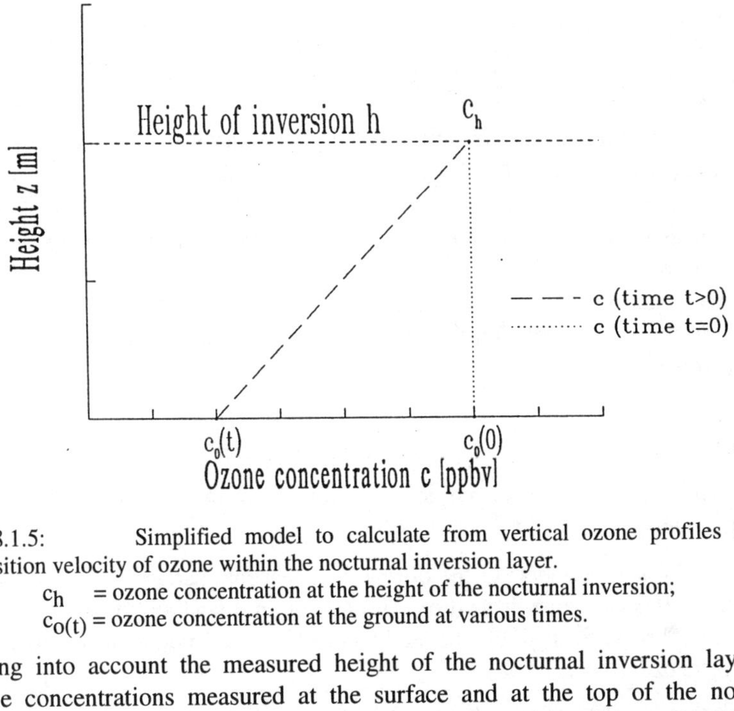

Fig. 8.1.5: Simplified model to calculate from vertical ozone profiles the dry deposition velocity of ozone within the nocturnal inversion layer.

c_h = ozone concentration at the height of the nocturnal inversion;

$c_{o(t)}$ = ozone concentration at the ground at various times.

Taking into account the measured height of the nocturnal inversion layer, the ozone concentrations measured at the surface and at the top of the nocturnal inversion as well as an exponential decay of the ozone concentration from the onset of formation of the nocturnal inversion, a simple model allows the dry deposition of ozone within the nocturnal inversion layer to be calculated:

$$C_o(t) = C_h e{-}2/h\ V_d t.$$

Thus, the surface ozone concentration under the nocturnal inversion layer undergoes an exponential decay with a first order rate constant equal to $-2V_d/h$. In Table 8.1.2 the dry deposition velocities obtained by this balloon-borne method of determination are compared with the ground based measurements of the dry deposition velocity simultaneously measured in the nested area of TRACT at the Gamshurst site, 5 km away from the site of measurements with the tethered balloon. Both sites were covered with short grass during TRACT. In Table 8.1.2 the results are compiled of the dry deposition velocities obtained by the two independent determination methods [17].

Table 8.1.2: Comparison of the dry deposition velocities of ozone obtained by the balloon-bourne method and by the eddy correlation technique.

Date [1992]	Time* [CET]	h [m]	O_3C_h [ppb]	$\overline{V}_d$ (eddy) [cm s^{-1}]	V_d (balloon) [cm s^{-1}]
Sept. 11	16.30–19.15	70	65	0.12	0.34
Sept. 16	18.15–20.15	90	50	0.14	0.70
Sept. 17	18.00–18.45	30	40	–	0.25
Sept. 18	17.00–18.30	30	27	0.14	0.30

The $\overline{V}_d$ data from the eddy correlation technique are derived from the half hourly averaged vertical turbulent flux measurements which are then averaged, yielding the mean value for the time given in Table 8.1.2. Since the ground based measurements of the dry deposition velocity of ozone cover only 20–50 % of the balloon-borne determination method, it is concluded that in addition to the dry deposition process also chemical reactions with the oxides of nitrogen are responsible for ozone depletion within the nocturnal inversion layer in the Upper Rhine Valley [16, 17].

Conclusions

- A novel fast-response ozone sensor proved to be successful in various environmental applications.
- Long-term measurements of the diurnal and seasonal variations of the vertical turbulent flux of ozone deposition onto agricultural land and other ecosystems can best be performed by the eddy correlation technique using this fast-response ozone sensor.
- Turbulent daytime ozone deposition fluxes need to be corrected due to microturbulent density fluctuations caused by the concomitant fluxes of heat and water vapour.
- The novel light-weight ozone sensor proved to be very effective for measuring vertical ozone profiles with a tethered balloon.
- The height of the nocturnal inversion layer can be determined from the vertical ozone profiles. Below the nocturnal inversion layer in the Upper Rhine Valley ozone is rapidly depleted by dry deposition onto the ground and by chemical reactions with the nitrogen oxides.

Acknowledgements

This project has been supported in part by funds of the German Federal Ministry for Research and Technology, reference No. 07EU811/2. The responsibility for the contents of this publication is assumed by the authors.

References

1. D. Fowler, J.H. Duyzer, in: M.O. Andreae and D.S. Schimel (eds), *Exchange of Trace Gases between Terrestrial Ecosystems and the Atmosphere,* John Wiley and Sons, Ltd. 1989, pp. 189–207.
2. S.B. Verma, *Remote Sens. Rev.* **5** (1990) 99–115.
3. H. Güsten, *GIT Fachz. Chem.* **36** (1992) 544–551.
4. J.A. Businger, A.C. Delany, *J. Atmos. Chem.* **10** (1990) 399–410.
5. H. Güsten, G. Heinrich, *Atmos. Environ.*, **30** (1996) 897–909.
6. H. Güsten, G. Heinrich, R.W.H. Schmidt, U. Schurath, *J. Atmos. Chem.* **14** (1992) 73–84.
7. H. Güsten, G. Heinrich, R.W.H. Schmidt, U. Schurath, in: R.D. Hudson (ed), *Ozone in the Troposphere and Stratosphere*, Part 1, NASA Conference Publication 3266, Goddard Space Flight Center, Greenbelt, MD, USA 1994, pp. 127–129.
8. H. Güsten, N. Ernst and G. Heinrich, *Ann. Meteorol.* **27** (1992) 84–85.
9. N. Ernst, H. Güsten, G. Heinrich, E. Mönnich, J. Weppner, in: P. Borrell, P. M. Borrell, W. Seiler (eds), *Proc. EUROTRAC Symp. 1990*, SPB Academic Publ. bv, The Hague 1991, pp. 127–129.
10. H. Güsten, G. Heinrich, E. Mönnich, D. Sprung, A.B. Ramadan, M.R.M. EzzEl–Din, D.M. Ahmed, G.K.Y. Hassan, *Atmos. Environ.*, **30** (1996) 911–918.
11. E.K. Webb, G.I. Pearman, R. Leuning, *Quart. J. Roy. Meteorol. Soc.* **106** (1980) 85–100.
12. K. Bernhardt, H. Piazena, *Z. Meteorol.* **38** (1988) 234–245.
13. E. Pattey, R.L. Desjardin, F. Boudreau, P. Rochette, *Boundary layer Meteorol.* **59** (1992) 195–203.
14. H. Güsten, N. Ernst, G. Heinrich, in: P.M. Borrell, P. Borrell, T. Cvitas, W. Seiler (eds), *Proc. EUROTRAC Symp. 1992*, SPB Academic Publ. bv, The Hague 1993, pp. 739–742.
15. D. Sprung, G. Heinrich, H. Güsten, to be published.
16. H. Güsten, G. Heinrich, E. Mönnich, and D. Sprung, in: P.M. Borrell, P. Borrell, T. Cvitas, W. Seiler (eds), *Proc. EUROTRAC Symp. 1994*, SPB Academic Publ. bv, The Hague 1994, pp. 639–642.
17. H. Güsten, G. Heinrich, E. Mönnich, D. Sprung, in: A.J. Ranzieri and P.A. Solomon (eds), *Regional Photochemical Measurements and Modeling Studies*, Air and Waste Management Association, Pittsburgh, USA 1995, pp. 221–226.

8.2 Real-time and Non-intrusive Detection of Ambient Ammonia using the Photothermal Deflection Technique

H.S.M. de Vries[1], F.J.M. Harren[1], G.P. Wyers[2], R.P. Otjes[2], J. Slanina[2] and J. Reuss[1]

[1]Department of Molecular and Laser Physics, University of Nijmegen, Toernooiveld, 6525 ED Nijmegen, The Netherlands.
[2] Netherlands Energy Research Foundation ECN, P.O. Box 1, 1755 ZG Petten, The Netherlands.

Summary

The recent development of a non-intrusive photothermal deflection (PD) instrument allows the quantitative detection of local ammonia concentrations in air to be measured continuously, sensitively and fast. Ammonia is vibrationally excited by an infrared CO_2 laser in an intracavity configuration. A HeNe beam passes over the CO_2 laser beam in a multipass arrangement and is deflected by a thermal, CO_2 laser induced, refractive index gradient. This deflection, due to the mirage effect, is a measure for the local ammonia concentration. In ambient air the detection limit is 0.5 ppb and the spatial resolution is 2.5 cm times π (0.282 mm^2). The time resolution is 0.1 s for single line operation and 15 s for multi-line operation. The system is fully automated and suited for measuring periods of at least one week. Results are compared with those of a continuous-flow denuder system.

Aims of the research

Ammonia and ammonium contribute to soil acidification and disturb the nutrient cycle. In the Netherlands dry deposition of ammonia, mainly originating from emission sources (livestock wastes) in the Netherlands itself, is responsible for approximately 33 % of total acidifying deposition [1]. Fast and sensitive detection methods for ammonia are needed to quantify air-to-surface exchange fluxes [2]. Attempts toward quantitative detection of ammonia employing (isotopic) CO_2 laser based photoacoustics and (less sensitive) direct absorption detection systems recently received attention [3–6]. Ammonia possesses high absorption coefficients and also a clear fingerprint spectrum in the $^{12}CO_2$ and $^{13}CO_2$ laser wavelength region [7]. Photoacoustics allows detection of ammonia at sub-ppb levels, however a serious drawback is its necessity of using a flow-through system which leads to sticking of ammonia to tubing and photoacoustic cell walls on account of its permanent dipole moment. This impedes fast time responses during measurements.

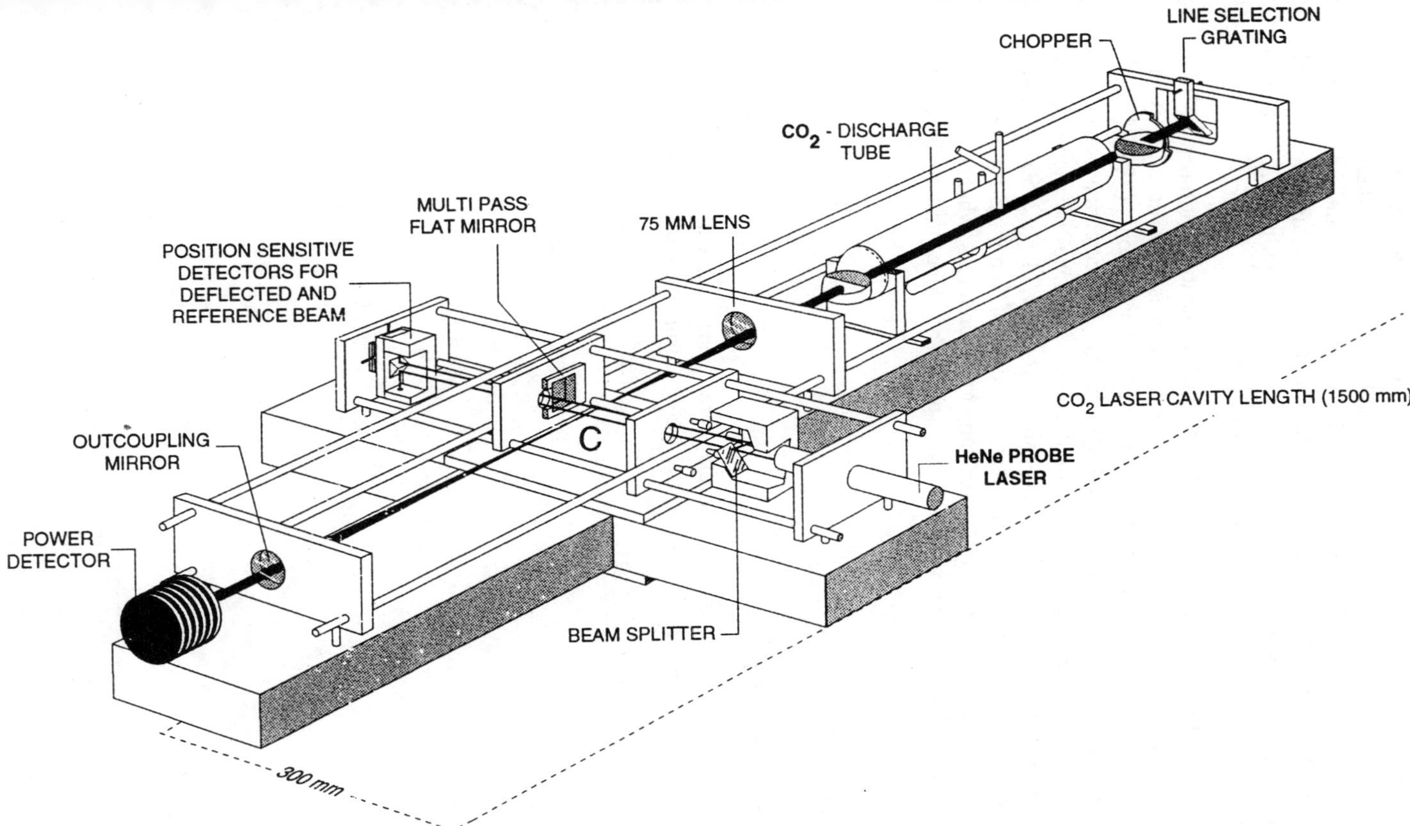

Fig. 8.2.1: The photothermal deflection rigid body frame, consisting of an intracavity CO_2 laser setup (including line selecting grating, chopper, 7.5 cm lens and out coupling mirror) and dual HeNe probe laser beams, which are obtained by a beam splitter unit. Above the multipass area the air is sampled by the rotating annular denuder unit. The interaction region of probe beam and CO_2 laser beam is presented in detail. The probe beam influenced by the mirage effect is passing over the CO_2 laser beam at a distance of about 1 mm; the reference beam at a distance of 7 mm under the CO_2 laser beam.

Here, the photothermal deflection (PD) technique is applied to perform fast measurements of ammonia concentrations in ambient air. The method is closely related to the photoacoustic technique in which a change in pressure is detected by microphones [10]. In PD a change in temperature induces a change in refractive index which is probed with a second laser beam [11]. The photothermal deflection set-up is shown in Fig. 8.2.1. and described elsewhere [12].

We have calibrated the laser-based photothermal detection instrument with a relatively inexpensive and thoroughly tested continuous-flow denuder (CFD) system; the latter is sufficiently selective and sensitive for ammonia (8 ppt) [8]. To compare estimated ammonia concentrations in the PD detection zone with continuous-flow denuder (CFD) values, a polyethylene tube was positioned above this area to sample the air. This system is calibrated with a blank, a 50 ppb and a 500 ppb NH_4^+ solution. However, the response time of 1 minute for the CFD system is not sufficiently fast for eddy correlation measurements, which rely on the covariation of fluctuations in vertical wind speed and concentration to derive ammonia fluxes [9].

Principal scientific results

Comparison of absolute values

In our laboratory ambient ammonia levels were determined with PD and CFD around the clock during periods of up to one week. During the night levels were on

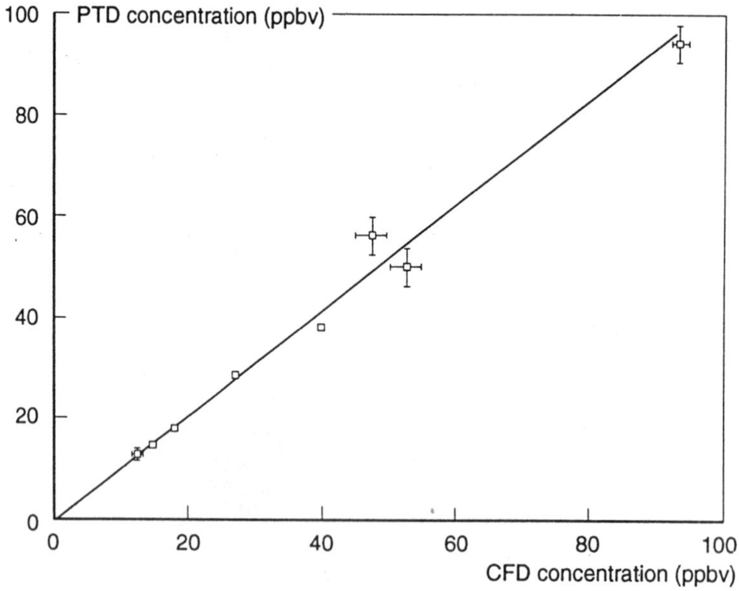

Fig. 8.2.2: Correlation of ammonia concentrations obtained with CFD and PD (slope of 1.02(5)).

the average 3 ppb lower than daytime levels, which slowly fluctuated between 10 and 20 ppb. The denuder tubing was positioned 100 mm above the PD sampling area. Ammonia sources were prepared from 0.25 %–2.5 % ammonia solutions in demineralized water to obtain concentrations of about 10, 30, 60 and 120 ppb in the sampling zone. The location of the source was at a distance of 50 cm from denuder tube and PD detection area. The correlation of CFD and PD signals is shown in Fig. 8.2.2.

For CFD the calibration was based on reference solutions. For PD the calibration was based on C_2H_4 measurements, corrected for saturation effects [10]. The absorption strengths (20.5 cm^{-1} for NH_3 on the 10R8 and 13.4 cm^{-1} for C_2H_4 on the 10P14) [13] and the laser power were taken into account.

Fast ammonia fluctuations

A double flow system, either of normal air containing ambient ammonia concentrations, or of air containing above-ambient ammonia levels, was used to switch within seconds between different ammonia concentrations. The flow was directly led into the detection zone, by positioning the tubing 20 mm below the crossing point of the two lasers. The denuder tubing was installed 20 mm above the multipass system. These short distances allowed to investigate rapidly changing ammonia concentrations arriving quasi-simultaneously at both detection systems.

The results are presented in Fig. 8.2.3. The arrows indicate the times at which higher levels of ammonia were flushed into the detection region, during 1, 2 or 3 seconds. The CFD technique demonstrates a smeared-out effect (dashed curve), whereas the PD setup (single line operation at the ammonia absorbing CO_2 laserline 10R8) reveals the ammonia passing in pulses through the detection zone. The signal on the non-absorbing laser line (10R10; lower trace Fig. 8.2.3) remains constant. It was impossible to measure simultaneously on the 10R8 and 10R10 CO_2 laser line with the actual PD system. Therefore, these control measurements were performed directly after the last 10R8 measuring period of 1 minute, *i.e.* 9 minutes after the start of this experiment, see Fig. 8.2.3. The flushing periods of 2, 3 and 3 seconds, respectively were similar for the 10R8 and 10R10 laser line experiments.

The signal is not disturbed by air turbulences themselves or variable water concentrations due to the ammonia solution used, as witnessed by the steady signal on the non-absorbing laser line (10R10). Note that the ammonia concentration can be obtained by subtracting both signals on the 10R8 and 10R10. In this way the influence of water vapor and carbon dioxide is minimized.

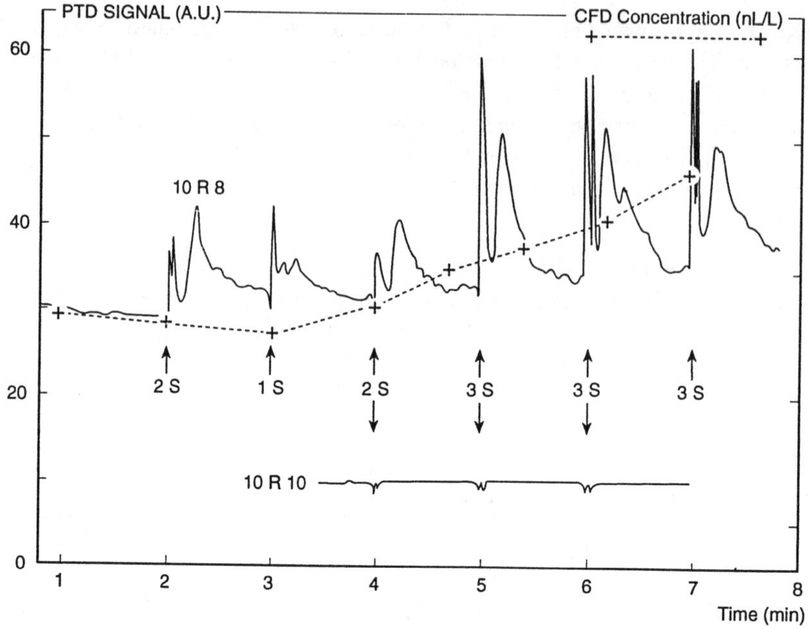

Fig. 8.2.3: Fast ammonia changes, by flushing ammonia enriched air through the detection area during 1, 2 or 3 seconds. CFD demonstrates a slowly increasing curve (dashed), while PD monitors fast changes on the 10R8 ammonia absorbing laser line and no change on the 10R10 non absorbing laser line.

Conclusions

The advantage of PD over currently existing techniques is the possibility to determine local ammonia concentrations in an open air environment, without disturbing the immediate environment. It also avoids the problem of gases sticking to wall material in conventional flow-through systems thus eliminating memory effects and time delays. The PD method is particularly suitable for eddy correlation measurements, where wind speed, wind direction and gas concentration are simultaneously determined. Injections of diluted ammonia in an air stream through the detection area show the fast time response of PD. A strongly absorbed CO_2 laser line by ammonia revealed ammonia entering the detection zone "in waves". The signal on a reference CO_2 laser line, where ammonia shows almost no absorption, was not affected. This indicates that other absorbing trace gases like water vapor and CO_2 did not disturb this experiment. In parallel with our PD system a relatively slow but very sensitive continuous-flow denuder (CFD; response time 60 s, detection limit 8 pL/L) monitored absolute ammonia concentrations. CFD demonstrated a kind of integrated ammonia concentration profile after the injections. CFD had also been utilized to show that PD permits easy absolute calibration for ammonia concentrations in the range of 10 to 120 nL/L.

PD detection is especially suited for reactive/adsorbing gases like ammonia and ozone; field measurements on ammonia could be carried out with a $^{13}CO_2$ laser to increase the sensitivity and to better discriminate against CO_2 background.

Acknowledgements

This work has been made possible through the financial support of the University of Nijmegen, The European Union (Science-project) and the Dutch Technology Foundation (STW).

References

1. G.J. Heij, J. Schneider, Report 200-09, Dutch priority programme on acidification, Bithoven 1991.
2. J.W. Erisman, A.W.M. Vermetten, W.A.H. Asman, A. Waijers-Ijpelaan, J. Slanina, *Atmos. Environ.* **22** (1988) 1153.
3. A. Olafsson, M. Hammerich, J. Bülow, J. Henningsen, *Appl. Phys.* **B49** (1989) 91.
4. W. Meienburg, H. Neckel, J. Wolfrum, *Appl. Phys.* **B51** (1990) 94.
5. R.A. Rooth, A.J.L. Verhage, L.W. Wouters, *Appl. Opt.* **29** (1990) 3643.
6. J.J.A.M. Sauren, D. Bicanic, K. van Asselt, *Infrared Phys.* **31** (1991) 475.
7. S.A. Trushin, *Ber. Bunsenges. Phys. Chem.* **96** (1992) 319.
8. G.P. Wyers, R.P. Otjes, J. Slanina, *Atmos. Environ.* **27** (1993) 2085.
9. D.W. Stocker, D.H. Stedman, K.F. Zeller, W.J. Massman, D.G. Fox, *J. Geophys. Res.* **98** (1993) 12619.
10. F.J.M. Harren, F.G.C. Bijnen, J. Reuss, L.A.C.J. Voesenek, C.W.P.M. Blom, *Appl. Phys.* **B50** (1990) 137.
11. W.B. Jackson, N.M. Amer, A.C. Boccara, D. Fournier, *Appl. Opt.* **20** (1981) 1333.
12. H.S.M. de Vries, N. Dam, M.R. van Lieshout, C. Sikkens, F.J.M. Harren, J. Reuss, *Rev. Sci. Instrum.* in press.
13. H.S.M. de Vries, Local trace gas measurements by laser photothermal detection; physics meets physiology, Thesis, University of Nijmegen, 1994.

8.3 Development of Instrumentation for Measurement of Concentrations and Surface-Exchange Fluxes of Air Pollutants

G.P. Wyers, R.P. Otjes, A. Wayers, J.J. Möls, A. Khlystov, P.A.C. Jongejan and J. Slanina

Netherlands Energy Research Foundation ECN,P.O. Box 1, 1755 ZG Petten, The Netherlands

Summary

Automated denuder systems have been developed for measurement of concentrations and exchange fluxes of ammonia and for simultaneous measurement of concentrations of nitric acid, nitrous acid, hydrochloric acid and sulfur dioxide. A new technique was established for sampling and analysis of aerosol, based on growth of particles under conditions of supersaturation and collection of the droplets in a cyclone. Surface exchange of ammonia has been studied over wheat, heathland and coniferous forest. Deposition to leaf surfaces is very efficient under humid conditions. Emission of ammonia from semi-natural vegetation was observed under dry conditions during the day. The dry deposition velocity of different size classes of particles was measured over a coniferous forest. The deposition velocity for sulfates showed a near-linear increase with u_* and was twice as high as previously inferred, 1.2–1.5 cm s^{-1} on average.

Aims of the research

- The development of analytical-chemical techniques for the measurement of concentrations and surface-exchange fluxes of acidifying and eutrophying air pollutants
- The application of these techniques to study surface-exchange of air pollutants

Principal scientific results

Continuous-flow denuder for ammonia

An automated denuder was developed for the measurement of atmospheric ammonia [1]. Ammonia is absorbed in a solution of $NaHSO_4$ on the walls of a rotating annular denuder (Fig. 8.3.1). The solution is removed from the denuder at a constant rate. After addition of NaOH this solution passes a semi-permeable membrane. Ammonia permeates the membrane and dissolves at the other side in a stream of double-demineralised water. The ammonium concentration in the water is measured by conductometry. The detection limit of this instrument is 6 ng NH_3 m^{-3}, the response time is 1.5 min and a relative precision better than 2 % can be obtained. ECN has participated with this instrument in a field intercomparison experiment for ammonia and ammonium in Rome (30 April–4 May 1990), organised by I. Allegrini on behalf of the CEC.

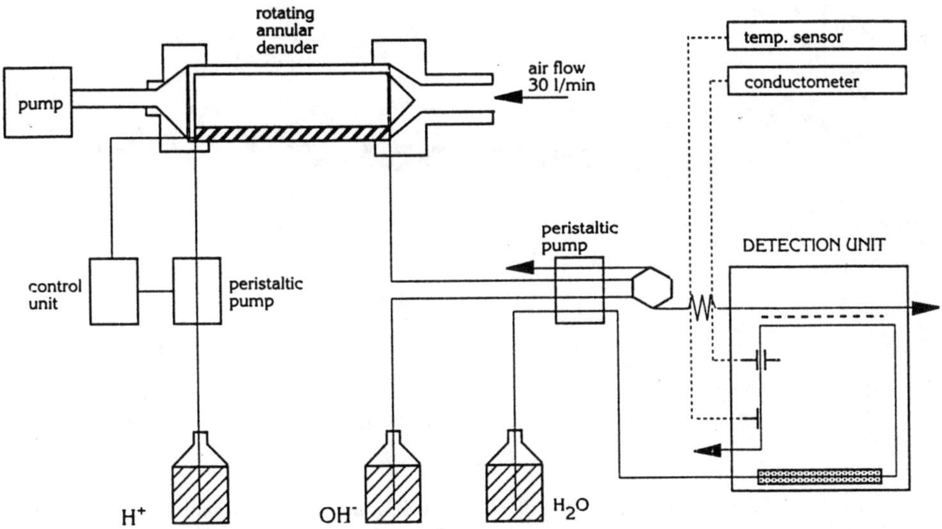

Fig. 8.3.1: Schematic diagram of the continuous flow denuder.

Surface-exchange of ammonia

Fluxes of ammonia over a wheat field were measured in Manndorf, Germany, in July 1990. Two V_2O_5 coated thermodenuders [2] were used to measure the vertical concentration gradients. A very consistent diurnal pattern was observed with deposition of NH_3 to the canopy at night and emission from the canopy during the day.

With the continuous-flow denuder surface exchange of ammonia was investigated over heathland at two sites in the Netherlands, Leenderheide (April/May 1991) and Elspeetsche Veld: (September–November 1991). At both sites it was observed that at a relative humidity (RH) above 80 % ammonia is deposited very efficiently with often a negligible surface resistance, indicating that deposition to the leaf-surfaces is the major pathway for ammonia under humid conditions. Sometimes a clear anti-correlation between the surface resistance and RH could be detected (Fig. 8.3.2), showing the influence of thin water layers at the canopy on the deposition of ammonia. Under dry (low RH) conditions during the day ammonia was emitted from the canopy (Fig. 8.3.3). No evidence was found for SO_2-NH_3 codeposition. However, after prolonged periods with large deposition fluxes of ammonia or

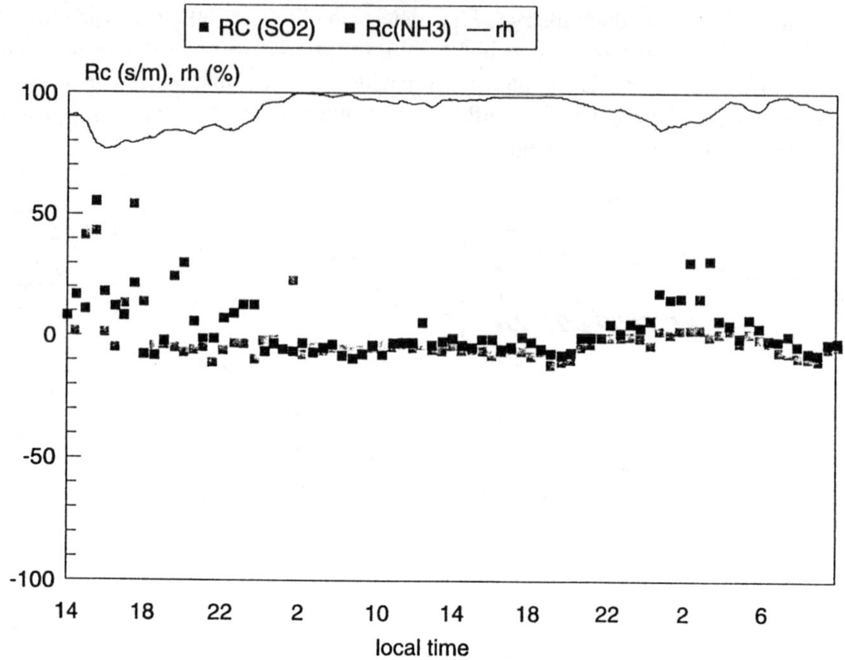

Fig. 8.3.2: Surface resistances (R_c) of NH_3 and SO_2 and relative humidity (RH) for the Elspeetsche Veld heathland, 6–7 November 1991.

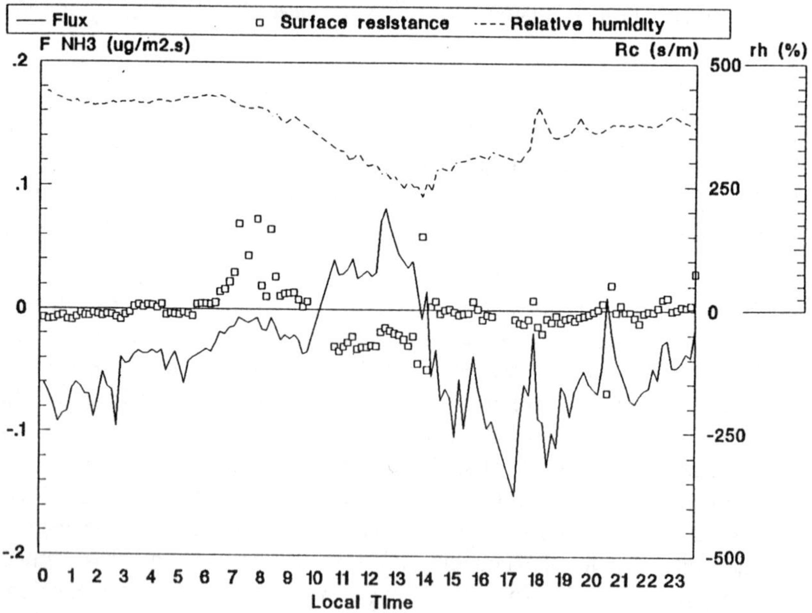

Fig. 8.3.3: Flux and surface resistance of ammonia and relative humidity for the Leende heathland on 1 May 1991. Strongly negative R_c at 10:00–14:00 represents ammonia emission.

sulfur dioxide it was sometimes observed that the surface resistance increased due to saturation of the leaf-surfaces with NH_3 or SO_2. The results of the measurements at Leenderheide and Elspeetsche Veld are discussed extensively in Erisman and Wyers [3].

From November 1992–November 1994 fluxes of ammonia over Speuld, a coniferous forest, were continuously monitored. The surface exchange of ammonia over this forest was largely similar to that observed over heathland: efficient deposition under humid (RH > 80 %) conditions, increased surface resistance at lower RH and emission during the day under dry conditions. Time series of NH_3 fluxes could be well described with a model developed by Sutton, based on reversible exchange with the stomata and irreversible deposition to leaf surfaces [4].

Rotating denuder for nitric acid, nitrous acid, hydrochloric acid and sulfur dioxide

A rotating annular denuder was connected to a Metrohm 690 ion chromatograph (IC) for simultaneous on-line analysis of SO_4^{2-}, SO_3^{2-}, NO_3^-, NO_2^- and Cl^- in the denuder solution. HNO_3, HNO_2, HCl and SO_2 are absorbed in a K_2CO_3 solution on the walls of a rotating annular denuder. A 5 mL loop is slowly filled with the denuder solution during 30 min. Bromide is added as an internal standard. The content of the loop is preconcentrated on a collection column and injected onto the analytical column of the IC. During the IC analysis a second loop is filled with a new denuder sample. Repeated analysis of standard solutions yielded relative standard deviations of 0.6–3.1 % for these anions. Thirty minute average

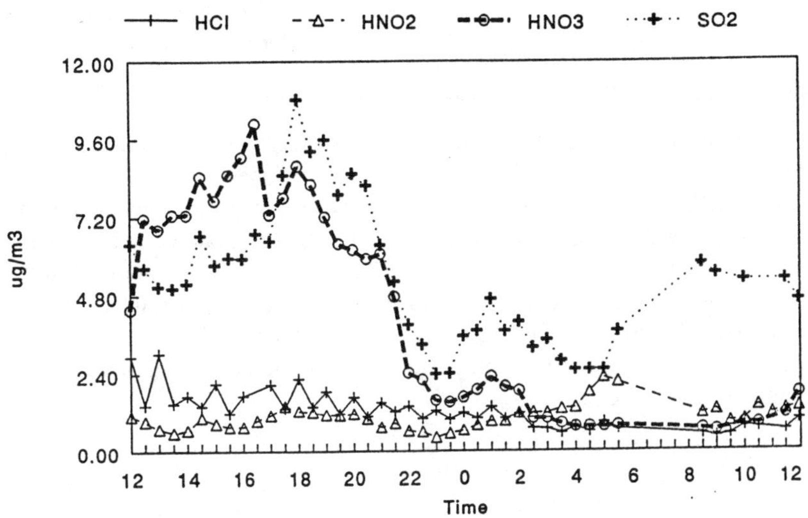

Fig. 8.3.4: Concentrations of HNO_3, HNO_2, HCl and SO_2 , measured simultaneously with a rotating annular denuder with on-line IC. Concentrations are measured during a period with photochemical smog on 30 June and 1 July 1992 near Petten.

concentrations of HNO_3, HNO_2, HCl and SO_2 measured with this system at the ECN site during a photochemical smog episode (30 June–1 July 1992) are shown in Fig. 8.3.4.

Steam jet aerosol collector (SJAC)

A new principle of sampling aerosol particles has been established which is based on forced growth of the particles and collection of the grown droplets in a cyclone [5]. An air flow containing particles, from which soluble gases are previously removed by a denuder, is rapidly mixed with steam (Fig. 8.3.5). The resulting supersaturation causes aerosol particles to rapidly grow into droplets.

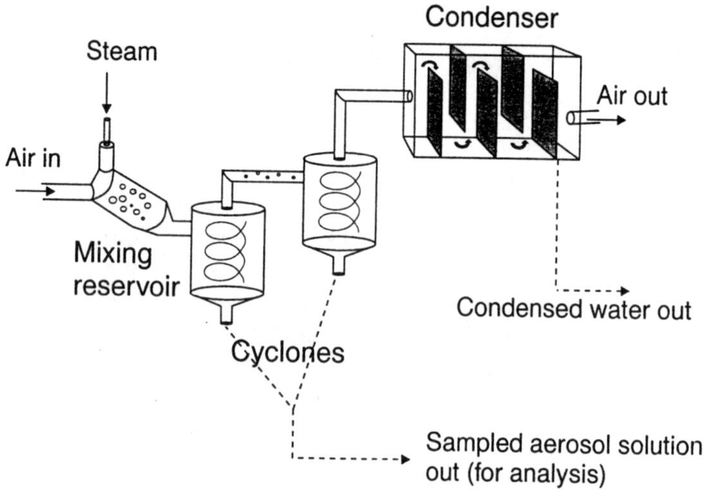

Fig. 8.3.5: Schematic diagram of the steam jet aerosol collector (SJAC).

The droplets containing the dissolved aerosol species are then collected by two cyclones in series. The solution from the cyclones is analysed by ion chromatography or flow-injection analysis. By DMPS measurements the prototype of this sampler was shown to quantitatively collect particles down to a diameter of several nanometers. The mass sampling efficiency is 99 % for ambient aerosol. Good agreement was obtained between this technique and denuder-filter pack measurements for sulfate, ammonium, nitrate and chloride. The sensitivity of this instrument is such that aerosol composition can be measured with a 15 min time resolution (Fig. 8.3.6).

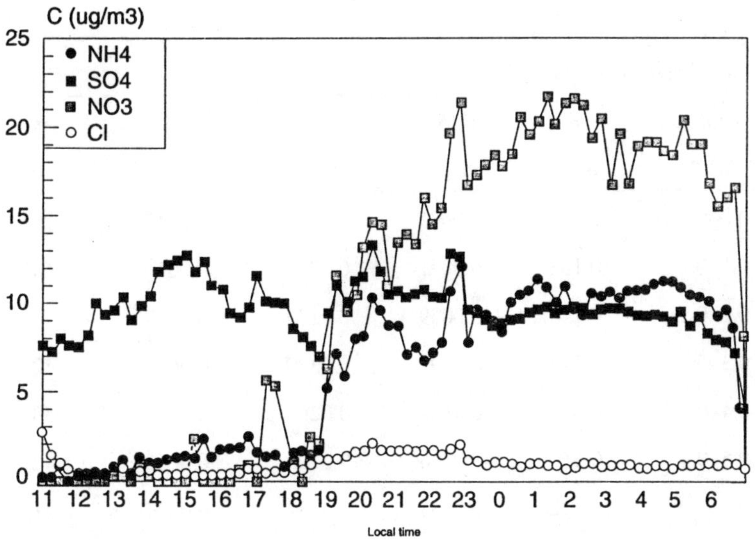

Fig. 8.3.6: Time series of SO_4^{2-}, NH_4^+, NO_3^- and Cl^- in aerosol measured with the SJAC.

Dry deposition of aerosol on a coniferous forest

To improve estimates of the contribution of dry deposition of particulate material to acidification and eutrophication of semi-natural ecosystems, experiments were carried out above a Douglas fir forest at Speuld. These experiments served to quantify the deposition velocity of different size classes of aerosol [6].

The deposition velocity of ammonium(bi)sulfate (MMD 0.8 mm, σ_g 2) was derived from vertical concentration gradients measured with CuO coated thermodenuders [7] using flux-profile relationships. The deposition velocity has a near-linear dependency on u_* (Table 8.3.1) and an average value of 0.012–0.015 m s^{-1} (for the period May–August 1993). This V_d exceeds values observed in previous micrometeorological studies over forests and is more than twice as high as the V_d predicted from a parameterisation which is currently used to estimate aerosol deposition in the Netherlands. The annual flux of sulfate to this site is estimated at 200–240 mol SO_4 ha^{-1} a^{-1}, implying that sulfate aerosol contributes for approximately one third to the dry deposition of SO_X at Speuld. If the substantial contribution of dry deposition of sulfate aerosol is taken into account, the discrepancy between deposition estimates by inference and from throughfall will be considerably reduced.

The dry deposition of ^{214}Pb was investigated by measuring airborne activity and activity accumulated on Douglas fir branches. This radioisotope is a natural tracer for submicron aerosol (AMD 0.4 mm, σ_g 3). The correlation of V_d with u_* is poor. The average deposition velocity ranges from 0.0061–0.0110 m s^{-1}, with a best estimate of 0.0073 ± 0.0049 m s^{-1}, a factor 1.6–2.1 smaller than the deposition

Table 8.3.1: Deposition velocities for different u_* ranges. Selection on u (> 1m s^{-1}) and sufficient fetch

u_* range	u_*	$V_d(50)$	V_d	s	n
m s^{-1}	m s^{-1}	m s^{-1}	m s^{-1}	m s^{-1}	
0.0–0.1	0.058	0.0005	0.0009	0.0012	28
0.1–0.2	0.153	0.0021	0.0049	0.0016	64
0.2–0.3	0.245	0.0059	0.0139	0.0437	71
0.3–0.4	0.348	0.0092	0.0300	0.0600	50
0.4–0.5	0.449	0.0184	0.0261	0.0349	53
0.5–0.7	0.610	0.0196	0.0306	0.0462	42
0.7–1.0	0.822	0.0365	0.0584	0.0827	31
all data	0.357	0.0062	0.0216	0.0480	339

velocity for sulfate. The difference is presumably due to the smaller size of ^{214}Pb as compared to sulfates.

The turbulent deposition of fog water to the canopy was directly measured by eddy correlation. The deposition by sedimentation was calculated from measured droplet size distributions. The turbulent deposition velocity is proportional to u_*^2 and is approximately half the value of the deposition velocity for momentum. The contribution of deposition of pollutants via fog water to the total input of acidifying compounds on forests is limited to approximately 5 % on an annual basis.

Conclusions

The techniques that have been developed in this project have proven very useful tools in the study of surface exchange of air pollutants. The use of these new techniques has contributed considerably to improvement of the understanding of the surface exchange of ammonia and aerosol.

Acknowledgements

This project was part of the EUROTRAC BIATEX programme and the Dutch Priority Programme on Acidification. Funding of this research by the Ministry of Housing, Spatial Planning and Environment and by the Ministry of Economic Affairs is gratefully acknowledged.

References

1. Wyers, G.P., Otjes, R.P., Slanina, J., *Atmos. Environ.* **27** (1993) 2085.
2. Keuken, M.P., Wayers, A., Möls, J.J., Otjes, R.P., Slanina, J., *Atmos. Environ.* **23** (1989) 2177.
3. Erisman, J.W., Wyers, G.P. *Atmos. Environ.* **27** (1993) 1937.
4. Sutton, M.A., Schjoerring, J.K., Wyers, G.P. *Phil. Trans. R. Soc. Lond. A*, **351** (1995) 261.
5. Khlystov, A., Wyers, G.P., Slanina, *J. Atmos. Environ.* **29** (1995) 2229.
6. Wyers, G.P., Veltkamp, A.C., Vermeulen, A.T., Geusebroek, M., Wayers, A., Möls, J.J. Report ECN-C--94-051 (1994) Netherlands Energy Research Foundation, Petten.
7. Slanina, J., Schoonebeek, C.A.M., Klockow, D., Niessner, R. *Anal. Chem.* **57** (1985) 1955.

Chapter 9

Exchange of Nitrogen Compounds and Oxygen

9.1 Ozone Deposition on Various Surface Types

S. Cieslik and A. Labatut

Joint Research Centre, Ispra , Italy

Introduction

It is now well established that the ozone molecules present in the lower troposphere generally undergo a downward flux through the surface layer, as a consequence of a dry deposition process at the surface. This fact was first described by Regener [1]. Rich *et al.* [2] studied the removal of ozone from the air by bean leaves under controlled conditions and found that its rate follows a pattern that is very similar to that of evaporation. Using the resistance analogy, they inferred from these observations that both processes are controlled by the stomatal aperture. Galbally [3] and Galbally and Roy [4] measured ozone deposition rates using enclosure techniques on various surface types, both bare and vegetated. Turner *et al.* [5] stated that both bare soils and vegetation are targets for ozone deposition.

Since it may be that enclosure methods do not provide a real picture of a process occurring in the atmosphere, micrometeorological methods have then been implemented. Vertical profile or gradient techniques have been used by Duyzer *et al.* [6], Broder and Gygax [7] and Colbeck and Harrison [8], and led to a number of reliable results that confirmed the role of stomata, but not excluding an ozone destruction process on bare soils, however. The development of the eddy correlation method greatly helped in understanding ozone deposition [9–13]. More recently, a device has been developed by Güsten *et al.* [14] for the measurement of ozone concentration fluctuations, using a chemiluminescent reaction with an organic dye. This sensor has a response time of about 0.2 s and, since it shows a good capability, it has been used for the present measurements in an extensive programme on various types of surface.

What kind of physical or chemical processes ozone deposition really means is not yet completely clear. As already stated, the most widely accepted hypothesis is that ozone deposition consists in its uptake by vegetation by diffusion through the stomata which are present on the leaves. But ozone deposition is also observed on some bare soil surfaces, by means of processes little is known about. A chemical destruction by contact with surface elements, similar to the wall effects observed in the laboratory when ozone is stored in glass vessels, is likely to occur. The problem is even more complicated when the surface is partly covered by vegetation elements, leaving portions of bare soil. In such quite frequent situations, deposition of ozone is the result of the superposition of those effects. Attempts to discriminate between stomatal processes and destruction at the soil have been made by Massman [15]. The measurements presented here have been made as well on bare soils as on active vegetation surfaces, including mixed situations. This work is an attempt to improve our understanding of the mechanisms governing the generation of downward fluxes of ozone in the atmospheric surface layer.

Theoretical background and method used

The vertical flux of ozone obeys the laws of the turbulent atmospheric transport of any conserved scalar quantity. It is generated by the interaction processes occurring at the air-surface interface and driven by atmospheric turbulence. Ozone is considered as a conserved quantity to a first approximation, the part participating in chemical conversion being limited with respect to the total amount of ozone. The processes involving ozone at the air-surface interface all represent sinks for ozone, and are commonly termed as "deposition". It should be pointed out here that the use of this word is somewhat inappropriate. If ozone is "deposited" on vegetation, in reality it penetrates into the tissues through the stomata and is metabolised. Obviously, such a process is not deposition. On the other hand, in presence of bare soil, ozone is probably destroyed by some kind of wall reaction. Again, this is not deposition.

The use of the term "deposition" is justified when the deposited substance, after removal from the atmosphere, remains on the surface. This is the case *e.g.* for aerosol particles and for some gaseous substances which are adsorbed on the surface. But certainly not for ozone, which is too reactive. For the latter, the use of the general term of "sink" is more appropriate. We can also speak of a "downward flux", which results from the action of turbulent vertical mixing, which tends to counterbalance the disappearance of the molecules at the surface.

The mathematical formulation of turbulent vertical fluxes through of a substance the atmospheric surface layer can be found in any textbook on micrometeorology (*e.g.* [16, 17]). Here, we just briefly state the main features of the formalism used.

The time-averaged continuity equation for an conserved (*i.e.* not subject to chemical production or loss) atmospheric scalar A is

$$\frac{\partial \bar{A}}{\partial t}+\bar{u}_i\frac{\partial \bar{A}}{\partial x_i}=\frac{\partial}{\partial x_i}\left[k\frac{\partial \bar{A}}{\partial x_i}-\overline{u'A'}\right] \tag{1}$$

where t is the time, x_i and u_i are the three space co-ordinates and components of the wind speed, respectively, the i subscript referring to the conventional x, y, and z directions. The overbars and primes refer to the time averages and the instantaneous fluctuations, respectively, of the variables, following the well-known Reynolds decomposition:

$$A=\bar{A}+A' \tag{2}$$

$$u=\bar{u}+u' \tag{3}$$

Since we are interested in vertical fluxes only, it is necessary to make sure that horizontal fluxes and advection processes are negligible, when

$$\frac{\partial \bar{A}}{\partial x}=\frac{\partial \bar{A}}{\partial y}=0 \tag{4}$$

When this horizontal homogeneity condition is fulfilled, Eq. (1) reduces to

$$\frac{\partial \bar{A}}{\partial l}=\frac{\partial}{\partial z}\left[k\frac{\partial \bar{A}}{\partial z}-\overline{w'A'}\right] \tag{5}$$

where w is the vertical component of the wind vector. By introducing a coefficient K_z defined as

$$K_z=\frac{\overline{w'A'}}{-\partial A/\partial z} \tag{6}$$

Eq. (6) becomes

$$\frac{\partial \bar{A}}{\partial l}=\frac{\partial}{\partial z}\left[(k+K_z)\frac{\partial \bar{A}}{\partial z}\right] \tag{7}$$

Observational evidence shows that $K >> k$ in the major part of the atmospheric boundary layer (the turbulent layer), except in a few millimetres in contact with the surface elements, called for this reason "laminar" or "viscous" sublayer.

We consider that, in the turbulent layer, the flux ∅) can be approximated by

$$\phi=-K_z\frac{\partial \bar{A}}{\partial z} \tag{8}$$

and, in the laminar layer by

$$\phi=-k\frac{\partial \bar{A}}{\partial z} \tag{9}$$

A further step is the introduction of the resistance representation, which is a simplified scheme, analogous to an electric circuit, permitting the separation and the study of the relative contributions of the different processes involved in vertical transport.

Here the scalar A under consideration is a conserved (*i.e.* not subject to chemical production or loss) depositing substance, quantified by its concentration c. The resistance scheme is introduced by integrating Eq. (7) from z = 0 (the theoretical ground level) to $z = z_R$, where z_R is a reference level, or measurement level (*e.g.* 8 m over low vegetation):

$$\phi\left[\int_0^{Z_L}\frac{dz}{k}+\int_{Z_L}^{Z_R}\frac{dz}{K_z}\right]=\overline{c}(Z_R)-\overline{c}(0) \tag{10}$$

where z_L is the upper limit of the laminar level. Now we define the aerodynamic resistance as

$$r_a \equiv \int_{Z_L}^{Z_R}\frac{dz}{K_z} \tag{11}$$

the laminar layer resistance as

$$r_b \equiv \int_0^{Z_L}\frac{dz}{k} \tag{12}$$

and the surface resistance as

$$r_c \equiv \frac{\overline{c}(0)}{\phi} \tag{13}$$

and Eq. (10) becomes

$$\overline{c}(Z_R)=(r_a+r_b+r_c)\phi \tag{14}$$

Further we introduce the following variable:

$$V_d \equiv \frac{1}{r_a+r_b+r_c} \tag{15}$$

we can also write

$$\phi = V_d\overline{c}(Z_R) \tag{16}$$

We see V_d has the dimension of a velocity. It is therefore called the deposition velocity since its introduction by Chamberlain and Chadwick [18] and is widely used in modelling work because of the extreme simplicity of Eq. (16). However, one must not forget that this parameter is not a real velocity of the molecules. Rather, it is a mathematical quantity that has the dimensions of a velocity.

Sites and instrumentation

In this contribution we present the results of five measuring campaigns which took place in three countries of Europe (Italy, Germany and France). The main features of these exercises are shown in Table 9.1.1.

Table 9.1.1: The different campaigns presented in this work

Place	Country	Co-ordinates	Abbreviation in text	Dates	Type of surface
Klippeneck	Germany	48°10' N; 8°45' E	K 92	Sept. 10–21, 1992	Grass (active)
Castelporziano	Italy	41°43' N; 12°24' E	BEMA93	June 2–12, 1993	Grass (dead), little vegetation
S. Pietro Capofiume	Italy	44°39' N; 11°37' E	SPC93	June 16–22, 1993	Maturing beet field
Castelporziano	Italy	41°43' N; 12°24' E	BEMA94	May 11–24, 1994	Grass (almost dead) and low vegetation
Le Dézert	France	44°05' N; 0°43' W	LD94	June 9–15, 1994	Pine forest with undergrowth

Klippeneck site is a little airfield on a high plateau (about 1000 m a.s.l.) in Southern Germany. The measurements were a contribution to the EUROTRAC TRACT–BIATEX campaign focused on air pollution transport in a complex topography area. The airfield is covered by grass and the fetch was about 1 km in the prevailing wind direction, however, with a slight slope.

The site of S. Pietro Capoflume is located in the Po river plain in Italy, in an extended agricultural area. The site is managed jointly by the Regional Agrometeorological Service of the Emilia-Romagna Region, providing' thus some organisational support. The area is completely flat up to distances of several tens of kilometres and the fetch conditions are excellent. the site was covered with beet plants in an active growth stage, thus providing good opportunities of studying the stomatal influence on deposition.

The two campaigns labelled "BEMA" took place in the nature reserve of Castelporziano near Rome. BEMA stands for "Biogenic Emissions in the Mediterranean Area" and is a project funded by the European Commission aimed at improving our knowledge of the influence of organic compounds emitted by Mediterranean vegetation on oxidants formation in the atmosphere. The project included measurements of deposition of some smaller molecular species. The measuring site was covered with short vegetation which was partially dry due to the season.

The "Le Dezert" site located in an extended pine forest in SW France (Landes area), with flat topography. The forest is very uniform and the fetch is good in all directions. The forest canopy extended from 15 to 24 m; undergrowth is present. Here, the measurements were made under the canopy, at 14 m a.g.l.

Measurements were carried out by means of a Kaijo-Denki Mod. Dat-300 3-D ultrasonic anemometer-thermometer coupled with a GFAS OSG-2 fast-response ozone analyser. A Campbell Scientific KH20 krypton hygrometer was used as fast response moisture sensor. Additional observations were made in order to complete the surface energy balance: a radiometer recorded the surface net radiative flux, and four Pt-100 type temperature probes were used to monitor the soil temperature at four depths, then converted into soil heat fluxes.

Observational results

If vegetation is seen as a probable sink for ozone, information is essential about its physiological state, and, in particular, about the activity of stomata, through which ozone is likely to penetrate into the plant tissues. Over a surface covered with vegetation, the efficiency of stomatal processes influences turbulent transfer of sensible and latent heat through the surface layer and, in particular, the Bowen ratio (sensible to latent heat flux ratio). These heat fluxes are measured by the eddy correlation system and a compared examination of the data from the different campaigns is the first step in the interpretation. A low Bowen ratio (corresponding to a high latent heat flux or evaporation) indicates that the stomata are active. Figs. 9.1.1 to 9.1.5 show the evolution of the turbulent heat fluxes during the different campaigns.

The following conclusions can be drawn on the state of vegetation:

- At K92 (Fig. 9.1.1), the Bowen ratio was close to unity (sensible and latent heat fluxes nearly equal) throughout the campaign. This indicates fairly active vegetation, which consisted of 10 cm high grass. The ground underneath was wet, supplying enough water for physiological activity. Here an important part of ozone deposition is expected to be due to the stomata.
- At BEMA93, the sensible heat fluxes exceeded by far the latent heat fluxes (Fig. 9.1.2). This is explained by the dry state of vegetation and soil. Evaporation was very low but not zero.
- At SPC93 (Fig. 9.1.3), the latent heat flux exceeded the sensible heat flux, at least for certain days (June 17, 20 and 21). For some other days, both heat fluxes were of comparable magnitude. In any case, the vegetation, which consisted of maturating beets, was very active. The interesting point here was that the soil surface was completely dry, indicating that evaporation could only occur through the leaves and stomata. Water supply to the plant probably came from lower soil levels. It can thus be concluded that all the evaporative flux is fully controlled by the stomatal processes, and that the latent heat flux data may be used for the calculation of stomatal resistances without any correction for possible evaporation from the soil.
- At BEMA94, the sensible heat fluxes exceeded the latent heat fluxes (Fig. 9.1.4), due to the dry state of vegetation and soil, but the difference was not as important as in BEMA93, because the campaign took place earlier in the year, and the vegetation was less dry. Heavy rain occurred on June 19, and, after this date, the latent heat fluxes were almost of the order of the sensible heat flux.

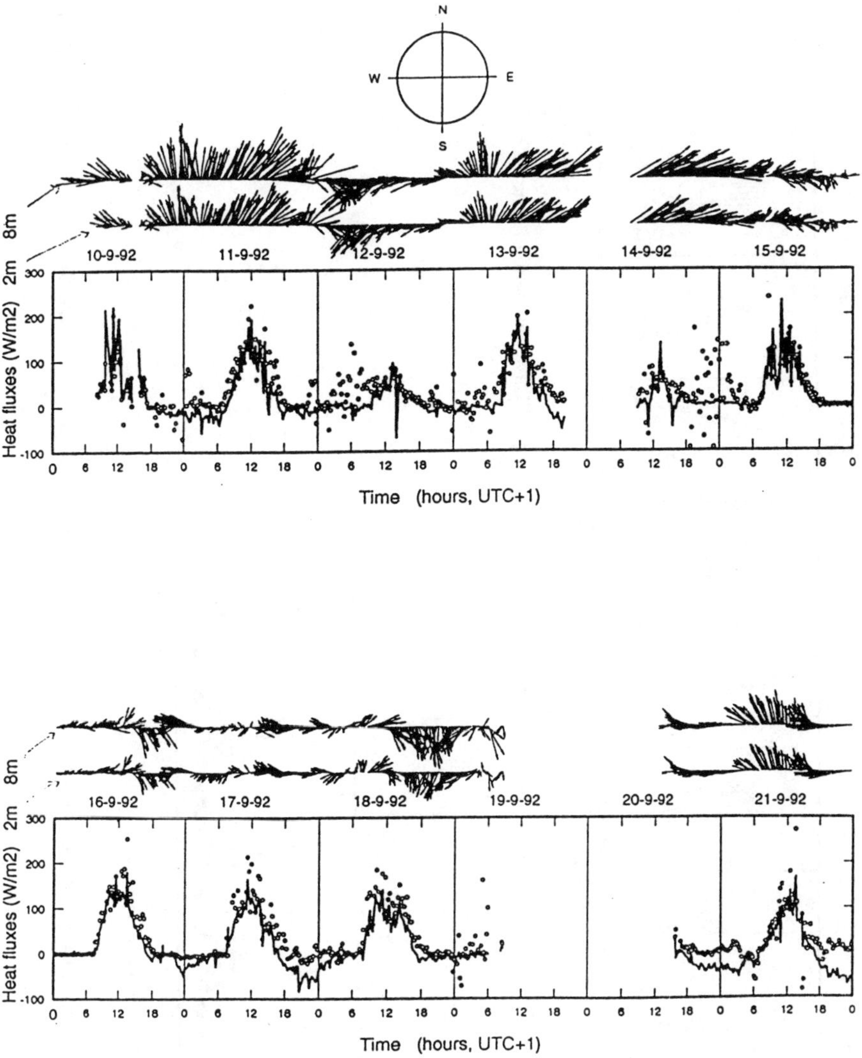

Fig. 9.1.1: Vertical fluxes of sensible heat (solid lines) and latent heat (circles) for the K92 campaign at Klippeneck in September 1992. The wind speeds and directions are also shown.

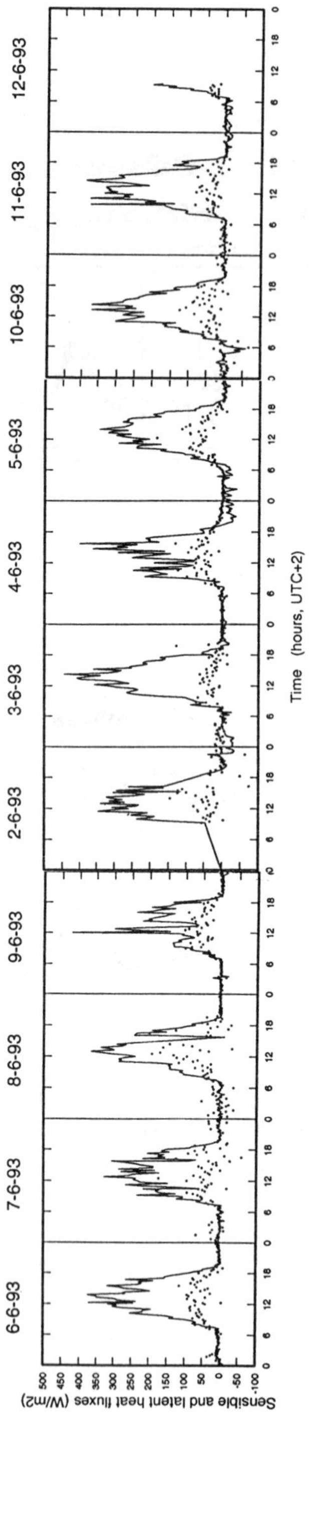

Fig 9.1.2: Vertical fluxes of sensible heat (solid lines) and latent heat (dots) for the BEMA93 campaign at Castelporiazno in June 1993.

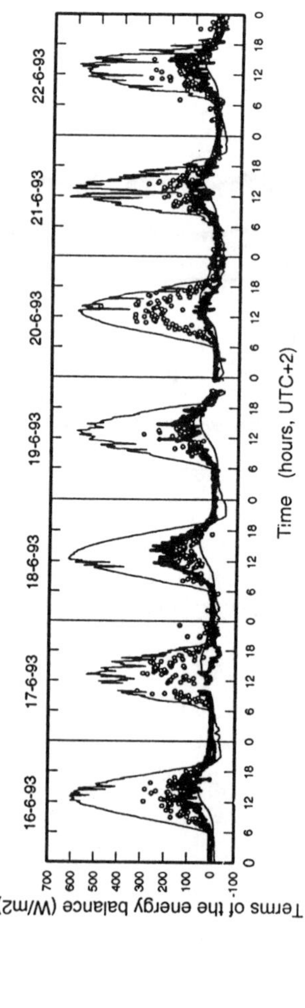

Fig. 9.1.3: Net radiative fluxes (thin lines), sensible heat fluxes (bold lines) and latent heat fluxes (circles) for the SPC93 campaign at S. Pietro Capofiume in June 1993.

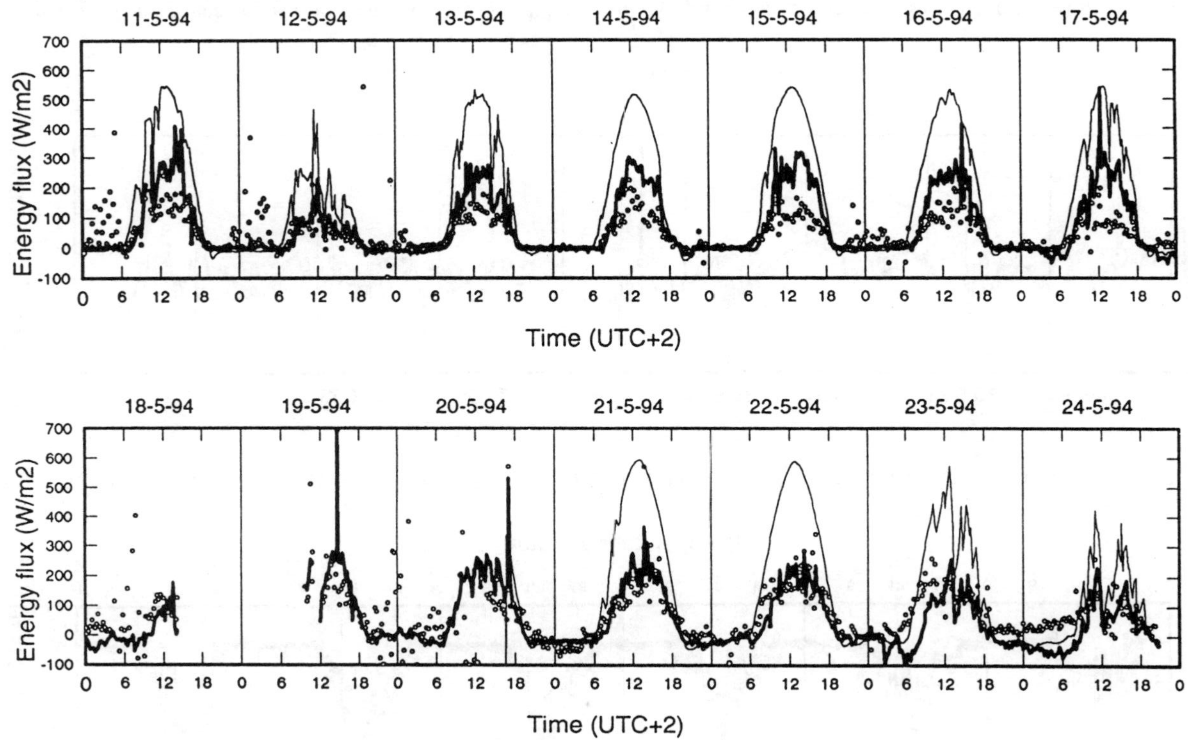

Fig. 9.1.4: Net radiative fluxes (thin lines), sensible heat fluxes (bold lines) and latent heat fluxes proportional to evaporation (circles) at Castelporziano in May 1994.

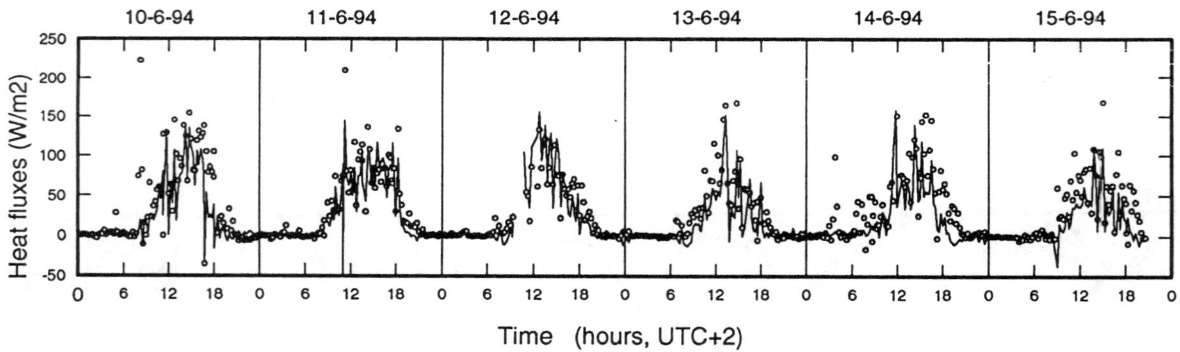

Fig. 9.1.5: Vertical fluxes of sensible heat (solid lines) and latent heat (circles) as observed during the Landes campaign in June 1994.

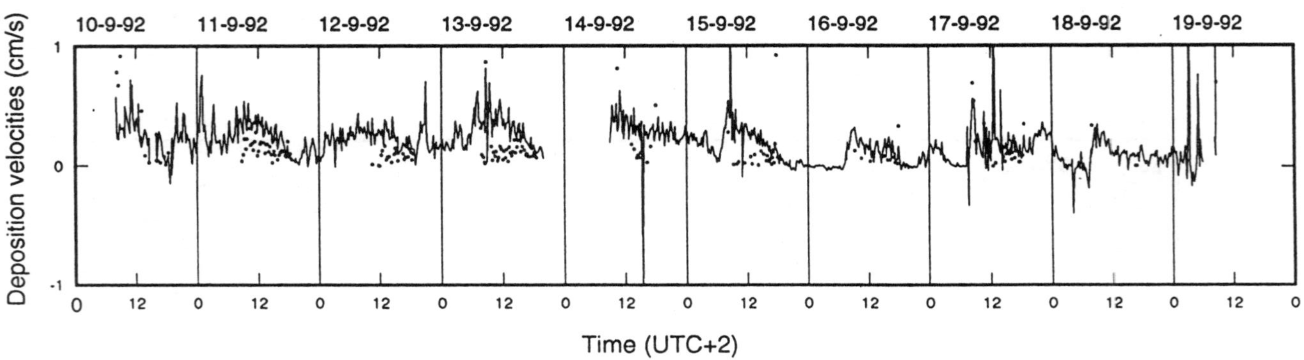

Fig. 9.1.6: Total ozone deposition velocities, V_d, (solid lines) and the contribution of the non-stomatal processes to ozone deposition, v_{NS}, (dots) for the K92 campaign at Klippeneck, September 1992.

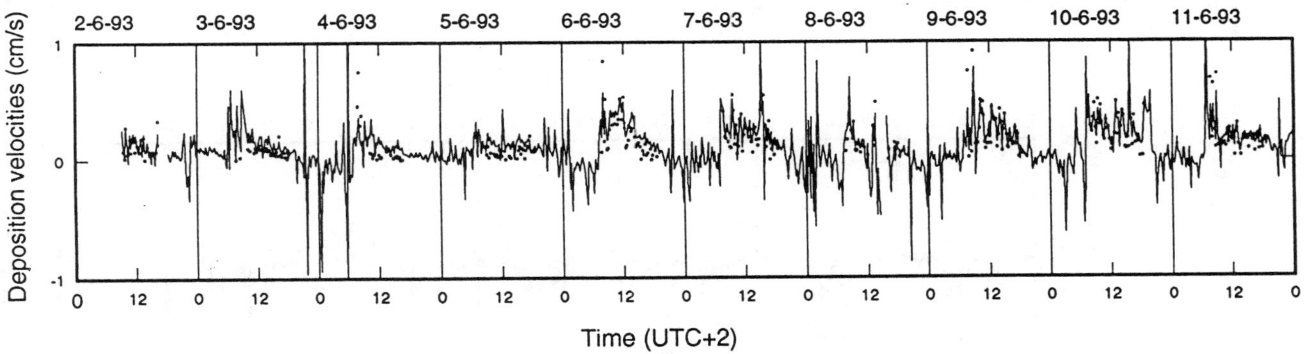

Fig. 9.1.7: Total ozone deposition velocities, V_d, (solid lines) and the contribution of the non-stomatal processes to ozone deposition, v_{NS}, (dots) for the BEMA93 campaign at Castelporiazno, June 1993.

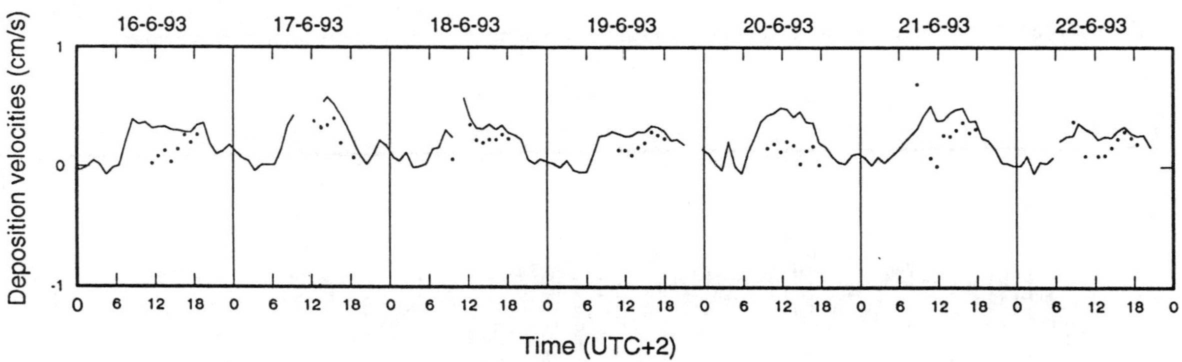

Fig. 9.1.8: Total ozone deposition velocities, V_d, (solid lines) and the contribution of the non-stomatal processes to ozone deposition, v_{NS}, (dots) for the SPC93 campaign at S. Pietro Capofiume in June 1993.

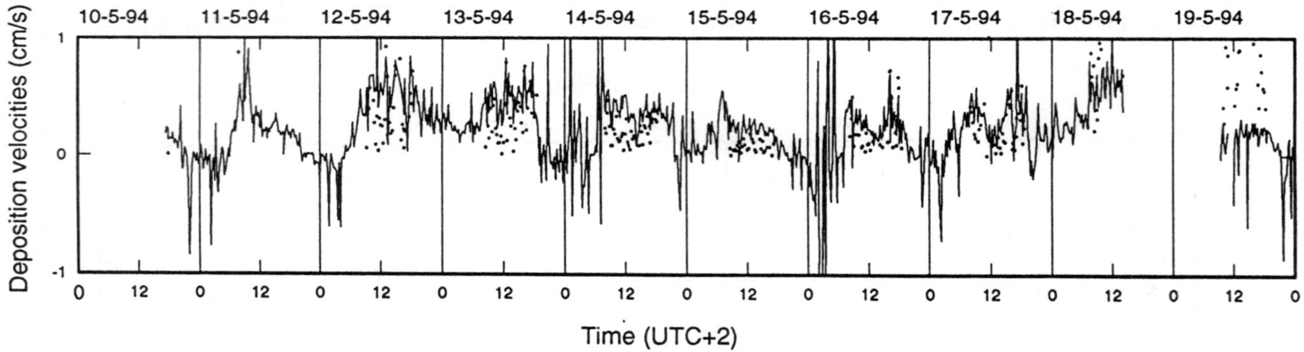

Fig. 9.1.9: Total ozone deposition velocities, V_d, (solid lines) and the contribution of the non-stomatal processes to ozone deposition, v_{NS}, (dots) for the BEMA94 campaign at Castelporiazno, June 1994.

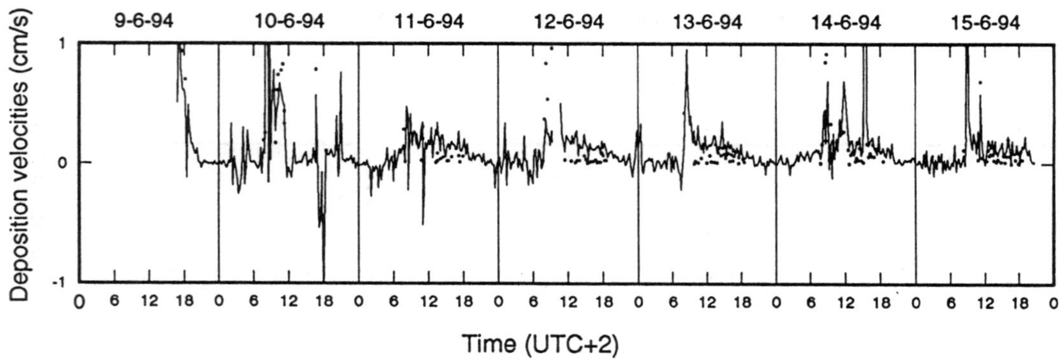

Fig. 9.1.10: Total ozone deposition velocities, V_d, (solid lines) and the contribution of the non-stomatal processes to ozone deposition, v_{NS}, (dots) for the LD94 campaign at Dezert in June 1994.

- At LD94 (Fig. 9.1.5), where the measuring point was situated under the pine tree canopy, the Bowen ratio was close to unity, due to the activity of the low vegetation in the undergrowth.

The results of the measurement of ozone deposition velocities by eddy-correlation for the different campaigns is represented in Figs. 9.1.6 to 9.1.10 by the solid lines.

All data show a similar behaviour, with an increasing deposition velocity during morning hours, a maximum value at noon and a decrease in the afternoon. The night-time data are scattered around zero. The daily cycle of ozone deposition roughly follows that of heat fluxes. The daily maximum generally ranged between 0.3 and 0.5 cm/s. Scatter is important, due the some limitations inherent to the eddy-correlation method, which requires horizontal homogeneity, not always respected. Examination of these results gives only a global picture of the ozone deposition phenomenon and it is practically impossible to detect the influences of different vegetation types. Further analysis, making use of the resistance analogy, is necessary.

Resistance study and partitioning

The surface resistances for a depositing substance (r_c), as defined in Eq. (15) are calculable from the knowledge of the measured deposition velocities and of the calculated aerodynamic and laminar layer resistances. But they do not discriminate between the different processes governing deposition. If we want to know which part of the deposition phenomenon is due to stomatal activity, it is necessary to determine the stomatal resistance, which is assumed to be a function of evaporation. The Penman-Monteith approach [19] gives an algorithm to calculate stomatal resistances r_s from the sensible and latent flux measurements. This method has been used here for the different campaigns.

The calculation of the aerodynamic and laminar layer resistances is based on Eqs. (11) and (12). The coefficient K_z has been calculated from the Monin-Obukhov theory, using the measured values of turbulence parameters (friction velocity and sensible heat flux). The coefficient k is kept constant and has its value for air ($0.135 \times 10\text{-}4\ m^2/s$). The value of z_R is taken equal to the height where $K_Z = k$.

The results of this procedure are shown in Figs. 9.1.11 to 9.1.14 for the different campaigns.

In the cases where the values of r_c and r_s are close to each other, it means that the most part of ozone deposition is due to the stomatal pathway. But the investigation may be conducted further in the following way.

The surface resistance r_c to ozone deposition may be decomposed as follows [15]:

$$\frac{1}{r_c} = \frac{1}{r_S\ D(H_2O) / D(O_3)} + \frac{1}{r_{NS}} \tag{17}$$

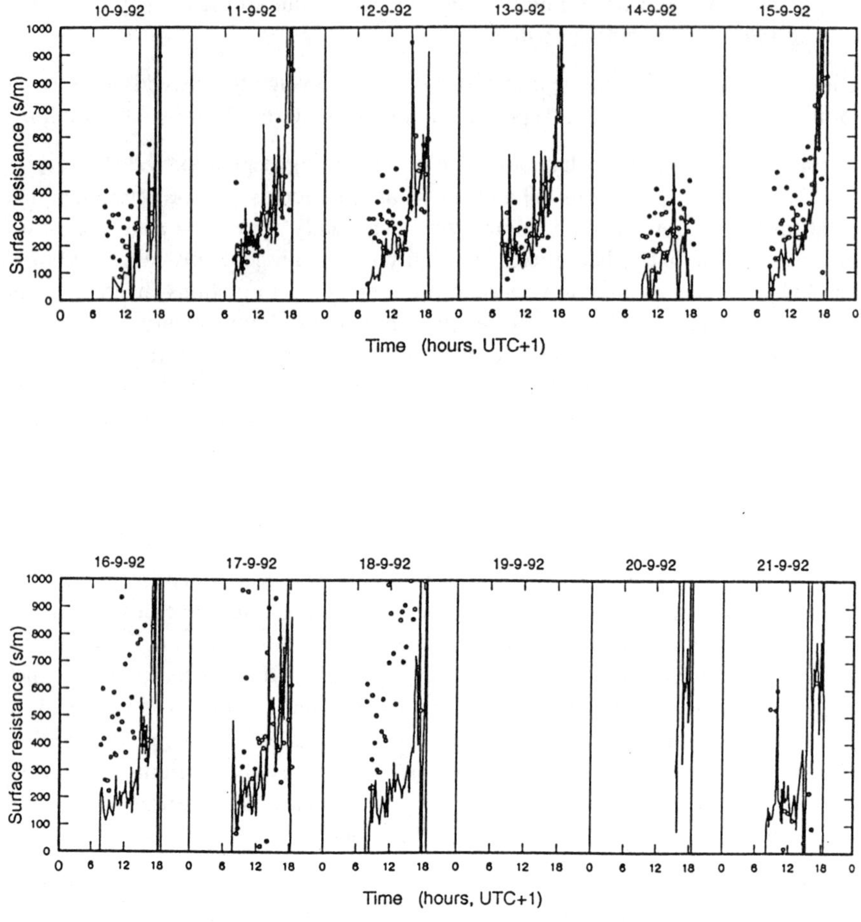

Fig. 9.1.11: Surface resistances for evaporation (solid lines) and for ozone deposition (circles) for the K92 campaign at Klippeneck in September 1992.

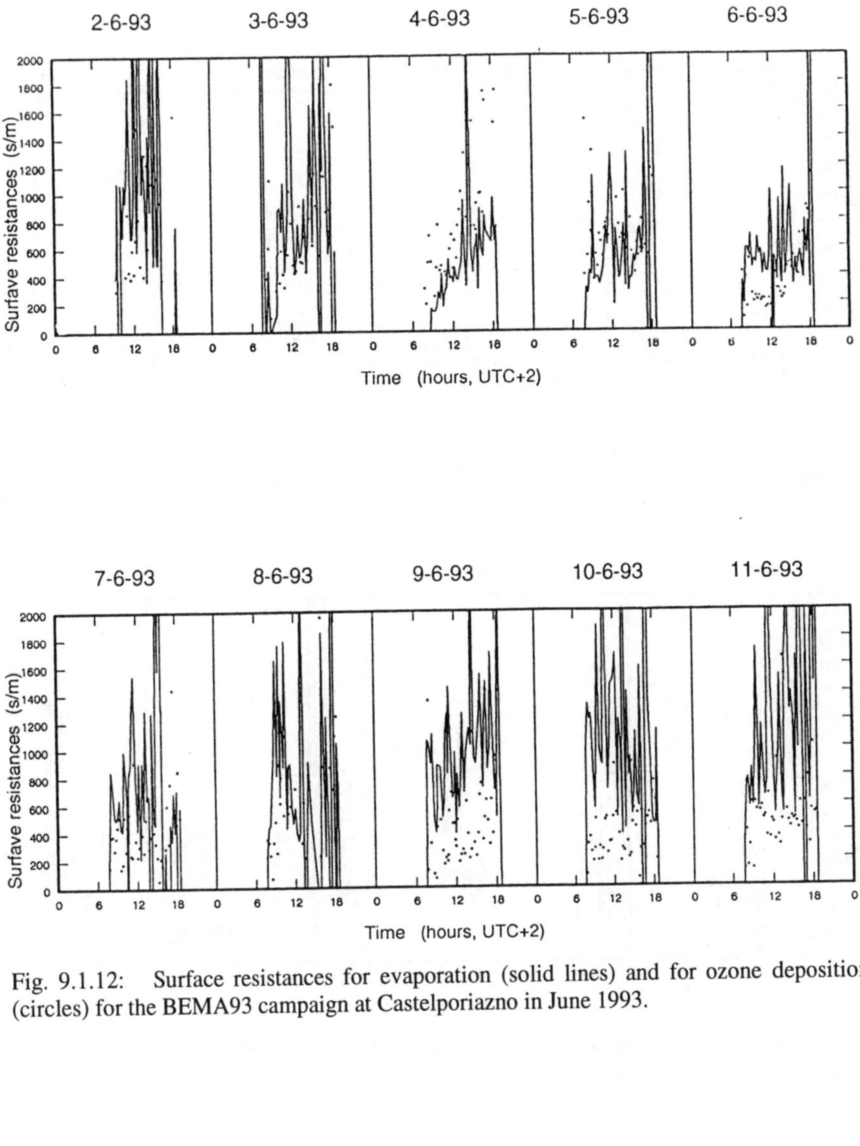

Fig. 9.1.12: Surface resistances for evaporation (solid lines) and for ozone deposition (circles) for the BEMA93 campaign at Castelporiazno in June 1993.

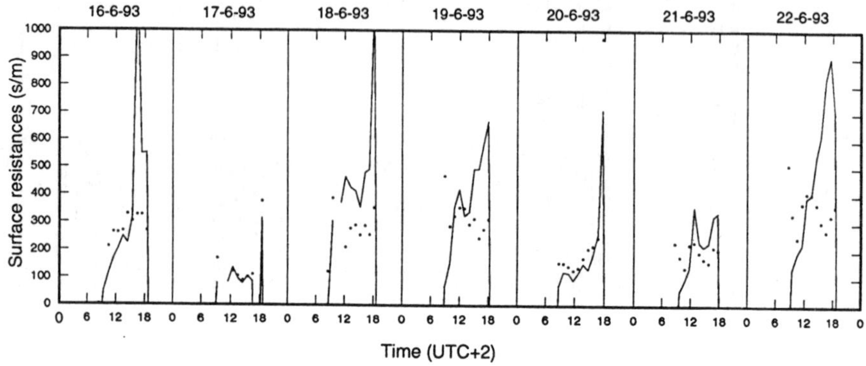

Fig. 9.1.13: Surface resistances for evaporation (solid lines) and for ozone deposition (circles) for the SPC93 campaign at S. Pietro Capofiume in June 1993.

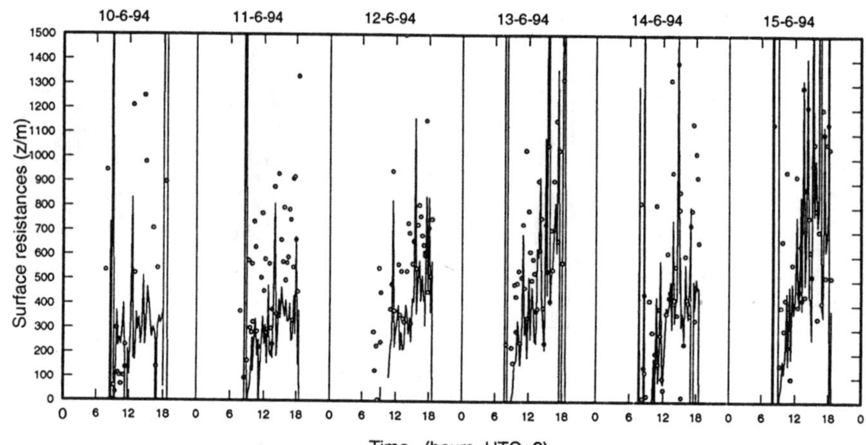

Fig.9.1.14: Surface resistances for evaporation (solid lines) and for ozone deposition (circles) for the Landes campaign at Dezert in June 1994.

where $D(H_2O)$ and $D(O_3)$ are the molecular diffusion coefficients through air for water vapour and ozone, respectively. Their ratio is equal to 1.65; r_{NS} is the resistance to ozone deposition corresponding to the non-stomatal part of the process. The calculation of this resistance can be made using Eq. (17). The final step is the calculation of a deposition velocity

$$v_{NS} = \frac{1}{r_{NS}} \tag{18}$$

which also corresponds to the non-stomatal contributions to deposition.

This procedure has been applied to the data from the different campaigns and is shown in Figs. 9.1.6 to 9.1.10, where the v_{NS} values are represented by the dots.

The results show interesting features. For the K92 measurements, where vegetation was active and no bare soil was apparent, the non-stomatal deposition velocity was low compared with the total deposition velocity. The major part of ozone deposition was thus apparently due to the stomatal pathway.

In the first BEMA campaign (June 1993), the non-stomatal part of the deposition velocity was almost equal to the total one, indicating the quasi-absence of stomatal processes. Ozone "deposited" directly on dead vegetation (or probably was destroyed by contact with it). For BEMA94, the non-stomatal contribution was still important, but was lower than in 1993, since some vegetation was still active. It is worth remembering that the measurement were carried out earlier in the year (May).

The data from SPC93 are interesting because stomatal activity was particularly important (strongly evaporating beet leaves), but, between the plants, very dry soil portions were visible. Deposition thus occurred through both pathways.

Finally, for LD94, the non-stomatal part was very low. Since the measuring facility was installed below the pine trees canopy, deposition could occur on the undergrowth only. This part of the vegetal cover was physiologically active. The total deposition velocity was lower than for the other campaigns, since the major part of deposition probably occurred on the pine canopy; less ozone was thus available for deposition at lower levels.

Conclusions

With the help of heat flux measurements including evaporation, a procedure is developed for discriminating between the contributions to ozone deposition by different pathways, *i. e.* stomatal and non-stomatal. It was generally admitted in the past that ozone deposits on plants by diffusing through the stomata, making thus this process directly related to evaporation, which is regulated by stomatal aperture. The measurements presented here and their analysis shows that this is true only in part. In the case of a full vegetation cover with strong stomatal activity, most of ozone deposition occurs through the stomata. When the vegetation is not active, ozone deposition also occurs by some other process, which is poorly known. When the situation is mixed, *e.g.* when there is active vegetation, but portions of bare soil are present, then both stomatal and non-stomatal processes contribute to ozone dry deposition.

References

1. V.H. Regener, The vertical flux of ozone, *J. Geophys. Res.* **62** (1957) 221.
2. S.E. Rich, P.E. Waggoner, H. Tomlinson, Ozone uptake by bean leaves, *Science* **169** (1970) 79.
3. I.E. Galbally, Ozone profiles and ozone fluxes in the atmospheric surface layer, *Quart. J. Roy. Met. Soc.* **97** (1971) 18.
4. I.E. Galbally, C.R. Roy, Destruction of ozone at the earth's surface, *Quart. J. Roy. Met. Soc.* **106** (1980) 599–620.
5. N.C. Turner, S. Rich, P.S. Waggoner, Removal of ozone by soil, *J. Environ. Qual.* **2** (1973) 259–264.
6. J.H. Duyzer, G.M. Meyer, R.M. Van Aalst, Measurement of dry deposition velocities of NO, NO_2 and O_3 and the influence of chemical reactions, *Atmos. Environ.* **17** (1983) 2117.
7. B. Broder, H.A. Gygax, The influence of locally induced wind systems on the effectiveness of nocturnal dry deposition of ozone, *Atmos. Environ.* **19** (1985) 1627.
8. I. Colbeek, R.M. Harrison, Dry deposition of ozone: some measurements of deposition velocity and of vertical profiles to 100 metres, *Atmos. Environ.* **19** (1985) 1807.
9. M.L. Wesely, J.A. Eastman, D.H. Stedman, E.D. Yalvac, An eddy-correlation measurement of NO_2 flux to vegetation and comparison to O_3 flux, *Atmos. Environ.* **16** (1982) 815.
10. N.H. Neumann, G. Den Hartog, Eddy-correlation of atmospheric fluxes of ozone, sulphur, and particulates during the Champaign Intercomparison Study, *J. Geophys. Res.* **90** (1985) 2097.
11. J.G. Droppo, Concurrent measurements of ozone dry deposition using eddycorrelation and profile flux methods, *J. Geophys. Res.* **90** (1985) 2111.
12. A.C. Delany, D.R. Fitzjarrald, D.H. Lenschow, R. Pearsonjr., G.J. Wendel, B. Woodruff., Direct measurements of nitrogen oxides and ozone fluxes over grassland, *J. Atmos. Chem.* **4** (1986) 429.
13. B.B. Hicks, D.R. Matt, R.T. McMillan, A micrometeorological investigation of surface exchange of O_3, SO_2 and NO_2: a case study, *Boundary Layer Meteorol.* **47** (1989) 321.
14. H. Giisten, G. Heinrich, R.W.H. Schmidt, U. Schurath, A novel ozone sensor for direct flux measurements, *Ozone Symposium,* Charlotteville, 1992.

15. W.J. Massman, Partitioning ozone fluxes to sparse grass and soil and the inferred resistances to dry deposition, *Atmos. Environ.* **27A** (1993) 167–174.

16. A. Poggi, *Introduction à la micrométéorologie*, Masson, Paris 1977.

17. R.B. Stull, *An introduction to boundary layer meteorology*, Kluwer, Dordrecht 1988.

18. A.C. Chamberlain, R.C. Chadwick, Deposition of airborne radioiodine vapour, *Nucleonics* **11(8)** (1953) 22.

19. J.L. Montheith, Evaporation and surface temperature, *Quart. J. Roy. Met. Soc.* **107** (1981) 1.

9.2 Measurements of Dry Deposition Fluxes of Nitrogen Compounds and Ozone.

Jan Duyzer, Hilbrand Weststrate, Henk Verhagen, Geurt Deinum and Jaap Baak

TNO Institute of Environmental Sciences, P.O. BOX 6011, 2600 JA Delft, The Netherlands

Summary

The deposition flux of ammonia was determined over a coniferous forest. An uptake rate significantly higher than can be explained by stomatal uptake was observed. On the basis of these measurements the annual averaged deposition velocity of ammonia to forest was estimated to lie between 20 and 33 mm s^{-1} the estimated flux lies between 21 and 44 kg N ha^{-1} yr^{-1}.

Simple procedures to correct measurements of the flux of NO, NO_2, and O_3, by micrometeorological methods were developed, tested and compared with existing procedures using a dataset generated from calculations with a 1-D surface layer model. The results showed that existing correction procedures can give biased results. A new, simple correction procedure was developed which derives fluxes that differ by less than 5 % from the true flux during the day and less than 20 % at night.

Measurements of the uptake of NO_2 by grass and heathland showed a close link with stomatal uptake. A long time series of measurements of O_3 gradients and fluxes (by eddy correlation) above forest showed that the flux gradient method can be used quite well provided locally derived flux profile functions are used. During a large field experiment organised by TNO in Speulderbos fluxes of NO, NO_2 and O_3 were measured above and below the canopy. Fluxes of NO_2 above the canopy were often away from the surface. A detailed canopy multi-layer model was developed to assist in the interpretation of the experimental results. Calculations with the model indicate that the observed fluxes are probably related to emissions of NO from the forest floor and conversion of this NO in the trunk space.

Aims of the research

The aims of the TNO research are :

- to compare and evaluate several techniques to measure trace gas fluxes.
- to determine dry deposition fluxes of all relevant nitrogen species and identify mechanisms that regulate these fluxes.

Principle scientific results

The TNO part of the project consists of the participation and organisation of joint experiments, measurements in the Netherlands and the interpretation and generalization of the results using models. The experimental work consists of measurement of surface fluxes of NH_3, NO, NO_2 and O_3 with micrometeorological methods and the measurement of PAN concentration.

Exchange of reactive trace gases the atmosphere and forest

The flux of ozone and heat was determined continuously over the forest by the eddy correlation method together with the temperature and concentration gradient of ozone over the canopy. Flux profile functions over the forest could be derived for heat. These showed systematic deviations from classical functions reported in the literature. Temperature gradients appear to be smaller than could be expected on the basis of the classical formulations. Especially during the day these gradients are very small due to the intense mixing over the canopy. However using robust statistics even for ozone a relation between the ozone flux and the concentration gradient could be found.

Fig. 9.2.1 shows that the flux of ozone calculated from the derived flux profile functions and the concentration gradient of ozone compares quite well with the flux derived from eddy correlation measurements [1]. This shows that the gradient method can be used over forests, provided locally derived flux profile functions are used.

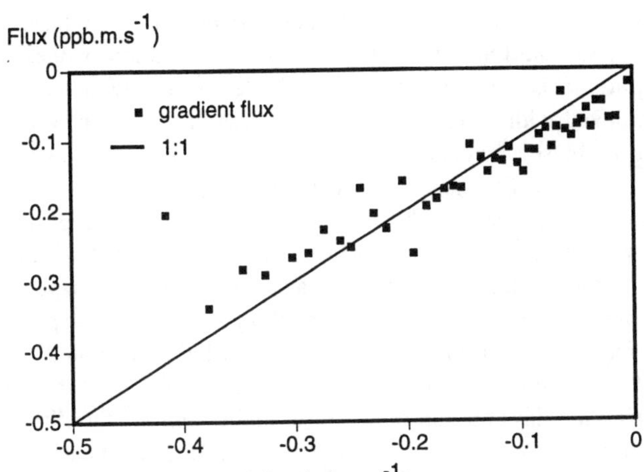

Fig. 9.2.1: The flux of O_3 for the Speulderbos calculated using the gradient method against the flux observed using the eddy correlation flux.

In June 1993 TNO organised a large joint field campaign in the Speulderbos. UMIST and RISOE participated in this campaign which lasted for three weeks. Fluxes of ozone and nitrogen oxides were measured at two levels above and one below the canopy. TNO instruments were positioned at the top level of the tower 36 meter above ground. The tower was located some 50 meters away from another tower (used by RIVM). The analysis of the campaign showed the measurements on the 'campaign' mast were compared with the continuous measurements carried out at the RIVM tower. The sensible heatflux, ozone flux and friction velocity measured on both towers during the campaign compared quite well. This indicates a fairly homogeneous flow field over the canopy.

NO and NO_2 fluxes from the forest floor were also determined using a box method. Deposition of NO_2 and emission of NO was observed. A net emission rate around 15 ng N NO was observed.

In line with earlier observations in the Speulderbos fluxes of NO_2 above and below the canopy were often directed upwards. This, in contrast with the generally accepted idea that forests absorb NO_2. In order to interpret these observations a comprehensive canopy exchange model was constructed at TNO in cooperation with UMIST. The results of the measurements during the campaign were compared with calculations with this model [2]. From these calculations it was concluded that the upward fluxes of NO_2 above the canopy are caused by emissions of NO from the forest floor. In the trunk space NO is converted to NO_2 by O_3. Some of this NO_2 is absorbed in the canopy the rest is vented into the atmosphere.

Correction of measurements of fluxes of reactive trace gases

Procedures to correct for the effect of chemical reactions on measurements of fluxes of NO, NO_2 and O_3 have been described. Verification of these procedures however requires flux measurements at multiple heights with an accuracy that cannot be achieved with current methods. The results of field measurements may therefore still be biased by chemical reactions [3, 4]. In Duyzer *et al.* [5] several procedures to correct flux measurements are compared and the magnitude of possible errors is discussed. The procedures are tested using the results of calculations with a comprehensive surface layer model which includes detailed descriptions of turbulence in the surface layer, a detailed description of surface exchange of trace gases and a scheme describing relevant atmospheric chemical reactions. It may be shown that fluxes of NO and NO_2 calculated from gradient measurements without a correction differ from the actual flux at the surface by up to 20 % (during the day) and up to 40 % (at night). For O_3 the difference is less than 7 % during the day and less than 20 % at night. It also appears that correction procedures described in the literature may give erroneous results because important assumptions are not valid and boundary conditions are not satisfied. A simple method (a correction factor) in which these assumptions are not necessary has been developed which calculates the flux from the gradients and provides systematic errors less than 5 % during the day and less than 20 % at night [5]. Fig. 9.2.2 gives an example of the various correction procedures applied on simulated data. LD2 is the procedure described by Lenshow and Delany [6] and JD91 the procedure described in Duyzer [7]. It is obvious that the classical correction procedures can lead to significant errors in the flux derived from observations. The different correction procedures are also applied on data from the Halvergate experiment [8].

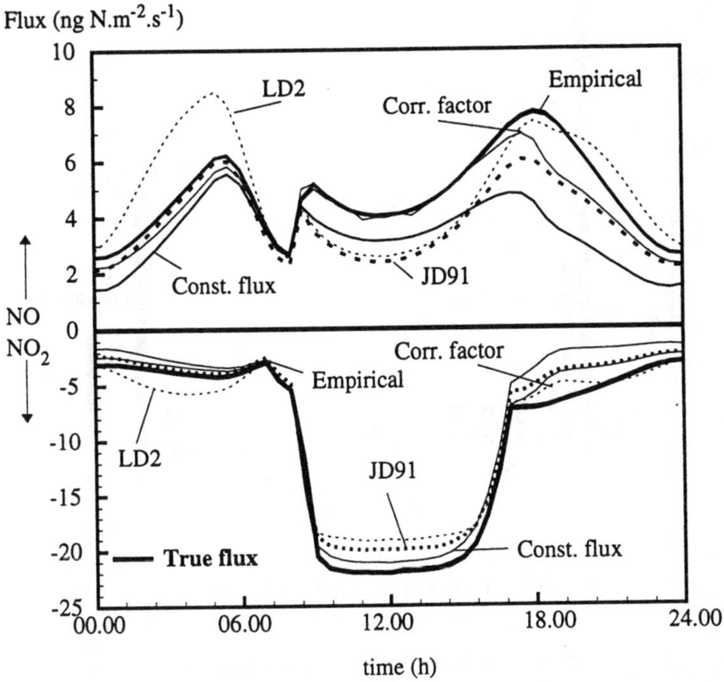

Fig. 9.2.2: The true surface flux of NO and NO_2 in model runs including hydrocarbons and the surface fluxes calculated from the concentrations at 0.5, 1, 2 and 4 meter using the classical gradient approach ('uncorrected fluxes') and the correction procedures described in the text.

NH_3

The measurements made during several campaigns from 1988 to 1990 above Speulderbos, a mature Douglas fir forest in the Netherlands are described in Duyzer *et al.* [9]. Using a more comprehensive instrumentation than that used over heathland [10], smaller gradients could be detected. An uptake rate significantly higher than can be explained by stomatal uptake was observed. Fig. 9.2.3 shows the frequency distribution of canopy resistances calculated from the measurements. Only at high water vapour pressure deficits (above 10 g m^{-3}) is absorption less efficient. Most of the variance in the observed fluxes could be explained using a resistance layer model assuming a strong cuticular and stomatal uptake under wet and moist conditions and a less efficient, probably stomatal, uptake under dry conditions.

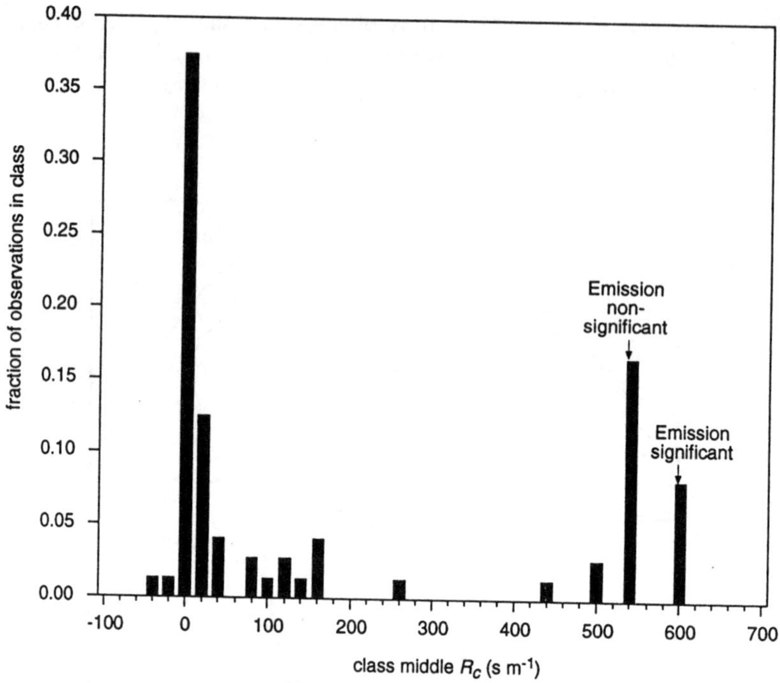

Fig. 9.2.3: Frequency distribution of the observed canopy resistance R_c for NH_3 over Douglas fir. Emission fluxes not differing significantly from zero are counted in the 550 s m^{-1} class, those differing significantly from zero are plotted in the 600 s m^{-1} class.

In order to generalize the results of the measurements over the Speulder forest a parameterisation scheme was used to estimate the annual dry deposition to the forest. The annual average deposition velocity was estimated to lie between 20 and 33 mm s^{-1}. The estimated flux lies between 21 and 44 kg N ha^{-1} yr^{-1}. These estimates are roughly equal to the throughfall fluxes observed at the site and slightly larger than previous model estimates. In any case it is clear that the estimated flux is substantially larger than the critical load of nitrogen to coniferous forest which was estimated to lie between 10 and 20 kg ha^{-1} yr^{-1} total N [11].

In a few cases above forest as well as above heathland fluxes of ammonia away from the surface were observed. These are quite common above recently fertilized crops. Farquahar [12] showed how a compensation point determines the equilibrium atmospheric concentration above which deposition and below which emission occurs. Over the forest as well as over heathland emission fluxes and high surface resistances were found in few cases when the temperature was high and the wind speeds were low. With those wind speeds the micro-meteorological method may be less applicable and the data are less reliable. Nevertheless these results are indications that at high temperatures a compensation point slightly under 1 µg m^{-3} could be present.

Evaporation of particulate ammonium nitrate caused by the higher temperatures near the canopy may lead to artefact gradients of ammonium nitrate. This evaporation may however at the same time affect ammonia and nitric acid fluxes and gradients. Recent experimental results suggest that evaporation of ammonium nitrate aerosol in air near the forest may also cause 'apparent' emission of NH_3 [13] at even higher concentrations as a consequence of this artefact.

Conclusions

The following conclusions can be drawn from the work carried out by TNO in the framework of the BIATEX programme :

- The micrometeorological gradient method can be applied over forest with good results provided local flux profile functions are used.
- The deposition of NH_3 by coniferous forest is higher than can be explained on the basis of stomatal uptake alone. An additional pathway, most probably deposition on the leaf cuticle, is more important. Estimates of the dry deposition of NH_3 to the Speulderbos show that the critical load for nitrogen is exceeded.
- Measurements of the deposition flux of nitrogen oxides by micrometeorological methods may be biased because of photolysis of NO_2 and chemical reactions between NO and O_3. Correction procedures from the literature may give incorrect results with differences as large as 100 % between calculated and true fluxes. A new, simple, correction procedure was developed which derives fluxes that differ by less than 5 % from the true flux during the day and less than 20 % at night.
- Measurements of the flux of NO_2 by micrometeorological methods have indicated that the deposition of NO_2 to grassland and heathland is most probably via stomata. NO is usually emitted from the soil.
- Fluxes of NO_2 above some forest canopies are often away from the atmosphere which is in contrast with current assumptions. Therefore current, model based, estimates of the input of nitrogenoxides into forest may be in error. This is caused by the emission of NO from the forest floor. The NO is converted in the trunk space by reactions with O_3. More research is needed into, in and below, canopy turbulence and advection and the emission of NO from the forest floor.

Acknowledgements

The work described here was funded by the Commission of the European Communities DG12, the Dutch Priority Program on Acidification (through RIVM), the Dutch Ministry of Housing Physical Planning and the Environment and TNO.

References

1. Duyzer, J.H, J.H. Weststrate, The use of the gradient method to monitor trace gas fluxes over forest, IMW TNO report R94/274, Delft 1995.
2. Duyzer, J., S. Walton, M. Gallagher, K. Pilegaard, A multi-layer model to describe the exchange of NO_x and O_3 above and below forest canopies. IMW TNO report R95/113, Delft 1995.
3. Hargreaves K.J., D. Fowler, R.L. Storeton-West, J.H. Duyzer, The exchange of nitric oxide, nitrogen dioxide and ozone between pasture and the atmosphere. *Environ. Pollut.* **75** (1992) 53–59.
4. Duyzer, J.H., D. Fowler Modelling land atmosphere exchange of gaseous oxides of nitrogen in Europe *Tellus* **46B** (1994) 353–372.
5. Duyzer, J.H., G. Deinum, J. Baak, The interpretation of measurements of surface exchange of nitrogen oxides; corrections for chemical reactions. *Phil. Trans. Royal Soc. London* **A 351** (1995) 1–18.
6. Lenschow, D.H., A.C. Delany, An analytic formulation for NO and NO flux profiles in the atmospheric Ltd. surface layer. *J. Atmos. Chem.* **5** (1987) 301–309.
7. Duyzer, J.H., The influence of chemical reactions on surface exchange of NO, NO_2 and O_3: results of experiments and model calculations. in: Schwartz and Slinn, (eds) *5th Int. Conf. on Precipitation Scavenging and Atmosphere–Surface Exchange processes,* Washington 1992, pp. 1105–1114.
8. Duyzer, J.H., Dry deposition of nitrogen compounds to semi-natural ecosystems. Thesis, University of Utrecht 1995.
9. Duyzer, J.H., H.L.M. Verhagen, J.H. Weststrate, F.C. Bosveld, A.W.M. Vermetten, The dry deposition of ammonia onto a Douglas fir forest in the Netherlands. *Atmos. Environ.* **28** (1994) 1241–1253.
10. Duyzer, J.H., Dry deposition of ammonia and ammonium aerosols over heathland, *J. Geophys. Res.* **99 D9** (1994) 18757–18763
11. Nilsson, J., Critical loads for sulphur and nitrogen. in : P. Mathy (ed), *Air pollution and ecosystems,* D. Reidel Publishing Company Dordrecht 1987, pp. 85–91.
12. Farquhar, G.D., P.M. Firth, R. Wetselaar, B. Weir,. On the gaseous exchange of ammonia between leaves and the environment: determination of the ammonia compensation point, *Plant Physiol.* **66** (1980) 710–714.
13. Duyzer, J.H., J.H. Weststrate, K. Beswick, M. Gallagher, Measurements of the dry deposition flux of sulphate and nitrate aerosols to the Speulderbos using micrometeorological methods. IMW-TNO report R94/255, Delft 1994.
14. Duyzer, J.H., J.H. Weststrate, H.L.M. Verhagen, M. Gallagher, H.Coe, D. Fowler, M.A. Sutton, The joint experiment on surface exchange of trace gases over the Leende heathland. IMW-TNO report PU 92/015, *EUROTRAC Newsletter No.10* (1992) pp. 30–36

9.3 Dry Deposition Monitoring of SO_2, NH_3 and NO_2 over a Coniferous Forest

Jan Willem Erisman[1], Marcel Mennen[1], Jos Hogenkamp[1], Erik Kemkers[1] Daniel Goedhart[1], Addo van Pul[1], Geert Draaijers[2], Jan Duyzer[2] and Paul Wyers[3]

[1]RIVM, P.O.Box 1, 3720 BA Bilthoven, the Netherlands,
[2]TNO, P.O.Box 6011, 2600 JA Delft, the Netherlands,
[3]ECN, P.O.Box 1, 1755 ZG Petten, the Netherlands

Summary

Since November 1992 vertical concentration gradients of SO_2, NO_2 and NH_3 as well as relevant meteorological parameters have been continuously measured at a Douglas fir forest site in the Netherlands. It was found that for SO_2 and NH_3 routine application of the gradient method is possible and annual average fluxes can be estimated. Routine application for NO_2 is not yet possible because of frequent instrumental failure and because the influence of chemical reactions on the fluxes can not yet be quantified. Eddy correlation measurements of $u*$ and H at 30 m height by TNO and those at 36 m height were compared to evaluate the constant flux layer assumption. It is obvious from these measurements that during this month the constant flux layer assumption for momentum and sensible heat fluxes was valid. Flux profile relationships for heat and ozone were derived from simultaneous eddy correlation and gradient measurements by TNO. Conventional relationships can be applied, provided a correction is introduced. From the results of the hourly average measurements, dry deposition parameters have been derived. Surface resistances for SO_2 and NH_3 show strong diurnal variations during dry conditions. Then lowest values occur during the daytime. When the leaf surface is wet, R_c values are generally low and independent on the time of the day. An R_c parameterization for SO_2 derived from analogous measurements over a heathland was tested at the forest site. Good agreement between measured and modelled dry deposition velocities was obtained with no systematic deviations. A first order surface resistance parameterization for NH_3 has been derived from the measurements at Speulder forest. When the NH_3 flux is directed towards the vegetation, there is reasonable agreement between deposition velocities obtained from the parameterisation and from the measurements. However, when NH_3 is emitted (~ 20 % of the time), modelled and 'measured' V_d can differ considerably.

Introduction

In 1990 a large project was initiated to develop a monitoring method for measurement of deposition of acidifying components onto forests. This project is the continuation of the successful development of monitoring methods for SO_2,

NO_2 and NH_3 deposition to low vegetation [1–6]. From January to August 1992 tests were done with the equipment over low vegetation at the heathland Elspeetsche Veld [7]. These tests included experiments determining effects of obstacles (monitor housing) to be installed in the mast above the forest on the momentum and heat flux measurements, optimisation of the gradient system for NO_2 and SO_2, and tests of eddy correlation measurements of the NO_2 flux. At the end of 1992 the optimised monitoring systems for SO_2, NO_2 (RIVM) and NH_3 (ECN) were installed at the forest site. Since then continuous vertical concentration gradient measurements of these components are available. In June 1992 also eddy correlation measurements of NO_2 were started. These data, however, have not yet been validated and evaluated. In this paper, the first results of the measurements made between November 1992 and September 1993 are presented.

Experimental procedure

The measuring site consists of a homogeneous 2.5 ha monoculture of Douglas fir, 35 years old with a stem density of 785 ha^{-1}. The mean tree height is 20 m. The canopy is well closed with the maximum leaf area density at a height of 10–14 m. The one-sided LAI varies through the year from about 8 in early spring to 11 at the end of the summer. A schematic of the monitoring system at the Speulder forest is shown in Fig. 9.3.1. The system is extensively described in [4] and [7].

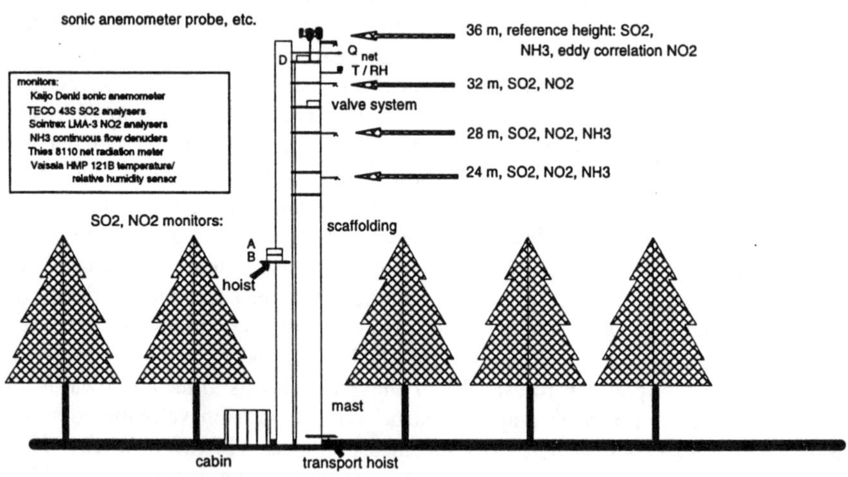

Fig. 9.3.1: Schematic of the monitoring system at the Speulder forest.

Flux profile relationships and the constant flux layer

During experiments in 1989 at the Speulder forest, KNMI compared gradients of T and u above the forest with fluxes of heat and momentum measured directly in order to test the commonly used flux profile relations [8]. From these experiments systematic differences were found between the Dyer and Hicks relations and those obtained above the Speulder forest [9]. This can be accounted for by introducing α_h, a height dependent correction factor. Flux profile relations for ozone and heat were evaluated from simultaneous eddy correlation and gradient measurements by TNO in 1993. The forest had grown about 3 m in height between 1989 and 1993. The displacement height in 1993 was assumed to be 15 m. The α_h factors found by TNO for O_3 and heat are similar to those found by [9]. It is concluded that the conventional flux-profile relations can be used for the estimation fluxes of heat and ozone using gradient measurements. It is assumed that these relationships can be applied for other trace gases (SO_2, NO_2 and NH_3).

The eddy correlation measurements of u_* and H at 30 m height by TNO and those at 36 m height by RIVM were compared to evaluate the constant flux layer assumption. Figs. 9.3.2 and 9.3.3 show scatter plots of hourly average u_* and H values measured at the two heights in December–June. There is a reasonable agreement between the 30 m and 36 m u_* values; most scatter is due to local influence on the fetch and to scatter in the data as a result of differences in averaging time.

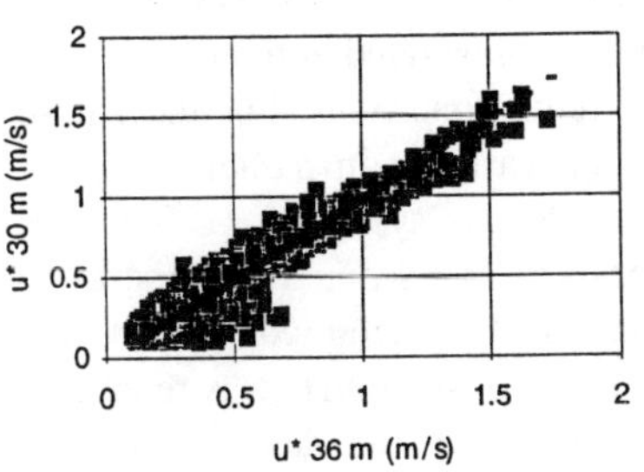

Fig. 9.3.2: Eddy correlation u_* measurements at 30 m compared to those measured at 36 m height.

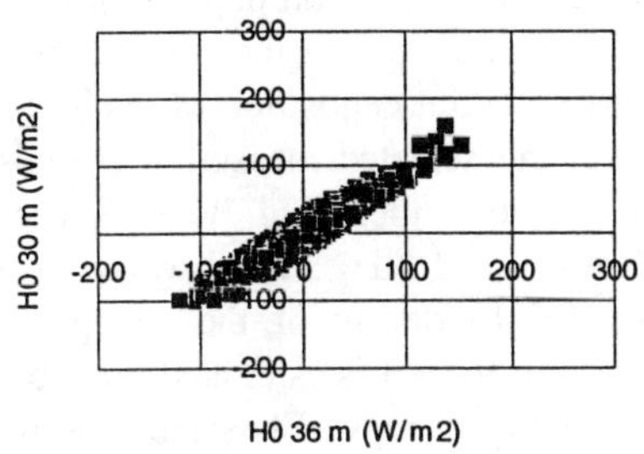

Fig. 9.3.3: Eddy correlation H measurements at 30 m compared to those measured at 36 m height.

The difference between the heat fluxes measured at the two levels was much larger. The reason for this large difference is that during and after rain or fog, heat flux measurements might be contaminated as a result of droplets on the sensors. Furthermore, at very low heat flux values (very stable conditions) the change on temperature inversions and thus decoupling of the air in and above the forest increases. From the two figures it is obvious that during these months the constant flux layer assumption for momentum and sensible heat fluxes is valid.

Average deposition parameters for selected periods

For each hour three vertical SO_2 and NO_2 gradients were averaged and c_*, F, V_d, R_a, R_b and R_c were calculated. The ten NH_3 gradients measured within one hour were also averaged and the deposition values were calculated accordingly. The dataset thus obtained has to be 'cleaned' by selection of hours during which the theoretical demands for the gradient technique were fulfilled, during which the concentrations were well above the detection limit, and during which there was no loss of necessary measurements due to technical problems. R_c values show strong variations with time; when the surface becomes wet, R_c values drop to zero, whereas at very dry conditions R_c values can easily increase up to values of 1000 s m^{-1} at night. An R_c parameterisation derived from analogous measurements over a heathland during three years [3, 10] was tested. This parameterization can be applied for routinely measured components; it is based on literature values for the stomatal resistance, on empirical values for wet surfaces (due to rain or at high relative humidity) and for snow covered surfaces.

Fig. 9.3.4 shows 'measured' and parameterised SO_2 R_c values and measured NH_3 R_c values for four succeeding days. Also shown in this figure are the concentrations of the two gases, the relative humidity and information on surface wetness. Parameterised and 'measured' R_c values for SO_2 show reasonable agreement, with high values during dry periods at night, low values during daytime and values approaching zero during wet surface conditions. R_c values for NH_3 show similar variations. For these days, as well as for the whole dataset no influence of NH_3 on deposition parameters for SO_2 and vice versa was observed. This in contrast to the observations of [6] who demonstrated influence of both gases on the deposition of each other under extreme conditions. For the dataset with the selected measuring periods, 40 % of the variance in parameterised V_d versus the 'measured' V_d was accounted for, with no systematic differences. Emission of NH_3 was observed on several occasions and seemed to be strongly related to drying of the canopy. For the four month period considered here (wintertime), it is estimated that about 20 % of the deposited NH_3 was re-emitted from the canopy. Emission was observed mainly during the day at a relative

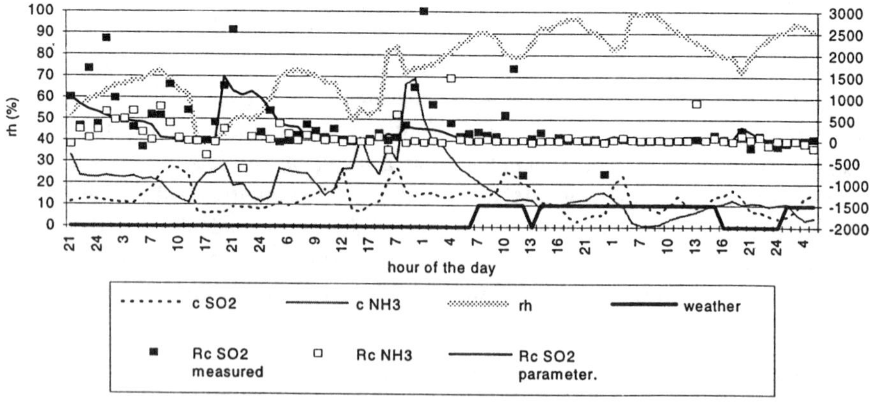

Fig. 9.3.4: R_c values for SO_2 and NH_3 from continuous measurements of vertical gradients over the Speulder forest ("Weather" provides information on precipitation: 0=dry; 10=rain).

humidity decreasing below 80 %. So far no relation is apparent between occurrence of emission and low or decreasing ambient NH_3 concentrations. When the canopy is dry and the flux is towards the surface, this flux is higher than the inferred stomatal flux, indicating that the external leaf surface is also an important receptor for NH_3.

A start was made with the derivation of a parameterised description for NH_3 surface exchange [5]. The prediction of occurrence of emission is still inadequate, as expected from the poor parameterization for drying of the canopy. Under conditions of deposition there is reasonable agreement between modelled and "measured" V_d values, with 25 % of variance accounted for.

References

1. Erisman J.W., PhD Thesis, University of Utrecht, Utrecht 1992.
2. Mennen M.G., *et al.* 1992
3. Erisman J.W., Elzakker B. van, Mennen M., Hogenkamp J., Zwart E., Beld L. van den, Römer F.G., Bobbink R., Heil G.,Raessen M., Duyzer J., Verhage H., Wyers G., Otjes R., Möls J., *Atmos. Environ.* **28** (1993) 487–496.
4. Erisman J.W., Mennen M., Hogenkamp J., Goedhart D., Pul A. van and Boermans J., in: Angeletti G., Pio C., Slanina J. (eds), *Air Pollution Report,* CEC, Brussels 1993.
5. Wyers G.P., Otjes R.P., Erisman J.W., in: Angeletti G., Pio C., Slanina J. (eds), *Air Pollution Report,* CEC, Brussels 1993.
6. Erisman J.W., Wyers G.P. *Atmos. Environ.* **27A** (1993) 1937–1949.
7. Zwart H.J.M.A., Hogenkamp J.E.M., Mennen M.G., *Report No. 722108001*, National Institute of Public Health and Environmental Protection, Bilthoven 1994.
8. Dyer A.J., Hicks B.B., *Quart. J. Roy. Met. Soc* **96** (1970) 715.
9. Bosveld F.C., *Report WR–91-02*, KNMI, de Bilt 1991.
10. Erisman J.W., Versluis A.H., Verplanke T.A.J.W., Haan D. de, Anink D., Elzakker B.G. van, Mennen M.G., Aalst R.M. van, *Atmos. Environ.* **27A** (1993) 1153–1161.

9.4 Flux of NO_X between Atmosphere and Vegetation – Parameterisation of Surface Resistance/Emission

Lennart Granat, A. Rondon, C. Johansson and R. Janson

Department of Meteorology, Stockholm University, S-106 91 Stockholm

Summary

Fluxes of gases including NO, NO_2, O_3 CO_2 and water vapour were studied with the enclosure technique in three different coniferous forest ecosystems with different air pollution loads and in the laboratory on seedlings, also with different exposures of air pollutants. The results have been expressed in the form of resistances (conductances) which will serve as a basis for a more general estimate to any given forest ecosystem in northern Europe.

The flux of NO at the forest floor (both up and down) is insignificant and can be neglected in Sweden and probably also at similar northern locations. The flux of NO to or from forest canopy is also small and can be neglected in comparison with the flux of NO_2. The flux of NO_2 to the forest canopy passes only through stomata and is, in southern Sweden, determined by stomatal resistance no additional internal resistance. This is valid for the whole concentration range investigated, from about 0.2 to more than 10 ppb. At a northern location this is true only at NO_2 concentrations above about 5 ppb. At this location and at lower NO_2 concentrations, an additional (internal) resistance is found. In model calculations, this additional resistance can be expressed as a compensation point, usually between 0.3 and 0.7 ppb, although the real mechanism can also be an internal resistance with no emission. Since the ambient summertime concentration of NO_2 in the region is about 1 ppb, the rate of uptake by the northern canopy is quite small. It is not yet well understood if the higher resistance is a geographical feature related to the lower pollution load, although there is some indirect evidence that this is the case. Deposition of NO_2 to the forest floor is important. Compared to the deposition to the canopy it is lower in southern Sweden but substantially higher at the northern site. The flux of O_3 to the canopy of Scots pine and Norway spruce can be described by stomatal conductance with no internal resistance, plus an external deposition.

The conductance for the external deposition is linearly related to the PAR flux with a coefficient of 0.0015 (mm s^{-1}) / (µE m^{-2} s^{-1}), valid for projected needle area and was determined up to 2000 µE m^{-2} s^{-1} in laboratory experiments.

The flux of HNO_3 and NO_2 (p) to a coniferous forest was measured with a foliar extraction method using branches distributed throughout the canopy. The deposition to individual branches was fund to be highly variable depending on their locations in the canopy. Comparison with simultaneously measured air concentrations will give an estimate of the deposition velocity.

Aims of the research

The aim of the project has been to find ways of quantifying fluxes of nitrogen compounds, in particular NO and NO_2 between the atmosphere and a coniferous forest ecosystem, including spatial and temporal variations. To this end, fluxes should be measured in the field under different environmental conditions, and the factors that regulate the flux evaluated. Trees of pine and spruce and lower vegetation should be investigated separately. In late 1993 the focus was shifted to more reactive compounds for which the aerodynamic resistance is most important. The aim is here to study fluxes by foliar extraction method and compare with common model formulations. In a parallel to this programme regular measurements are being made of air concentration of HNO_3(g), NO_3(p), NH_3(g) and NH_4(p) as an input for calculation of deposition flux.

Selection of coniferous forest ecosystem as the subject for this project was motivated by the lack of knowledge of the deposition flux of NO_x and other N components, especially in a northern location with somewhat lower air pollution levels and also of the large areal extension of forest in northern Europe, the fact that deposition might be an important sink for NO_2 (and other nitrogen compounds) and that the forest can be affected by this deposition.

Principal scientific results

Fluxes of gases including NO, NO_2, O_3, CO_2 and water vapour were studied in three coniferous forest ecosystems, two at Simlångsdalen located in southern Sweden (56 N, Scots pine, LAI = 3), both moderately affected by regional scale air pollution, and one at Jädraås (61 N, Scots pine, LAI = 4, Norway spruce, LAI = 8.5), which is a clean site receiving about one third of the pollution load (including N) found at Simlångsdalen. Fluxes were studied with the enclosure technique using several different chambers. Measurements of the flux to or from the ground were made with the static type chamber (where the change in concentration was followed for some 7 minutes). The flux to or from the canopy was measured with transparent, climate controlled chambers under a variety of conditions. Corrections were made for photolysis of NO_2, reaction of NO and O_3 in the chamber air, as well as for wall effects. Measurements were made mainly in May and June during three consecutive years in addition to previous studies. Extensive measurements were also performed with seedlings of pine in the lab under conditions that resembled those both in northern and southern rural Sweden. Surface resistances (or in some cases conductances) were evaluated from the measured fluxes, as this provides a way to generalise the findings and enables flux calculations to a variety of surfaces . NO_2 concentrations obtained with the photolytic and ferrous sulphate converters in the forest of Jädraås shoved good agreement (despite potential influence of interferences), an important finding since different equipment was used at the two sites and during different times [1–6].

Many measurements of the flux of NO from 5 plots on the forest floor in Jädraås showed an emission that varied from 0.04 to 0.20 ng N m^{-2} s^{-1}. The flux at Simlångsdalen, based on a few measurements, was quite similar, 0.08 to 0.38 ng N m^{-2} s^{-1}. These fluxes can be considered negligible if compared with the total deposition of N to the forest ecosystem (wet deposition of NO_3–N for instance corresponds to 7 ng N m^{-2} s^{-1}).

The flux of NO_2 at the forest floor was always downwards. Due to the short time of coverage, measurements should represent undisturbed conditions even though the chambers used were impermeable to light. In day-time measurements, the deposition velocity (relative to ground area) varied from 3 to 5 mm s^{-1} with a mean of 4 mm s^{-1} corresponding to a surface resistance of 250 s m^{-1} (well-mixed air in the chamber). At the rather unpolluted site in Jädraås, the flux to the forest floor is thus larger than to the forest canopy (see also below).

The flux of NO to or from individual branches was found to be less than 0.1 ng N m^{-2} s^{-1}, the detection level of the method. Previous measurements with elevated NO concentrations showed a deposition velocity at least 10 times lower than stomatal conductance.

Considering the ratio of NO to NO_2 in ambient air, the flux of NO to vegetation is negligible, as is the emission.

Factors controlling the flux of NO_2 to the canopy showed important differences between the two sites. At Jädraås, the northern site, the deposition of NO_2 was controlled by the stomatal conductance, as calculated from water transpiration, indicating no external uptake and essentially zero internal resistance at concentrations usually above 5 ppb. At lower concentrations, the resistance increased and at 1 ppb or lower, the uptake was usually around zero. This can be interpreted either as an increasing internal resistance or as the existence of a compensation point. During a few single cases, an emission of NO_2 was in fact observed, but it is still an open question whether or not there are mechanisms in the needles that produce a true compensation point and emission at lower concentrations. From a modelling point of view, it is convenient to infer a compensation point. Flux measurements could then be quantitatively explained as governed by stomatal conductance and a compensation point ranging from 0.3 to 0.7 ppb for Scots pine in a northern location. This may however vary with time. Earlier measurements on the same vegetation gave values of about 3 ppb.

The results from Simlångsdalen were less complicated to interpret. Fluxes could be well described with stomatal conductance and no additional deposition. A compensation point would have to be about 0.2 ppb which is nether statistically significant nor of any importance for the flux calculations, and can be therefore considered zero. Experiments with seedlings in the laboratory gave results which are very similar to those from Simlångsdalen. While the different behaviour between Jädraås and Simlångsdalen is well supported, it is less clear whether the difference depends on a general regional difference or is due to different stand parameters. No similar measurements in the northern boreal forest ecosystem have been done at ambient NO_2 concentration. However, during an experiment with an open air exposure system (6 ppb NO_2) during the vegetation period, no extra N compounds were found, either in the canopy or in the throughfall. This, to some extent, corroborates the view that the coniferous forest canopy is a small or negligible sink for NO_2 in northern (> 61 °) locations.

The rate of O_3 uptake by the canopy (spruce and pine) is rather well correlated to stomatal conductance but is quantitatively larger than can be explained for this pathway even if the internal resistance is zero. A deposition to external surfaces is therefore inferred. The additional uptake was found to be well related to PAR intensity, both in the field and during laboratory experiments with two year old seedlings of Scots pine.

In the lab experiment conductance for external deposition increased from zero in the dark almost linearly to 3 mm s^{-1} at PAR equal to 2000 $\mu E\ m^{-2}\ s^{-1}$. The mechanism is suggested to be photochemically induced reactions on the needle surface. Rather modest differences in the specific surface resistance were found at the different field sites. The minimum resistance for external uptake was: Jädraås, pine 330 s m^{-1}, Simlångsdalen, pine 250 s m^{-1} and Simlångsdalen spruce 330 s m^{-1} which can be compared to 330 s m^{-1} for seedlings in the lab experiment. It should be noted, however, that there are large differences in the LAI between the sites. During summer conditions (when the measurements were made), the flux of NO_2 at Simlångsdalen would make up some 35 % of the total flux of oxidised nitrogen, while at Jädraås, it would be about 25 %, in this case almost exclusively to the forest floor.

The foliar extraction method for deposition measurements of ammonia and nitrate compound was tested in the lab and used in the field. It was found that the surface resistance for HNO_3 was very low, as expected, and that nitrate could be recovered quantitatively, but also that deposition per leaf area differed between branches of different density. From the regular air concentration measurements compiled during the year, it was seen that the fraction of gas phase to total concentration varies considerably between sites for both nitrate and ammonium (5 to 27 % and 8 to 20 % respectively) and at one site over the year (maximum in spring and summer respectively). Of the measured compounds, HNO_3 concentration shows the largest depletion during the night (50 %). Using these concentration data and combining them with estimated dry deposition velocities, (that are to be verified in the program), yields a dry deposition flux of up to 50 % for nitrate and 80 % for ammonium, as compared with wet deposition.

Conclusions

In line with the original aims, the project has provided data on emission fluxes and surface resistances that can be applied in general flux calculations in the following cases: Flux estimates for NO to forest floor and canopy, which appear to be generally valid in coniferous forest ecosystems in Sweden. Parameterisation of canopy resistance for the flux of NO_2. Some uncertainty remains as concerns how to generalise the findings in northern Sweden, but a tentative recommendation is given. The flux of NO_2 to the forest floor was larger than expected. A tentative magnitude of the resistance is given but no systematic study was made on how to describe the resistance. The ozone measurements were included only to give supporting information but turned out to give new quantitative information on the processes during external deposition. A scheme to estimate this additional flux is described. The parameterisation of surface resistances will be combined with models for aerodynamic and stomatal resistances.

A new system to measure the flux of ammonium and nitrate compounds to the coniferous forest has been tested in the lab and used in the field. Evaluation is in progress.

Acknowledgements

During the whole project support has been obtained from the National Swedish Environment Protection Board, while Stockholm University contributed with basic resources. Support for one year periods have been obtained from Swedish Council for Forestry and Agricultural Research and The Natural Science Council.

References.

1. A. Rondon, C. Johansson, L. Granat, Flux of NO_2 between the atmosphere and a pine stand in central Sweden at low NO_2 concentrations. in: P.M. Borrell, P. Borrell, T. Cvitaš, W. Seiler (eds), *Proc. EUROTRAC Symp. '92*, SPB Acedemic Publishing bv, The Hague 1993, pp. 173–174.
2. A. Rondon, L. Granat and C. Johansson, in: G. Angeletti, S. Beilke, J. Slanina (eds), *Field Measurements and Interpretation of Species related to Photo-oxidants and Acid Deposition, Air Pollution Research Report* **39**, CEC, Brussels 1992, pp. 311–318.
3. A. Rondon, C. Johansson and L. Granat, Dry deposition of nitrogen dioxide and ozone to coniferous forests. *J. Geophys. Res.* **98** (1993) 5159–5172.
4. A. Rondon, Photoinduced deposition of ozone on the plant leaf cuticle. in: Atmosphere-surface exchange of nitrogen oxides and ozone. PhD thesis, Department of Meteorology, Stockholm University 1993.
5. A. Rondon, C. Johansson and L. Granat, Comparison of stomatal and cuticular uptake of ozone by Scots Pine and Norway Spruce. in: P.M. Borrell, P. Borrell, T. Cvitaš, W. Seiler (eds), *Proc. EUROTRAC Symp. '92*, SPB Academic Publishing bv, The Hague 1993, pp. 723-726.
6. A. Rondon and L. Granat: Studies on the dry deposition of NO_2 to coniferous pecies at low NO_2 concentrations, *Tellus* **46**B (1994) 399–352.

9.5 First Emission Inventory of Ammonia in Timis County

Eugen Lorinczi, Ana Ritivoiu and Tamara Camaroschi

Environmental Protection Agency Timisoara, Bl. Mihai Viteazul No. 32,
1900 Timisoara, Romania

Summary

The Environmental Protection Agency Timisoara is responsible for the environmental protection activity in Timis county, the biggest in the country, situated in western Romania. The complexity of the environmental problems is enhanced because of this border position. As a characteristic of the Timis county is the presence of a large livestock sector.

Aims of the research

The Environmental Protection Agency Timisoara accomplished the first NH_3 emission inventory regarding Timis county, using the formulae and coefficients developed at IIASA, in order to know the share of this pollutant in the frame of all pollutants [1–4].

Principal scientific results

The percentage of pigs is the highest from all the country: 11.5 %, while other categories of animals represent, as follows: cattle: 3.5 %, sheep: 4.6 %, goats: 2.3 %, horses: 3.1 %, poultry: 2.6 %. The calculations indicate that the major part of the NH_3 emissions is due to livestock farming (86.4 %). In Timisoara pigs are the main pollutant factor (56.8 %) as compared with the whole country where dairy cows play this part (30.3 %). The calculated values of the NH_3 emissions in Timis county are the greatest in Romania. They are even greater than in some European countries, Fig. 9.5.1.

Looking at the NH_3 emission dynamic during the 4 years period, the level of emission is decreasing. This is due to the abatement of livestock farming and to a reduced application of fertilisers. The same tendency can be observed in the study referring to the whole country. Regarding the emission density, the Timis county values exceeds permanently the value calculated for Romania. From this point of view, our county is the third place in the country. A weak point of this study is that the calculated data could not be compared with field measurements.

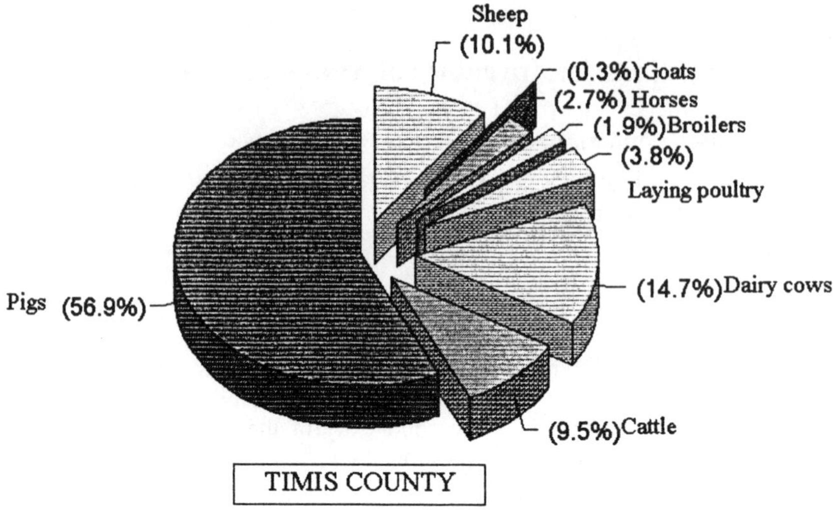

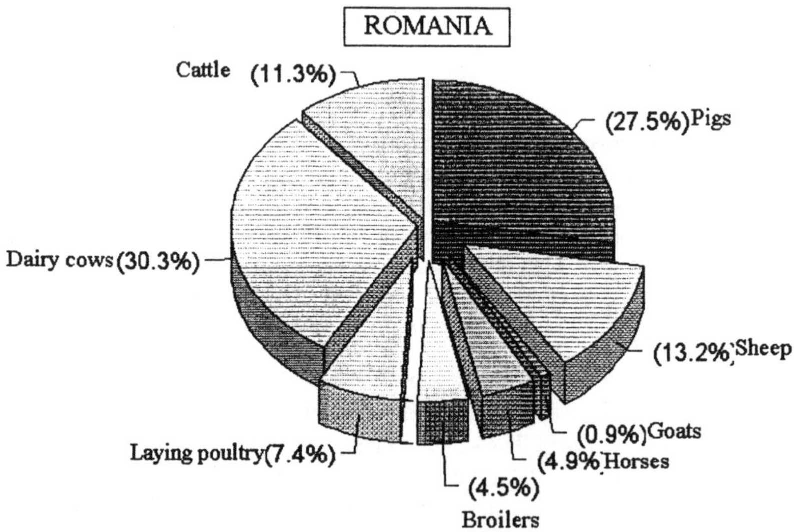

Fig. 9.5.1: Comparison of livestock farming. NH_3 emissions in Timis County and Romania.

Conclusions

The study of the NH_3 emissions in Timis county shows a very high level of this environmental pollutant. The study of its negative effects is not a simple job, it must involve a lot of work and specialists collaborations. The main source of NH_3 emissions in Timis county is the livestock units. The first step that can be done for the abatement of NH_3 emissions is the restriction of the great livestock units. At least it would be necessary to reconsider the structure of the livestock population and the relocation of these farms. This can be carried out only through joint efforts of the competent authorities. In this regard, EPA Timisoara does his job.

In the future, it would be useful and interesting to complete this study with values based on administrative surfaces. In this way the NH_3 emission reduction measures could be taken more easily.

References

1. National Commission for Statistics, *Romania Statistical Yearbooks* 1990, 1991, 1992, 1993.
2. Ammonia Emissions in Europe: Emission Coefficients and Abatement Costs, *Proc. IIASA Workshop* 4–6 February 1991, Laxenburg, Austria.
3. Wolf, J.M., Romania – Agriculture and National Resources: Environmental Concerns, March 1992, Maryland, USA.
4. IIASA, Status Report: *Past and Future Emissions of Ammonia in Europe*, March 1991, Laxenburg, Austria.

9.6 Surface Exchange of Nitrogen Oxides over Different European Ecosystems

Franz X. Meixner[1], Jörg Ludwig[1], Hans Müller[2], Franz Böswald[2] and Birgit Terry[2]

[1]Max-Planck-Institut für Chemie, Abteilung Biogeochemie, Postfach 3060, D-55020 Mainz, Germany
[2]Fraunhofer-Institut für Atmosphärische Unweltforschung, Kreuzeckbahnstr.19, D-82467 Garmisch-Partenkirchen, Germany

Summary

The surface exchange of nitric oxide (NO) and nitrogen dioxide (NO_2) was investigated over three European ecosytems (grassland, farmland, heathland) during extended field campaigns in 1989–1991. The applied flux measuring method (dynamic chamber technique) was successfully intercompared to micrometeorological and other techniques. In general, NO emission (from soils) and NO_2 deposition (to vegetation) was found for the different ecosystems which could be quantified for different time scales. Influencing factors of both, NO emission and NO_2 deposition were identified and studied by field as well as by laboratory experiments.

Aims of the Research

The overall aim of the present contribution was to increase knowledge on the processes that determine the exchange of nitrogen oxides (NO_x) at the atmosphere/biosphere interface. For that, the contribution's activities were focused:

(a) on intercomparisons of different flux measuring methods (micrometeorological and enclosure techniques) during joint CEC/BIATEX field campaigns,

(b) on individual field measurements over a representative European ecosystem (wheat field) to study diurnal and seasonal variations of NO_x fluxes as well as their dependence on physiological and soil parameters, and

(c) on mechanistic laboratory studies to parameterise environmental, biological and nutritional factors which influence NO_x exchange, and to investigate metabolic processes responsible for the consumption of NO_x in wheat plants.

Principal scientific results

Joint and individual field experiments.

Joint CEC/BIATEX/SANA intercomparison field experiments were performed over marshland (Halvergate, U.K., 1989), over a wheat field (Eisdorf, Germany, 1991), and over heathland (Grote Heide, the Netherlands) [1–3]. All exercises have demonstrated specific advantages and disadvantages of the particular methods (aerodynamic gradient, eddy correlation, dynamic box techniques, as well as a laboratory method using individual soil samples). A basic problem, common to all field methods, namely the interference of fast chemical reactions of the NO–NO_2–O_3 triad with (turbulent) transport during the exchange process was identified and corresponding correction procedures could be developed; under/overestimation of the corresponding fluxes was quantified and range between a factor of 0.2 – 2 [4–8].

Along with the joint intercomparison experiments, a so-called BIATEX individual experiment was performed at Manndorf (Lower Bavaria, Germany) [9, 10]. Applying a dynamic chamber system [10], measurements during four experimental periods (29 May–01 June, 26–28 June, 10–07 July, 07–15 August, 1990) addressed the NO_x exchange over wheat at eight plots on a commercial farm. The wheat was managed according to the usual farming praxis. On March 20, May 8 and 15, June 1 and 12, and August 7, nitrate and ammonium fertilizers were applied to the field with a total amount of 218 kg N $ha^{-1.}$ Three of the measurement plots received a reduced amount of 88 kg N ha^{-1} (March 20, May 15, June 1) to account for the influence of fertilizer application on the NO_x vertical net flux.

In total, we investigated 3 European ecosystems, namely grassland, heathland and farmland at different geographical location (52°36' N, 01°37' E; 51°20' N, 05°30' E; 48°43' N, 12°57' E), and of different utilisation (pasture; natural; wheat), different soil type (gleysol; podsol; luvisol), different texture (silt loam; sand; silt loam), and different vegetation (*Lolium perenne* and *Poa trivialis*; *Calluna vulgaris* (partially dead); *Triticum aestivum*).

NO_x *and* NO_2 *fluxes*

A summary of observed net vertical NO, NO_2, and NO_x fluxes is given in Fig. 9.6.1. For NO_x (= NO + NO_2) a net uptake was observed for the heathland and during most parts of the farmland experiments. Net NO_x release was found for the pasture ecosystem and during the last period of the Manndorf experiment. Both cases can be attributed to very recent farming activities just prior to the experiments (cattle grazing and harvest/fertilization, respectively) which favoured (temporally limited) enhanced NO emission from soil. On average, NO_2 deposition (negative net vertical NO_2 fluxes) was observed for all investigated ecosystems. Significant NO_2 emission was observed during extended periods of the MA4 experiment after recent fertilization (August 7), and occasionally during the HA,

MA1, MA3 and GH experiments. In general, NO_2 deposition was observed to be dependent on photosynthetic active radiation and the growth state of corresponding vegetation indicating stomata control of the NO_2 uptake. Average surface resistances $r_c(NO_2)$ are given in Table 9.6.1. Seasonality of $r_c(NO_2)$ is best demonstrated for the wheat field experiments MA1..MA3 (during MA4 (after harvest), NO_2 uptake was no longer controlled by stomates). Before and after the period of strongest growth rate (June 1990, MA2), higher values of $r_c(NO_2)$ were observed. This could be easily explained by the temporal variation of corresponding biomass owing functional stomates and the corresponding physiological conditions of the wheat plants. Compared to the wheat plants, $r_c(NO_2)$ is higher for the heathland and the pasture sites. While the pasture data may be biased by instrumental artefacts (see [10]), heathland data are in good agreement with those derived by independent methods (*e.g.* [11]).

Table 9.6.1: Surface resistance of NO_2 (in s m^{-1}) versus photosynthetic active radiation as observed during field experiments

		Photosynthetic active radiation (PAR) in µmol $m^{-2}s^{-1}$		
		< 1	10–1000	> 1000
HA	18–21 Sep, 1989	931	303	659
MA1	29 May–01 June, 1990	355	261	98
MA2	26–29 June, 1990	101	53	20
MA3	10–20 July, 1990	127	97	129
MA4	07–15 Aug, 1990	370	97	emission
GH	29 Apr–10 May, 1991	460	630	1418

NO *fluxes*

As far as NO fluxes are concerned, the agricultural managed ecosystems "pasture" (mean: +10.6 ng N $m^{-2}s^{-1}$) and "wheat field" (mean of particular experiments: +0.2 to +8.9 ng N $m^{-2}s^{-1}$) exhibit higher NO emissions than the natural heathland (mean: +0.8 ng N $m^{-2}s^{-1}$). The large seasonal variation of NO emission measured at the Manndorf wheat field site (MA1..MA4) is mainly attributed to the addition of N containing fertilizers. Experiments performed after recent application of fertilizer (June, August) show remarkably enhanced levels of NO emissions. NO emission is low during the experiment of July with no fertilizer application prior to the measurements. The dominant influence of fertilizer addition is also reflected in the difference between plots that are managed at different fertilizer regimes. Throughout the season, NO emission from regularly fertilized plots (218 kg N ha^{-1} yr^{-1}) is higher (up to a factor 10 and more) than the NO emission from plots that received a reduced input of nitrogen (88 kg N ha^{-1} yr^{-1}).

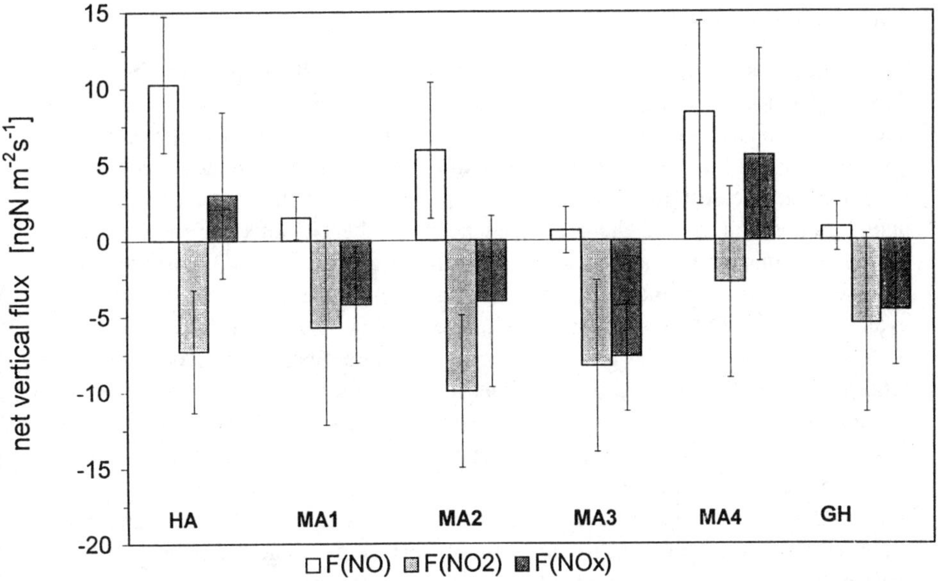

Fig. 9.6.1: Mean net vertical fluxes of NO, NO_2 , and NO_x (=$NO+NO_2$) observed during the field experiments over grassland (HA), farmland (MA1..MA4), and heathland (GH). HA and GH refer to the CEC/BIATEX intercomparison experiments at Halvergate (U.K., 1989) and Grote Heide (The Netherlands, 1991). MA1...MA4 refer to the BIATEX individual experiments at Manndorf, Germany, 29 May–01 June, 26–29 June, 10–20 July, 07–15 Aug. 1990.

By the simultaneous operation of three to five dynamic chambers (each enclosing 0.302 m^2 of soil) on an area of 10–400 m^2 of the corresponding ecosystem, we have some evidence on the spatial representativeness of the results. NO net fluxes vary about a factor of two between the different chambers. This spatial variability is quite similar for all sites and shows the reliability of flux measurements performed by a chamber technique. As indicated by the Manndorf experiments, where seasonal variability of NO net fluxes is more than one order of magnitude, the extrapolation to larger time scales from a limited number of measurements is a matter of substantial uncertainty. Nevertheless, we assume that mean fluxes measured at the pasture and the heathland site are rather representative for the corresponding ecosystem type, as variation due to human influence is of minor importance. This assumption may be supported by comparison to results reported elsewhere [12]: at several grassland sites mean NO emission was found between 1.5–7.3 ng $N\,m^{-2}s^{-1}$, with higher values for grazed pastures than for natural grasslands. To our knowledge, no measurements of NO exchange over heather have been published by now, but the observed mean value of 0.8 ng $N\,m^{-2}s^{-1}$ is similar to those results which are reported for other natural ecosystems like temperate forest soils (0.2–0.8 ng $N\,m^{-2}s^{-1}$, see also [12]).

Compensation points

The simultaneous occurrence of NO emission and NO deposition has been described before (*e.g.* [13]) and the mixing ratio at which NO emission and NO deposition balance equal to yield a net flux of zero has been defined as the compensation point. The compensation point varies considerably as it depends on environmental factors that influence those processes that finally lead to emission and/or deposition. While the calculated mean of the NO flux for the particular periods of our measurements always result in net emission of NO, observation of NO deposition were made as well during several parts of our studies. The examination of NO fluxes as a function of the ambient NO mixing ratio shows that net deposition is measured if a certain level of ambient NO is exceeded (see Fig. 9.6.2). During the Manndorf experiments, net NO deposition was found for ambient mixing ratios ≥ 1–2 ppb in July 1990, whereas net NO emission occurred at lower mixing ratios. A similar change from net emission to net deposition is seen at an ambient mixing ratio of about 1–4 ppb in the results of the Grote Heide experiment (Fig. 9.6.3). If conditions prevail which favour NO emissions, then there exists no relation between NO net flux and ambient mixing ratio as demonstrated in Figs. 9.6.2 and 9.6.3 for the August period of the Manndorf and the entire Halvergate experiment. As deposition is low compared to emission, the variation of factors that control emission processes is of greater importance than

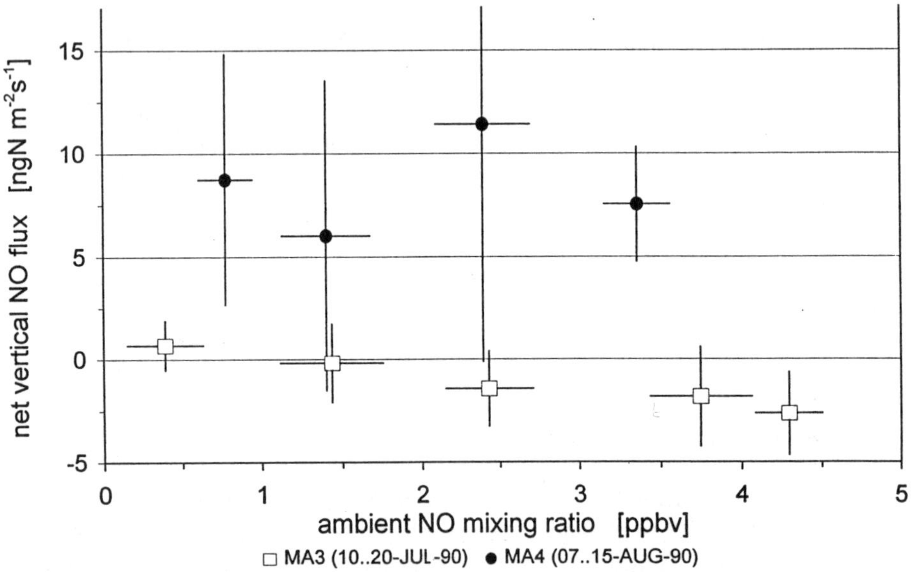

Fig. 9.6.2: NO net fluxes as a function of ambient NO mixing ratio for the experiments MA3 and MA4 (Manndorf, FRG, 1990). Measurements are summarized in discrete intervals of 1 ppb. Each data point represents the average vertical net NO flux (± standard deviation) at the average ambient mixing ratio (± standard deviation) observed in each interval. Note that data points at low ambient mixing ratios generally represent higher numbers of individual measurements.

the variation of ambient mixing ratio. Laboratory studies on soil samples from the Manndorf wheat field site which were performed under controlled environmental conditions [14] have shown that a compensation point exists for those situations as well. Compensation points up to 60 ppb with a high diurnal variability have been observed for those samples which showed also considerable NO productivity. Therefore, it is reasonable that under corresponding field conditions (MA4 and HA, where NO mixing ratio was always below 10 ppb) NO deposition could not be observed. Significant (individual) NO_2 emission could be occasionally observed during most field experiments (see above). However, emission fluxes were always very small. With the exception of the MA experiment (where NO_2 emission could clearly attributed to recent fertilisation) we were not able to deduce significant NO_2 compensation points from corresponding plots of net NO_2 flux versus NO_2 mixing ratio due to the high scatter of corresponding data. However, our laboratory experiments on wheat plants from the Manndorf site clearly indicate NO_2 compensation points of ≈ 2 ppb [15] and 0.2–3 ppb [10], respectively.

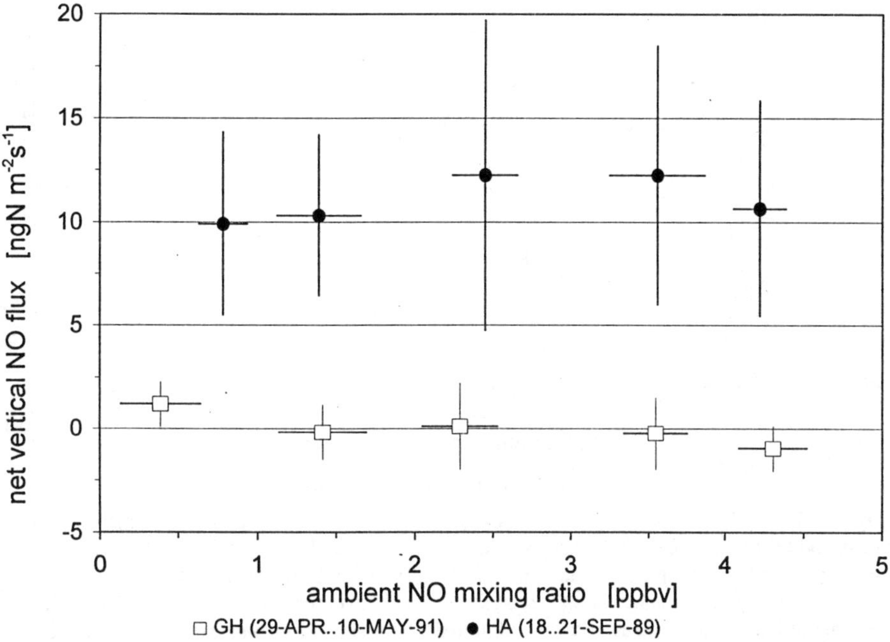

Fig. 9.6.3: same as Fig. 9.6.2, but for the experiments at Halvergate (U.K., 1989) and Grote Heide (The Netherlands, 1991).

Influence of soil nitrogen.

Enhanced levels of NO emission after the application of nitrate and ammonium fertilizers to the Manndorf wheat field indicate that NO exchange is strongly related to the nitrate and ammonium contents (availability) in the soil. Direct measurements of soil nitrate and ammonium were performed at the end of all particular experimental periods mentioned above. The relationship of NO emission to the soil nitrate content is shown in Fig. 9.6.4. Over the wide range of observed nitrate levels, the NO net flux correlates quite well to the nitrate content of the soils. No significant correlation could be found between ammonium content and NO fluxes (see [10]). The spatial variability of net NO fluxes as observed by the individual chambers (see above) can not be explained neither by the variation of the corresponding nitrate nor the ammonium content of the soil. Considering that the results shown in Fig. 9.6.4 summarise all our measurements which were taken under very different environmental conditions, we suggest that at first the nitrate content of the soil should be used as a suitable parameter to estimate the magnitude of the NO net emission flux of a given site. The influence the soil ammonium content seems to be more complicated.

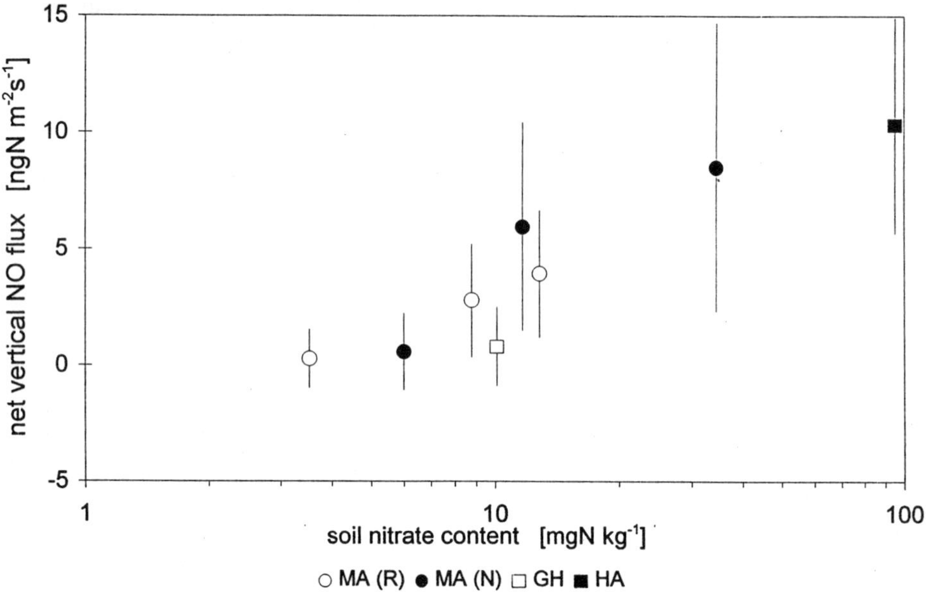

Fig. 9.6.4: NO net flux versus nitrate content of soil for the field experiments over grassland (HA), heathland (GH) and farmland (MA). "N" and "R" refer to measurements (at MA) performed on plots which received normal and reduced fertilizer amounts, respectively. Each data point represents the average vertical net NO flux (± standard deviation) measured by individual chambers during an experiment's period. Nitrate content was determined in soil samples of 0–30 cm depth taken close to the chambers at the last day of each experiment.

Influence of soil temperature and soil moisture.

Although the importance of soil temperature and soil moisture for the NO production in soils is apparent, previous studies came to conflicting results concerning their influence in regulating the NO flux (see [12]). The complex relationship between soil temperature, soil moisture content and NO exchange can be demonstrated by the data of the third field experiment at Manndorf (MA3, July 10–20, 1990). Analysing the total data set, no significant correlation between soil temperature and net NO flux could be obtained. After classifying the data into cases of wet and dry surface conditions (as determined by a surface wetness sensor), the effects of soil temperature and moisture become evident. Fig. 9.6.5 shows a pronounced increase of net NO flux with increasing soil temperature above dry (soil) surface. In the contrary case of wet surface, there exists no significant correlation. Most likely, the slight variations of the absolute soil moisture content at the surface of the wet soil seem to influence the net NO exchange and upset the influence of soil temperature during this particular measurement period. Results shown in Fig. 9.6.5 also indicate a higher net NO flux at a given soil temperature if the surface is wet. Even the small amount of water, which may be deposited to the soil surface by dewfall, causes obviously an increase of the net NO flux. It is not clear yet, whether this effect is due to the direct influence of water or due to the mobilisation of nutrients. Nevertheless, this fact again suggests that the NO emission may originate in the uppermost soil

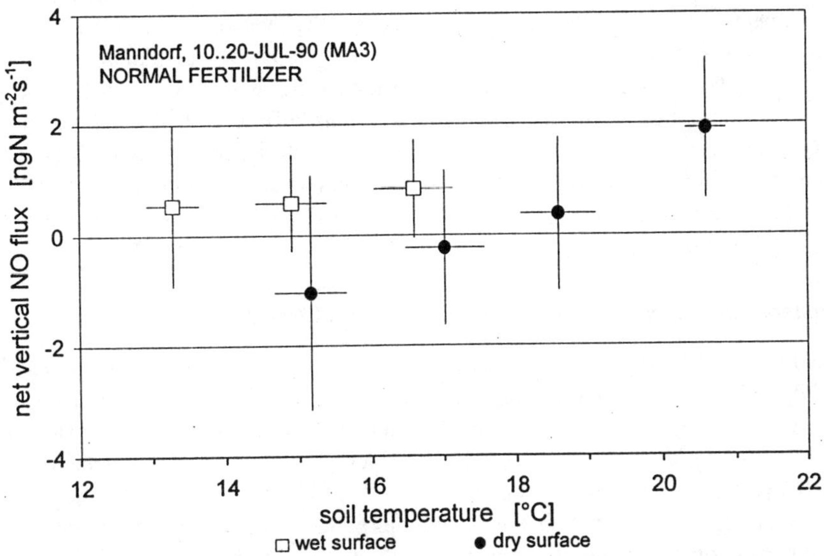

Fig. 9.6.5: Dependence of NO net flux on surface wetness and soil temperature (at d=1 cm). Data points represent the average NO net flux (± standard deviation) for discrete temperature intervals of 2 K at the average (± standard deviation) soil temperature observed in this interval.

layer. The hypothesis, that (microbial) processes responsible for NO production may be located in the first centimetres/millimetres of soil depth was first published by Galbally and Johansson [16] and was confirmed by Remde *et al.* [6] from their study on soil samples from the Halvergate marshland site.

Laboratory studies

Wheat plants of different growth state were regularly transferred from the Manndorf wheat field site to a growth chamber during the vegetation period 1991. There, they were exposed to different NO and NO_2 mixing ratios, different light intensities and different relative humidities as well as a set of wheat plants which were separately grown from the very first in the growth chamber. The exchange of NO and NO_2 between atmosphere and the wheat plant-soil system was found to consist generally in net NO emission from the soil and NO_2 deposition to the leaves of the wheat plants; for both processes a strong diurnal variation was observed exhibiting maximal values during day and minimal values during night. NO_2 deposition was shown (a) to be linearly dependent on light intensity (photosynthetic active radiation) and (b) to be mainly under stomatal control (in some cases cuticular uptake was also evident). Within the range of 0–60 ppb the NO_2 net flux is also linearly correlated with NO_2 mixing ratio. Zero NO_2 net flux was observed around 2 ppb (NO_2 compensation point) and significant NO_2 emission from the leafs of the wheat plants could be found below that mixing ratio. Experiments addressing the relation of ambient NO_2 versus nitrate reductase as well as nitrite reductase activity in the leaves of the wheat plants were not conclusive. By fumigating wheat plants (35 days old) with $^{15}NO_2$ and following the $^{14/15}N$ content of particular amino acids through different plant compartments, it could be directly shown that NO_2 is taken up by the leaves and is used for the synthesis of glutamate, glutamine and further amino acids and is transported to other plant compartments (*e.g.* the roots). Further details are described by Weber [15] and Ludwig [10].

A general algorithm to estimate* NO *emission from soils?

Based on their field studies over different American ecosystems Williams *et al.* [17] developed a general algorithm for the estimation of NO emission for a variety of soil categories. The NO emission flux is described by $F(\text{NO}) = \text{A} \exp(\text{B}^* T_{soil})$, where T_{soil} is the soil temperature (°C), B is a dependency coefficient which is assumed to be constant across biomes, and A is a biome fitting parameter (ng N $m^{-2}s^{-1}$) which is deduced/interpolated from observed relationships of NO emission versus nitrate content of soils. Using "A" parameters of Williams *et al.* (grazed pasture, wheat field, heather = forest soil) and mean soil temperatures observed during our field experiments we calculated corresponding NO emission fluxes. These are shown in Fig. 9.6.6 together with the experiment averaged NO emission data for the three ecosystems investigated during this study. Whereas for the pasture and the heathland sites the NO flux is always underestimated (at least by a factor of 4), the calculated wheat field emissions overestimate the observed NO emissions up to a factor of 15. The obvious discrepancy is certainly due to the application of "A" parameters which are constant with time and may be also due to the statistically not very significant

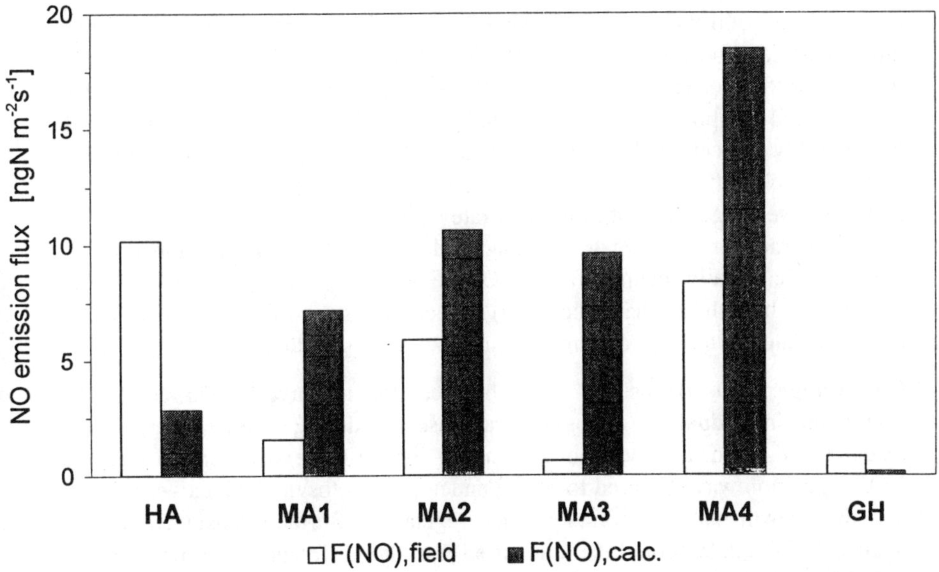

Fig. 9.6.6: Mean NO emissions observed during the experiments over grassland (HA), farmland (MA1..MA4), and heathland (GH) versus estimates of the NO emission flux obtained by the generalisation scheme of Williams et al. (1991) using actual soil temperature data from the experiments. For details, see text.

deduction of "A" parameters (see [17] and references therein). Furthermore, it should be recalled, that the field data leading to the development of the algorithm of Williams *et al.* [17] have been obtained by a dynamic chamber technique which uses so-called "zero-air" (synthetic air, free of NO) for flushing the chambers. It should be noted, that the observation of any possible NO deposition is basically excluded by this technique.

Conclusions

The intercomparison of our flux measuring method, the dynamic chamber technique, with micrometeorological and other techniques resulted in satisfying agreement. It should noted, that the interference of fast chemical reactions of the NO–NO_2–O_3 triad with (turbulent) transport regimes is a common (and partially serious) problem to all field methods which address NO_x surface exchange. However, corresponding correction procedures could be developed.

The investigated ecosystems are normally sources of nitric oxide. The emission of NO is controlled by a variety of environmental variables and human activities. The importance of individual factors is matter of reasonable complexity and can change on short time scales (*e.g.* weeks, season). Highest NO fluxes can be observed in case of nitrogen input into a given ecosystem (by fertilization or grazing), high soil temperatures and a sufficient (not yet quantified) amount of soil moisture. In the contrary case of low soil nitrogen, low soil temperatures or dryness of the soil the net NO emission from soil is reduced to a large extent and nitric oxide can be deposited from the atmosphere at ambient NO mixing ratios higher than the corresponding NO compensation point mixing ratio under these conditions (1–5 ppb). Regional NO_x budgets of polluted areas in Europe may be considerably concerned by this fact. Due to their complexity, a rank order of factors which influence NO emission from soils can presently not be given. However, we suggest that at first the nitrate content of the soil should be used as a suitable parameter to estimate the magnitude of the NO net emission flux of a given (agriculturally managed) site. Our field and laboratory results strongly support the hypothesis that (microbial) processes responsible for NO production may be located in the first centimetres/millimetres of soil depth.

On average, NO_2 deposition was observed for all investigated ecosystems. Significant individual NO_2 emission was observed during extended periods after recent fertilization, and occasionally during most field experiments. In general, NO_2 deposition was observed to be dependent on photosynthetic active radiation and the growth state of corresponding vegetation. Surface resistances $r_c(NO_2)$ against NO_2 uptake could be quantified for the different ecosystems, clearly indicating diurnal (as well as seasonal) variation. As indicated by field experiments, the NO_2 exchange may be also bi-directional. Laboratory experiments on wheat plants clearly indicated NO_2 compensation points of 0.2–3 ppb.

A simple generalised scheme (algorithm) for the estimation of biogenic NO emissions from soils [17] failed when applied to our field data and compared to observed experiment averaged NO emission data from the different ecosystems. It is suggested, that a simple "biome-fitting" factor as used by Williams *et al.* [17] is by far not sufficient to describe temporal and geographical variation of the behaviour of individual biomes as long as this parameterization is not based on specific (microbial) processes which produce and consume NO in soil.

Acknowledgements

The presented work was supported by Bundesminister für Forschung und Technologie (contract 07EU713-7) and by the Commission of the European Communities (contracts EV4V-0167-C, EV4V-00883-C).

References

1. J.H. Duyzer, D. Fowler, F.X. Meixner, G. Dollard, C. Johanson, M. Gallagher, The Halvergate trace gas experiment on surface exchange of oxides of nitrogen: preliminary results, in: Beilke, S., Millan, M., Angeletti, G. (eds) *Field measurements and interpretation of species derived from* NO_x, NH_3, *and* VOC *emissions in Europe, Air Pollution Research Report* **25**, Brussels, CEC, Directorate-General for Science, Research and Development 1990, pp. 267-276.

2. J.H. Duyzer, J.H. Weststrate, H.L.M. Verhagen, M. Gallagher, H. Coe, D. Fowler, M. Sutton, The joint experiment on surface exchange of trace gases over the Leende heathland, in: Borrell, P. and Cvitas, T. (eds), *EUROTRAC Newsletter 10/92*, EUROTRAC ISS, Garmisch-Partenkirchen 1992, pp. 30–36.

3. M. Nestlen, H. Müller, F. X. Meixner, Measurements of vertical fluxes of CO_2, H_2O, NO, NO_2, and O_2 by eddy correlation: First results from The BIATEX/SANA experiment "Eisdorf" March 1991, in:Borell, P. M., Borell P., Cvtaš, T., and Seiler, W. (eds) *Proc. EUROTRAC Symp. '92,* SPB AcademicPublishing bv, The Hague 1993, pp 727–731.

4. G. Kramm, H. Müller, D. Fowler, K.D. Höfken, F.X. Meixner, E. Schaller, A modified profile method for determining the vertical fluxes of NO, NO_2 , ozone, and HNO_3 in the atmospheric surface layer, *J. Atm. Chemistry* **13** (1991) 265.

5. F.X. Meixner, F. Böswald, K. Eiblmeier, G. Kramm, J. Ludwig, S. Leimbach, H. Müller, M. Nestlen, P. Weber, Measurements of vertical fluxes of nitrogen oxides, *EUROTRAC Ann. Rep. 1991, Part 4 BIATEX*, EUROTRAC ISS, Garmisch-Partenkirchen 1992, pp. 60–66.

6. A. Remde, J. Ludwig, F.X. Meixner, R. Conrad, A study to explain the emission of nitric oxide from a marsh soil, *J. Atmos. Chem.* **17** (1993) 249.

7. J.H. Duyzer, PhD Thesis, Univ. Utrecht, The Netherlands (1994).

8. G. Kramm, R. Dlugi, G.J. Dollard, T. Foken, N. Mölders, H. Müller, W. Seiler, H. Sievering, On the dry deposition of ozone and reactive nitrogen compounds, *Atmos. Environ.* submitted.

9. F.X. Meixner, J. Ludwig, A field experiment on surface exchange of trace gases : Ecosystem "wheatfield", Manndorf/ Lower Bavaria, in: Angeletti, G., Beilke, S. and Slanina, J. (eds) *Field measurements and interpretation of species related to photooxidants and acid deposition, Air Pollution Research Report* **39,** Brussels, CEC, Directorate-General for Science, Research and Development 1992, pp. 325–337.

10. J. Ludwig, Untersuchungen zum Austausch von Stickoxiden zwischen Biosphäre und Atmosphäre, PhD thesis, Univ. Bayreuth, Germany 1995.

11. H. Coe, M. Gallagher, Measurements of dry deposition of NO_2 to a Dutch heathland using the eddy-correlation technique, *Quart. J. Roy. Meteor. Soc.* **118** (1992) 767.

12. C. Johansson, Fluxes of NO_x above soil and vegetation, in: Andreae, M.O., and Schimel, D.S. (eds) *Exchange of trace gases between terrestrial ecosystems and the atmosphere*, New York: John Wiley & Sons 1989, pp. 229-246.

13. C. Johansson, L. Granat, Emission of nitric oxide from arable land, *Tellus* **36B** (1984) 25.

14. J. Ludwig, P. Weber, F.X. Meixner, H. Rennenberg, Surface fluxes of NO and NO_2 by a dynamic chamber technique - Laboratory studies on wheat, in: Angeletti, G., Beilke, S. and Slanina, J. (eds) *Field measurements and interpretation of species related to photooxidants and acid deposition, Air Pollution Research Report* **39**, Brussels, CEC, Directorate-General for Science, Research and Development 1992, pp. 257-265

15. P. Weber, Aufnahme und Metabolisierung von atmsophärischem Stickstoffdioxid (NO_2) in Weizen (*Triticum aestivum*, L.), PhD thesis, Techn. Univ. München, Wiss. Verlag Maraun, Frankfurt am Main 1993.

16. I.E. Galbally, C. Johansson, A model relating laboratory measurements of rates of nitric oxide production and field measurements of nitric oxide emissions from soils, *J. Geophys. Res.* **94** (1989) 6473.

17. E.J. Williams, A. Guenther, F.C. Fehsenfeld, An inventory of nitric oxide emissions from soils in the United States, *J. Geophys. Res.* **97** (1991) 7511.

9.7 Measurement of Ambient Concentration and Surface Fluxes of O_3, SO_2 NO_x and TSP in Uludag National Park.

Ulviye Özer[1] and Semra G. Tuncel[2]

[1]Chemistry Department, Uludag University, Bursa, Turkey,
[2]Chemistry Department, Middle East Technical University, 06531 Ankara, Turkey

Summary

Uludag station is situated in the northwest part of the Turkey at longitude $40^\circ\ 10^1$ N, and latitude $20^\circ\ 08^1$ E (Fig. 9.7.1). The station is operated by the Middle East Technical University and Uludag University. Based on the fact that fluctuation of ambient concentration results in the changes in the vertical fluxes, first ambient concentration data were obtained. Starting in March 1993, measurements of O_3, NO_x, SO_2 and TSP concentration at ground level have been made continuously. Meteorological parameters wind direction, wind speed, temperature and humidity were also measured continuously. Hard winter conditions at the Uludag mountain site (average snow-fall is about 1.5m) resulted in some technical problems with the instruments, which in turn resulted in some gaps in the data.

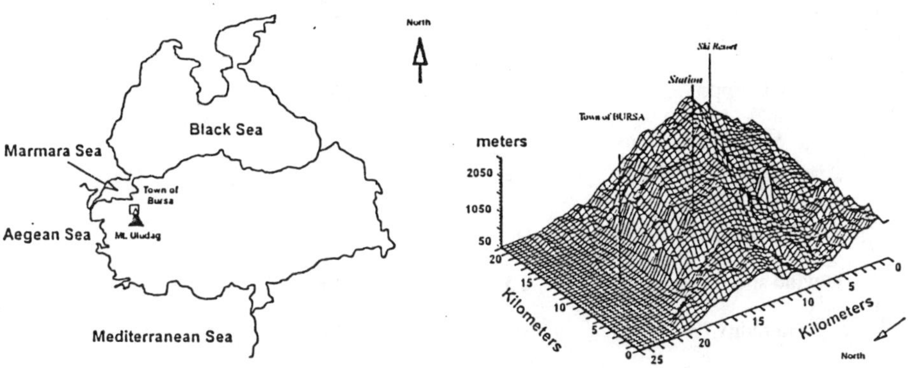

Fig. 9.7.1: Location and topographical features of the station.

Measured concentrations of the O_3, NO, NO_2, SO_2, and TSP were discussed concerning meteorological parameters, air quality standards and contributing source regions through wind sector analyses. By now we have ambient concentration data of O_3, NO_x, SO_2 and TSP for more than two years period. This data enabled us to recognise levels of trace gases and their contributing source regions and photochemical smog scenario in Uludag National Park which is suffering from forest decline. In order to control atmospheric exchanges of O_3 and its precursor gases to natural vegetation, one has to also measure vertical fluxes.

We were planing to start the experiments for vertical flux measurement of O_3 and its precursors using gradient methods. But, because of the insufficient funding in the project we were not able to start up flux measurements.

Aims of the research

To asses the impact of O_3, NO, NO_2, SO_2, and TSP and their contributing source regions.

To study the atmospheric exchanges of O_3 and its precursor gases to natural vegetation in a rural area of northwesten Turkey, through to measurements of ambient concentrations and surface fluxes.

Principal scientific results:

Annual averages of measured parameters at Uludag mountain is given in Table 9.7.1. These are the averages for 1993 and 1994. Ozone decreased for about 13% although there is an increase in the NO and NO_2 concentrations in 1994. Sulfur dioxide concentrations increased for about a factor of ten in 1994. There is not a substantial differences in the annual averages of the other parameters from one year to another.

Table 9.7.1: Annual averages of measured parameters at Uludag Mountain

Parameter	1993	1994
NO(ppb)	0.4	0.7
NO_2(ppb)	1.5	2.2
O_3 (ppb)	32	23
SO_2 (ppb)	2.5	26
TSP (mg/m^3)	18	15
wind speed (m/s)	2.5	2.8
Humidity (%)	61	70
Temperature (°C)	7.3	2.8

Further statistics of the average O_3 concentration at Uludag mountain are as follows. The mean of 1 h average 9800 values was 28 ppb with a standard deviation of 16 ppb. The mean hourly O_3 concentration for a 24 month period were above the US Air Quality Standard of 15 ppb. The highest recorded 1 h average concentration of O_3 was 98 ppb which is below the US air quality standards of two hourly values of 120 ppb per year. Ozone data collected at Uludag mountain were distributed nearly normally. The range was 3–98 and the median was 28. About 45% of the 1 h average O_3 values fell between 20 and 40 ppb. During a 24 month measurement about 11 of 9800 1 h average values exceeded 50 ppb.

Monthly mean concentrations for NO, NO_2, O_3 and TSP are given in Table 9.7.2. These findings indicate that O_3 levels we observed in Uludag mountain is basically from the photochemical production from the precursor gases of urban and industrial emissions with subsequent transport to our sampling area [5].

Table 9.7.2: Monthly averages of two year daily measurements at Uludag Mountain. Units are ppb unless otherwise specified

Months	NO	NO_2	O_3	SO_2	TSP(mg/m^3)
January	0.5	0.2	18	13	5.1
February	–	–	21	38	25
March	0.6	1.2	26	14	22
April	0.7	2.6	27	10	17
May	1.2	1.8	18	6.2	16
June	1.0	2.9	25	--	14
July	0.5	4.6	31	1.2	22
August	0.45	1.9	44	5.1	22
September	0.40	0.9	35	8.1	14
October	0.52	1.4	22	7.2	13
November	–	0.94	23	13	6.3
December	–	1.6	19	14	4.0

All variables showed a seasonal trend. Concentration of O_3, NO_2 and SPM are higher during summer months and decrease substantially in the spring and autumn. NO do not show significant seasonal variation. Concentrations of SO_2 on the other hand, are lower during the summer period. The observed trends in the concentrations of pollutants are probably due to variation in the source strength and transport patterns. The photogeneration of O_3 and NO_2, both in the vicinity of the station and in source regions (the city of Bursa and industries around it), determines the seasonal behaviour of those gases. Higher photochemical production of O_3 during the summer months, due to increased solar flux, results in the higher concentrations observed during the summer months. The summer maxima of O_3 is most pronounced in August with a 1 h peak value of 56 ppb. The highest 1 h peak value of 98 ppb was observed in September 1993. The more vigorous conversion of NO to NO_2 during the summer months can also explain the high NO_2 concentrations in the summer period [1, 2]. Concentrations of both O_3

and NO_2 observed at Uludag mountain were comparable with rural concentrations reported elsewhere [3].

The NO concentrations do not show any seasonal variation and are quite low. The depletion of NO and lack of any long and short term trends indicate that the transport from the Bursa city and industries is more important than local generation from the vehicular activities in the vicinity of the station, not only for NO and also for the other parameters.

There are no important sources of SO_2 and SPM around the station. Consequently, the observed concentrations of these pollutants are explained by transport from the urban and industrial areas where fossil fuel is burned. Low concentrations of SO_2 in the station during the summer months are probably due to the lack of residential heating during this period.

The observed concentration of SPM may be due to airborne dust particles in the summer when the soil is dry, and to the transport of anthropogenic emissions in the spring and winter when the soil in the vicinity of the station is either damp or snow covered.

Concentrations of all measured parameters, with the exception of NO show well-defined diurnal variations especially for summer months. Fig. 9.7.2 shows diurnal variations of O_3, NO, NO_2 and SO_2 for summer. Ozone concentration peaks between 3 and 4 p.m. Concentration of NO stays constant throughout the day, but NO_2 and SO_2 shows a substantial diurnal variation of 2.3 ppb.

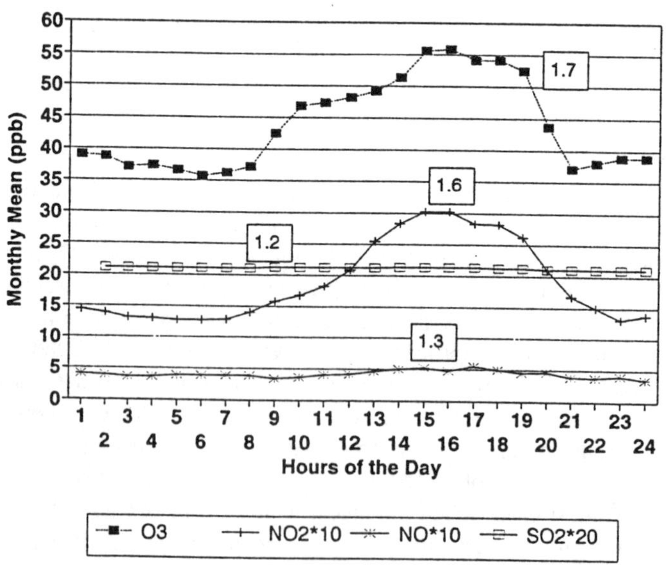

Fig. 9.7.2: Average diurnal variation of O_3, NO, NO_2 and SO_2 in summer 1993 at Uludag mountain. The numbers given on the curves are peak or mean ratios for the given parameter.

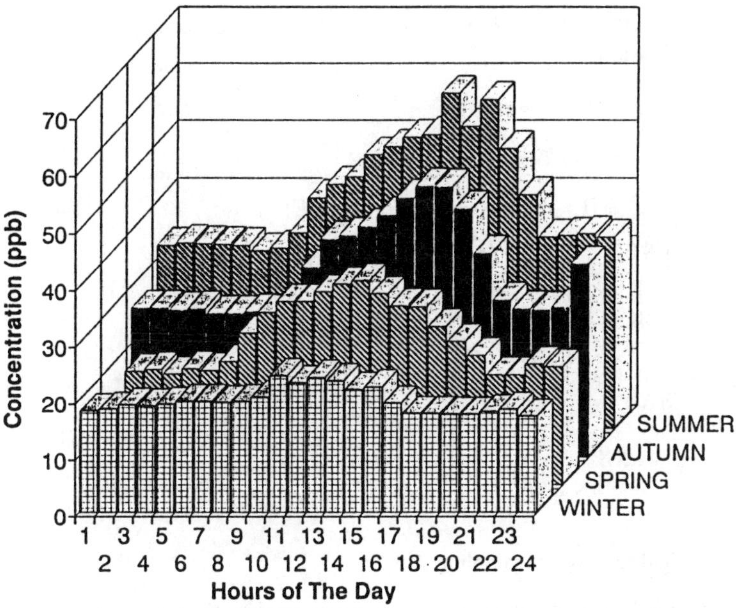

Fig. 9.7.3: Seasonal variation of the hourly average O_3 concentration.

The diurnal variations of NO, NO_2 and O_3 reveals a typical photosmog cycle with oxidant production (O_3, PAN, etc.) in the afternoon by photochemical reactions of the primary pollutants NO_x and unburned hydrocarbons from industry and traffic [4]. The diurnal variation of hourly O_3 concentrations in different seasons are shown in Fig. 9.7.3. The diurnal variations shows a late afternoon O_3 peak around 16:00 in the summer shifts toward earlier hours and become less pronounced in winter because of the solar influences. The diurnal variations of daily 1 h O_3 concentrations was 17 ppb in summer and 3.2 ppb in winter.

Low and constant NO concentrations together with relatively strong diurnal patterns in the NO_2 and O_3 concentrations may indicate that the NO is largely consumed during the transport but transport distance is not long enough for the complete consumption of NO_2. This scenario suggests a transport time of approximately 1–6 hours. With an average wind speed of 2.5 m s^{-1}, the sources which accounts for observed concentrations at the station site should be 9–50 km from the station. The two most important emission sources in the region are the city of Bursa which is located 20 km to the north and a coal-fired power plant which is located 40 km the southwest of the station. Since these two major sources are located at different wind sectors, their plumes are intercepted at the station when winds are from different directions. The north wind sector includes not only the town of Bursa but also a number of industrial emissions, particularly emissions from various textile plants. The prevailing winds in Uludag mountain were from the southeast to southwest direction, with daytime wind speed being much higher than at night-time. Fig. 9.7.4 is the wind rose observed at the station and wind sector identification.

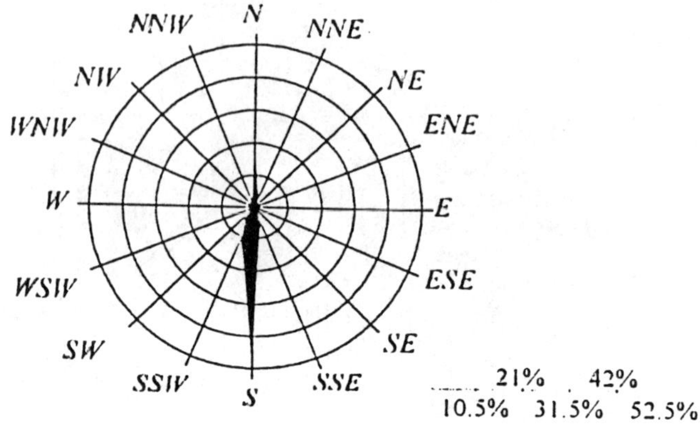

Fig. 9.7.4: Wind rose for the sampling site.

Table 9.7.3: Hourly average concentrations (ppb) of measured parameters and wind sector at Uludag Mountain

Wind direction	NO	NO_2	O_3	SO_2	SPM($\mu g/m^{3)}$
N	-	0.90	47.3	2.02	13.3
NNW	-	-	20.1	8.2	10.0
SE	0.35	0.15	49.2	3.8	17.2
SSE	-	0.95	22.8	9.7	9.3
S	0.6	1.22	25.4	5.2	17.1
SSW	0.39	1.56	34.8	2.5	17.8
SW	0.30	1.9	43.7	1.5	33
WSW	0.25	1.25	35.1	2.7	31.3

Concentration of gaseous pollutants at different wind direction are summarised in Table 9.7.3. It can be clearly seen that the highest O_3, NO_2 and SO_2 concentrations are associated with wind sectors between N and NE and between S and SW. The first sector between N and NE includes the city of Bursa and various small scale industries around the city and the second sector which is between S and SW includes the thermal power plant.

Since the Uludag mountain is a forested sensitive area contribution of each of these sources on the observed concentrations is more important than the average concentrations at each wind sector, as it is a direct measure of the concentration of sources on the deposition of pollutants like SO_2 and O_3 on the forest ecosystem.

Concentration of each wind sector on the observed concentrations of measured parameters are given in Table 9.7.4. As it is clearly seen from the table, the largest contribution of O_3 concentration is observed from S and SSW sectors. Approximately 80 % of the observed O_3 and SO_2 concentrations are accounted by these two sectors. Since the coal-fired power plant is the only source in that sector, it is the most important source of O_3 and SO_2 at the Uludag mountain. Although the same sector also have the highest contribution for the NO, NO_2 and SPM, this is purely due to very high frequency of winds from that sector and can not be attributed to the power plant in that sector. If the power plant is the main source of these pollutants one would expect not only high fractional contribution from S and SSW sectors but also high concentrations when the winds blow from these directions.

These findings indicate that O_3 levels we observed at the Uludag mountain site are basically from the photochemical production from the precursor gases of urban and industrial emissions with subsequent transport to our sampling area [5].

Table 9.7.4: Percent contribution to the average parameter value for different wind sectors

Wind sector	NO	NO_2	O_3	SO_2	SPM(μg/m^{3})
N		3.5	3.6	0.83	2.0
NNW	-	-	0.9	3.4	0.80
SE	5.6	4.3	2.3	12	2.2
SSE	-	3.8	6.3	22	4.5
S	4.8	18	44	58	52
SSW	78	62	39	15	33
SW	7.9	5.7	2.7	0.20	3.4
WSW	4.0	2.5	1.7	0.60	2.6

Conclusions

The O_3, SO_2, NO, NO_2, SPM concentrations and meteorological parameters were measured at the Uludag mountain site in Turkey for a 24 month period.

The mean hourly O_3 concentration of 29 ppb was above the US standard of 15 ppb. The US standard of two hourly values per year above 0.12 ppm was not exceeded.

The diurnal variation of O_3 was 27 ppb for summer and 3.2 ppb for winter seasons. Although the diurnal variations was small for NO, a relatively strong diurnal variation of 2.3 ppb was observed for NO_2.

Short and long term variations of O_3, NO_x, and SO_2 indicate that the O_3 levels observed at Uludag mountain are basically from the photochemical production from the precursor gases of urban and industrial emissions with subsequent transport to the Uludag station.

The highest percent concentrations of O_3, SO_2, NO_x and TSP we observed under southerly winds which is the major wind pattern in the area.

Major source regions were Bursa city and industries around, and a thermal power plant.

We had hoped to measure dry deposition fluxes of O_3 and NO_x, but because of insufficient funding, it was not possible to accomplish this goal.

Acknowledgements

We would like to thank to Turkish Scientific and Research Council (TUBITAK) for the financial support for this project.

References

1. N.A.Kelly, G.T., Wolf and A.M. Ferman, *Atmos. Environ.*, **15** (1984) 561.
2. J.A. Logan, *J. Geophys. Res.*, **90** (1985) 10463.
3. A.Martin and F.R. Barber, *Atmos. Envir.,* **15** (1981) 567.
4. G.T.Wolff. P.S. Lioy, *Environ. Sci. Tech.* **14** (1994) 1257.
5. S.G. Tuncel, A.Baykal, G.Tuncel and U.Özer, Measurement of NO_X, SO_2, SPM and O_3 at a High Altitude Station in Northwestern Turkey, *Israel J. Chem.,* **34** (1995),403.
6. NAPAP, The National Acid Precipitation assessment Program III, (1987), **4**, Atmospheric Processes.

9.8 Surface Exchange of Ammonia over Grazed Pasture

M.A.H.G. Plantaz, A.T.Vermeulen and G.P. Wyers

Netherlands Energy Research Foundation ECN, P.O. Box 1, 1755 ZG Petten, The Netherlands

Summary

From July 1992 until July 1994 continuous measurements of the exchange of ammonia between surface and atmosphere over cultivated pasture were carried out at Zegveld in the west of the Netherlands. Fluxes were determined according to the profile method, using thermodenuders and continuous flow denuders for NH_3 measurements. Concentrations showed a dominating diurnal trend of high values at night and lower values during the day, due to differences in mixing intensity with clean air from higher layers of the atmosphere. In the grazed season average concentrations were higher than in the ungrazed season, due to grazing, fertilizer and manure application and plant activity in the region. In the grazed season emission dominated during daytime, while at night and also during daytime in the ungrazed season deposition was usually observed.

Exchange at this location can be roughly described with a model combining bi-directional exchange with the stomata, and deposition to (water layers on) the leaf surface.

Additional SO_2 measurements showed efficient deposition to the outer leaf surface, probably due to uptake in water layers on this surface, according to a correlation with humidity. No clear evidence of co-deposition with NH_3 was found, although the slightly lower average SO_2 deposition in winter compared to summer could indicate the effect of a lower NH_3/SO_2 ratio.

Aims of the research

The aim of the research project was the study of exchange of ammonia over grazed and fertilized pasture in a region of intensive livestock breeding and its dependence on:

- seasonal and diurnal variations in meteorological and surface conditions;
- canopy wetness;
- eventual relations in exchange behaviour with other air pollutants, such as SO_2 and NH_4NO_3.

Principal scientific results

A description of the automated continuous measurement system, of data treatment and of principal results of the measurements of NH_3 by continuous flow denuders and of SO_2 were given by Plantaz *et al.* [1, 2]. Further analysis has been carried out with respect to the NH_3 concentration and surface exchange, now including all thermodenuder measurements. Generally, this larger and more complete dataset confirmed the patterns described in the earlier reports.

Diurnal concentration variations generally consisted of high concentrations at night, followed by decreased values during the day, with minima in the afternoon. Table 9.8.1 gives a summary, subdivided per season. The intensity of vertical mixing within the atmosphere probably is one of the causes for this pattern. Low mixing layer height and low turbulence intensity prevent ammonia emitted at low level at relatively short distance from being mixed with clean air from higher levels, as occurs during the day. For SO_2 a contrary pattern had been found; a large part of the SO_2 is emitted at higher levels, at larger distances and poor mixing at night in this case fails to compensate for deposition, resulting in lower concentrations.

Table 9.8.1: Mean diurnal variations of concentrations of NH_3 ($\mu g m^{-3}$) at Zegveld, July 92–July 1994, per season.

Time (hour)	Grazed		Ungrazed		All seasons	
	avg	sd	avg	sd	avg	sd
0:00–5:00	12.1	8.2	8.3	9.1	10.2	8.9
6:00–11:00	10.2	7.0	7.7	8.5	9	7.9
12:00–17:00	7.8	5.9	6.6	8.7	7.2	7.3
18:00–23:00	9.6	8.0	7.7	8.7	8.7	8.4

Table 9.8.1 also shows that in the grazed season, which roughly spans the growing season of the grass or the summer half-year (March/April to November), higher concentrations are found than in the ungrazed season (winter half-year). In winter, direct volatilisation from cattle excretes is reduced as cattle are kept housed, and produced manure is stored in tanks, of which only 1/3 to 1/2 is applied at the end of the same winter. The rest is applied in summer. Also by cattle grazing and fertilizer application enables some degree of volatilisation of NH_3. Furthermore, plant metabolism may result in stomatal emission of NH_3 [4], bringing, as a net result, NH_4^+–N or NO_3^-–N from the soil and soil solution into the atmosphere.

Typical diurnal and seasonal patterns of NH_3 exchange, mentioned in the earlier reports were also confirmed by the thermodenuder measurements. In all seasons deposition dominated at night, but in the grazed season during daytime emission occurred, resulting in net emission in this season (Table 9.8.2). In the ungrazed season daytime emissions were much less common, less intensive and short, (except for some periods after manure application), resulting in net deposition during this season. A typical example of diurnal trends of concentration and exchange in the grazed season is presented in Fig. 9.8.1.

Table 9.8.2: Mean diurnal variations of fluxes of NH_3 ($ngm^{-2}s^{-1}$) at Zegveld, July 92–July 94, per season.

	Grazed		Ungrazed		All seasons	
	avg	sd	avg	sd	avg	sd
Night-time	–8.4	65	–27	47	–20	55
Daytime	62	151	–2.7	66	37	130
Total day	41	135	–17	57	12	108

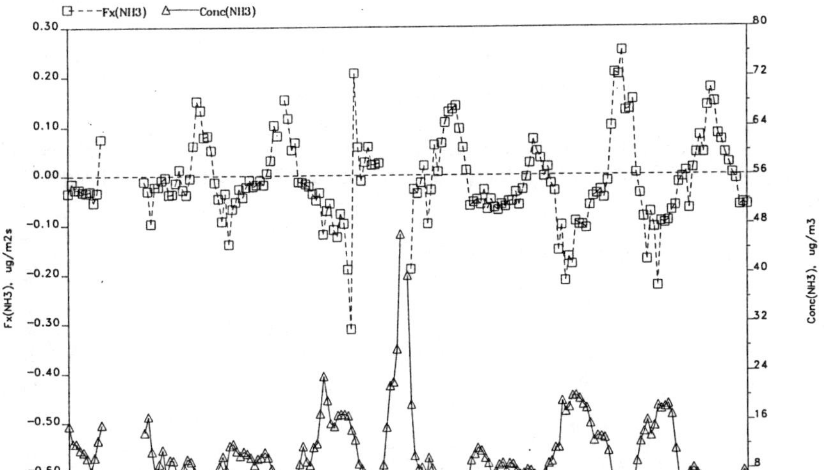

Fig. 9.8.1: Concentration and flux of ammonia at Zegveld, from continuous flow denuder measurements during the grazed period (October 1993).

An attempt has been made to get more insight into the mechanisms involved in the bi-directional exchange, of which the basis consists of reversible equilibria of gaseous NH_3 with dissolved NH_4^+ or with NH_4^+ in salts. On the surface, NH_4^+ is found *e.g.* in manure, in water, in water layers or salts on leaf surfaces and in the intercellular fluid of stomata. Because of the equilibria, low or decreasing air concentrations will favour emission from the surface. Therefore a comparison was made between fluxes and concentrations in both seasons, subdivided per class of u_* to assure identical turbulence conditions (Fig. 9.8.2). The diagram reflects the different trends of fluxes in summer and winter; also the effect of turbulence to concentration is illustrated, for both seasons. However, the concentration level as an isolated factor obviously does not account for the change in flux direction in the grazed season, as the generally lower concentrations and stronger decrease of concentration with growing u_* might be expected to favour more frequent emission

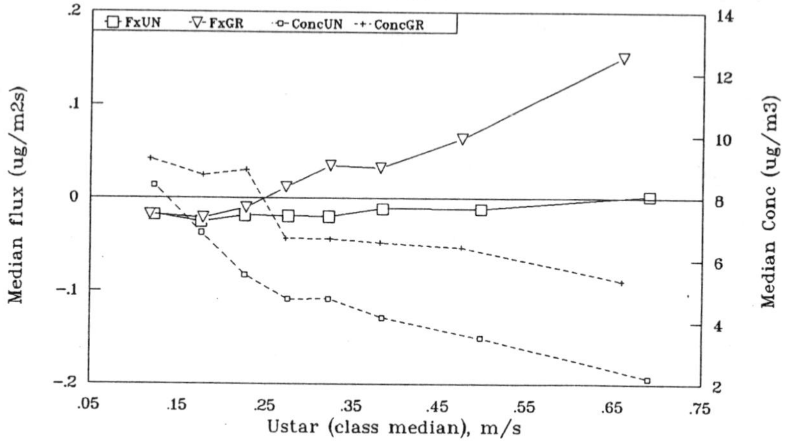

Fig. 9.8.2: Variation of ammonia flux and concentration with u_* during the grazed (GR) and ungrazed (UN) season, from thermodenuder measurements at Zegveld from 1992 to 1994.

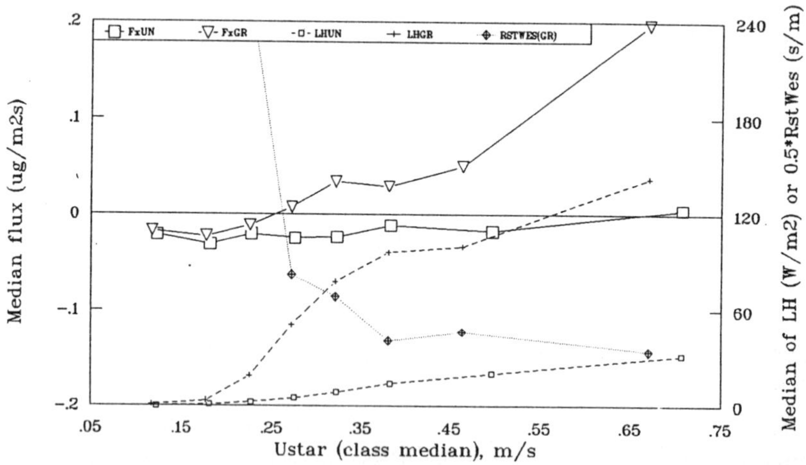

Fig. 9.8.3: Variation of fluxes of ammonia and latent heat with u_* during the grazed (GR) and ungrazed (UN) season, at Zegveld from 1992–1994, and stomatal resistance parameterised according to Wesely [3], in the grazed season (in ungrazed season > 1000 sm^{-1}).

in winter. So probably other mechanisms are (also) involved. Fig. 9.8.3 gives a similar comparison between seasons for latent heat. It shows that emission fluxes roughly start when the latent heat flux is larger than about 40 Wm^{-2}, from whereon transpiration becomes a significant part of total evaporation, which is also reflected in the decreased stomatal resistance (parameterised according to Wesely [3]).

The relationship with transpiration and stomatal resistance suggests that stomatal processes play an important role in NH_3 emissions. Also, analysis of time series showed that emissions generally cover well the period of the day that significant opening of the stomata is expected, including the variation with the time of the year. Combination with data on land treatment made clear that significant daytime emissions even occurred if more than a week had passed since cattle grazing or liquid manure application to the pasture, making direct volatilisation from these sources unlikely as the cause of the emissions.

Comparing the diurnal cycles in summer and the dominance of deposition in winter, a two-way route for exchange with the vegetation is likely. Night-time deposition during all seasons points to the leaf surface as sink. During the day in the growing season stomatal exchange in principle may concern either emission or deposition. A few examples were found of deposition at daytime in the growing season, when stomata were open. In most cases they coincide with much higher NH_3 (air) concentrations than usual at daytime, suggesting that the concentration within the substomatal cavity acted as a compensation point. Leaf surface deposition may proceed independently of stomatal flux direction, but is probably limited when the surface is dry at low relative humidities. In addition to these routes and the patterns involved, less regular (soil)emissions take place from excretes from cattle, manure application and eventually fertilizer application.

Because of the bi-directional character of NH_3 fluxes, simple application of the resistance analogy for inferring fluxes, by expressing the sink inefficiency of the surface only in a residual surface or canopy resistance, R_c, is not valid, as this supposes a zero surface concentration. The different pathways and locations of exchange prohibit a simple adaptation by the introduction of a compensation point. This problem is dealt with by a parameterization proposed by Sutton and Fowler [5], in which a "canopy compensation point", ($C_{zo'}$) is defined (located at the theoretical bottom of the laminar boundary layer), given by:

$$F_g = (C_{zo'} - C_{zref})/(R_a+R_b)$$

The transport between $z_{o'}$ and the actual sink and source locations of the vegetation is divided into a flux to the leaf surface (F_w, probably uptake in water layers), and a flux through the stomata (F_{st}). Applying again the resistance analogy yields:

$$F_g = F_w + F_{st} = -C_{zo'}/R_w + (C_{zo'} - C_{st}) / R_{st}$$

Here C_{st} is the stomatal compensation point for NH_3, being the concentration in the air inside the stomatal cavity.

This parameterisation was tested for the Zegveld data.

R_{st} was estimated both from bulk canopy resistance (R_{stb}) for water vapour and by the parameterisation by Wesely (R_{stWes}), assuming a minimal stomatal resistance of 50 sm^{-1}.

R_w was derived from nocturnal R_c values measured by the continuous flow denuder system, taking into account low or high humidity and grazed or ungrazed season (Table 9.8.3).

Table 9.8.3: R_w estimates (leaf surface resistance, in sm^{-1}) used in parameterisation of fluxes according to "canopy compensation point" model of Sutton and Fowler, [5].

	RH<80	RH>80
Grazed	150	100
Ungrazed	150	80

For estimation of C_{st} a dataset was selected with dominating stomatal flux to be expected (global radiation > 350 Wm^{-2}, dry > 5 hours, RH < 80 %). By calculating R_w for different values of C_{st} (at measured F_g, C_{zref}, R_a, R_b and R_{st}), a value of C_{st} = 20 $\mu g\ m^{-3}$ resulted in R_w values within reasonable limits for most of the selected time periods.

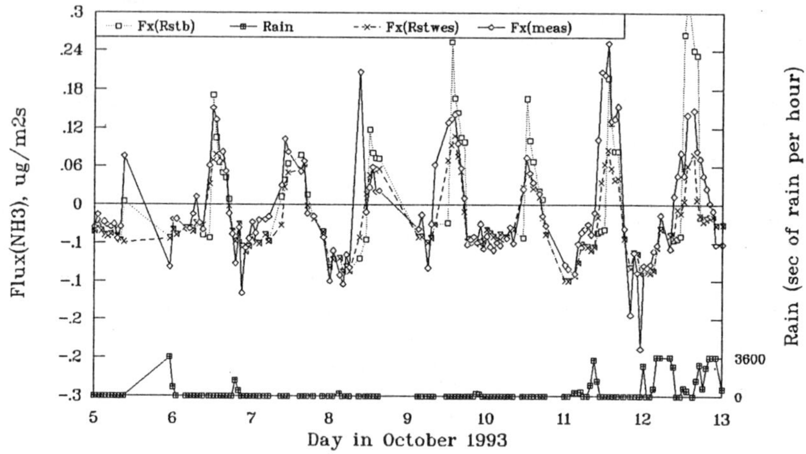

Fig. 9.8.4: Measured and parameterised fluxes of ammonia at Zegveld, October 1993. Parameterization based on a model by Sutton and Fowler [5]. Input stomatal resistances calculated according to Wesely [3] (R_{stWes}) and to bulk surface resistance for water (R_{stb}). Stomatal compensation point of 20 $\mu g\ m^{-3}$.

Results of fluxes inferred according to this parameterization are shown in Fig. 9.8.4, for both estimates of R_{st}. Given the uncertainty of the input values for R_w and C_{st} the model gives a relatively good prediction of flux direction and magnitude reproducing the observed diurnal patterns.

However, still large improvements of the model and of the estimates of input parameters are necessary. Both estimates of R_{st} sometimes result in large errors in predicted flux. Probably a more sophisticated estimate of C_{st} is necessary, introducing *e.g.* a temperature dependency. R_w parameterisation needs improvement, with respect to periods during or after rain, and probably with respect to saturation and/or growth or evaporation of water layers. The evaporation of water layers most probably accounts for several periods of nocturnal fluxes alternating between deposition and emission, while the model predicts clear deposition.

Nevertheless the performance of the model supports the validity of the mechanistic approach that the exchange with this grazed, cultivated pasture is largely determined both by bi-directional stomatal exchange and by deposition to the leaf surface.

Conclusions

The project has provided insight into the patterns and magnitudes of concentration and exchange flux of NH_3 over grazed pasture in a region of intensive livestock breeding. The dataset will enable further refinement to modelling of the process, which will be reported separately.

Acknowledgements

This study was part of the EUROTRAC BIATEX programme. Funding of the BIATEX programme by the Ministry of Economic Affairs and the Ministry of Housing, Spatial Planning and Environment is gratefully acknowledged.

References

1. Plantaz, M.A.H.G., A.T.Vermeulen and G.P.Wyers, in: *EUROTRAC Ann. Rep.1993, Part 4*, EUROTRAC ISS, Garmisch-Partenkirchen 1994, p. 66.
2. Plantaz, M.A.H.G., A.T. Vermeulen and G.P. Wyers, in: P.M. Borrell, P. Borrell, T. Cvitaš, W. Seiler (eds), *Proc. EUROTRAC Symp. '94,* SPB Academic Publishing bv, The Hague 1994, p. 661.
3. Wesely, M.L., *Atmos. Environ.* **23** (1989) 1293.
4. Schjørring, J.K., in: T.D. Sharkey *et al.* (eds) *Trace gas emissions by plants*, New York 1991, p.267.
5. Sutton, M.A., D. Fowler, in: WMO/GAW No 91, World Meteorological Organization, Geneva, 1993, p.179.

9.9 The Speulderbos Experiment 1993 - Experimental and Modelling Results

Sam Walton[1], M. Gallagher[1], T.W. Choularton[1], J. Duyzer[2], and K. Pilegaard[3].

[1]UMIST, Department of Atmospheric Physics, P.O. Box 88, Manchester, U.K.
[2]TNO, Institute of Environmental Sciences, P.O. Box 6011, 2600 JA Delft, The Netherlands
[3]RISØ National Laboratory, Denmark

Summary

Conflicting results in NO_x and O_3 dry deposition literature necessitates the requirement for understanding the effects of chemistry on their flux profiles. This is particularly important over forests where the micrometeorology and plant physiology is more complex than for other vegetation.

In 1993 research groups from the United Kingdom (UMIST), Holland (TNO) and Denmark (RISØ) co-ordinated in a field campaign which was designed specifically to measure the flux of O_3, NO_2 and NO over coniferous forest. The experimental site was located in a mature Douglas fir stand (22 m high) at Speulderbos in central Holland. Flux measurements were carried out from a 37 m tower at more than one height above the forest canopy (25 m and 35 m), in order to determine the flux divergence between observational levels. A further measurement height beneath the canopy (7 m) was included to gain a fuller understanding of processes involved in gas-surface exchange and the important role which the canopy plays in these processes. Measurements were made on a continuous basis using both the eddy correlation and gradient techniques from the 19th June to 2nd July 1993. Soil emission and deposition was measured during the final period of the experiment using the chamber technique. Detailed data analysis of the experimental results has been carried out at UMIST. The main results are presented below along with a tabulated summary of two case study periods.

A computer deposition model was devised in collaboration with TNO to reproduce the experimental results and describe some of the chemical and micrometeorological mechanisms involved.

The basis for the model was also used by Baldocchi [1] for the estimation of SO_2 dry deposition to a deciduous oak forest and adapted for a coniferous forest by Deinum [2]. The *revised* model considers NO_x, O_3 and hydrocarbon reactions both above, within and below a coniferous forest canopy. It also accounts for an NO soil emission flux at the forest floor. Data form the Speulderbos experiment was used as direct model input to try and reproduce the observed NO_x and O_3 fluxes both above and below the canopy. Model results along with a sensitivity study were used to interpret the observed measurements made at Speulderbos and obtain a clearer picture of the chemistry and processes involved at the atmosphere-forest interface.

Aims of Research

The aim of the research was to look at the role which chemistry plays, together with other influential factors, in perturbing the NO_x and O_3 flux over coniferous forest. It was also hoped that some explanation could be given as to why particular differences and anomalies in measurements of NO_2 fluxes over forest have previously been observed.

The Speulderbos experiment aimed to achieve these goals by making NO_x and O_3 flux measurements at several heights above and below the canopy. The deposition model was aimed at trying to replicate the Speulderbos data and explaining the complex results. From the research it was hoped that a better estimation of the flux and total N budget may be derived which can be used in larger scale deposition models.

Experimental Results

The results from Speulderbos suggested that both NO_2 and O_3 were depositing through leaf stomata for the entire period, although a further sink for O_3 was present during the early hours of the morning. This enhanced O_3 deposition was associated with soil and cuticle uptake, together with a chemical sink via reaction with hydrocarbons and NO emitted from the soil.

O_3 fluxes showed almost entirely deposition with the greatest fluxes in the morning and early afternoon at both heights. The 25 m and 35 m O_3 fluxes were in good agreement with each other and exhibited little variation with height. Fluxes were lower during the evening when O_3 concentrations were high, which suggested that some degree of stomatal closure was present. The stomatal resistance to O_3 indicated that the needle stomata were opening just after sunrise at 05:00 and remaining open until late evening. Stomatal aperture was at its greatest between 08:00 and 09:00 in the morning and reduced gradually after 12:00. The stomatal resistance rose to above 1000 sm^{-1} before 20:00.

NO_2 fluxes were much harder to interpret because resistance analogies were made useless since an emission of NO_2 was observed from the canopy. The NO_2 fluxes only showed some brief periods of deposition during the mornings, with emission seen for the rest of the time. Flux perturbation and divergence between heights was caused by the complex photochemistry above and below the canopy. Nocturnal fluxes showed a mixture of emission and deposition, with little divergence between heights.

The 26th June showed a greater degree of NO_2 deposition at 25 m than on previous days, when NO_2 concentrations were at their highest, peaking around

33 ppb at 08:00. This gave an indication that NO_2 deposition was more likely to occur with a sizable NO_2 concentration aloft.

Extreme radiation levels and temperatures which occurred from 00:00 29th June to 00:00 2nd July created a high degree of photochemical activity above and below the forest canopy. O_3 fluxes in the morning were enhanced by the increased O_3 concentrations and PAR levels throughout this final period, which stimulated stomatal opening. However, the degree of stomatal closure in the afternoon was more severe during such prolonged periods of intense solar radiation.

The average NO soil flux taken between two enclosure chambers was 22.2 ng $N\,m^{-2}\,s^{-1}$. Previous enclosure measurements made in unfertilised forest soils in Sweden [3] indicated an NO emission level two degrees of magnitude lower than measured at Speulderbos, with a median value of 0.3 ng $N\,m^{-2}\,s^{-1}$. The application of nitrate fertilizers on two types of the Swedish forest soils led to a marked increase in the NO emission rate. Largest fluxes were up to 60 ng $N\,m^{-2}s^{-1}$.

At Speulderbos however, the total N deposition amounts to around 50 kg $N\,ha^{-1}\,yr^{-1}$ [4]. Although this cannot be considered to be a fertilized soil, the amount of Nitrogen input is comparable to that of a fertilized area of vegetation. Williams *et al.* [5] observed fluxes of up to 338 ng $N\,m^{-2}\,s^{-1}$ in a corn field which had recently been fertilised with 15 kg urea ha^{-1} and 15 kg $NH_4\,NO_3\,ha^{-1}$.

The levels of NO soil emissions at Speulderbos indicate that NO chemistry has a important role to play on the observed fluxes both above and below the canopy and certainly contributes in some way to the large emission fluxes of NO_2 which were measured.

Conclusions

1. During periods of little water stress and high NO_2 concentration (> 20 ppb), NO_2 deposition was observed.

2. NO_2 emission from canopy with was prevalent during the campaign which was linked to a strong NO soil emission.

3. *Daytime* O_3 canopy resistances followed the stomatal resistance closely, suggesting stomatal uptake plus cuticular and soil uptake.

4. An enhanced deposition for O_3 was seen during the early hours of the morning. Possible sinks include reaction with NO and hydrocarbons or an increased cuticle deposition due to high degree of needle wetness.

5. Stomatal opening was observed to be greater in the morning than during evening periods when PAR levels were similar, suggesting an asymmetric response to PAR. Stomatal closure in the afternoon was increased by prolonged, elevated radiation levels.

Table 9.9.1 summarises the experimental results and shows the typical observed behaviour during two case study periods.

Table 9.9.1: Case study summary table

	Case study 1 00:00 24–27/6/93	Case study 2 00:00 29/6/–2/7/93
Meteorology	Cloudy, overcast some periods of rain	Hot, dry with high radiation levels
PAR Temperature	Low	High
NO_2 concentrations	High 0.5–42 ppb	Low 1–12.5 ppb
O_3 concentrations	Low 8–60 ppb	High 15–90 ppb
Stomatally controlled NO_2 and O_3 deposition?	Yes	Yes, although further sink for O_3 detected early morning
	Typical daytime resistances	
R_c H_2O	50–150 s/m	100–400 s/m
R_c NO_2 *	–	–
R_c O_3	150–400 s/m	10–300 s/m
NO ground emission	–	9–37 $ngN/m^2/s$
Daytime O_3 flux	–0.48 to –0.73 $\mu g/m^2/s$	–0.6 to –0.97 $\mu g/m^2/s$
Daytime NO_2 flux	–0.12 to +0.12 $\mu g/m^2/s$	–0.012 to + 0.12 $\mu g/m^2/s$
NO_2 flux divergence?	Some divergence	Yes, large divergence seen in morning

* R_c NO_2 – no canopy resistance is given owing to the observed upward flux of NO_2 making resistance analogy useless.

Model Results

Figs. 9.9.1 and 9.9.2 show the model concentration and flux variations with height for 12:00 on the 29th June, along with the corresponding measurements at 7 m, 25 m and 35 m.

Above the canopy height ($z/h > 1$) all the gases exhibit a small concentration gradient. On passing below the canopy the O_3 concentration decreases rapidly, indicating a high level of canopy uptake. NO_2 concentrations do not show the same percentage decrease as O_3, even though there is still NO_2 canopy uptake in the model. This is caused by an NO_2 production below canopy via the NO + O_3 reaction, elevating NO_2 concentrations.

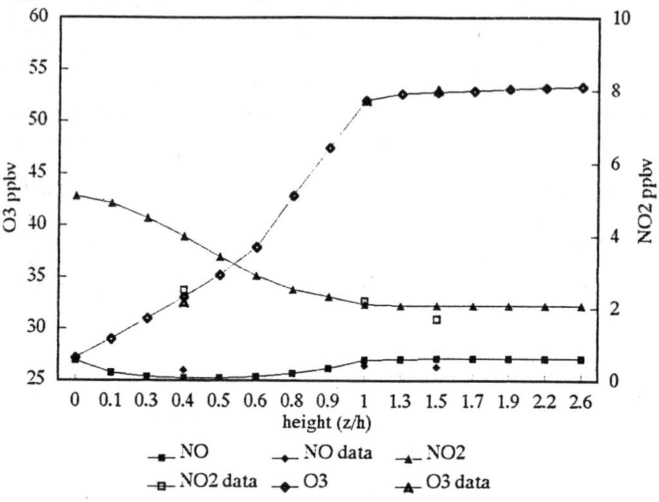

Fig. 9.9.1: Model and measurement concentrations vs. height 12:00 29/6/93, Speulderbos

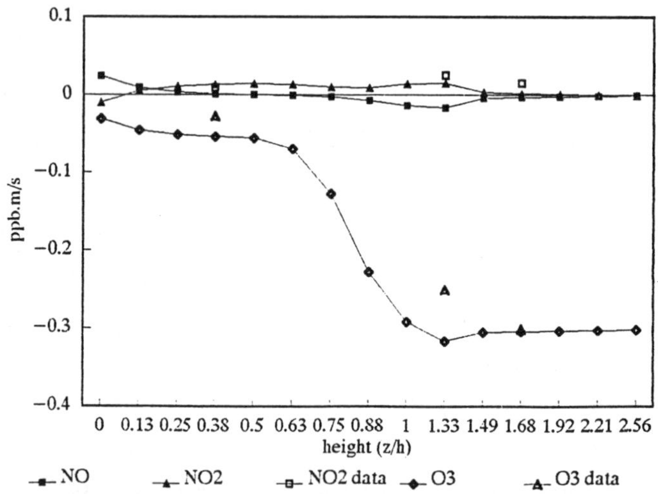

Fig. 9.9.2: Model and measurement fluxes vs. height 12:00 29/6/93, Speulderbos

O_3 fluxes are more or less constant with height above the canopy but show a sharp reduction in magnitude on passing below the canopy level into the trunk space (z/h < 0.5). O_3 fluxes in the trunk space display little deposition, which corresponds to the measurements made at 7 m.

NO_2 fluxes show emission which becomes more constant with increasing height above the canopy. The observed emission is created by a reversal in the NO_2 concentration gradient.

At the surface both NO_2 and O_3 are depositing with NO being emitted.

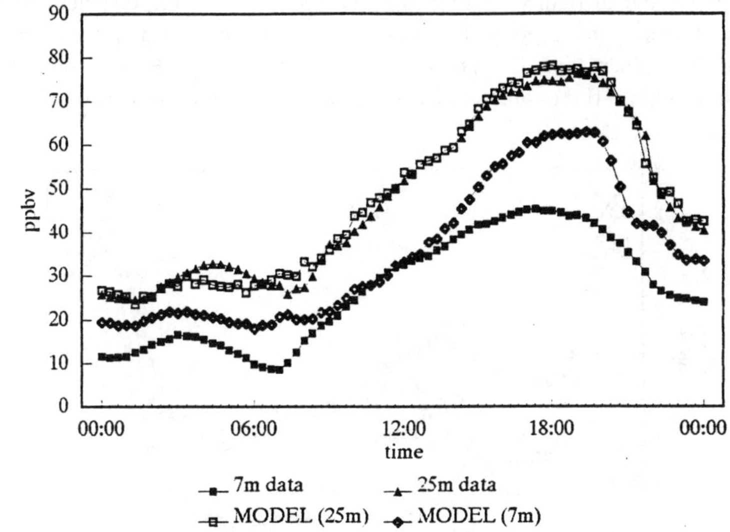

Fig. 9.9.3: Model and measurement O_3 concentrations 29/6/93, Speulderbos

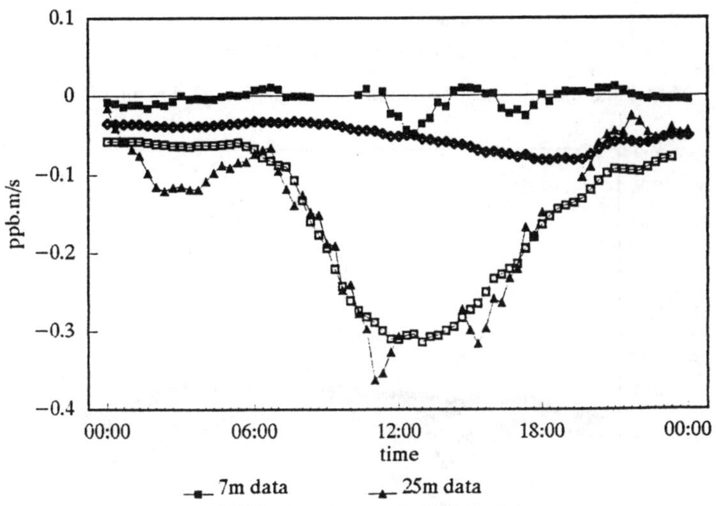

Fig. 9.9.4: Model and measurement O_3 fluxes 29/6/93, Speulderbos

Figs. 9.9.3 and 9.9.4 present the measurements and model's O_3 concentration and flux at the 7 m level and the 25 m level for the 29th June. The O_3 concentrations at the 25 m level are in very good agreement with the data. The model is able to reproduce the O_3 flux observed at 25 m to within 5 % of the observed flux for the majority of the time.

At the 7 m level the model concentrations indicate more deviation from the measurements. The O_3 flux below the canopy is slightly overestimated. This is partly due to the difficulty in assessing the magnitude of the turbulent diffusion coefficient below the canopy. For this reason, the below canopy concentration gradient is steeper than the gradient above the canopy. Hence, both the O_3 concentrations and fluxes are very hard to predict accurately for every situation.

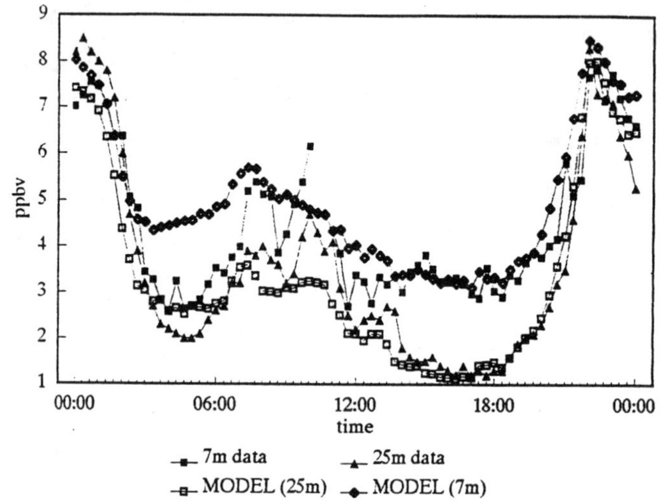

Fig. 9.9.5: Model and measurement NO_2 concentrations 29/6/93, Speulderbos

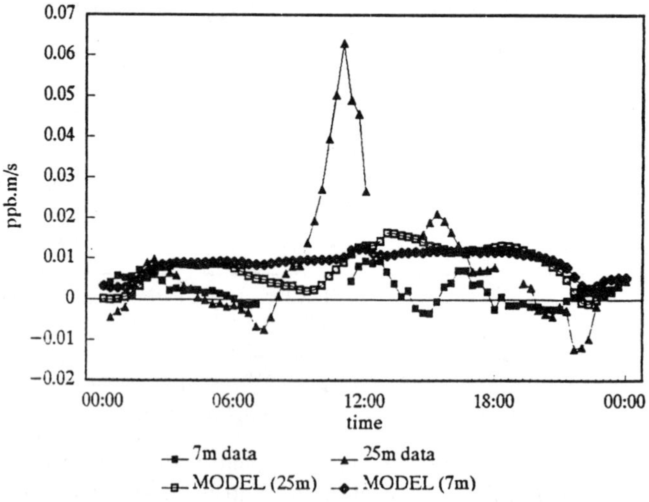

Fig. 9.9.6: Model and measurement NO_2 fluxes 29/6/93, Speulderbos

Figs. 9.9.5 and 9.9.6 give the corresponding NO_2 concentrations and fluxes for both the 7 m and 25 m level for the 29th June. The NO_2 concentrations at the 25 m level show a very close match with the data set. However, at the 7 m level the NO_2 concentration is sometimes overestimated. One possible reason behind this large overestimation of NO_2 below the canopy is the constant NO soil emission which is defined in the model as being 23 ng N m^{-2} s^{-1}. The model does not simulate the large spatial variation in NO soil flux which was observed at Speulderbos and therefore produces a higher concentration of NO_2 than would normally be observed below the canopy.

Model Sensitivity Results

No upward flux of NO_2 is observed in the model when zero NO emission is present at the forest floor. Secondly, the concentration of NO_2 aloft determines not only the size of NO_2 deposition to the canopy but also the extent to which the NO soil emission perturbs the above canopy NO_2 flux. Thirdly, the canopy resistance to uptake is also important when considering both emission and deposition events. NO_2 emission from the canopy is enhanced when the canopy uptake is low (*i.e.* when stomata are shut) and *visa versa* for deposition events. This leads to an equation for the estimation of the observed NO_2 flux above canopy which is a function of the NO soil flux, the NO_2 concentration and the canopy resistance:

$$F_{NO_2} = f(F_{NOsoil}, -C_{NO_2}, (-R_{cO_3})^{-1})$$

The NO soil flux contributes towards the observed NO_2 emission. The negative terms ($-C_{NO_2}$ and $-R_{cO_3}$) denote a contribution towards observed NO_2 deposition.

The canopy resistance is calculated for O_3, since the uptake of NO_2 to the canopy is almost identical to that for O_3 in the model and almost all NO_2 R_c values are negative and therefore rendered useless.

Large NO_2 emission ($F_{NO_2} >> 0$) is observed when the NO soil flux and the canopy resistance are high and the NO_2 concentration is low (< 1–2 ppb).

Deposition events ($F_{NO_2} << 0$) are seen when there is both a low soil flux and canopy resistance, coupled with a high NO_2 concentration (> 10 ppb).

The most common case is when the NO_2 flux is low and positive or close to zero. This occurs when F_{NOsoil} is marginally larger than the deposition terms C_{NO_2} and R_{cO_3} *i.e.* moderate NO_2 concentration and canopy resistance.

A canopy compensation point is observed when both the emission and deposition terms are equivalent. This can cover a wide range of concentrations for a changing canopy resistance and NO soil emission rate.

Typical nocturnal compensation points for a soil flux of 20 ngN $m^{-2}s^{-1}$ and for a canopy resistance of between 500 sm^{-1} and 1000 sm^{-1} range from 17 to 35 ppb, which is sometimes common place at night.

For an identical soil emission rate the daytime compensation points range from between 5 and 10 ppb for resistances of 150 to 300 sm^{-1}. Such concentrations rarely occurred during the day at Speulderbos throughout the experimental period. At Speulderbos the maximum nocturnal concentrations of NO_2 reached about 25 ppb, hence deposition was much more likely to be seen at night than during the day.

General Conclusions

The Speulderbos experiment offered some interesting results and gave the opportunity to compare data from different measurement levels. The model is able to reproduce both the measurement concentrations and fluxes fairly accurately. Modelling and making measurements below the canopy poses more problems than above the canopy because of the low levels of turbulence and mixing which in turn increases the errors in results.

In respect of the initial aims of the research the model is able to offers some explanation of the observed O_3 and NO_2 flux along with the canopy behaviour and gas chemistry. Specific meteorological conditions can also be reproduced to give an estimate of NO_x and O_3 flux to the canopy.

Acknowledgements

This work has been supported by the Commission of the European Communities, the National Environmental Research Council and the European Environmental Research Organisation.

References

1. Baldocchi D.D., A multi layer model for estimating sulfur dioxide deposition to a deciduous oak forest canopy, *Atmos. Environ.* **22** (1988) 869–884
2. Deinum, (1993)
3. Johansson C., Field Measurements of Emission of Nitric Oxide from Fertilized and Unfertilized Forest soils in Sweden, *J.Atmos. Chem.* **1** (1984) 429–442
4. Erisman, Heij, (1991)
5. Williams E.J., Parrish D.D., Buhr M.P., Fehsenfeld F.C., Measurement of Soil NO_x Emissions in Central Pennsylvania, *J. Geophys. Res.* **93D8** (1988) 9539–9546

Chapter 10

Exchange of Sulfur Compounds and Multi-compound Measurements

10.1 Measurements of Wet and Dry Deposition in Timis County, Romania

A. Anghelus, M. Lörinczi, E. Lörinczi and A. Ritivoiu

Environmental Surveillance and Protection Agency Timisoara, B-dul Mihai Viteazul No.32, 1900 Timisoara, Romania

Summary

Our agency situated in the western part of Romania, was founded in August 1990. The activity area of the agency in Timis county is based on the following data: total area 869700 hectares; population 726000; 82 cities and villages; forest covered area 108184 hectares; 9 natural reservations. A characteristic of the Timis county is the great number of livestock (*ca.* 2000000 animals) which induce a high pollution level. Thus NH_3 is one of the most important air pollutants in this area. The complexity of the problems is also seen on the legislative side, since Timis county is close to the border and the agency has to take into account the international laws and conventions.

Precipitation measurements, performed periodically, include: volume, pH and conductivity measurements. The atmospheric pollutant concentrations are measured by means of a stationary installation, while the wet and dry depositions are determined by photo-colorimetry and gravimetry, in several places. Heavy metal depositions were monitored in collaboration. Unfortunately the anthropogenic and biogenic emissions are not evaluated.

Aims of the research

The completion and systematisation of some air quality indicators in order to know better the air pollution in Timis county.

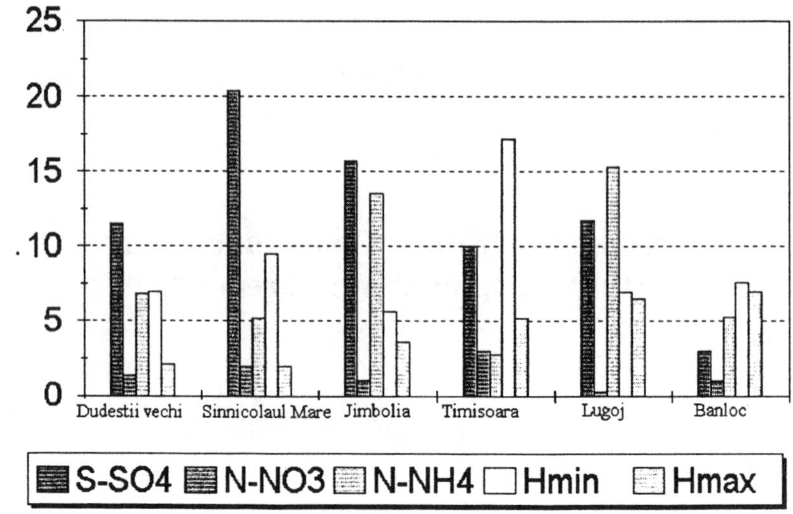

Fig. 10.1.1: Wet Depositions, in $mg/m^2/day$

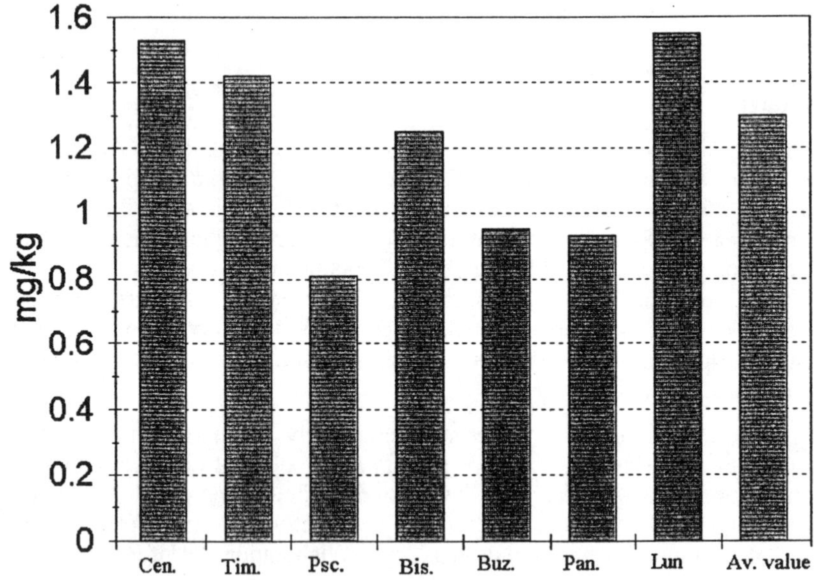

Fig. 10.1.2: Cadmium analyses form mosses

Principle scientific results

It was the first air quality study from this point of view. The sampling points were distributed all over the county (25 points) including, expecially the main city,

Timisoara. The monthly values of sedimentable powders excedeed the limit value – in some industrial zones of the county (Deta) and in Timisoara, due to the traffic, mainly in the periods without precipitations. Regarding the chemical analysis of the precipitations, sampled in points illustrative for the whole county, they were, also, performed monthly. The values, determinated in 1991, were compared with multiannual values, taking, also into account the average quantity of the precipitations. These results were correlated with moss analysis, performed in the frame of an European heavy metal deposition programm with the Plant Ecology Department University of Lund, Sweden.

Conclusions

The agency's laboratory studied and analysed the quality of the wet an dry depositions in Timis county.

The results of these determinations indicate a high pollution level, confirmed by the analyses performed through biomonitoring, by mosses (*Hypnum cupressiforme*). An extension of a monitoring system, is developing. In order to obtain the results required by the European level, it is recommended to improve the endowment of our laboratory with high-tech equipment.

10.2 Measurement of Dry Deposition Velocity of Ozone, Sulfur Dioxide and Nitrogen Oxides above Pine Forest and Low Vegetation in Different Seasons by the Gradient Method

László Horváth[1], Tamás Weidinger[2], Zoltán Nagy[1] and Ernó Führer[3]

[1]Institute for Atmospheric Physics, H-1675 Budapest, P.O.Box 39, Hungary,
[2]Eötvös Loránd University Budapest, Dep. for Meteorology, H-1083 Budapest, Ludovika tér 2, Hungary,
[3]Forest Research Institute, H-1023 Budapest, Frankel Leó u. 44, Hungary

Summary

Results of the field campaigns carried out in Hungary in different seasons above different types of surfaces show that deposition velocities are higher in instable cases and for forest in contrast with stable cases and grass surfaces. There is no regular variation of deposition velocities as a function of season, very probably other meteorological and chemical parameters determine the rate of deposition rather than the seasonal variation.

Aims of the research

Dry deposition velocity of gases depends on meteorological conditions and on the physiological state of plants. The main aim of our work was to determine the dry flux and deposition velocity of the trace gases above tall and low vegetation in Hungary, in different seasons, under different meteorological conditions. Between 1990 and 1994 nine field experiments have been carried out in the four seasons in a Norway spruce forest and above grass.

Principal scientific results

Among the nine field campaigns between 1990 and 1994 five were organized at the Nyírjes station in a spruce forest in winter, spring, summer (2) and autumn. Four expeditions were carried out in spring, summer and autumn (2) at the Hortobágy station above an uncultivated grass surface.

A gradient method was applied to determine the dry flux and deposition velocities [1] based on the simultaneous measurements of profiles of wind speed, air temperature and mixing ratio of different trace gases as ozone, nitric oxide, nitrogen dioxide and sulfur dioxide. The arrangement of meteorological sensors and air inlets for monitors has been described by some previous reports [2-13]. In and above the 14-15 m height forest canopy the measured parameters were collected by data acquisition system with 3–5 minute averaging time interval at 4-7 levels. Upper level was at the height of 21–25 m. The calculated average displacement height and roughness length ($d + z_0$) was about at 13 meter. Average leaf area index was 3.4. Measurements above grass (dominated by fescue; *Festuca pseudovina* with an approximate 20 cm height) were carried out between 0.5–4 m levels. Some information about field experiments can be seen in Table 10.2.1. The methodology of the first pilot campaign in Hortobágy, in the autumn of 1990, was different from the following ones so it is not discussed in this summary paper. The results were reported elsewhere [5].

Supplementary measurements were parallel performed for wind direction, global radiation, radiation balance, radiation spectrum, Bowen ratio, relative humidity, leaf area index, soil moisture, concentrations of anthropogenic and biogenic hydrocarbons. Sonic anemometer and krypton hygrometer measurements were also performed at a group of campaigns.

The average dry deposition velocity of gases are compiled in Table 10.2.1. for instable (Richardson's number < 0) and stable (Richardson's number > 0) cases. As it can be seen, dry deposition velocity of ozone and sulfur dioxide are much higher in instable cases, as it was expected. For the forest, the deposition velocity is higher compared with the results for grass surface. It is due to the one order of magnitude difference between the calculated eddy-diffusivity figures, measured at Nyírjes and Hortobágy stations.

Deposition velocity of nitric oxide and nitrogen dioxide are negative in some cases indicating that in the function of meteorological, biological and chemical circumstances emission processes are expected. Because our theory supposes that concentration at the surfaces equals to zero, the effect of emission makes the calculated deposition velocities for nitrogen oxides to be questionable.

In Figs. 10.2.1 to 10.2.8 we collected the daily variations of dry deposition velocities for these gases. These figures and Table 10.2.1. show that no regular variation of dry deposition velocities can be observed as a function of season. More probably the dry deposition processes are controlled by other meteorological and chemical factors.

Table 10.2.1: Summary of the field campaigns between 1991-1994

site	date of experiment	measuring heights (m)	height of plants (m)	average dry deposition velocities ($cm\ s^{-1}$)							
		w: wind velocity T: air temperature M: mixing ratio		$R_i<0$				$R_i>0$			
				O_3	SO_2	NO	NO_2	O_3	SO_2	NO	NO_2
Nyírjes (forest)	5-30 September, 1991	w (21,19,17,16,15) T (21,17,15,6,0.5) M (19,15,2.5,0.5)	14	1.15	n.a.	0.70	0.52	0.08	n.a.	0.39	0.03
	20 March–16 April, 1992	w (21,19,17,16,15) T (21,17,15,6,0.5) M (19,15,6,0.5)	14	1.09	0.60	1.09	0.22	0.21	0.20	0.07	0.09
	28 January–17 February, 1993	w (25,21,19,17,15) T (25,21,19,17,15,6,0.5) M (25,19,15,2.5,0.5)	14	0.24	1.19	0.16	n.a.	0.01	0.25	-0.01	n.a.
	8 July-29 August, 1993	w (25,21,19,17,15) T (25,21,19,17,15,6,0.5) M (25,19,15,2.5,0.5)	15	0.74	1.57	–0.17	–0.29	0.18	0.34	–0.15	–0.02
	15 July-6 August, 1994	w (25,21,19,17,15) T (25,21,19,17,15,8,0.5) M (25,19,15,3,0.5)	15	1.33	0.92	0.26	–0.38	0.23	0.09	–0.04	-0.10
Horto-bágy (grass)	19-30 June, 1993	w,T (4,2,1,0.5) M (4,0.5)	0.2	0.51	0.19	-0.03	n.a.	0.15	0.05	0.01	n.a.
	5 May-21 June, 1994	w,T,M (4,2,1,0.5)	0.2	0.53	0.44	0.15	–0.11	0.28	0.03	–0.04	–0.04
	13-30 September, 1994	w,T (4,2,1,0.5) M (4,0.5)	0.2	0.28	0.35	0.23	0.00	0.18	0.17	0.04	–0.03

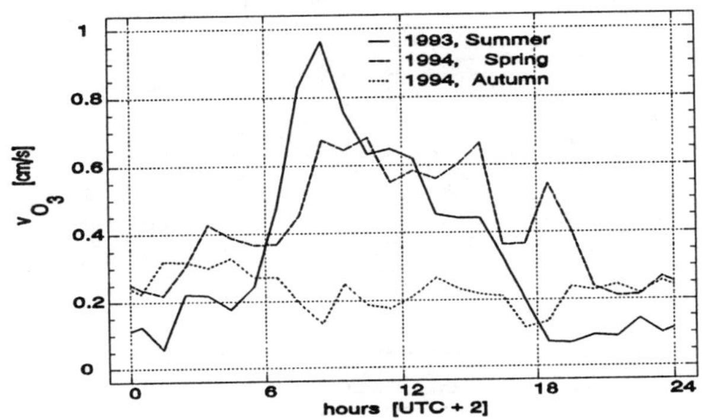

Fig. 10.2.1: Diurnal variation of dry deposition velocity of ozone (Hortobágy).

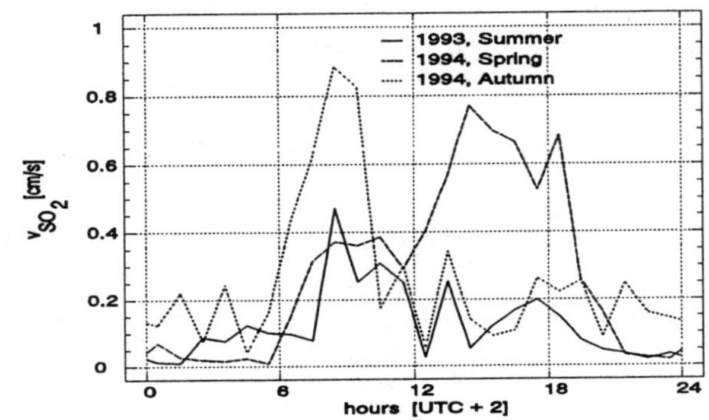

Fig. 10.2.2: Diurnal variation of dry deposition velocity of sulfur dioxide (Hortobágy)

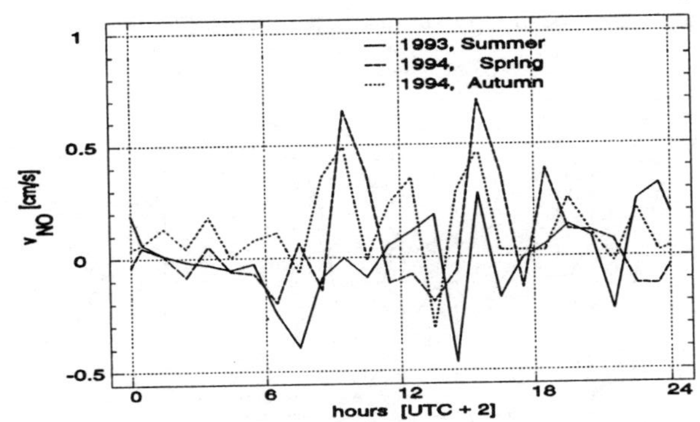

Fig. 10.2.3: Diurnal variation of dry deposition velocity of nitric oxide (Hortobágy)

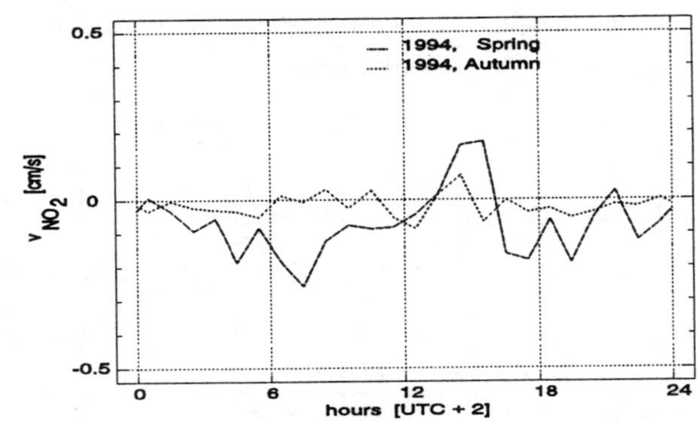

Fig. 10.2.4: Diurnal variation of dry deposition velocity of nitrogen dioxide (Hortobágy).

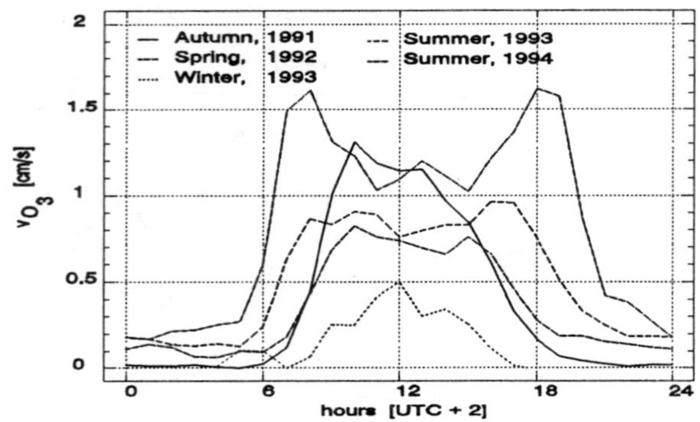

Fig. 10.2.5: Diurnal variation of dry deposition velocity of ozone (Nyírjes)

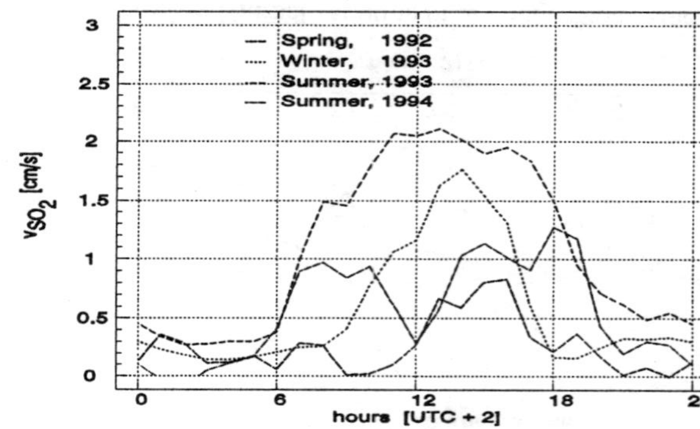

Fig. 10.2.6: Diurnal variation of dry deposition velocity of sulfur dioxide (Nyírjes)

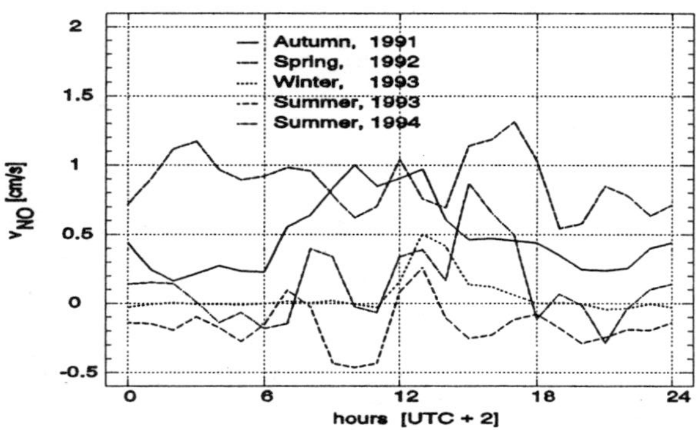

Fig. 10.2.7: Diurnal variation of dry deposition velocity of nitric oxide (Nyírjes).

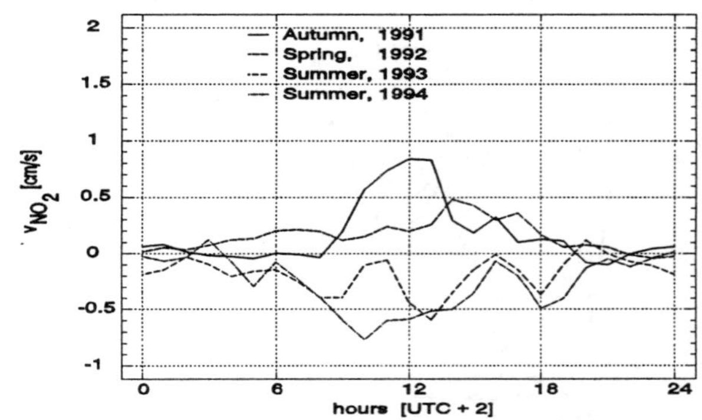

Fig. 10.2.8: Diurnal variation of dry deposition velocity of nitrogen dioxide (Nyírjes).

Acknowledgements

The following organizations supported our contribution to BIATEX subproject in the last few years:

- National Committee for Technological Development
- National Scientific Research Fund (OTKA), contract No. 341 91/94
- European Communities, PHARE/ACCORD project, contract No. H-9112-0399
- European Communities, PECO project, contract No. ERBIV 5V CT 920060
- US Hungarian Joint Research Fund, contract No. 161/91

We would like to thank them for the funding.

References

1. Weidinger, T., Iványi, Zs., Horváth, L., Pinto, J.,Valigura, R., Application of Monin-Obukhov similarity theory for calculation of surface layer fluxes. Part 1.: Sensitivity analysis of surface layer parametrization for universal functions, roughness length and displacement height, *Boundary Layer Meteorol.* submitted.
2. Horváth, L., Measurement of dry deposition of nitrogen compounds and oxidants. *EUROTRAC annual report, BIATEX*, EUROTRAC ISS, Garmisch-Partenkirchen 1992.
3. Horváth, L., Gradient measurements of dry deposition flux of nitric oxide, nitrogen dioxide, ozone and sulfur dioxide above forest and low vegetation. Brief summary of the five field campaigns carried out in 1990-1993. In: J. Slanina, G. Angeletti and S. Beilke (eds), *Air Pollution Research Report* 47 CEC 1993.
4. Horváth, L., Bozó, L., Haszpra, L., Kopacz, J., Molnár, Á., Nagy, Z., Führer, E., Sitkey, J. and Weidinger, T., Gradient measurement of the dry flux of nitrogen dioxide and ozone over a Pine forest in Hungary, in: P. M. Borrell et al (eds), *Proc. EUROTRAC Symp. '92*. SPB Academic Publishing bv., The Hague 1993.
5. Horváth, L., Bozó, L., Haszpra, L., Kopacz, J., Molnár, Á., Práger, T., Weidinger, T., Gradient measurement of air-soil exchange of gases, in: S. E. Schwartz and W.G.N. Slinn (eds), *Precipitation Scavenging and Atmosphere-surface exchange* 2, Hemisphere Publishing Corporation, Washington, Philadelphia, London 1992, pp. 637-648.
6. Horváth, L., Weidinger, T., Nagy, Z., Gradient flux measurement of nitric oxide, ozone and sulfur dioxide over short vegetation during the TRACT experiment, September, 1992. *EUROTRAC annual report: TRACT*, EUROTRAC ISS, Garmisch-Partenkirchen 1993.
7. Horváth, L., Weidinger, T., Nagy Z., Führer, E., Gyuricza, A., Kopacz, J.,Lovas, K., Gradient measurement of the dry flux of atmospheric trace gases over short vegetation and forest. *EUROTRAC annual report: BIATEX,* EUROTRAC ISS, Garmisch-Partenkirchen 1993.
8. Horváth, L., Nagy, Z., Führer, E. and Weidinger, T., Field experiments for estimation of turbulent fluxes over different types of surfaces in Hungary. *Annales Geophysicae, Part II.* Supplement **12** (1994) C335.

9. Horváth, L., Nagy, Z., Weidinger, T., Measurement of nitric oxide, ozone and sulfur dioxide flux by gradient method over short vegetation during the TRACT experiment, September 1992, *Geophysica et Meteorologica,* 1994.

10. Horváth, L., Führer, E., Gyuricza, A., Lovas, K., Nagy, Z. and Weidinger, T., Gradient flux measurements of gases above short vegetation and forest, in: P.M. Borrell, P. Borrell, T. Cvitaš, W. Seiler (eds), *Proc. EUROTRAC Symp. '94*, SPB Academic Publishing bv, The Hague 1994, pp. 477-478.

11. Horváth, L., Nagy, Z., Weidinger, T., Artz, R., Luke, W.T., Valigura, R., Pinto, J., Womack, J., Measurement of flux of trace gases (O_3, NO_x, SO_2, CO_2, HNO_3), particulate sulfate and nitrate and water vapor over short vegetation by gradient and eddy-correlation techniques in Hungary. *Annales Geophysicae*, Supplement **13** (1995).

12. Horváth, L., Nagy, Z. and Weidinger, T., Concentration and deposition velocity measurements of nitric oxide, sulfur dioxide and ozone by gradient method during the TRACT campaign. *Atmos. Environ.* submitted.

13. Luke, W.T., Draxler, R.R., Horváth, L., Springtime Measurements of photochemically active trace species (NO, NO_x, NO_y, O_3, CO) in Central Europe. 1994 Fall Meeting of *American Geophysical Union.*

10.3 Biogenic and Anthropogenic Trace Gas Fluxes between the Atmosphere and a Polluted Mixed Forest of Central Europe

W. Jaeschke, Th. Dietrich, U. Schickedanz and M. Klauer

Zentrum für Umweltforschung (ZUF), Georg-Voigt-Straße 14,
D-60325 Frankfurt am Main, Germany

Summary

In order to investigate the exchange of trace gases above a forest canopy in the densely settled area of Frankfurt, a tower of 51 m height was erected and a constant flux model for the determination of trace gas fluxes has been developed. The reliability of the gradient model was examined under different meteorological conditions by means of two fast response systems. These investigations showed us that the constant flux assumption is fulfilled if the wind passes over the fetch. Furthermore an intercomparison between the gradient model and the eddy correlation method was performed with respect to $O_{3\text{-}3}$.

Beside ozone the concentrations of NO and NO_2 are also continuously measured and the appropriate fluxes subsequently calculated. The availability of detection methods for COS, DMS, H_2S, PAN, SO_2 and VOCs enables an application of the gradient method to be made to these species. A simultaneous measurement of all these trace gases was performed during a measuring campaign in August 1994. The goal of these efforts is to obtain information on the amount of pollutant uptake in the mixed forest ecosystem investigated for the measured species.

In order to observe the influence of air pollutants on the monitored forest, their concentration together with the meteorological parameters are compared with the CO_2 deposition. Investigations made so far seem to indicate that the daytime CO_2 uptake by the forest canopy is diminished during periods of high air pollution. An intercomparison of CO_2 fluxes measured with those calculated by a global biosphere model, which actually does not consider the influence of pollutants and biotic interferences, is accomplished at present.

Aims of the research

The main goal of our activities is to enlarge the knowledge about the amount and effects of air pollutant deposition on a typical mixed forest ecosystem in a polluted urban area of central Europe.

For this purpose fluxes of different trace gas species are determined. CO_2 deposition is monitored with the intention to investigate if an impairment of high pollutant doses on the CO_2 uptake by the ecosystem can be observed.

Further subjects of interest are the improvement and validation of our gradient model and a method intercomparison between eddy correlation and the gradient method.

Activities during BIATEX

In the Frankfurter Stadtwald, which is situated in the centre of the conurbation Rhine-Main, a tower of 51 m height was erected. This tower includes one movable (elevator) and four fixed measuring platforms in 22, 25, 33 and 50 m height. Each platform was equipped with measuring devices for wind speed, temperature and humidity as well as with Teflon tubes to suck air samples for trace gas measurements (CO_2, NO, NO_2 and O_3). Additionally two sonic anemometers (28 and 42 m) and a fast response O_3 analyser (28 m) were installed.

In order to calculate trace gas fluxes a gradient model has been developed. To ensure its applicability extensive investigations were carried out to validate the model by means of the two fast response systems. Permanent efforts in ameliorating the model quality are undertaken. An extension of the model on other parameters such as the surface resistance has been made. The measurement of ozone fluxes by means of a fast response system enables us to accomplish a method intercomparison between the gradient and the eddy correlation method.

An extension of the continuously monitored programme to other trace gases was made during a measuring campaign in August 1994. Gradients and fluxes of CO_2, COS, NO, NO_2, O_3, PAN, SO_2 and VOCs were determined simultaneously in co-operation with the IFU Garmisch-Partenkirchen (VOCs) and the meteorological institute of the Joh. W. Goethe Universität Frankfurt (COS and CS_2). For this purpose the gradient model applicability has been adapted for all trace gas measurements with a sampling rate up to 30 minutes. In early November another measuring period for reduced sulfur compounds (COS, CS_2, H_2S) took place with our own facilities. In January 1995 SO_2 gradients were measured in order to calculate fluxes during a winter period.

The influence of photosmog on the CO_2 uptake by the ecosystem Frankfurter Stadtwald was investigated by means of the measured CO_2, O_3 and meteorological data. A first comparison of modelled and measured CO_2 fluxes during stress episodes was performed.

Results and conclusions

The installation of a fast response system allows a continuous observation of heat and momentum flux at 28 and 42 m. As already described in the Aveiro report [1], the temporal evolution of heat and momentum flux is highly correlated for both heights in winter as well as in summer.

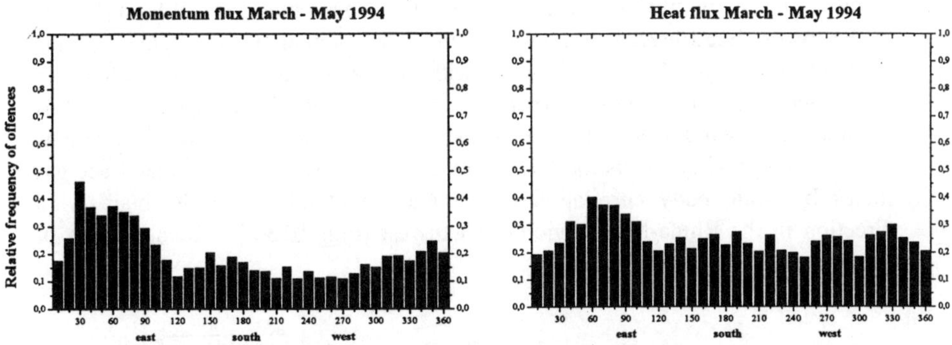

Fig. 10.3.1: Frequency of offences for momentum and heat fluxes against the constant flux assumption of the dependence on wind direction, March–May 1994.

To investigate this point in more detail, it is necessary to observe turbulent fluxes for a long time period at two heights and depending on the wind direction. Fig. 10.3.1 shows the frequency of offences against the constant flux assumption for a three month period in spring 1994. For heat as well as for momentum flux a high frequency of offences against the constant flux assumption during north-easterly winds can be seen. This can be explained by the fact, that the meteorological platforms are directed to the southwest, so that the air stream is disturbed by the tower before reaching the sonic anemometer in situations with prevailing winds from the northeast. The frequency of offences has its minimum during synoptic situations with southwest winds. This confirms the good horizontal homogeneity of the fetch in this direction. More offences can be seen during periods with winds from the northwest, because the homogeneity is disturbed by a nearby forest aisle with a railway line. Additionally the fetch is not as long as it is in the southwest.

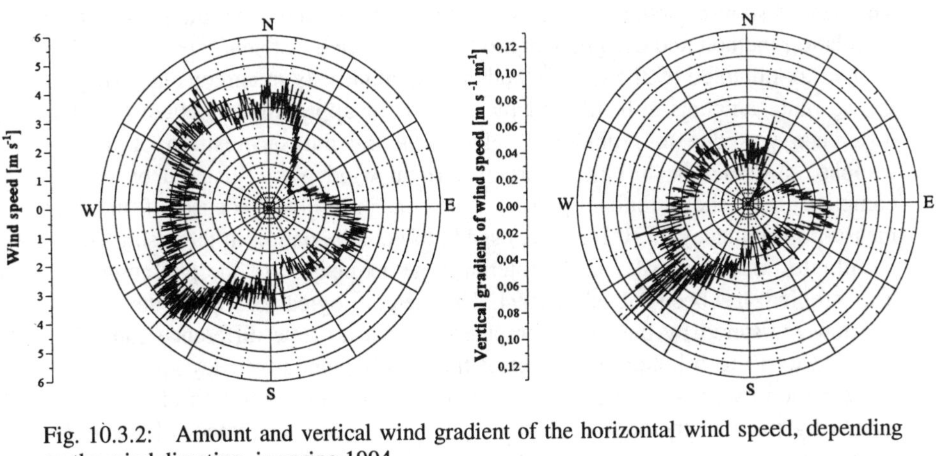

Fig. 10.3.2: Amount and vertical wind gradient of the horizontal wind speed, depending on the wind direction, in spring 1994.

Additionally to this fulfilment the horizontal wind speed must have at least 0.7 m/s in the lowest fixed level to allow gradient flux calculations. The quality of the gradient model results is increasing with an increasing vertical gradient of windspeed. Fig. 10.3.2 shows, that the vertical gradient of wind speed as well as the horizontal wind speed itself reach their highest value for synoptic situations with southwest winds. Thus for these conditions it is possible to calculate trace gas fluxes by both, eddy correlation and gradient method. Since the main wind direction in the Rhine-Main region is southwest (Fig. 10.3.3), the calculation of fluxes is possible for most cases.

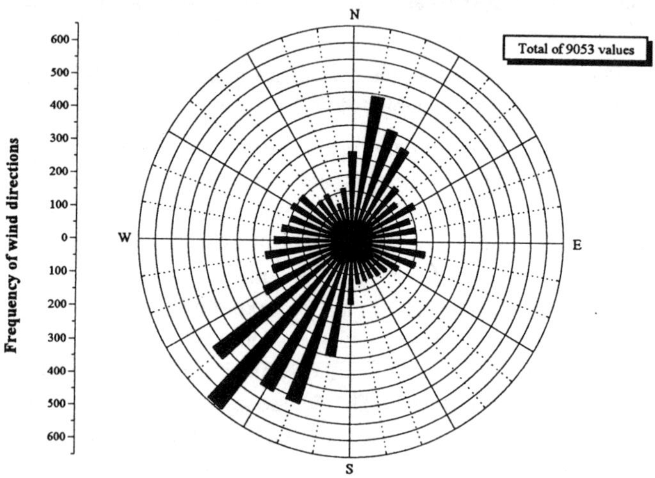

Fig. 10.3.3: The frequency of different wind directions at a height of 42 m on the ZUF tower between March and May, 1994. The wind direction is divided into classes, each of 10 degrees.

The fast response system for ozone allows an intercomparison of the eddy correlation method and our gradient model. As reported last year, the course of the ozone flux obtained by the two methods was qualitatively in agreement. However, the absolute amounts still differed by a factor of approximately three [2]. This difference was due to problems we had with the data acquisition of the fast response system. The turbulent part of vertical wind was registered by another computer than the turbulent part of gas concentration. Because of the high measuring frequency of 10 Hz it is possible to get a deviation in the results due to an infinitesimal inaccuracy in the internal time acquisition of the two computers. This problem has been solved in 1994. The meteorological as well as the gas data are now registered by the same computer, so that no time difference can occur. This led to a much better agreement of the results gained by eddy correlation to those gained by gradient method. In Fig. 10.3.4 the time evolution of ozone flux gained by the two methods is shown. There is a good agreement of the fluxes during daytime; in the showed night it was not possible to calculate fluxes by the gradient method, because the wind was to weak. However, all these investigations showed us, that our gradient model is reliable and enables us to calculate fluxes of every trace gas with a sufficiently high transfer velocity with regard to its chemical reactivity.

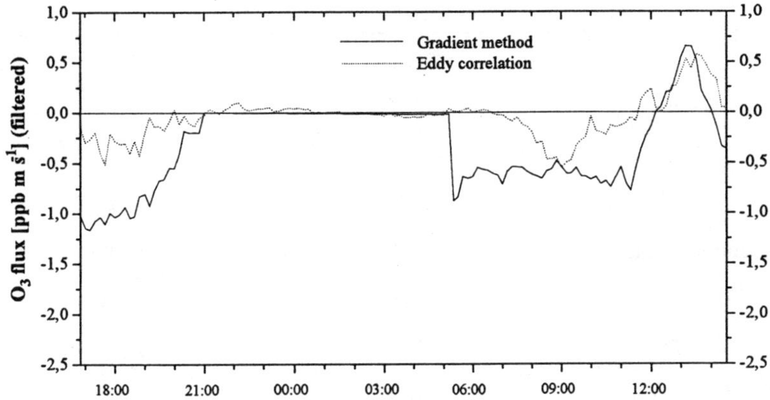

Fig. 10.3.4: Ozone flux at the ZUF tower on July 11th and 12th, measurements by eddy correlation and the gradient method.

To get a better idea of the gas exchange over the forest canopy, a measuring campaign was carried out at the ZUF tower in August 1994. After an extreme heat wave in late July and early August with temperature maxima of more than 35 °C the temperature during the campaign varied from 10 to 28 °C. Beside the scientists from the ZUF Dr. Rainer Steinbrecher (IFU-Garmisch-Partenkirchen) and Dr. Heinz Bingemer (Met. Inst. Univ. Frankfurt) did participate with their VOC respective reduced sulfur compound facilities. The simultaneously measured compounds were CO_2, COS, CS_2, NO, NO_2, O_3, PAN, SO_2 and VOCs. The results confirmed, that the measuring site is strongly influenced by traffic exhaust. The found concentration and flux range of some trace gases is shown in Table 10.3.1.

The reduced sulfur compounds were only measured between 3 and 4 p.m. on August 23rd. The belonging temperatures ranged from 27 to 28 °C. Anthropogenic as well as biogenic VOC fluxes show low correlation with wind direction and seem to be emitted on a more regional scale. In addition to the dense traffic in the Rhine-Main region aeroplanes do also influence our VOC measurements, since their height is only 250 m over the measuring tower while landing. In contrast to this behaviour other trace gases like SO_2, COS and CS_2 are probably dominated by more local anthropogenic sources, since they show an exhalation. Only during rush hour a deposition of SO_2 can be observed. SO_2 concentration is strongly correlated to wind direction. High concentrations can be measured if the wind comes from towns, highway crossings or the airport.

Table 10.3.1: Measured concentrations and fluxes during the measuring campaign August 1994.

	Concentration in 22 m [ppt]	Average flux [ppt m s^{-1}]
δ-3 carene	10–120	–0.08
1.8 cineol	10–430	0.42
COS	600–900	0.3
CS_2	42–73	0.04
ethylbenzene	20–590	–0.78
limonene	10–100	0.04
PAN	65–1100	–2
β-pinene	20–250	–0.16
SO_2	500–2000	2.5
Toluene	180–3250	–0.55
m-, *p*-xylene	20–1640	–0.36
o-xylene	30–630	–0.55

In early November there was a second measuring period concerning the reduced sulfur compounds, this time with our own facilities. This period was characterised by amazing high temperatures of 7–17 °C, which is very unusual for our region at that time of the year. This may have had an intensifying effect on the biogenic emission of reduced sulfur compounds in addition to the dominating anthropogenic emissions. The trace gas gradients were measured on November 3rd 2 to 5 p.m. and on November 4th 11 a.m. to 2 p.m. The temperature ranged from 13 to 17 respective 13 to 15 °C. However, the amounts of the calculated fluxes in [ppt m s^1] are given in Table 10.3.2. As expected the SO_2 concentration in January was higher than in summer, due to the emissions during the heat phase. The concentration in 22 m ranged from 0 to 23 ppb, the flux from –12 to +130 [ppt m s^{-1}].

Table 10.3.2: Measured trace gas fluxes in [ppt m s^{-1}] November 1994.

Date	COS	CS_2	H_2S
03.11.	4.8–8.5	0.53–0.69	3.2–6.35
04.11.	2.1–7.4	0.37–0.9	1.1–5.3

Efforts in finding an integral indication of the impairment of air pollutants on photosynthesis are undertaken at present. Therefore an evaluation of CO_2 flux, trace gas concentration and the prevailing meteorological condition is made. At a first view an increase of CO_2 uptake by the forest ecosystem with decreasing ozone concentration can be observed during a photosmog episode (Fig. 10.3.5).

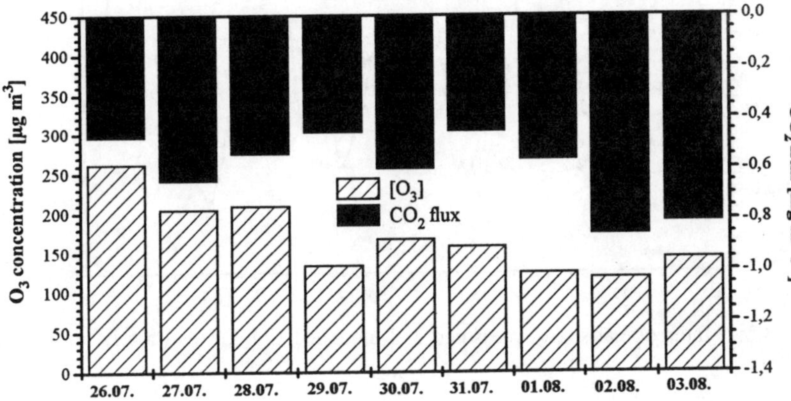

Fig. 10.3.5: Daytime average of CO_2 flux and O_3 concentration during an extreme heatwave in 1994.

Up to which extend the meteorological parameters took part in this trend is currently being investigated. For example July 29th was a very cloudy day with low global radiation. In consequence photosynthesis as well as ozone formation decreased. However, the decrease of O_3 concentration from July 26th to 28th as well as from July 30th to 31st is connected with two ozone alarms imposed by the local government. Those alarms were connected with speed limits in order to reduce traffic exhaust. The meteorological parameters do not show a significant change in the way, that the degree of O_3 depletion could be explained. This circumstance supports the hypothesis, that the increased CO_2 uptake may be due to decreasing O_3 doses.

The potential CO_2 flux over the forest canopy in 1994 is furthermore determined by means of the "Frankfurter Biosphären Model" (FBM) of the former BIATEX PI Prof. Dr. G. Kohlmaier [3]. The FBM does not consider the influence of pollutants and parasites. The really measured meteorological data was used as an input parameter. Driving variables are temperature, radiation and precipitation. As model output we got among others hourly means of CO_2 flux. A first comparison of measured and modelled data showed both a good qualitative and quantitative agreement. Even the fine structure of the daytime fluxes (especially on June 15th to 17th) is in amazing accordance (Fig. 10.3.6).

The lower amount of the measured CO_2 fluxes in June is possibly due to an impact by the timely correlated parasite infestation, which had led to a considerable loss of foliage in May. The increase of the measured CO_2 uptake during the shown period is in agreement with a resprout of the damaged plants in June. The modelled fluxes at night differ from the measured ones, because the FBM does not

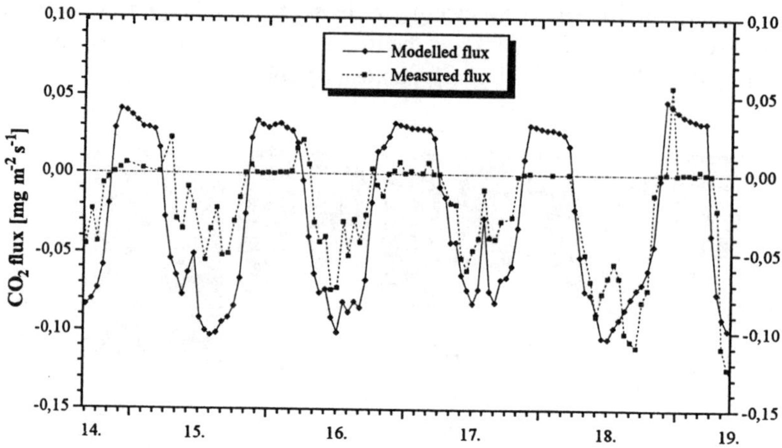

Fig. 10.3.6: Intercomparison of measured and modelled CO_2 fluxes (hourly means) for June, 1994.

consider transport processes. The typical inversion layer during summer nights leads to an accumulation of CO_2 near the ground as a consequence of plant and soil respiration. The accumulated CO_2 is either taken up by plants in the morning after sunrise or exhaled after the inversion has gone. Sometimes exhalation fluxes can also be observed in the evening after photosynthesis has become weaker but before the inversion layer is formed. This behaviour occasionally leads to a CO_2 exhalation peak in the morning respective in the evening.

An improvement of the biosphere model results is actually carried out by adapting its parameter more specifically to the ecosystem. A more intensive comparison of measured and modelled data concerning the photosmog episode in late July is not yet achieved. We hope to find some more indications about a diminished CO_2 uptake (measured) compared to the undisturbed potential uptake (modelled) due to unfavourable photo-oxidant doses.

Acknowledgements

Funds for this research were founded by the BMFT–Bundesministerium ,für Forschung und Technologie, Project 07EU766/0. Beyond that we like to thank Dr. Rainer Steinbrecher (IFU), Dr. Heinz Bingemer (Met. Inst. Univ. Frankfurt) and all scientists of the ZUF – especially Dr. Norbert Belz, Udo Krischke, Oliver Muth, Andrea Reinecke, Torsten Salkowski, Markus Sommerfeld and Jörg v. Trümbach – for their participation during the measuring campaign Aug.'94. We are also grateful to the group of Prof. Dr. G. Kohlmaier (IPTC Univ. Frankfurt) – namely Stefan Habermehl, Jürgen Kindermann and Ulrich Jäkel – for their co-operation with regard to the activities concerning the biosphere model (FBM).

References

1. W. Jaeschke, U. Schickedanz, W. Wobrock, Turbulent Fluxes of NO_x and O_3 between a Forest Ecosystem and the Atmosphere under high Pollutant Levels, in: J. Slanina, G. Angeletti, S. Beilke, S. (eds). *General assessment of biogenic emissions and deposition of nitrogen compounds, sulphur compounds and oxidants in Europe. Air pollution Research Report* **47**, CEC, Brussels 1993.
2. W. Jaeschke, Th. Dietrich, U. Schickedanz, Determination of CO_2, NO, NO_2 and O_3 fluxes between the Atmosphere and a Strongly Polluted Forest Ecosystem, *EUROTRAC Ann. Rep., BIATEX*, EUROTRAC ISS, Garmisch-Partenkirchen 1993.
3. M.K.B. Lüdeke, *et al.*, The Frankfurt Biosphere Model: A Global Process-Oriented Model of Seasonal and Long-Term CO_2 Exchange between Terrestrial Ecosystems and the Atmosphere. I. Model Description and Illustrative Results for Cold Deciduous and Boreal Forests. *Clim. Res.* **4** (1994) pp 143–166.

10.4 Exchange of Reduced Sulfur Compounds between the Biosphere and the Atmosphere

J. Kesselmeier[1], U. Bartell[1], S. Blezinger[1], W. Conze[1], C. Gries[3], C. Hilse[2], R. Hofmann[1], U. Hofmann[1], A. Hubert[1], U. Kuhn[1], F. Meixner[1], L. Merk[1], T.H. Nash III[3], G. Protoschill-Krebs[1], F. Velmecke[1], C. Wilhelm[2] and M.O. Andreae[1]

[1]Max Planck Institute for Chemistry, Biogeochemistry Dept., P.O.Box 3060, D-55020 Mainz, Germany,
[2]Botanical Institute, University Mainz, D-55099 Mainz; Germany,
[3]Dept. of Botany, Arizona State University, Tempe AZ 85287-1601, USA

Summary

In the course of our studies we developed a new sampling method for reduced sulfur compounds and used this method for investigations on the role of trees, crops, algae, lichens and leaf litter within the sulfur cycle. These studies were done under field conditions in conjunction with micrometeorological measurements as well as leaf or branch cuvettes. Laboratory studies gave insight into some basic physiological processes and regulation mechanisms.

Aims of the research

The subject of our research is the exchange of reduced sulfur compounds (H_2S, COS, CH_3SH, CH_3SCH_3 and CS_2) between the biosphere and atmosphere. The emission budgets still show significant uncertainties and field data as well as mechanistic work is needed. Within the compound's spectrum, carbonyl sulfide is an important S-species with climatic relevance.

Principal scientific results

Development of a sampling technique

A commercially available Nafion drier (Perma-Pure) was tested under laboratory and field conditions and shown to be suitable for measurements of reduced sulfur compounds such as H_2S, COS, CH_3SH, CH_3SCH_3 (DMS), and CS_2 in dry and humid air in the ppt range. For the scavenging of oxidants we combined the Nafion drier technique with a cotton oxidant scrubber. Results obtained under laboratory conditions with humidified samples and in the course of a small field experiment confirm the oxidant scrubber qualities of cotton and indicate that the cotton oxidant scrubber in combination with a Nafion drier is suitable for measurements of nearly all reduced sulfur compounds except CH_3SH. The technique was intercompared for quality assurance with a cryogenic trapping/thermal desorption technique used by the Institute for Atmospheric Research (Fraunhofer Institute, IFU, Garmisch-Partenkirchen) and found to be of high accuracy.

Enzymatic background for the consumption of carbonyl sulphide* (COS) *by higher plants and algae

Vegetation is believed to be an unquantified sink in the global cycle of COS. We investigated whether there is an enzymological background for the consumption of COS by higher plants in analogy to CO_2. Photometric measurements demonstrated that all enzymes involved in the CO_2 assimilation by higher plants (ribulose-1,5-bisphosphate-carboxylase, rubisco, and phosphoenolpyruvate-carboxylase, PEP-Co) can also metabolise COS. However, the key enzyme for COS metabolism in higher plants is carbonic anhydrase (CA), an enzyme which probably directly splits COS into CO_2 and H_2S. In the presence of CA the CO_2 fixation rate of both carboxylating enzymes was significantly increased. This means all enzymes involved in CO_2 assimilation participate in the COS metabolism.

Such a pathway is in accordance with the observed deposition of COS to vegetation. Carbonic anhydrase, isolated from pea leaves, was shown to consume carbonyl sulphide (COS) directly. The enzyme showed a very high affinity (Km 39 μM COS) towards this substrate. The CA model was supported by fundamental investigations with the green alga *Chlamydomonas reinhardtii*, grown under two different CO_2 regimes, thus developing different amounts of CA. The uptake of COS by this organism was directly dependent on the amount of CA, which was linked to the growth conditions. The experiments gave reason to believe that CA is the key enzyme for the consumption of COS in all plants. The high affinity for the substrate COS explains how CA is able to overcome the high concentration differences of COS (550 ppt) and CO_2 (350 ppm) in the atmosphere. The identification of this enzyme furthers our understanding of additional sinks for COS, which are needed to understand the balance of known global sources.

Exchange of carbonyl sulfide* (COS) *between plants and the atmosphere: the role of compensation points

Young corn, pea and rapeseed plants were exposed to compressed synthetic air containing varying COS concentrations. The results suggest that COS exchange depends highly on the ambient COS mixing ratios. Ambient COS mixing ratios larger than 150 ppt resulted in a deposition of COS to all plant species studied. Significant (confidence level > 95 %) COS emission was only detected from rapeseed leaves at COS mixing ratios lower than 90 ppt. We computed COS compensation points around 90 (57–135) ppt and 144 (0–328) ppt COS for rapeseed and corn. For both plant species we found a close correlation between the photosynthetic CO_2 assimilation and the COS uptake. In contrast to the gas exchange studies with corn and rapeseed, pea plants did not change in response to increased COS concentrations between 350 and 900 ppt COS, neither did they show correlation with photosynthesis. However, for all three plants studied we found indications that COS is taken up preferentially over CO_2 under normal ambient conditions. The work confirms our results and interpretations obtained during field work (developing and mature plants), which showed that deposition of COS to plants and vegetated land correlated closely with CO_2 assimilation. This close correlation to photosynthetic CO_2 assimilation and the observed compensation points around 150 ppt indicate that terrestrial plants might always consume COS especially when they are photosynthetically active. Furthermore, we know about the physiological background of the coupling of the assimilation of CO_2 and COS via carbonic anhydrase (see above). When CO_2/COS uptake ratios are compared to atmospheric ratios, the plants investigated even exhibit a preference for COS over CO_2. The amount of CO_2 fixation and the COS/CO_2 ratio can be used to calculate global COS deposition to vegetation. Using a figure of 100–120 × 10^{15}g (C) as CO_2 fixed per year by terrestrial vegetation on a global scale and assuming that the uptake ratio COS/CO_2 is just similar to that in the atmosphere, we calculate 0.86–1.0 Tg (COS) consumed globally by terrestrial vegetation which is in accordance with other estimates. However, the sink might be larger if all plants really prefer COS over CO_2 and thus might balance the known sources. In contrast, COS emissions in ambient air, as sometimes observed are not understood. Therefore, the biogenic production of COS by plants should be intensively investigated to further understanding of the exchange behaviour of this important trace gas. Until both the deposition and emission of COS are understood, we will have severe problems in determining global fluxes.

Exchange over fields (wet meadow, wheat field)

We measured the emission and deposition of the reduced sulfur compounds COS, CS_2, H_2S, and DMS over wheat in the Danube valley during the three vegetation periods: germination (April 1990), growth (May 1990) and seed ripening (June 1990). The variation of COS concentration gradients and fluxes showed that the developmental stage of plants is of high consequence on sulfur flux measurements. In spring H_2S gradients were mainly positive (deposition) whereas CS_2 gradients pointed to an emission from the soil/plant system to the atmosphere. In May and June we observed positive gradients for H_2S and CS_2 over the day (deposition). During the night, the H_2S gradient increased and the CS_2 gradients decreased. The daily DMS pattern indicated soil emission during dawn and night. Daytime DMS measurements, however, were low probably because DMS is rapidly oxidised under sunlight. Summarising, the wheat plants in May and June were responsible for a deposition of H_2S and CS_2 under daylight conditions. At the early stage of development in April only H_2S was found to be deposited to the soil plant system, whereas CS_2 was emitted by the soil/plant system, obviously driven by temperature. For COS the results clearly showed that the deposition of COS to plants is coupled to the actual photosynthetic CO_2 assimilation. The findings imply that measurement of sulfur compounds should be made at the same site during a complete vegetative cycle to enable regional and global scale fluxes to be calculated.

European and tropical tree species

In November 1991 we measured the exchange of reduced sulfur compounds between several tropical tree species in the rain forest of Cameroon. We measured the concentrations of reduced sulfur compounds in the atmospheric boundary layer over this equatorial African rain forest. In addition, we studied emission and deposition fluxes of volatile reduced sulfur compounds from tropical tree species by using cuvettes to directly measure the exchange behaviour of tree twigs. These cuvettes were operated in a forest clearing near ground level as well as on top of the forest canopy, employing a specially designed tree top jungle raft ("Treetop Raft III, Dirigible version", Operation Canopée) placed on the canopy crown. The results show qualitative and quantitative disparities between different tree species as well as between individuals of the same species near ground level and at the top of the canopy. We found a correlation between photosynthetic CO_2 assimilation and emission of sulfur compounds. Comparison between measurements at the ground and at the canopy top show that the studied tree species adapts its photosynthetic CO_2 assimilation in response to the extreme climatic conditions at the canopy top. This is accompanied by a quantitative change in trace gas emission. Lower CO_2 fixation rates are accompanied by an increase in the emission of reduced sulfur compounds. We propose the increase of DMS emission at the canopy top to be explained by a potential demand for nitrogen in the foliage resulting in an accumulation of sulfur. Such a sulfur accumulation in the presence of an intensive light-driven electron flow may be the reason for a high emission of reduced sulfur compounds. Thus, an excess of sulfur may result in an increased sulfur reduction. To consume the reductive force, the plant metabolism may switch to a reduction of, for example, sulfate and production of organic sulfur compounds, thus leading to an excessive emission of DMS. Therefore, it is of interest to look for correlation between the emission of sulfur and the photosynthetic capacity of the plant species or ecotype, especially as a decrease in photosynthetic yield, due to an adaptation to extreme conditions, seems to be accompanied by an increase in sulfur emission.

The Cameroon experiment gave evidence that DMS is the predominant sulfur compound emitted by the trees investigated. This conclusion may be transferred to trees of our mid-latitude regions. In the course of 1992, we investigated several mid-latitude tree species. Some European tree species tested showed a similar correlation between DMS emission and photosynthetic CO_2 assimilation as is found with tropical tree species. COS uptake/emission depended on the physiological status (assimilation or respiration) of the plant. Under active photosynthetic CO_2 assimilation we always found an uptake of COS. H_2S emissions were only observed in situations which may be regarded as stress (start of cuvette experiments or cutting branches).

Exchange of reduced sulfur compounds between marine phytoplankton species and the atmosphere

In laboratory experiments we studied the uptake of COS by marine phytoplankton. The algae *Mantoniella squamata, Prymnesium parvum and Amphidinium klebsii* were grown under low light conditions (10 µmol photons / m^2 s) in batch cultures and under high light conditions (100 µmol photons / m^2 s) in Kniese cultures. Algal samples were taken from cultures in growing phase conditions. The preliminary data show that all phytoplankton species took up COS from their surrounding medium. As shown for *Mantoniella*, the COS hydrolysis was dependent on the growing conditions. The uptake of COS was significantly higher when the algae were grown under low light conditions as compared to the uptake rate of algae grown under high light conditions. We assume that under low light conditions the algae are forced to synthesize higher amounts of carbonic anhydrase (CA). Since carbonic anhydrase is regarded to be responsible for the uptake of COS (see above) low light conditions followed by higher CA activities might be an explanation for the observed differences. Saturation of the uptake was reached at COS concentrations higher than 6 µM. The $K_{1/2}$ value for the COS uptake was found to be 2–3 µM. Thus the uptake of COS by these marine algae is similar to the COS uptake by the sweetwater alga *Chlamydomonas reinhardtii* (see above).

Changes in environmental conditions might also affect the production of DMS by marine phytoplankton. Up to now neither the conditions which lead to DMS emission nor the exact release rates of intracellular DMS have been fully investigated. In preliminary experiments we found that after centrifugation even at low speed (10 g), the algae emitted DMS for about 4 hours. Centrifugation might impose stress on the algae and thus induce emission of DMS. The DMS intracellular concentration and the related release is currently under investigation.

Leaf litter as a component of the sulfur exchange processes

We investigated the deposition and emission of reduced sulfur compounds to and from dead leaf material. Leaf litter was collected from beech forests near Mainz from two horizons: i) upper layer (litter horizon, LH) and ii) fermentation horizon (one year old, FH). Leaf litter samples were transferred into a cuvette system and investigated for their sulfur compound exchange behaviour while being constantly flushed with ambient air. It became obvious that leaf litter may switch between emission (mainly DMS) and deposition (only COS) of sulfur compounds, depending on the moisture content of the dead leaf material. The microbial activity (CO_2 emission by respiration) within the litter samples evolved in close correlation to the leaf litter moisture. Comparing the LH layer with the FH-layer we could show that the upper layer exhibits the dominant activity, both for the CO_2 emission and for the sulfur compound exchange. The CO_2 emission on a dry weight basis is three times higher and the DMS emission is higher by a factor of 10–15 in the upper LH than in the lower FH. Due to the change in the emission/deposition pattern, we may conclude that there are different regimes active and responsible for the sulfur compound exchange depending on moisture content. Either there are at least two microbial communities within the leaf material preferring different humidities or there is a shift from biogenic to physicochemical activities. Consequently, we conclude that leaf litter may serve as a source (DMS) and a sink (COS) for sulfur compounds. These results give reason to consider plant leaf litter as an additional sink for COS.

Lichens involved in the sulfur exchange

The role of algae in sulfur exchange in the case of oceanic phytoplankton is quite well known. No such data exist for fungi and lichens, though these organisms may be of significant relevance for some special ecosystems. For example lichens, symbionts of fungi and algae, may be the dominant autotrophs in a number of coastal deserts and polar systems and may be major components as epiphytes in some forests. It may be assumed that these symbionts significantly influence the exchange of trace gases. Especially, as the lack of stomates and cuticles prevent a biological control of deposition and emission processes. In the first approach, we investigated several lichen species for their uptake and emission of reduced sulfur compounds from ambient air. Experiments were conducted with 14 lichen species sampled from different locations in Europe and North America, 10 containing green algae as photobiont and 4 cyanolichens. Prior to trace sulfur gas measurements, the lichens were maintained in a growth chamber with a 18 h day: 6 h night cycle, a photoperiod chosen to match the southern Alaskan habitat and to approximate that of the German collections. Light conditions were maintained at 300 μmol photons m^{-2} s^{-1} at 23 °C and a relative humidity of 70 % and dark conditions were 18 °C and 75 % relative humidity.

All lichens, those containing green algae as well as cyanobacteria as their algal component, showed an emission of H_2S and DMS. Of special importance is the uptake of COS by these organisms. All lichens investigated showed an uptake of COS, independently whether they were illuminated or in the dark. Only a water content sufficient for the onset of physiological activity (photosynthesis or respiration), with an optimum around 50% thallus water content was required to start an uptake of COS. Such an uptake can only be explained by a light-independent step in the consumption of COS. Carbonic anhydrase (see above) is a light-independent enzyme and is found in nearly all organisms. Thus, a CA-induced consumption of COS would explain this uptake. In 1994 two field studies at an open oak woodland in California with a very high lichen load confirmed the significant uptake of COS as obtained in the laboratory. The data are under evaluation. We assume lichens to be a significant sink for COS in lichen dominated communities. Compared to the uptake by terrestrial higher vegetation with values between 1–6 mg/m^2 a the consumption of COS by Lichens may contribute significantly to the global budget of COS exchange.

Conclusions

The aims of our studies were to contribute to the knowledge on the exchange of reduced sulfur compounds between the biosphere and the atmosphere. With a highly sensitive method, dynamic cuvettes and field studies we were able to get new information about the production and consumption of sulfur compounds by higher plants, algae and symbionts such as lichens. In the course of these studies new significant sinks for COS could be identified. Furthermore, due to enzymatic studies, the physiological background for the consumption of COS by plants and algae is understood.

Acknowledgements

This project was funded by the Bundesminister für Forschung und Technologie (BMFT) as a project within the section Biosphere Atmosphere Exchange (BIATEX) of the European Environmental Programme EUROTRAC. However, without the fundamental support by the Max Planck Society this work would not have been possible. We are grateful to Renate Müller for technical assistance in growing *Chlamydomonas reinhardtii*. The assistance of Linda Geiser with lichen collections in Alaska, Jean Knops with collections in California and Dr. Volkmar Wirth with German collections are gratefully acknowledged. Special thanks to the Graduate Colloquium "Spurenstoffkreisläufe" of the University of Mainz (financed by the Deutsche Forschungsgemeinschaft) for their support of financing the stay of T.H. Nash in Mainz. In addition we acknowledge financial assistance from research funds from U.S. Environmental Protection Agency to Arizona State University.

Chapter 11

Biogenic VOC Emissions

11.1 Exchange of Short Chained Organic Acids between the Biosphere and the Atmosphere

J. Kesselmeier[1], C. Ammann[1], J. Beck[1], K. Bode[1], R. Gabriel[1], U. Hofmann[1], G. Helas[1], U. Kuhn[1], F.X. Meixner[1], Th. Rausch[2], L. Schäfer[1], D. Weller[1] and M.O. Andreae[1]

[1]Max Planck Institute for Chemistry, Biogeochemistry Dept., P.O.Box 3060, D-55020 Mainz, Germany
[2]Botanical Institute, Ruprecht Karls University, Im Neuenheimer Feld 360, Heidelberg, Germany

Summary

We have investigated the exchange of short chain organic acids between the biosphere and the atmosphere. Emission from different tree species under laboratory as well as field conditions was studied in order to get an estimate of the contribution of the release of organic acids into the atmosphere by the vegetation. Additionally we gathered information about the correlation of the emission with several plant physiological parameters such as photosynthesis, respiration, transpiration and stomatal conductance. Furthermore, a one year study (1994) supported by micrometeorological measurements over an agricultural site will give some insight into the fluctuations of acids in the air as well as the emission/deposition processes. Investigations on the role of leaf litter were done to discuss the role of litter layers in forests, and first experiments on the determination of the apoplastic concentration of acids and ions will help to discuss production, transport and release phenomena within and from plant tissues.

Aims of the research

The subject of our research is to contribute to the knowledge of the emission of short chained organic acids (formic and acetic acid) between the biosphere and atmosphere. The knowledge of the biogenic production and release of these oxygenated compounds still suffers from inadequate understanding.

Principal scientific results

Emission of short chained organic acids from spruce trees under field conditions

To investigate whether vegetation acts as a relevant source, we studied the emission of organic acids by spruce twigs using a dynamic cuvette system. The cuvette studies were carried out in spring 1990 on the Kleiner Feldberg, Taunus, and in autumn 1990 at the LfBAM tower of the Meteorological Institute, Munich, in the Bavarian forest. Within all experiments photosynthetic active radiation (PAR), temperature, relative humidity and CO_2 exchange were recorded. Within a typical diurnal emission pattern the emission ranged between 4 to 56 pmol/min (formic acid) and 10 to 100 pmol/min (acetic acid) with a maximum in the early afternoon. Organic acid fluxes correlated with CO_2 exchange. However, in fall the correlation between fluxes of acids and the CO_2 exchange showed only a weak dependence on the photosynthetic activity, whereas the measurements in spring exhibited a significant dependence on the CO_2 exchange. Using minimum and maximum emission rates of 7-70 pmol formic acid per μmol CO_2 assimilated and 10-100 pmol acetic acid per μmol CO_2 assimilated, we scaled up with the global net primary production rate of 6.5×10^{15} g.d.w./year (2.86×10^{15} g C) given by Whitaker and Likens [1] for a coniferous forest. Thus, we came up with a global acid release of $1.7–17 \times 10^{-9}$ mol/a for formic acid and $2.4–24 \times 10^{-9}$ mol/a for acetic acid. It is obvious that the coniferous forest may act as a significant source.

Emission of organic acids by different tree species under controlled climatic conditions and purified air

The aim of this research was to compare the emission of formic and acetic acid of different tree species under controlled conditions in a plant growth chamber with the exclusion of oxidative processes by atmospheric oxidants. The measurements were done using a dynamic cuvette system enclosing whole above ground parts of young trees. The cuvette was flushed with purified air, containing only low amounts (< 0.2 ppb) of formic and acetic acids. The experiments were performed with young spruce (*Picea abies* (L.) *Karst.*), beech (*Fagus sylvatica* L.), and ash (*Fraxinus excelsior* L.) trees and always lasted several days, to obtain a data set giving a protocol of the time of adaption. Measurements of the exchange of acetic and formic acid were combined with measurements of both transpiration and assimilation. All tree species investigated emitted both acids. The mean emission rates obtained with spruce on a leaf area basis ranged between 0.2 to 1.0 nmol m^{-2} min for formic and 0.1 to 0.4 nmol m^{-2} min^{-1} for acetic acid. A beech species showed a comparable emission in a similar range between 0.1 and 0.8 nmol m^{-2} min^{-1} for both acids. The highest emission rates with values ranging between 0.5 and 5.5 nmol $m^{-2} min^{-1}$ for formic acid and 0.3 and 1.6 for acetic acid was obtained with an ash species. All trees were measured several times during the year and it became obvious that there was a seasonal effect, especially observed with the ash tree which showed the highest emission rates in the autumn when the leafs were discoloured. In all cases highest emission rates were observed under light conditions with plants being photosynthetically active and showing open stomata. Lower emissions or even sometimes deposition of the acids were observed during the dark periods. A significant correlation was found between the acid emission and transpiration. In all cases this correlation was better for formic than for acetic acid. The emissions were always dominated by formic over acetic acid with a factor of 2.1 for ash and 1.7 for spruce, whereas it was nearly balanced for beech showing a ratio of 1.1.

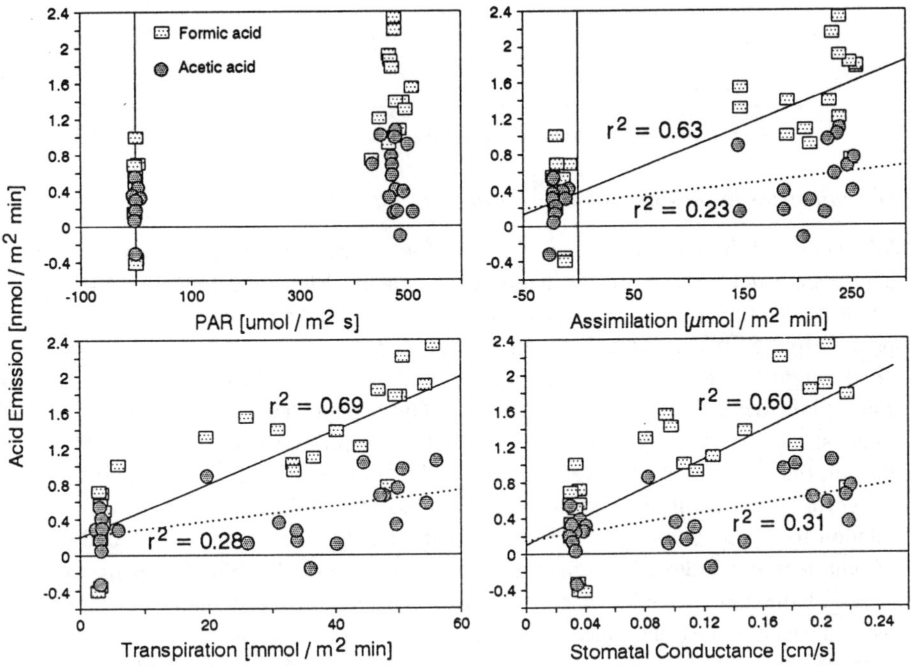

Fig. 11.1.1: Correlations of organic acid emissions by a young ash trees with light (PAR), assimilation, transpiration and stomatal conductance. Data were obtained in the course of a 3 day experiment after an adaption of 40 hours.

We used the data to check whether the observed emission rates might be important for the global budget estimating on a leaf area basis. For this rough calculation a global forest area of about 48.5×10^6 km^2 with a total leaf area of 413 m^2 m^{-2} was taken into account as given by Whittaker and Likens [1]. Assuming photosynthetically active plants for 12 hours a day and a vegetation period of 6 months a year, we come up with global annual emission rates ranging between 14 and 256 Gmol for formic and 9 to 94 Gmol for acetic acid. Following this crude approach the global forests, excluding other vegetation types as savannas, shrublands etc., might act as relevant continental sources (Table 11.1.1).

Table 11.1.1: Compilation of continental sources for formic and acetic acid in the atmosphere. Values given in Gmol a^{-1}.

Source	formic acid	acetic acid
biomass burning	55–202	182–650
vegetation	14–256	9–94
savanna soils	36	24
ants	10	–
vehicular traffic	0.4	1.3
ruminants	–	2.5

Data are based on Helas and Kesselmeier [2] and Bode [3].

Organic acids in the atmosphere over an agricultural site

We measured the concentrations of acetic, formic, pyruvic, propionic and oxalic acid over a field of oats in Bellheim (Rhine valley, Rheinland Pfalz) during the whole vegetation period: soil only (March, before sowing), germination (April, plant height 5 cm), growth (beginning of May, plant height 20 cm; end of May, plant height 50 cm), grain filling (beginning of June, plant height 80 cm; end of June, plant height 115 cm), seed ripening (July, plant height 120 cm), stubble (August, after harvesting). The aims of the field experiments were to learn about the diurnal and seasonal variations of the mixing ratios as compared to meteorological conditions and to calculate emission and deposition of the acids in relation to the developmental stage of a field of oats. Thus, plant's fresh and dry weight, leaf area index, the exchange of water and carbon dioxide, the fertilization status of the soil as well as meteorological and micrometeorological parameters were measured over the season. Organic acids, nitric acid and ammonia were cryogenically trapped with a specially designed siliconised glass trap and analysed by ion chromatography. First results show a strong increase of the mixing ratios from March until August and strong diurnal variations with minima in the night-time and maxima at the time of high solar radiation. In March the daytime averages were around 200 ppt for acetic and formic acid, 5 ppt for propionic acid, 1 ppt for oxalic acid. In August the maxima reached 5000 ppt for formic and acetic acid, 250 ppt for propionic and pyruvic acid, 35 ppt for oxalic acid. Pyruvic acid showed the highest mixing ratios during the morning time. All gradient data point to deposition.

Investigations on the role of leaf litter

In contrast to exchange measurements on vegetation and soils, no data exist about the role of microbial activities within leaf litter as a source or sink for short chained organic acids. Therefore we investigated the role of beech leaf litter

(*Fagus sylvatica*) from a forest near Darmstadt using a silanised glass cuvette flushed constantly with clean air. In addition to the exchange of acids water content and microbial respiration were measured simultaneously. Due to flushing, the leaf litter was constantly dried with time and an simultaneous increase of the emission rate was observed. From below a water content of 50 % we found a constant maximum emission rate which did not alter within the remaining period of the experiment, showing no linkage to microbial production. These constant emission rates observed changed depending on the seasons, the leaf litter was investigated. The maximum emission was measured during summer with 34.9 ± 0.4 nmol/100 g d.w.h for formic and 54.1 ± 0.6 nmol/100 g d.w.h for acetic acid. The lowest values were found during winter with 14.9 ± 0.7 nmol/100 g d.w.h for formic 6.2 ± 1.3 nmol/100 g d.w.h for acetic acid. Constant adapted emission rates being independent from microbial production over longer time periods gave reason to assume a large pool of organic acids within the leaves. To get an estimate of this pool we measured organic acids after homogenisation of leaf litter with highly purified water and found high amounts of both acids, up to 35 mmol/100 g d.w.h for formic and 17 mmol/100 g d.w.h for acetic acid. Thus, a big pool of organic acids is present in leaf litter, the release being dependent on the stage of decomposition, the break-down of the membrane structures and thus the exposure of the acids to the atmosphere. This interpretation is supported by exchange measurements after chloroform fumigation. We found a 100 times higher emission rate for formic and 20 times for acetic acid due to the disintegration of the cuticular barrier.

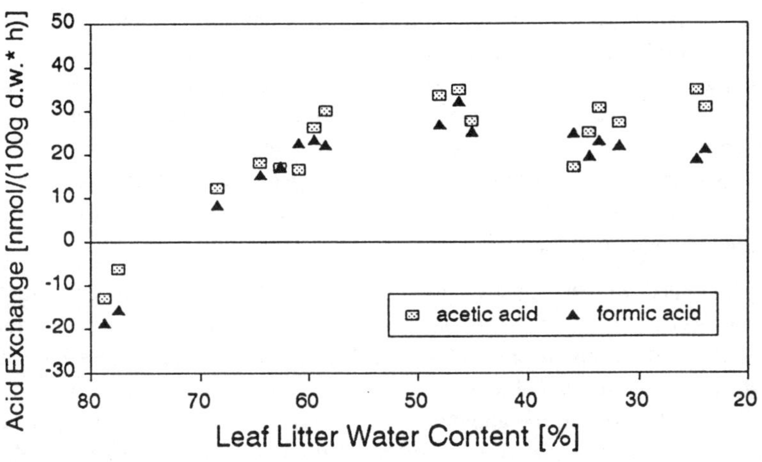

Fig. 11.1.2: Release of organic acids from leaf litter under ambient air conditions showing deposition as well as emission depending on the water content of the leaf litter.

Additional experiments were carried out under ambient air concentrations. With air containing 3.7 ± 1.3 ppb for formic acid and 5 ± 1.7 ppb for acetic acid the litter layer took up both acids, nearly independently from the water content of the leaves. Under more natural acid concentrations found in forests, 2.2 ± 0.1 ppb for formic and 1.7 ± 0.1 ppb for acetic acid, we found deposition of both acids to humid litter (water content above 70 %) and emission from dry litter (water content below 70 %) (see Fig. 11.1.2). Summarizing, we could show that the acid exchange depends on the leaf litter water content under ambient air conditions. With a high water content the leaf litter serves as a sink, whereas under drier conditions these compounds are released from an enormous pool of acids on and within these leaves. Actual microbial activities are obviously not necessary to explain the acids release and emission, though it may be assumed that the microbial metabolism might contribute to the overall pool. These results explain observed daily patterns of organic acids in air masses near soil level.

Apoplastic concentration of organic acids and inorganic ions in tree leaves

The cell wall represents the main part of the so called apoplast, consisting of cell walls, intercellular spaces and the lumina of nonliving cells such as the xylem vessels. By passing the cell wall, the trace gas leaves the liquid phase and enters the gas phase inside the leaf. Thus, the knowledge about the concentration of organic acids in the apoplast as the primary boundary layer between atmosphere and biosphere is a basic requirement for emission studies. Assuming that the apoplast is the origin of organic acid emission within the leaf, we investigated the concentration in the apoplastic sap.

The experiments were carried out with 4 years old beech trees (*Fagus sylvatica* L.) and 30 to 40 years old Mediterranean oak trees (*Quercus ilex* and *Quercus virgiliana*). Apoplastic fluid was obtained by the pressure chamber technique (Scholander Bomb). Organic acids found were acetate, formate, pyruvate, lactate. Additionally inorganic anions, such as chloride, nitrate, phosphate, sulfate were analyzed. Inorganic cations (potassium, calcium, magnesium, natrium) were determined by capillary zone electrophoresis. The determination of proteins (18.4–66 kD) was carried out with SDS polyacrylamide gel electrophoresis. The results obtained for the proteins and pyruvate indicated that the experimental techniques inflicted no damage upon the leaf cells.

The data obtained with beech give preliminary estimates of the concentration of formic and acetic acids in the apoplastic sap ranging between 3 and 110 μmol/l. The data obtained for apoplastic fluids from Mediterranean oak species indicated no clearly diurnal pattern. Table 11.1.2 summarises the concentration of organic acids and inorganic ions in the apoplastic sap averaged over a whole day. *Quercus ilex* and *Quercus virgiliana* showed different amounts and compositions of all analysed ions. The ion concentrations of lactate, chloride and phosphate were higher in *Q. virgiliana*. In contrast, formate and sulfate were higher in *Q. ilex*. Pyruvate was in the range of the detection limit for *Quercus ilex* but occurred in as high amounts as acetate and formate in *Q. virgiliana*.

Table 11.1.2: Ionconcentrationsof extruded sap of apoplastic compartments.

Species	*Q.ilex*	*Q.ilex*	*Q.ilex*	*Q.virgil.*	*Q.virgil.*
Date	17.05.	24.05.	25.05.	22.05.	26.05.
n	5	4	6	4	5
	[μmol/l] (min/max)				
lactate	28 (12/48)	31 (11/42)	15 (4.8/34)	44 (14/76)	42 (26/81)
acetate	10 (6.8/14)	12 (6.5/17)	12 (8.8/15)	32 (11/43)	6.3 (5.6/7.6)
formate	11(10.7/11.9)	31 (21/41)	19 (13/26)	5.9 (5.3/6.7)	5.1 (1.5/10)
pyruvate	0.3 (0/1.8)	0.0	0.0	7.0 (2.5/9.6)	16 (10/27)
chloride	92 (74/103)	102 (44/222)	82 (18/160)	150 (32/263)	165(107/224)
nitrate	3.6 (2.0/6.9)	2.9 (0/6.8)	5.2 (0/18)	5.0 (3.7/5.8)	8.5 (2.,2/19)
sulfate	28 (15/40)	34 (15/50)	24 (6/52)	6.3 (4.8/7.8)	23 (7/41)
phosphate	30 (15/58)	76 (25/126)	42 (21/84)	110 (40/197)	135 (70/184)

Values are means of two fractions obtained from each measurement averaged over n daily measurements.

Conclusions

The vegetation is involved in the direct emission of short chained organic acids into the atmosphere. The emission is related to plant physiological parameters as photosynthesis, transpiration and stomatal conductance. Rough estimates give reason to believe that the amount of acids released might act as a relevant continental source. Also leaf litter may contribute to the atmospheric budget.

Acknowledgements

This project was funded by the Bundesminister für Forschung und Technologie (BMFT) as a project within the section Biosphere Atmosphere Exchange (BIATEX) of the European Environmental Programme EUROTRAC. We acknowledge additional financial support by the Graduate Colloquium "Spurenstoffkreisläufe" of the University of Mainz (Deutsche Forschungsgemeinschaft) and by the Max Planck Society.

References

1. Whittaker, H., Likens, G.E.; The biosphere and man, in: Lieth, H., Whittaker, H. (eds), *Primary productivity of the biosphere. Ecological Studies 14*, Springer Verlag, Berlin 1995, pp. 305-328.

2. Helas, G., Kesselmeier, J., Estimates on sinks and sources of formic and acetic acid, in: Slanina, J., Angeletti, G., Beilke, S. (eds), *General Assessment of Biogenic Emissions and Deposition of Nitrogen Compounds, Sulphur Compounds and Oxidants in Europe*, CEC Air Pollution Research Report **47** E. Guyot SA, Brussels, 1993, pp 299-304..

3. Bode, K., Die Abgabe von organischen Säuren an die Atmosphäre durch Pflanzen verschiedener Entwicklungsstufen. Dissertation, Fachbereich Biologie, Universität Mainz 1994.

11.2 Forest Emissions of Hydrocarbons

Casimiro Pio, Teresa Nunes and Antonio Valente

Departamento de Ambiente e Ordenamento, Universidade de Aveiro,
3810 Aveiro, Portugal

Summary

Sampling and analytical techniques were developed for the simultaneous measurement of isoprene and terpenes emitted by vegetation. An evaluation of the hydrocarbon emission rates by common tree and shrub species in Portugal was performed applying a semi-quantitative method, (~ 50 different species were studied). Field and laboratory studies with the bag enclosure method were used to quantify the emission rate of monoterpenes and/or isoprene by *Pinus pinaster* and *Eucalyptus globulus*. In a pine patch forest tracer and bag enclosure methods were employed together to better access emission rates. Measurements were also done in the field during the year, for terpenes emitted as result of resin-tapping practices.

Aims of the research

* Development and optimisation of different sampling methods for the evaluation of hydrocarbon emissions by vegetation to the atmosphere;
* Determination of semiquantitative hydrocarbon emission rates for the most common tree and shrub species in Portugal;
* To access the importance of hydrocarbon emissions from resin-tapped trunks, by maritime pine trees;
* To evaluate isoprene/monoterpene emission rates from *Pinus pinaster* and *Eucalyptus globulus* trees by different field techniques;
* To establish hydrocarbon emission rates for the species refereed above as a function of meteorological variables and season of the year.

Principal scientific results

During the first part of the project most of the time was spent in developing sampling and analytical methods for the application in hydrocarbon compounds identification and quantification. Hydrocarbon compounds, namely isoprene and monoterpenes, were sampled in Tenax-TA glass tubes and then analysed by GC/FID after controlled thermal desorption. Details of the sampling and analytical methods are described in [1].

An experimental set-up was mounted in the laboratory for the quantification of the capability of more common species in Portugal for the emission of hydrocarbons to the atmosphere. A total of 27 tree, 15 shrub, 2 grass, 7 fruit tree and 2 agricultural species were investigated [2]. This survey pointed for the first time that some broad leave tree species, common in Mediterranean climates, do not emit isoprene, emitting monoterpenes instead. For example *Quercus suber* and *Quercus ilex* emit a variety of non-isoprene hydrocarbons with emphasis in α-pinene and β-pinene or sabinene, myrcene (*Q. ilex*) and limonene (*Q. suber*). Our results obtained in the laboratory with single leaves were confirmed later in the field, for *Quercus ilex*, in the BEMA project [3]. This information is very important for a correct inventory of biogenic emissions in southern Europe, where this two species are very common, being predominant in some regions, such as the south of Portugal.

During decades resin tapping has been an important industrial activity associated with forest management in Portugal. Together with China and the former USSR, Portugal is one of the biggest producer of gum from living trees in the world. Near 70 % of the national pine forest area is tapped, with 35 to 45 million trees worked annually. Measurements of VOC emissions from resin tapping were made with a continuous flow-through chamber technique [4]. α- and β-pinene are the main compounds emitted from resin, constituting approximately 88 % of volatile VOCs [5]. Resinated trees emit 3 to 150 times more VOCs than non-resinated trees. Emissions increase exponentially with temperature, (see Fig. 11.2.1) and are dependent on wind speed. Chemical reactions at the resin surface result in an enrichment of β-pinene in emissions, relative to resin

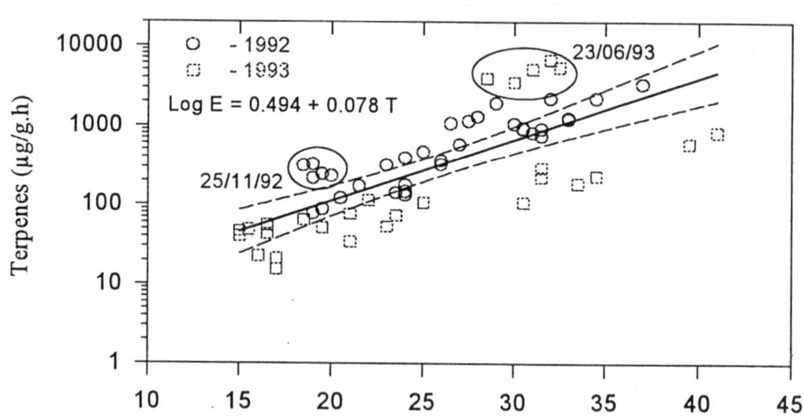

Fig. 11.2.1: Terpene emission to the atmosphere from *Pinus pinaster* trees as result of resin tapping activities.

composition. From the emission algorithm obtained, the number of resin tapped trees, the mass of resin produced and average ambient temperatures we have calculated that 8,000 to 12,000 tons of terpenic VOCs are emitted annually in Portugal, as result of resin tapping practices.

For two years field experiments were carried out in a pine forest with the objective of evaluating VOC emissions to the atmosphere. Emissions from *Pinus pinaster* species were determined with the bag enclosure method and an algorithm describing emission rate as a function of temperature was proposed, (see Fig. 11.2.2) [6]. VOC emission rates from pine plantules were also investigated in the laboratory. α- and β-pinene were again the main compounds emitted, but with higher emission rates than adult trees.

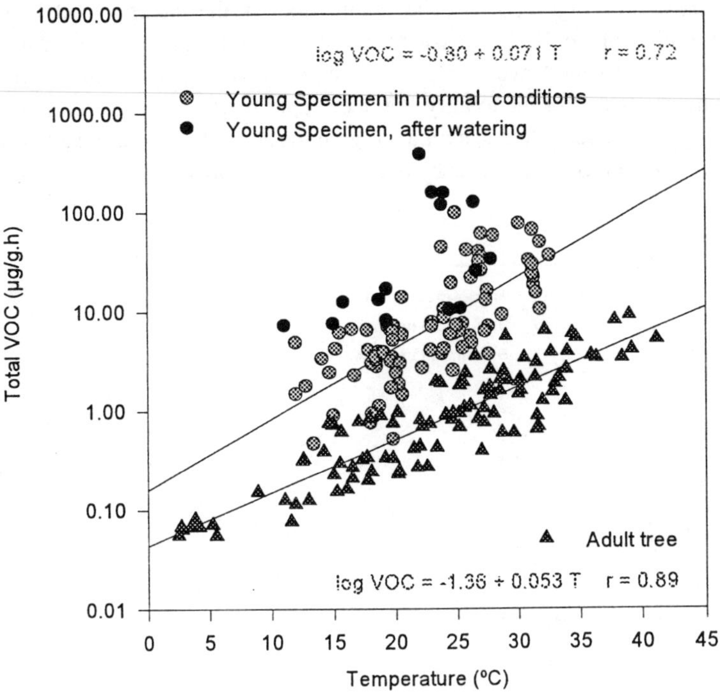

Fig. 11.2.2: Emission rates and emission algorithms for pine needles from young and adult *Pinus pinaster* specimen.

A total of seven tracer release experiments, using SF_6 as tracer gas, were done successfully in the field, some of them simultaneously with bag enclosure measurements, at temperatures ranging from 10 to 30 °C. The tracer gas was released in several points at canopy level with known flux rates. Dispersion and dilution of the detected VOCs, during transport between the forest and samplers, were evaluated from tracer release fluxes and tracer concentrations. From this and VOC air concentrations at sampling points upwind and downwind the forest, the total emission from forest trees was calculated [5].

The biometry of the forest patch was studied extensively through different techniques, including cutting and sampling of some trees, in order to determine total leaf biomass and leaf biomass distribution, necessary for a correct quantification of emission fluxes. The emission algorithm previously calculated was applied to the forest patch in association with the BEIS model, which gives leaf temperature distribution across the forest canopy as a function of solar radiation, wind speed and humidity [7], and VOC emission fluxes calculated for the forest patch during the tracer experiments.

For all VOCs, differences between tracer measurements and the algorithm calculations are always lower than 65 %, with an average of 31 % for the seven experiments, (see Fig. 11.2.3). The application of the canopy model gives an average excess of 16 % in emissions by comparison with tracer measurements, while by considering leaves temperature equal to air temperature, results in a deficiency of 10 %. For the totality of tracer experiments, the algorithm associated with the BEIS model calculates an average emission rate for the forest patch that differs from tracer measurements by only 2 %. VOC composition

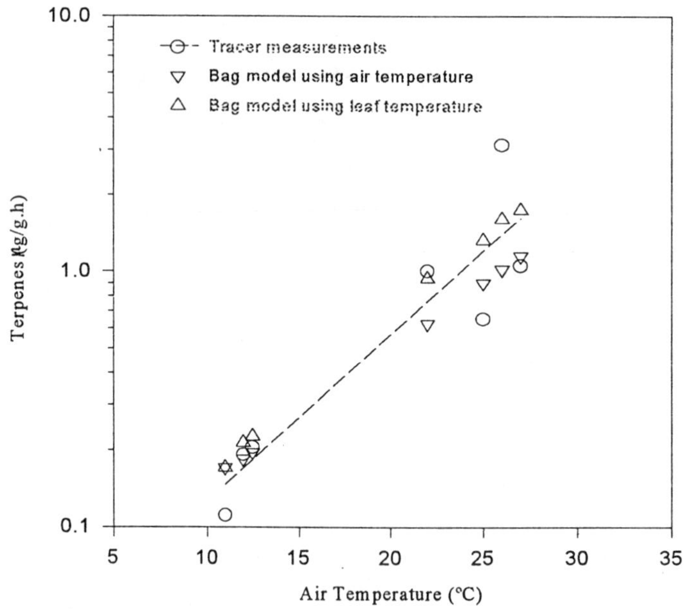

Fig. 11.2.3: Terpene emission rates measured for seven tracer experiments performed in a *Pinus pinaster* forest, and calculated by using the bag enclosure algorithm.

measured in the ambient air during tracer experiments doesnot agree so well with algorithms for individual terpenes as for total VOC data. Ambient β-pinene concentrations observed were lower than expected, principally during summer experiments, while limonene and myrcene had an opposite behaviour.

Field measurements with the bag enclosure technique were performed with *Eucalyptus globulus*. Eucalyptus leaves emit isoprene mainly and also smaller quantities of a variety of heavier hydrocarbons such as cineool and α-pinene. Laboratory experiments with young eucalyptus trees showed a similar behaviour. Some samples, analysed by GC/MS, permitted the identification of *trans*-ocimene as the more common emitted VOC after isoprene. Emission of isoprene is relevant only during the day at high temperatures and solar radiation intensity, (see Fig. 11.2.4). The rate of isoprene emission by eucalyptus leaves during daylight is two orders of magnitude higher than terpenic emissions. Isoprene emissions are strongly dependent from solar radiation, being affected also by ambient temperature, (see Fig. 11.2.5). Emission rates calculated from our field measurements are in a reasonable agreement with the algorithm developed by Günther *et al.* [8] that describes emissions as a function of temperature and solar radiation intensity, even taking into account that our PAR measurements are only an approximated average of the light intensity received by different leaves in the exposed tree branch.

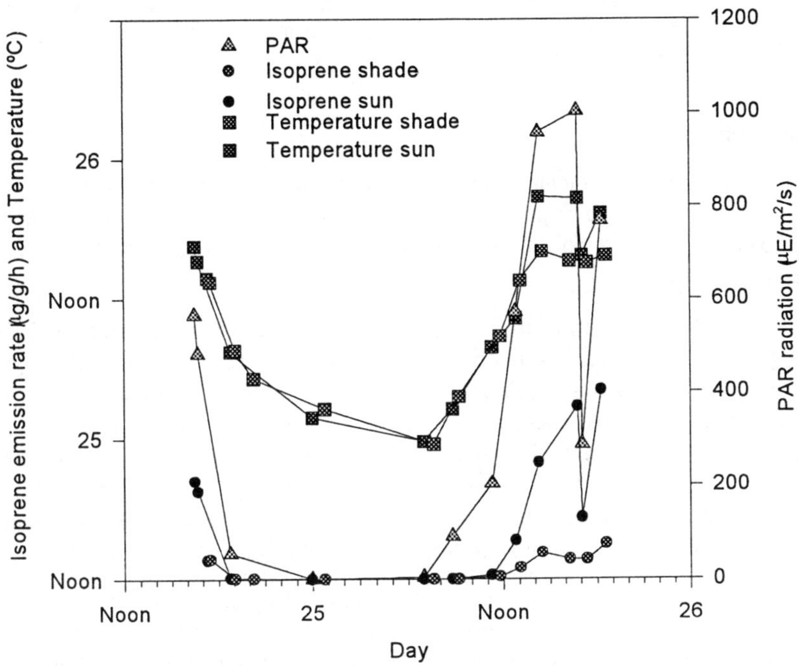

Fig. 11.2.4: Typical isoprene diurnal emission rates from eucalyptus leaves. The figure presents two parallel experiments with branches in direct sunlight and in the shadow. Temperature and PAR are shown for reference.

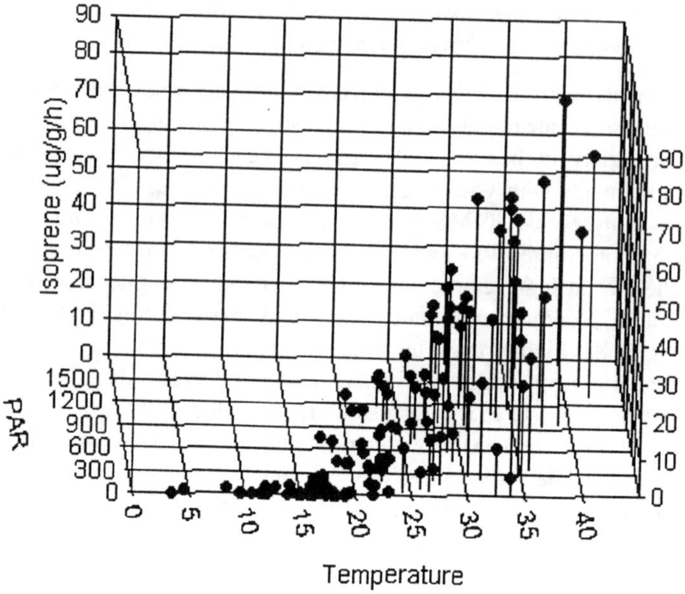

Fig. 11.2.5: Isoprene emission from *Eucalyptus globulus* as a function of leaf temperature and photosynthetically active radiation (PAR). Experiments were performed between 1993 and 1994 at Quinta das Corgas, Portugal.

Conclusions

From emission measurements and spatial distribution of vegetation it is possible to conclude that only a few species, including maritime pine, eucalyptus, cork and holm oaks, need to be considered, to produce a correct biogenic VOC emission inventory in Portugal.

The main VOC emissions from *Pinus pinaster* are α-pinene and β-pinene and only a temperature dependency was observed. Young pine trees seem to have higher emission rates than adult specimen possibly as result of the presence of a larger amount of new needles where the monoterpene syntheses activity is higher [9]. Tracer experiments applied to the quantification of VOC emission rates from *Pinus pinaster* showed that the algorithm developed from bag enclosure measurements gives correct flux information. Fluxes of monoterpenes in resinated pine forests are much higher than in forests where this practice is absent.

The main VOC emission from eucalyptus trees is isoprene. Monoterpenes were also emitted but at lower emission rates. Isoprene emission rates measured in the field are well described by the algorithm developed by Günther and collaborators.

References

1. Pio C., Measurements of biogenic non-methane hydrocarbon (BNMHC): Emissions from vegetation representative of European Ecosystems Final Report STEP CONTRACT CT90-0060, 1993.
2. Pio C. A., Brito, S. V., Nunes, T. V., Volatile hydrocarbon emissions from common and native species of vegetation in Portugal, in: *General assessment of biogenic emissions and deposition of nitrogen compounds, sulphur compounds and oxidants in Europe*. CEC/BIATEX workshop 1993, pp. 291–298.
3. Enders G., Biogenic Emissions in the Mediterranean Area BEMA-Project. *Report on the preliminary BEMA measuring campaign at Castelporziano*, Rome (Italy), June 1993. Environment Institute - Joint Research Centre-EC.EUR 15955 EN, 1994.
4. Pio C. A., Valente A. R., Terpene emissions from resin tapping practices, in: *General assessment of biogenic emissions and deposition of nitrogen compounds, sulphur compounds and oxidants in Europe*. CEC/BIATEX workshop 1993, pp. 243–249.
5. Valente A.R., Natural hydrocarbons in the atmosphere. Contribution for the knowledge of non methane hydrocarbons emission fluxes from Pinus pinaster in Portugal, (portuguese), PhD Thesis. Submitted, Departamento de Ambiente e Ordenamento, Universidade de Aveiro Portugal, 1995.
6. Pio C. A., Nunes T. V., Valente A. R., Measurement of VOC emissions from *Pinus pinaster* species by bag enclosure and tracer techniques, in: *IV Symposium CEC/EUROTRAC Project*, 1994.
7. Pierce, T., Waldruff, P., PC-BEIS: A personal computer version of the biogenic emissions inventory system *J. Air & Waste Management Association* 41 (1991) 937–941.
8. Günther A. B., Monson R. K., Fall R., Isoprene and monoterpene emission rate variability: observations with Eucalyptus and emission rate algorithm development *J. Geophys. Res.* **96D6** (1991)10799-10808.
9. Bernard-Dagan C., Carde J. P., Gleizes M., Etude des composés terpéniques au course de la croissance des aiguilles du pin maritime: comparaison de données biochimiques et ultrastructurales, *Can. J. Bot.* 57 (1979) 255–263.

11.3 Investigations on Emissions of Low Molecular Weight Compounds (C_2–C_{10}) from Vegetation

Rainer Steinbrecher[1], J. Hahn[1], K. Stahl[1], G. Eichstädter[2], K. Lederle[1], R. Rabong[1], A.-M. Schreiner[1] and J. Slemr[1]

[1]Fraunhofer-Institut für Atmospärische Umweltforschung (IFU), Kreuzeckbahnstr. 19, D-82467 Garmisch-Partenkirchen, Germany,
[2] Institut für Botanik der Technischen Universität München, Arcisstr. 21, D-80333 München, Germany

Summary

Biogenic hydrocarbons are known to act as important precursors in tropospheric photochemical ozone formation. Large uncertainties exist about the composition of the mix of volatile organic compounds emitted by various plant species, and the respective emission rates. A screening study of important woody and crop plants revealed that the portion of isoprene emitting species within deciduous tree species is small. But in contrast to the general assumption, all investigated deciduous tree species emitted traces of monoterpenes. As expected, oaks are strong isoprene emitters, however, some Mediterranean oak species emitted no detectable amounts. Isoprene emission is strongly dependent on temperature and insolation. Wheat and rape seed plants emitted only traces of isoprenoids. The emission and deposition behavior of wheat plants, as far as C_2 to C_9 hydrocarbons (NMHCs), formaldehyde, and acetaldehyde are concerned, was studied both in the field (BIATEX experimental site, Manndorf, Bavaria) and in the laboratory. Ambient temperature appears to control the fluxes of ethane, ethene, propane and propene, whereas acetaldehyde emission by wheat plants as well as by Norway spruce is controlled by light. Gradient measurements of NMHCs over a grassland area in the Rhine valley (TRACT Erlenhof I7, Brühl, Germany) showed that meadows have a negligible impact on ambient air mixing ratios of NMHCs. Over a spruce canopy (BIATEX experimental site Schachtenau Bayerischer Wald National Park, Germany) the twelve most abundant non-terpenoid NMHC exhibited no distinct diurnal cycle, and only small differences in mixing ratios were detected between the heights 31 and 51 m indicating that the influence of the canopy on the abundance of the non-terpenoid NMHCs present in the air above the canopy was small. Aldehyde mixing ratios above a spruce canopy, however, may significantly be influenced either by direct emission of aldehydes from spruce or by the production of aldehydes during photochemical degradation of precursors. Investigations of NMHC fluxes in strongly polluted forest ecosystems (BIATEX, Frankfurter Stadtwald, Frankfurt, Germany; BEMA, Castelporziano, Rome, Italy) revealed a complex flux pattern for selected biogenic and anthropogenic NMHCs, which is possibly a result of the heterogeneity of vegetation in the main wind direction.

Aims of the research

- Identification of neutral organic C_1–C_{10} compounds emitted by the vegetation,
- Determination of emission rates of selected compounds as a function of season and time of day,
- Determination of biological and meteorological factors controlling the emission rates, and
- Calculation of canopy fluxes of selected compounds.

Principal scientific results

Method development

Enclosure techniques

The measurement of the CO_2/H_2O gas exchange and the emission rate of selected hydrocarbons from plants were performed using three different types of open flow enclosure devices [1, 2, 3]. A small enclosure with quartz glass plates on the top and bottom of an aluminium cylinder was used for studying the emission of volatile organic compounds (VOCs), the CO_2 net uptake and the H_2O water vapour gas exchange by plant parts or leaves. Two fans mounted inside the cylinder ensured complete mixing of the air volume passing through the enclosure at a flow rate of 4 L (STP) min^{-1}. A larger teflon bag cylinder (50 cm length, 30 cm i.d.), fixed between two teflon coated stainless steel plates containing all the sensors and a fan, allowed the investigation of larger twigs. This enclosure system was operated at an air flow rate of 40 L (STP) min^{-1}. A third system was set up to enclose plants growing in soil or potted plants. It consisted of a stainless steel frame with a teflon collar, Duran glass cylinders of variable height, and a Duran glass cover containing a fan for mixing the box air volume. All sensors were mounted in the Duran glass cylinder. The air flow rate was set between 20 and 100 L (STP) min^{-1} depending on the volume of the enclosure. The three above mentioned enclosure devices were either flushed with ambient or purified air.

Sampling, GC and HPLC techniques

Isoprene, monoterpenes, and other C_2 to C_{10} hydrocarbons in air samples were collected alternatively in the mixed-bed adsorption tubes (Carbosieve III, Carbotrap and Carbotrap C, Carbosieve III and Carbotrap) of two on-line GC systems [3, 4, 5]. It was ensured by careful laboratory studies that ozone and NO in ambient air samples did not influence the recovery rate of isoprene trapped on the adsorbents [6]. TENAX TA and Carbosieve were used as adsorbents for off-line analysis of isoprene in air samples [7]. For analysing off-line the whole pattern of the C_2 to C_5 hydrocarbons in ambient air, the grab sample technique using electropolished stainless steel canisters was employed [8]. Monoterpene mixing ratios in air were determined after sample pre-concentration on TENAX TA [9].

Air samples were taken using a sample flow rate of 150 ml (STP) min^{-1} and a sampling time of 30 to 40 min. The adsorbent was filled in HCl deactivated glass tubes. In all cases, the trapped hydrocarbons were thermally desorbed, subsequently separated on capillary columns, and quantified using flame ionisation detectors. The details of the gas chromatographic analytical method are described elsewhere [1, 3–9, 11]. The quality of the NMHC analysis was checked by participation in several international intercomparison exercises [12, unpublished results E. Apel, NCAR, Boulder, Colorado].

Atmospheric C_1 to C_3 carbonyls were trapped on 2,4-dinitrophenylhydrazine coated silica gel after removing ozone with a KJ scrubber. Air was sampled at a flow of 2 liters (STP) min^{-1} for one hour. The hydrazone derivatives of the carbonyls were eluted from the cartridges using acetonitrile and analysed by HPLC [10].

Screening studies

In the climate chamber of the IFU, two coniferous, eight deciduous and two shrub species were studied to evaluate the NMHC emission potential under almost natural conditions [1, 2, 11]. CO_2/H_2O gas exchange measurements were performed to recognise possible stress effects on the plants during the emission study and to supervise the physiology of the plants. In order to correlate the emission of monoterpenes from the plants with the monoterpene pool in the plants, the amount of endogenous terpenes was determined. Among the plant species investigated, only oak (*Quercus robur* L.) showed a significant isoprene emission. Larch (*Larix decidua* Mill.) and pine (*Pinus sylvestris* L.) emitted traces of isoprene, sometimes not exceeding the detection limit of 0.1 ng C m^{-2} total leaf surface s^{-1}. As expected, the conifers emitted monoterpenes (for structures see Steinbrecher *et al.*, this volume). The main compounds in the emission pattern of *Pinus* were α-pinene, *ß*-pinene and myrcene, in the night and during the day. The monoterpene emission rates on total leaf area basis of *Larix* were less than those of *Pinus* reaching, *e.g.* in the case of α-pinene, only one fifth of the emission during the daylight hours. The higher emission during the day compared to night values is mainly due to the higher temperatures in the daytime. Surprisingly, all examined broad leaf plant species emitted traces of *p*-cymene (0.1–0.9 ng C m^{-2} s^{-1}). Limonene was also detected in the monoterpene emission pattern of nearly all investigated plant species except lime (*Tilia cordata* Mill.). Willow (*Salix cineria* L.) and hazelnut (*Corylus avellana* L.) exhibited a relatively broad emission spectrum with up to eight different monoterpenes. α-Pinene and traces of other monoterpenes could also be detected in pentane leaf extracts of the deciduous plant species (100–200 µg g^{-1} DW) [11]. In general, the terpenoid emission of rape seed (*Brassica napus* ssp. napus) and corn (*Zea mays* L.) plants were negligible. Only rape seed emitted up to 0.4 ng C m^{-2} s^{-1} sabinene and limonene at 25°C during the day. At night, no terpenoid emission was measurable. A screening study of Mediterranean oak species and typical plant communities of the pseudo steppe indicated that selected oak species (*Quercus frainetto* [Ten.], *Quercus petraea* L., *Quercus pubescens* Wild.) are strong isoprene emitters but the deciduous oak *Quercus cerris* L. and the evergreen oak *Quercus suber* L. emitted only negligible amounts of isoprene [3]. Monoterpenes were not detected in the VOC emission pattern. The investigated isoprene emitting oak species exhibited a great inter-species variation with a range of 1.6 nmol m^{-2} total leaf area s^{-1} (*Q. pubescens*) to 5.28 nmol m^{-2} s^{-1} (*Q. frainetto*) at comparable leaf temperatures (36 °C) and photon flux densities (1400 µE PAR). The investigated two pseudo steppe plant communities with *Asphodelus microcarpus* Viv. and *Cistus salvifolius* L. as dominating plant species emitted isoprene at maximum rates of 2.32–3.89 nmol m^{-2} soil s^{-1} at leaf temperatures around 39°C and photon flux densities of 1900 µE PAR (photosynthetically active radiation).

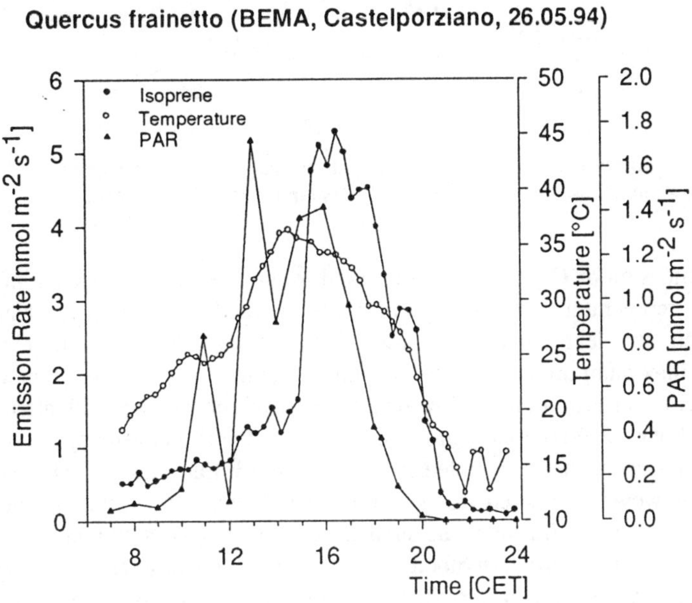

Fig. 11.3.1: Diurnal cycle of photon flux density of photosynthetically active radiation (PAR), air temperature in the vicinity of a leaf, and isoprene emission from the Mediterranean oak *Quercus frainetto* at Castelporziano, Rome, Italy.

Factors controlling the emission of VOCs from plants

The emission of VOCs from plants is thought to be controlled by various factors. The following parameters were examined in relation to the emission of VOCs: leaf temperature, PAR, net CO_2 assimilation, transpiration, leaf to air water vapour conductance, and water potential of the leaves. In general, isoprene emission of plants is well correlated to leaf temperature and PAR [3, 5]. A correlation to the other parameters is often not obvious. Even under certain unknown circumstances, a correlation between PAR and leaf temperature is not very distinct. The maximum

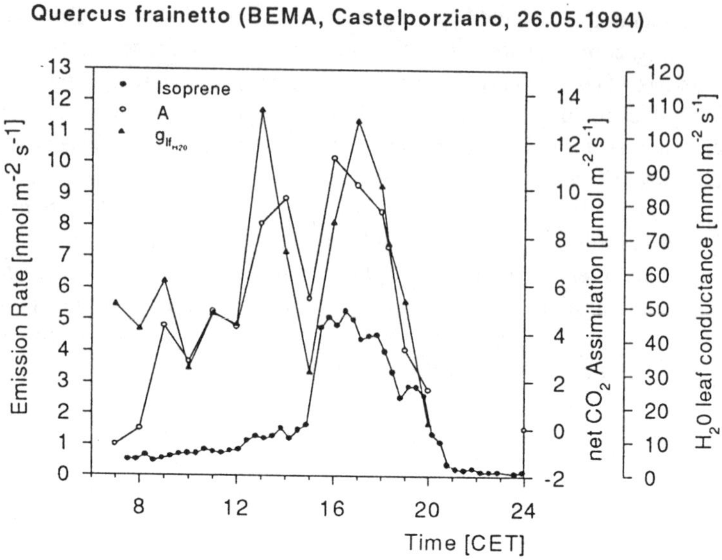

Fig. 11.3.2: Diurnal cycle of net CO_2 assimilation, leaf-to-air water vapour conductance g_{H_2O}, and isoprene emission of *Q. frainetto* at Castelporziano, Rome, Italy.

isoprene emission of *Q. frainetto* is observed in the afternoon, about 1 h after the maximum temperature and insolation were reached (Fig. 11.3.1). Net CO_2 assimilation and leaf conductance rose continuously during the morning and reached a maximum around noon, following the diurnal cycle of leaf temperature and PAR (Fig. 11.3.2). In the following 2 hours, leaf conductance and, after a time delay of 1 hour, net CO_2 assimilation were reduced due to a closure of the stomata, possibly as a consequence of intensive water loss by transpiration. The isoprene emission, however, increased slowly during the morning in contrast to the other parameters measured. But after the sudden drop in net CO_2 assimilation and leaf conductance in the early afternoon, isoprene emission dramatically increased. Later in the day, temperature, PAR, CO_2 assimilation, and leaf conductance were well correlated. It may be speculated that the biosynthesis of isoprene is not switched on before a primary pool of precursors is filled.

In an intensive laboratory study, the emission/deposition behaviour of nineteen C_1 to C_9 VOCs with regard to wheat plants (*Triticum aestivum* L.) was investigated as a function of temperature, air humidity, and insolation [1]. The following compounds were studied: (1) the hydrocarbons ethane, ethene, ethine, propane, propene, *i*-butane, 1-butene, *n*-butane, isoprene, *n*-hexane, benzene *n*-heptane, toluene, ethylbenzene, *o*-xylene, *m*-, *p*-xylene, chloromethane; (2) the carbonyls formaldehyde and acetaldehyde. The main factor influencing the flux behaviour of hydrocarbons was temperature (Fig. 11.3.3), whereas the carbonyl fluxes appeared to be controlled by light. The flux of ethane, ethene and propene varied from a deposition at low temperatures (10 °C) to an emission (30 °C) in light.

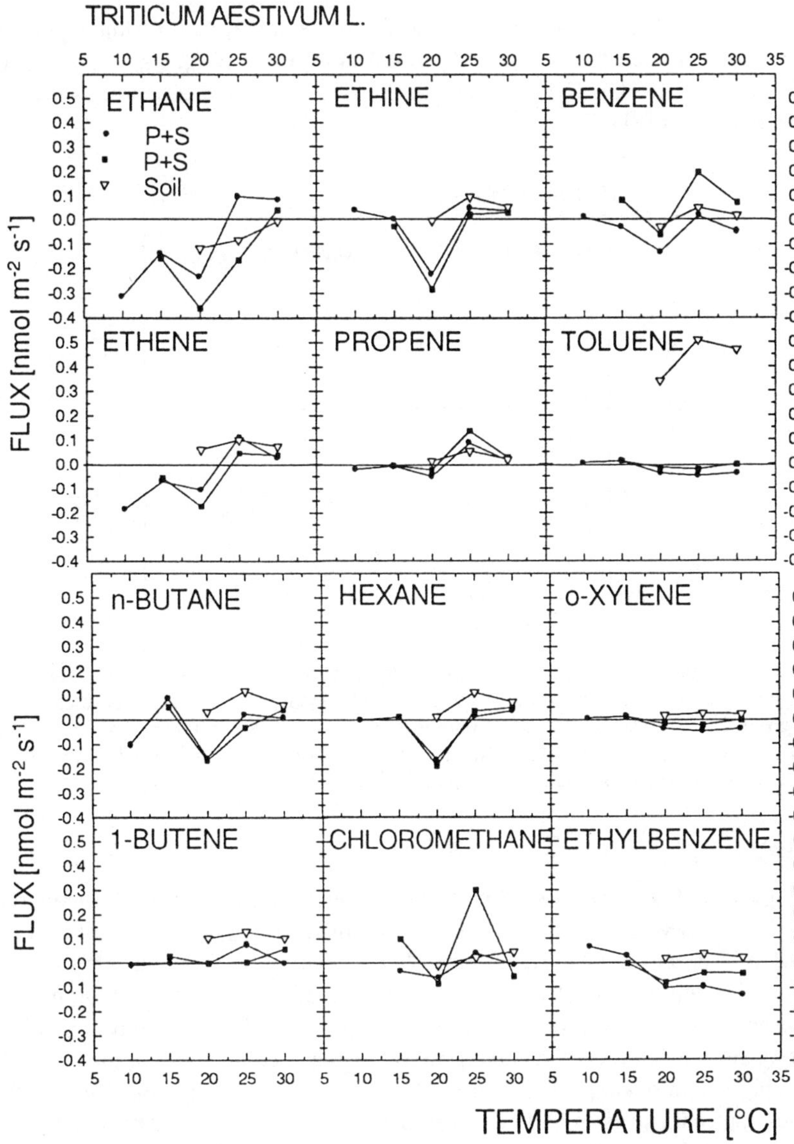

Fig. 11.3.3: NMHC fluxes from a wheat field and the relation to air temperature. The experiments were performed in climate chambers at the IFU at 600 µE PAR and a relative air humidity of 40–50 %. The flux rates are related to soil area at a leaf area index of 6.5. Abbreviations: P + S = plants + soil.

In order to specify the contribution of the soil and the wheat canopy to the total VOC flux from a wheat field, the plants were removed and the soil studied separately. It was found that ethane, ethene, and propene and the carbonyls formaldehyde and acetaldehyde were emitted by the wheat plants. Isoprene was not detected in the emission spectrum. No temperature effect on the carbonyl emission was observed but light enhanced the acetaldehyde emission significantly from zero emission during the night to 0.2 nmol m^{-2} soil area s^{-1} in the daytime (600 µE PAR, 20 °C). The acetaldehyde emission from Norway spruce showed the same behaviour [8] (Fig. 11.3.4).

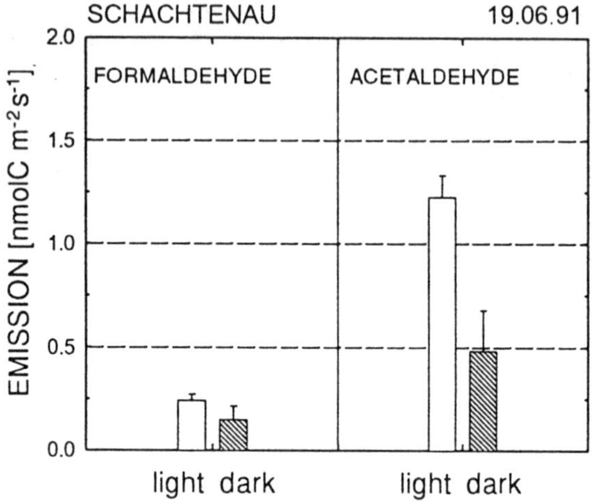

Fig. 11.3.4: Influence of light on aldehyde emission from a living spruce twig, referring to total leaf area in the sun crown of a spruce tree (enclosure measurements 28 meters above the ground at 800 ± 50 mE $m^{-2}s^{-1}$ photosynthetically active radiation, 15 ± 0.5 °C leaf temperature, 50 ± 1 % relative humidity; sampling time 1 hr). The stack bars represent mean values of five measurements, the bars indicate the standard deviation.

Canopy fluxes

For the calculation of VOC fluxes from plant canopies, the gradient approach GRAD was used [13, 14, Steinbrecher *et al.*, this volume]. Over a wheat canopy at the BIATEX site Manndorf, Danube Valley, Germany, the gradients of the six dominating atmospheric hydrocarbons (ethene, ethane, propane, chloromethane, benzene, and toluene) were not statistically significant. Thus, the impact of a wheat canopy on the hydrocarbon mixing ratios of these compounds in ambient air is probably insignificant [1, 15]. In the frame of the TRACT experiment in the Rhine Valley, diurnal cycles of C_2 to C_6 hydrocarbon mixing ratios at 0.8 and 6.8 meters above a grassland surface were recorded [4]. The TRACT field experimental site I7, 'Erlenhof' nested area, was classified as a meadow with a large number of species on low fertile soil of moderate moisture containing wet parts. The site was surrounded by forests containing medium to high isoprene emitters (*Quercus, Populus*). Mixing ratios of C_2 to C_6 hydrocarbons in ambient air ranged from non-detectable (50 pptC) to 7 ppb, with peak values of up to 9.5 ppb for *i*-pentane. Vertical gradients of the atmospheric hydrocarbons studied were either not observed or very small (± 17 ppt m^{-1}). No reproducible diurnal cycles in hydrocarbon mixing ratios were observed. Box measurements showed that the impact of the enclosed meadow on the hydrocarbon mixing ratios in ambient air was negligible. In the air above the spruce canopy of the BIATEX site Schachtenau, Bavarian Forest National Park, VOCs with chain lengths from C_1 to C_8 were analysed [8]. Up to sixteen compounds were quantified and their diurnal cycles recorded at 31 m and 51 m above ground (canopy height: 31 m). In general, the twelve most abundant NMHCs exhibited no distinct diurnal cycle. Only small differences between the two heights were detected, indicating that the impact of the spruce canopy on the NMHC mixing ratios in the air above the canopy was only small. Median values for the mixing ratios of both formaldehyde and acetaldehyde in ambient air above the canopy were found to be a factor of two higher during the day (1 ppb) than during the night (0.5 ppb). This difference suggests that the canopy may have acted as a source during the day, either through direct emission of aldehydes (Fig. 11.3.4) or emission of their unstable precursors (aldehydes are formed as relatively stable products of the photochemical degradation of unsaturated hydrocarbons such as isoprene and monoterpenes). Fluxes of monoterpenes at the same field site are given in a companion paper in this volume.

Over a pine/oak forest in the Mediterranean area (BEMA, Castelporziano), gradient measurements of selected atmospheric trace gases were performed [16]. The varying shape of the gradients reflected the complexity of transport and chemical processes in the air above the tall canopy at that site. The relatively high ethene and propene mixing ratios in the air above the canopy at certain times during the day may be caused either by emissions from the sea surface and/or emissions from traffic (roads and nearby airport). During a BIATEX field campaign in a mixed forest in a strongly polluted area [13] near Frankfurt/Main, Germany, canopy fluxes of selected anthropogenic (toluene, ethyl benzene, *m*-, *p*-xylene, *o*-xylene) and biogenic (α-pinene, β-pinene, Δ^3-carene, 1,8-cineol) hydrocarbons were determined in co-operation with U. Schickedanz and T. Dietrich, Zentrum für Umweltforschung, Universität Frankfurt [17]. During two sunny days in August, the flux behaviour of biogenic compounds changed from an emission of up to 80 pmol m^{-2} soil area s^{-1} on the first day to a deposition of up to 160 pmol m^{-2} soil area s^{-1} on the next day. Anthropogenic compounds were deposited at rates between 20 and 350 pmol m^{-2}, with significantly lower rates on the second day. This flux behaviour of selected VOCs over a forest may be explained by a non-homogenous distribution of NMHC sources in the main wind direction during the day. With easterly winds during the first day, the dominating trees in that direction were deciduous trees which are known as low monoterpene emitters. In the vicinity east of the tower, however, coniferous trees known as high monoterpene emitters were dominating. During the second day, westerly winds were prevailing. In this direction, conifers constitute the main portion of the trees. However, in the vicinity west of the tower, there is an oak plantation known to be a negligible monoterpene emitter. Large sources of anthropogenic VOCs (highway crossings, Frankfurt airport) are situated west of the tower. The results demonstrate that, in addition to other information, a detailed mapping of vegetation is a prerequisite to understand the VOC fluxes determined by gradient measurements over tall vegetation.

Conclusions

In the last years, the research performed in BIATEX within the framework of EUROTRAC reduced the uncertainties with respect to VOC emission/deposition from/onto vegetation. In different vegetational communities, typical for Europe, the emission spectrum of main VOC compounds was determined, and the main variables controlling the source strengths were identified. In summary: the aims set at the beginning of BIATEX have been reached. The data presented will help to improve the biogenic emission inventories of Europe and thus improve the input data for modeling tropospheric air chemistry. The improvement of the tools for studying exchange processes between biosphere and atmosphere and studies related to the physiology and biochemistry of the synthesis of VOCs in plants will further improve the construction of regional scale biogenic emission scenarios in a rapidly changing environment.

Acknowledgements

This work was funded by the German BMBF (EUROTRAC/BIATEX project no. 07 EU 720/2), and the CEC (STEP project no. 0060 and JRC BEMA project no. EV5V-CT93-0329, DG XII DTEE). We greatly acknowledge the technical assistance of M. Jocher, M. Wiedemann, I. Karbach, S. Tranos, numerous students, and the people of the park service of the Bavarian Forest National Park and of Castelporziano. We would like to especially thank G. Enders, Institute of Meteorology, University Munich, for his support during the intensive field campaign in the Bavarian Forest. The support of F. Meixner and co-workers, B. Karnbaum, and M. Schwaiger during the Manndorf wheat field experiment is also greatly acknowledged.

References

1. R. Steinbrecher, K. Stahl, J. Slemr, J. Hahn, *Abschlußbericht BMFT* 07 EU 720/2 (1992) pp. 73.
2. A.-M. Schreiner, Diplomarbeit, Lehrstuhl für Botanik der TU-München 1992.
3. R. Rabong, R. Steinbrecher, *BEMA annual report* 1994, Envirnoment Institute, JRC Ispra, Italy (in press).
4. R. Steinbrecher, W. Junkerman, M. Jocher, *EUROTRAC annual report 1992, Part 4, BIATEX*, EUROTRAC ISS, Garmisch-Partenkirchen 1993, pp. 135–144.
5. R. Steinbrecher, R. Rabong, in: P.M. Borrell, P. Borrell, T. Cvitaš, W. Seiler (eds), *Proc. EUROTRAC Symp. '94*, SPB Academic Publishing bv, The Hague 1994, pp. 502–506.
6. K. Lederle, Diplomarbeit, Fachhochschule Isny (1994) pp. 50.

7. R. Steinbrecher, I. Karbach, S. Nitz, H. Ziegler, in: P.M. Borrell, P. Borrell, T. Cvitaš, W. Seiler (eds), *Proc. EUROTRAC Symp. '92*, SPB Academic Publishing bv, The Hague 1993, pp. 284–288.
8. R. Steinbrecher, M. Jocher, M. Wiedemann, J. Slemr, J. Hahn, in: P.M. Borrell, P. Borrell, T. Cvitaš, W. Seiler (eds), *Proc. EUROTRAC Symp. '92*, SPB Academic Publishing bv, The Hague 1993, pp. 299–302.
9. R. Steinbrecher, G. Eichstädter, W. Schürmann, L. Torres, B. Clement, V. Simon, D. Kotzias, R. Daiber, J. Van Eijk, *Intern. J. Environ. Anal. Chem.* **54** (1994) 283–297.
10. J. Slemr, *J. Anal. Chem.* **340** (1991) 672–677.
11. G. Eichstädter, G. Dissertation, Technische Universität München, 1993.
12. J. Hahn, in: P.M. Borrell, P. Borrell, T. Cvitaš, W. Seiler (eds), *Proc. EUROTRAC Symp. '94*, SPB Academic Publishing bv, The Hague 1994, pp. 399–403.
13. W. Jaeschke, Th. Dietrich, U. Schickedanz, in: *EUROTRAC annual report 1993, Part 4, BIATEX*, EUROTRAC ISS, Garmisch-Partenkirchen 1994, pp. 102–108.
14. R. Dlugi, T. Anderson, Th. Foken, C. Frank, H. Lößlein, G. Roider, H. Schroers, J. Schween, Th. Seiler, B. Wichura, M. Zelger, in: *EUROTRAC annual report 1993, Part 4, BIATEX*, EUROTRAC ISS, Garmisch-Partenkirchen 1994, pp. 109–128.
15. R. Steinbrecher, K. Stahl, J. Hahn in: *The Chemistry of the Atmosphere: Its impact on global change*. CHEMRAWN VII World Conference, Baltimore, December 2–6, 1991, International Union of Pure and Applied Chemistry and The American Chemical Society, (1991) A9.
16. R. Steinbrecher, W. Junkermann, J. Steinbrecher, J. Slemr, in: P.M. Borrell, P. Borrell, T. Cvitaš, W. Seiler (eds), *Proc. EUROTRAC Symp. '94*, SPB Academic Publishing bv, The Hague 1994, pp. 506–511.
17. R. Steinbrecher, S. Tranos, *Tätigkeitsbericht 1994*, IFU Garmisch-Partenkirchen, (in press).

11.4 Monoterpene and Isoprene Emission in Norway Spruce Forests

Rainer Steinbrecher[1], H. Ziegler[2], G. Eichstädter[2], U. Fehsenfeld[1], R. Gabriel[2], Ch. Kolb[1], R. Rabong[1], R. Schönwitz[2] and W. Schürmann[2]

[1]Fraunhofer-Institut für Atmosphärische Umweltforschung (IFU), Kreuzeckbahnstr. 19, D-82467 Garmisch-Partenkirchen, Germany
[2]Institut für Botanik und Mikrobiologie der Technischen Universität München, Arcisstr. 21, D-80333 München, Germany

Summary

The isoprenoids isoprene and monoterpenes are hydrocarbons with important ecological functions. In conifers monoterpenes are synthesised mainly for protection against insects. In a *ca.* 90 year old Norway spruce stand in the Bavarian Forest National Park, Germany, and on the Wank mountain near Garmisch-Partenkirchen, Germany, the formation of isoprene and monoterpenes in needles of *Picea abies* [L.] *Karst.*, the emission into the atmosphere, the diurnal and seasonal cycles of emission and immission, the horizontal and vertical distribution of the source strength in and above the canopy and the fate of isoprenoids in the atmosphere were studied. The work was performed together with meteorologists and air chemists. Controlling factors in both, isoprene emissions and monoterpene emissions from intact needles are leaf temperature and photosynthetic active radiation (PAR). On the basis of these two parameters emission factors and emission algorithms were calculated. The sun crown of the trees is the dominating source for isoprenoids in the forest during the physiologically active period of the year. In 1991, 88 % of the monoterpens were emitted in July and August. The contribution of the trunk compartment to total forest emission ranges from 1 % to 64 % and is not clear yet. The contribution of the soil is negligible (3 %). The emission pattern and the emission rates of monoterpenes dramatically change after injuring of needles and trunks by opening the resin ducts. The α-pinene fluxes of the forest calculated using the gradient approach and the enclosure technique are similar and indicate that the sun crown of the forest is the main source of the isoprenoids transported in upper air layers. Around noon, however, the gradient approach resulted in deposition of α-pinene, although this compound is emitted from the twigs in large amounts due to high temperature and high solar insulation. At the site the ozone forming potential through photochemical isoprenoid degradation should be of minor importance, since the terpene/NO_x ratio in ambient air is mostly < 1 during the day. The interdisciplinary investigations of reactive trace gases released into the atmosphere by Norway spruce ecosystems demonstrate the complexity of the interactions soil, tree, canopy, atmosphere, man and animal.

Aims of the research

- Method development of enclosure techniques for the investigation of isoprene and monoterpene emissions from twigs, trunks and soil under natural conditions.
- Method development of GC techniques for the analysis of isoprene and monoterpenes in ambient air samples.
- Investigation of the source strength of isoprene and monoterpenes in the crown, trunk and soil compartment of Norway spruce forests.
- Calculation of emission factors and emission algorithms for monoterpenes from Norway spruce.
- Evaluation of the enclosure methods and the gradient approach for determining the monoterpene emission from Norway spruce forests.

Principal scientific results

Method developments

Enclosure techniques

A climate controlled cuvette system, flushed with purified air was used to study the emission of isoprene, monoterpenes and the CO_2/H_2O gas exchange of a living attached spruce twig [1,2,3]. Stainless steel, aluminium, glass and Teflon were used in the construction of a cuvette (volume 0.5 L) in order to avoid contamination of the air passing through it. The cuvette was flushed with purified air at a flow rate of 1 L min^{-1}. All air flows were controlled by mass flow meters. A part of the cuvette outlet air was passed through the sampling device for isoprene and monoterpene analysis. Differences in mixing ratios of CO_2 and H_2O between inlet and outlet air of the cuvette were measured with an infrared gas analyser. After A/D transformation the various signals were used as input variables in a program calculating the parameters: net CO_2 assimilation A, transpiration E, leaf to air water vapour pressure deficit ΔW, leaf conductance for water vapour g_l, internal CO_2 concentration c_i [4]. Emission rates and gas exchange parameters were calculated on a whole surface area basis. Total leaf area of the spruce twig was determined by multiplying the projected area by 2.74 [5].

For studying the cone emission and trunk emission of Norway spruce in the laboratory, the open-flow gas exchange chamber system as described above was modified [6]. For studies of the undisturbed trunk emission of spruce in the field, a special open-flow Teflon enclosure was designed [7]. An aluminium ring coated with a silicon band was attached carefully to the bark of trees. A Teflon cylinder was fixed at the aluminium frame and an aluminium plate with a fan ensuring complete mixing of the enclosure volume at an air flow of 6 L min^{-1}. In order to stabilise the enclosure system the aluminium plate was mounted on a tripod.

Soil emission measurements were performed in a modified open-flow gas exchange device with a specially developed soil enclosure [8]. A soil area of 0.07 m^2 was covered by the enclosure which consisted of two aluminium rings connected by a viton o-ring and covered by a Duran glass cylinder to allow light incidence. Complete mixing of the cuvette volume was achieved by a ventilator resulting in a wind velocity of *ca.* 1 m s^{-1} at a cuvette flow of 12 L min^{-1}.

Sampling techniques

Monoterpene mixing ratios in air samples were analysed after preconcentration on TENAX TA. Isoprene mixing ratios were determined after preconcentration of air samples on mixed adsorbent beds (TENAX TA/Carbosphere, Carbosieve III/Carbotrap/Carbotrap C). The tubes were either mounted in an automated sampling device for subsequent loading of 12 adsorption tubes using defined sampling times and mass controlled sampling flows or mounted directly on the gaschromatograph for on-line analysis. Details are described elsewhere [1–3, 7–10]

GC and GC/MS techniques

Adsorbed compounds were thermally desorbed and flushed with oxygen free helium into a liquid nitrogen cooled pre-column situated in a programmed temperature vaporiser (PTV). Cryo-focused compounds were separated on different capillary columns mounted in various gas chromatographs (GC) and detected by flame ionisation (FID). Details of the analysis are described elsewhere [1–3, 7–10].

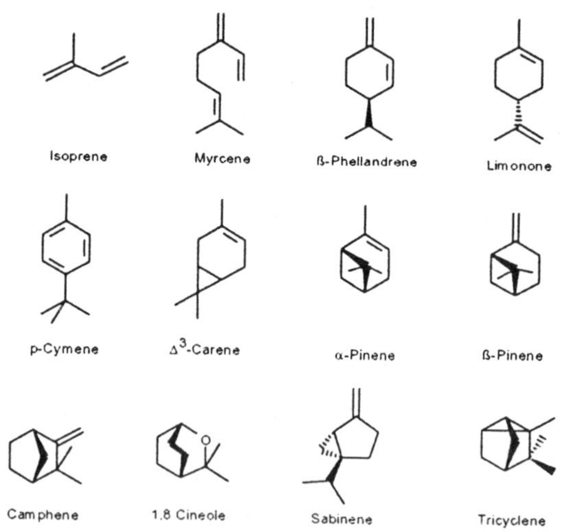

Fig. 11.4.1: Structures of selected isoprenoids present in air samples of a Norway spruce forest.

Isoprenoids were identified by mass spectroscopy (MS) and coinjection of reference substances detected by FID [1, 3, 9, 10]. For structures see Fig. 11.4.1. For preparation standards of monoterpenes, a permeation device was used [1, 3].

The variation coefficients for the calibration of the different monoterpenes were better than 15 % (n = 10). The quality of quantification was recently ensured by an European intercomparison experiment for monoterpenes in air samples [3]. The quantification of isoprene was accomplished with external gaseous standards (10 ppb) [10]. The accuracy of the whole sampling and analysis device was ± 2 % (standard deviation of the mean mixing ratio in the ppb range; n = 10). The quality of the isoprene analysis was checked by several international intercomparison exercises (unpublished results Eric Apel, NCAR, Boulder, Colorado; Paolo Cicioli, CNR, Rome).

Isoprenoid emission in the crown compartment

Isoprene emissions of attached spruce twigs were studied in the laboratory and in the field at Schachtenau in the Bavarian Forest National Park [9] and on the Wank mountain, Garmisch-Partenkirchen [10]. Climate chamber experiments revealed a strong correlation of the isoprene emission rate with PAR and leaf temperature. Increasing PAR and leaf temperature levels enhanced emission rates until saturation, followed by a decline in the emission under extreme conditions (PAR = 2560 μE, leaf temperature = 38 °C). Field measurements confirmed the strong correlation between the isoprene emission rate and PAR and leaf temperature (Fig. 11.4.2). The isoprene emission rates exhibited a distinct diurnal cycle very similar to daily PAR and leaf temperature variation with minimum levels near the detection limit (< 2 pmol m^{-2} s^{-1}) during the night, a continuous increase starting at sunrise to maximum values of 650 pmol m^{-2} s^{-1} in the early afternoon (under favourable PAR and leaf temperature conditions on summer days), followed by a decline back to the night levels. Similar runs with much lower emission rates were observed on cloudy or hazy summer days or cold and sunny autumn days. The mean isoprene emission rate observed in the field under standard conditions (900 μE PAR, 23 °C leaf temperature) was 237 pmol m^{-2} s^{-1} ranging from 108 pmol m^{-2} s^{-1} to 364 pmol m^{-2} s^{-1}. Under field conditions the isoprene emission showed no tendency to saturate. Furthermore no decline in the emission rate was observed for increased light intensities and temperature [10]. Plotting the isoprene emission rate versus net CO_2 assimilation demonstrates that there is no obvious correlation between the two parameters. No correlation between isoprene emission plotted versus leaf conductance for water vapour was observed [10]. These observations underline the importance of leaf temperature and PAR as controlling variables for the isoprene emission of Norway spruce needles.

Monoterpene emission from Norway spruce twigs was intensively studied at the field site Schachtenau in the National Park of the Bavarian Forest on 85–122 year old trees [1, 2, 11]. As expected the emission rate of monoterpenes is controlled through numerous endogenous and external factors.

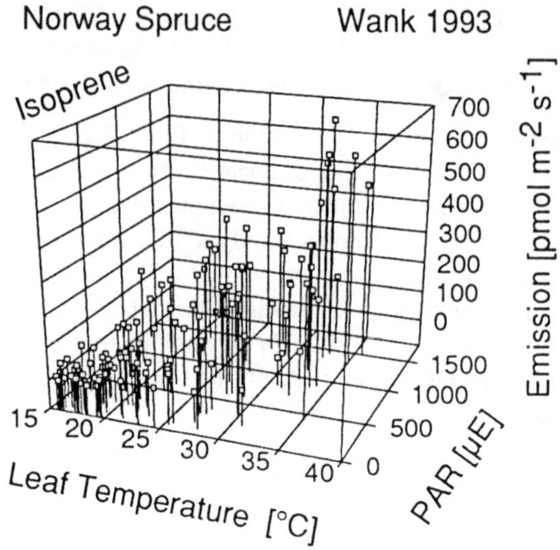

Fig. 11.4.2: Isoprene emission of Norway spruce twigs under various environmental conditions. From August to October 1993 seventeen diurnal cycles of isoprene emission were recorded from ten different southwesterly exposed current year twigs of a 180 year old Norway spruce tree. The emission rates are related to total leaf area and plotted against leaf temperature and photosynthetically active radiation (*PAR*) given as photon flux density E (mol photons m^{-2} s^{-1}).

The variation of monoterpene emission between three trees and 14 twigs of the same age of field grown spruce showed coefficients of variation between 11 % for α-pinene and 44 % for β-pinene. Both compounds account for 42 % of the total monoterpenes emitted. See Fig. 11.4.1 for other compounds emitted. The coefficients of variation of the relative amounts of the different monoterpenes emitted range from 10 % for α-pinene to 29 % for β-pinene.

The seasonal needle age is of importance only for new needles. Both absolute and relative amounts change during needle development. Before bud burst monoterpene emission is low. Limonene and 1,8 cineole are the main compounds. When the needles are fully developed the emission remains more or less constant, α-pinene being the main compound in the emission spectrum.

Twigs in the shadow crown may emit only a tenth of the α-pinene peak values observed for sun twigs around noon at similar leaf temperatures (Fig. 11.4.3). The diurnal cycle of the α-pinene emission clearly follows the course of the PAR. The effect of light on other monoterpenes is not so well developed. In Fig. 11.4.4 all field emission data of α-pinene are plotted against leaf temperature and PAR.

It is obvious that α-pinene emission is controlled not only by leaf temperature but also by light (PAR: Spearman rank correlation coefficient r_s of 0.638; n = 123; leaf temperature: r_s 0.534; n = 121). Light and temperature may control the monoterpene emission through the pool size and/or the stomatal opening, through the isoprenoids are released from the leaves. A change in stomatal conductance, however, induced by water stress or abscisic acid fed through the xylem does not change the source strength of α-pinene from needles [2, 12].

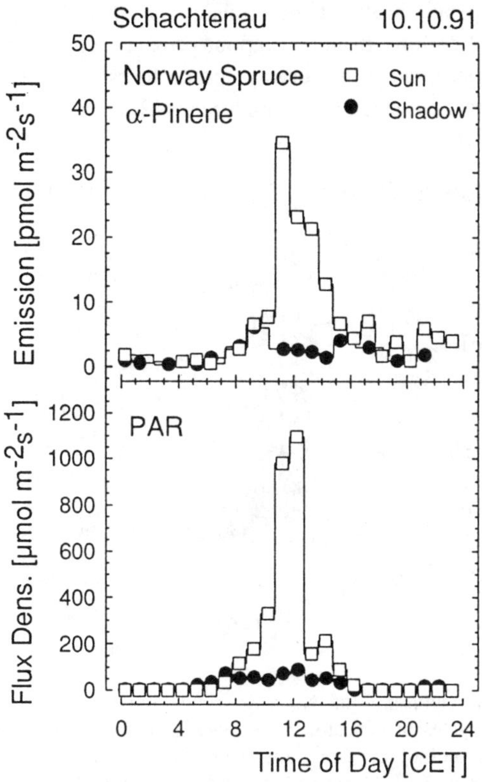

Fig. 11.4.3: Diurnal cycle of α-pinene emission rates of twigs in the sun crown (26 m above ground level) and shadow crown of a ca. 90 year old Norway spruce tree. The PAR levels in the sun crown and shadow crown are also shown. The temperatures of needles from the investigated twigs in the sun crown and shadow crown were similar.

A direct link between endogenous needle monoterpenes and the monoterpenes emitted was not obvious, except when the needles were damaged [1, 2]. This observation was proven using the ^{13}C isotope technique. After fumigation with $^{13}CO_2$, labelling was only found in monoterpenes emitted and not in the endogenous monoterpene pool. Further, this experiment indicates that the monoterpenes emitted are synthesised directly from photosynthetic products [13].

The carbon loss through a monoterpene emission of 209 pmol m^{-2} s^{-1} during a photosynthetic C-fixation of 3.3 µmol CO_2 m^{-2} s^{-1} amounts to 0.2 %.

In the emission spectrum of cones of Norway spruce, eight monoterpenes were detected and quantified [6]. Isoprene could not be detected. The emission rates are related to cone surface area. In the studied temperature range (15–35 °C), *β*-pinene (25–125 pmol m^{-2} s^{-1}), α-pinene (10–60 pmol m^{-2} s^{-1}) and limonene (5–30 pmol m^{-2} s^{-1}) were found as major compounds, followed by 1,8-cineole, myrcene, *β*-phellandrene, camphene and sabinene in minor amounts. For all detected monoterpenes a direct relationship of the emission rates from the temperature, and, therefore, from the vapour pressure, could be proven. The cones in the crown compartment of a Norway spruce forest may only contribute to *ca.* 2 % to the total forest emission of monoterpenes [2].

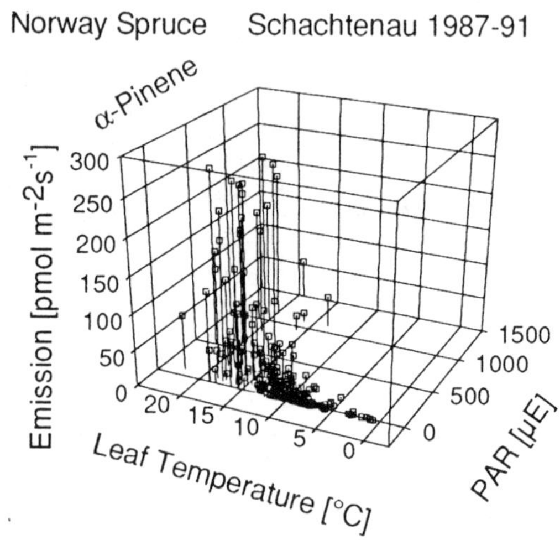

Fig. 11.4.4: α-pinene emission of Norway spruce twigs under various environmental conditions. Diurnal cycles of emission rates of twigs from different trees and at different position on the tree and of different age were recorded at the field site Schachtenau, Bavarian Forest National Park, during the years from 1988 to 1991. The emission rates are related to total leaf area and plotted against leaf temperature and PAR.

Isoprenoid emission in the trunk compartment

The emission of monoterpenes through the cortex of unwounded trunks showed considerable variation from tree to tree [6, 7]. Isoprene was not detected in the emission spectrum. Eight year old cloned spruce trees emitted α-pinene, the major compound in the trunk emission pattern of monoterpenes, at a rate of 0.01 nmol m^{-2} trunk surface s^{-1} in the summer (Fig. 11.4.5). The emission rates of the other monoterpenes, limonene, β-pinene, and myrcene were below 4 pmol m^{-2} s^{-1}. In general, the relative amount of monoterpenes emitted from trunks of field grown spruce are comparable to those of cloned trees in the climate chamber. Over 75 % of the total monoterpenes emitted are attributed to α-pinene and β-pinene. Emission studies on undisturbed trunks in the field in spring, however, revealed emission rates which were higher by a factor of 28–185 compared to the rates of cloned tress. The significantly higher emissions rates in the field may be due to different genetic disposition and/or differences in the morphology of the cortex and thus different diffusion resistances for monoterpenes. The monoterpene emission *M* is well correlated to the bark temperature *T*. The coefficients of determination for the linear relationship between ln*M* and 1/*T* of α-pinene and β-pinene are generally in the range of 0.57–0.87. Diurnal cycles of the monoterpene emission recorded at similar bark temperatures for sunny and cloudy days revealed comparable emission rates of 2 nmol total monoterpenes m^{-2} s^{-1} in the morning ($T = 9.5$ °C) and 4 nmol m^{-2} s^{-1} in the evening (18.9 °C) but a factor of two higher emission rates in the early afternoon (cloudy: 3 nmol m^{-2} s^{-1}; T = 18.4 °C; sunny: 7 nmol m^{-2} s^{-1}; T = 18.9 °C). This result may be explained by the fact that especially during sunny days in the early afternoon more assimilatory power and other products of the photosynthesis are available compared to cloudy days. It may be speculated that enhanced precursor levels and/or enhanced levels of assimilatory power are partly used for an increased synthesis of monoterpenes which in turn may also influence the source strength of monoterpene emission from trunks through still unknown processes. The contribution of undisturbed trunk emission of α-pinene to total canopy emission ranged from 1 % to 64 % (20 °C; 1,000 μE PAR) and is not yet clear.

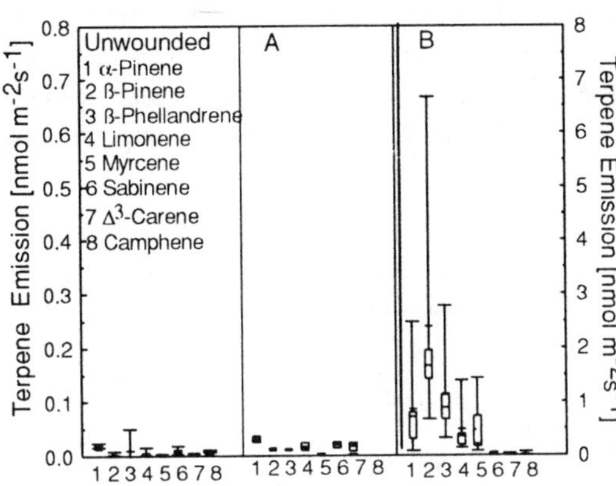

Fig. 11.4.5: Monoterpene emission of five year old trunk segments of eight year old cloned Norway spruce trees. The emission rates are related to trunk surface area and trunk surface area + unjured area, respectively, eight hours after wounding at 20 °C surface temperature. A: injured trunk without resin flow. B: injured trunk with resin flow. Shown are the box plots of seven trees with the 65 and 95 percentile, respectively, and the median.

Injury to the bark, resulting in resin loss, dramatically increases the monoterpene emission and changes the relative amounts of the different monoterpenes emitted significantly (Fig. 11.4.5). Eight hours after wounding, causing a small resin droplet, the level of β-pinene emission was still a factor of 80 higher; the emission of β-phellandrene, however, was higher by a factor of 366 compared to non-injured trunks. Therefore, injuries to the bark inflicted by lumbering or deer biting change the source strength and the spectrum of the monoterpenes emitted in the trunk compartment of a Norway spruce forest.

Isoprenoid emission in the soil compartment

The soil emission in Norway spruce forests was investigated on soils from the Bavarian Forest and the Högel Forest, near Augsburg, Germany [8, 14, 15]. Laboratory studies were performed on soils from the Högel Forest. In the field, the soil emission was studied at the site Schachtenau in the Bavarian Forest National Park. The soil of the investigated plots was covered mainly with brownish spruce needles. Living vegetation was not present. It is assumed that in a dense Norway spruce forest underbrush contributes only a minor fraction of the total biomass density to the forest. Therefore, underbrush may contribute only negligible amounts to the total forest emission of monoterpenes. Ten monoterpenes, which are also present in discarded needles were detected in emission from the soil (for structures see Fig. 11.4.1). The main compounds are, limonene (25 %), α-pinene (22 %), and camphene (16 %). The chromatogram pattern from the soil emission is comparable to the endogenous monoterpene content of the upper soil layer but differs significantly from that of a physiologically active needle. Isoprene was not detected in the emission spectrum of the soil. The sum of monoterpenes emitted by the soil ranges from 4–340 pmol m^{-2} soil area s^{-1} (n = 288) at a temperature range from 1 °C to 21 °C. The main controlling variables are assumed to be plot structure and temperature. In the field, monoterpene emission rates M are well correlated with soil temperature T. For the major compounds the coefficients of

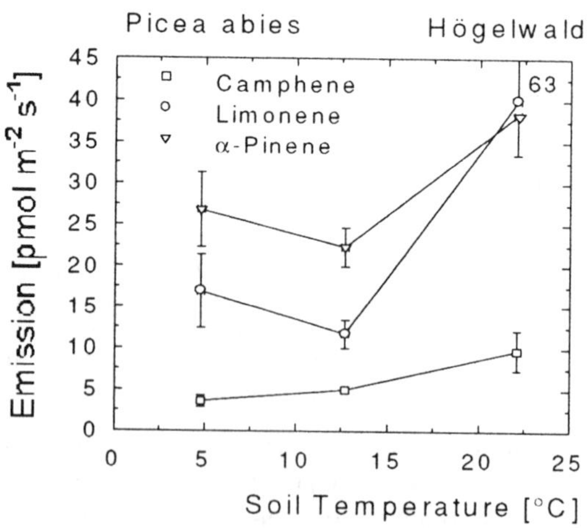

Fig. 11.4.6: Monoterpene emission and soil temperature in 1 cm depth of soils from a Norway spruce forest. The soil humidity ranged from 62 % and 74 %, the air humidity ranged between 47 % and 78 %, and the PAR was 232 μE. Given are the mean emission rates and the standard deviation of five determinations of four plots.

determination in the linear relationship of ln*M* and 1/*T* range from 0.61–0.80 (n = 288). At a soil temperature of 20 °C a mean α-pinene emission rate of 41 pmol m^{-2} soil area s^{-1} is calculated. On soil plots from the Högel Forest the dependence of monoterpene emission from various environmental factors was investigated in a climate chamber (Fig. 11.4.6). Soil humidity (52 %–74 %) and ambient air humidity (24 %–74 %) did not have an important influence on the monoterpene emission rates from the soils. A marked increase of monoterpene emission was observed at temperatures between 12 °C and 23 °C. The expected exponential increase of the emission rate between 5 °C and 12 °C soil temperature, however, was not observed. This result points to the fact that in addition to temperature other factors like the activity of micro organisms and macro organisms may also trigger the monoterpene emission rate from the soil. On a sunny and warm day (20 °C soil surface temperature; 1,000 µE PAR), the soil in a Norway spruce forest may contribute less than 3 % to total forest emission of α-pinene.

Evaluation of the enclosure methods and the gradient approach for determining monoterpene emission rates of Norway spruce forests.

The monoterpene flux of a forest was calculated using the gradient approach GRAD [2]. The terrain fullfils the micro meteorological requirements for GRAD trace gas flux calculation at the Schachtenau site [11]. Gradient measurements and enclosure measurements were performed at the same time in order to compare the different methods used for forest flux calculation of α-pinene (Fig. 11.4.7). GRAD fluxes at a mean height of 31m and 41m were similar in their diurnal cycle. Emission rates of up to 700 pmol α-pinene m^{-2} soil area s^{-1} were calculated during the day and no flux was observed at night. The GRAD emission rates were generally comparable to the emission rates calculated on the basis of enclosure measurements from twigs in the sun crown of trees. During the day around noon, however, the GRAD showed deposition of α-pinene whereas the α-pinene was emitted in large amounts by the twigs. This observation was possibly due to chemical processes at the upper edge of the canopy leading to an enhanced degradation of the emitted compound or due to a complex meteorological situation at the site resulting in a violation in the GRAD requirements at that time. The results indicate that α-pinene, which is transported in upper air layers of the troposphere originates mainly from the sun crown of the canopy. Monoterpenes emitted from the soil or the trunks do not seem to be transported in air layers above the forest. It should be noted that the emission rates of the two twigs in the sun crown of two different trees are more or less similar, pointing to the fact that the genetic variation may not be as important as believed and that environmental factors like temperature and PAR are the dominating variables controlling α-pinene emission from Norway spruce.

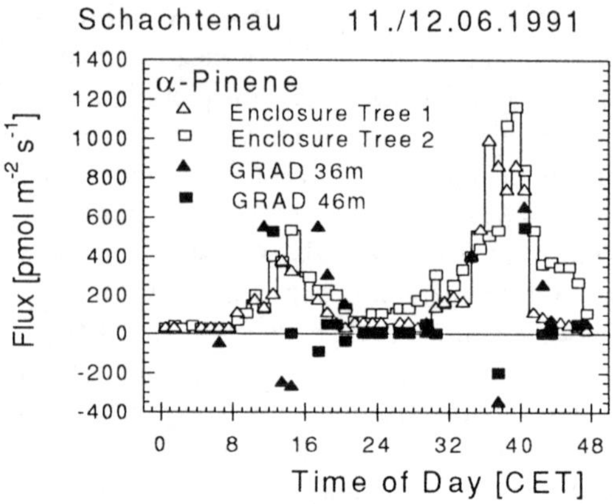

Fig. 11.4.7: α-pinene fluxes of a Norway spruce forest. Fluxes were calculated either using the gradient approach GRAD or using enclosure measurements. GRAD: two flux values were calculated on the basis of gradients in ambient air concentration between 31 m and 41 m and 41m and 51m above ground, respectively. Enclosure: the flux values were calculated on the basis of the emission of two twigs in the sun crown of two different trees. Enclosure fluxes are related to soil area (LAI 16.7). It is assumed that only the sun crown fraction (80 % of total leaf area) of the canopy is responsible for the total forest emission.

Calculation of emission factors and emission algorithms for Norway spruce

The field measurements of monoterpene emission rates in the Bavarian Forest performed between 1987 and 1991 were used to test a recent emission algorithm for monoterpenes from plants [16–18]. In the emission model MT1 the monoterpene emission rate M at a given leaf temperature T [K] is expressed as a function of the monoterpene standard emission rate M_S at the standard temperature T_S [K] and an empirical coefficient $\beta = 0.09$.

$$\text{MT1:} \qquad M = M_S \exp(\beta\,(T - T_S)) \tag{1}$$

Based on the field emission data, a standard emission of 17.15 pmol α-pinene m^{-2} total leaf area s^{-1} at a standard temperature of 15 °C was calculated. With the MT1 model (Eq. 1) only 68 % of the total variation of the α-pinene emission from Norway spruce could be explained. The summarised field α-pinene emission data from Norway spruce at various environmental conditions between 1987 and 1991 (Fig. 11.4.4) clearly showed a light effect in addition to the temperature effect, especially at leaf temperatures above 10 °C. On the basis of the field measurement data, a two parameter multiple regression model named LR2 (Eq. 2) with leaf temperature T [°C] and the natural logarithm of PAR [μE] was established in order to increase the predictability of α-pinene emission.

$$\text{LR2:} \qquad \ln M = -2.40 + 0.22\,T + 0.41 \ln PAR \tag{2}$$

The LR2 model is able to explain up to 77 % of the observed variation within the α-pinene emission data from Norway spruce. More variables in Eq. (2), like net

CO_2 assimilation, leaf conductance for water vapour, transpiration, or leaf to air water vapour pressure deficit, did not significantly increase the predictability of the model.

Both models, MT1 and LR2, were used to estimate the monthly mean potential emission rates of α-pinene for the Norway spruce trees in the Bavarian forest. In Fig. 11.4.8, both model estimates are shown. At leaf temperatures near 15 °C (used as standard temperature in the MT1 model) both models give similar estimates. But at leaf temperatures above and below the standard temperatures in summer or winter, the results of the two different models exhibit considerable deviations, due to the limitations of the MT1 model. In July the mean α-pinene emission rate is estimated at 34 pmol m^{-2} leaf area s^{-1} by the LR2 model, whereas

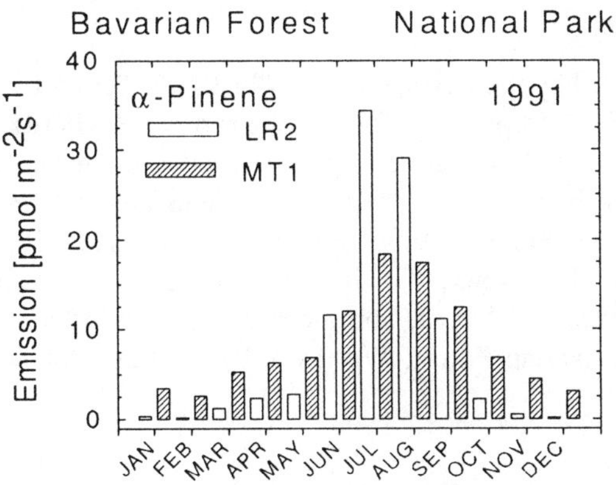

Fig. 11.4.8: Estimates of α-pinene emission related to leaf area for Norway spruce in the Bavarian Forest. Input parameters for the emission models MT1 and LR2 were mean monthly day temperature and mean monthly day *PAR* for 1991 [2]. It is assumed that the mean monthly day leaf temperature equals the mean monthly day ambient air temperature.

the MT1 model predicted only 19 pmol m^{-2} leaf area s^{-1}. In January the LR2 model predicted 0.5 pmol α-pinene m^{-2} s^{-1}, whereas the MT1 model estimated 3.5 pmol α-pinene m^{-2} s^{-1}. The new algorithm for α-pinene emission from Norway spruce which is based on field data containing PAR and temperature as controlling factors significantly improves the emission estimates of α-pinene for Norway spruce compared to the results of a model incorporating only temperature.

Conclusions

In the last years during EUROTRAC work in the subproject BIATEX a lot of progress has been made with respect to our understanding how isoprenoid emission of Norway spruce forests, one of the dominant forest types in northern Eurasia, is controlled and the original aims set at the beginning for BIATEX have been reached. Several internal and external factors have been intensively studied in the laboratory and the field, leading to the formulation of an emission algorithm for the dominating monoterpene α-pinene emitted by Norway spruce forests. The presented emission data and emission algorithm should help to reduce the large uncertainties in present biogenic VOC estimates for Europe used in oxidant models [19–21]. Further research will help to clarify the genetic, physiological, and biochemical basis of the isoprenoid biosynthesis possibly leading to revised emission algorithms. Ongoing and future field work in a variety of ecosystems of the world will help to establish more accurate vegetation specific standard emission factors and revised emission algorithms for monoterpenes. With the progress made in canopy modelling and remote sensing, the uncertainties in isoprenoid emission inventories on local, regional and global scales will further be reduced.

Acknowledgements

This work was funded by the German BMBF No. 07 EU 720/2, 07 EU 722, 07 EU 812, the Bavarian State Department for Science and Art BayFORKLIM B II 3 and the CEC STEP No. 0060. We gratefully acknowledge the technical assistance of M. Jocher, I. Karbach, S. Tranos, numerous students and the people of the park service of the Bavarian Forest National Park. We would especially like to thank G. Enders, W. Breit, P. Dreier, G. Jacoby and U. Teichmann, Institute of Meteorology, University Munich, for their support during the intensive field campaigns in the Bavarian Forest. We thank Dr. P. Haug for comments on the manuscript.

References

1. R. Steinbrecher, Dissertation TU-München, Fakultät Chemie, Biologie, Geowissenschaften, 1989.
2. W. Schürmann, Dissertation TU-München, Fakultät Chemie, Biologie, Geowissenschaften, 1993.
3. R. Steinbrecher, Eichstädter G., Schürmann W., Torres L., Clement B., Simon V., Kotzias D., Daiber R., Van Eijk J., *Intern. J. Environ. Anal. Chem.* **54** (1994) 283–297.
4. T.J. Ball, in: Zeiger *et al.* (eds), *Stomatal Function*, Stanford University Press, California 1987, pp. 445–467.
5. M. Riederer, Kurbasik K., Steinbrecher R., Voss A., *Trees* **3** (1988) 165–172.
6. R. Gabriel, Diplomarbeit, Lehrstuhl für Botanik, TU-München, 1991.
7. Ch. Kolb, Ch. Diplomarbeit Fachhochschule Isny, 1994.
8. G. Eichstädter, Dissertation, TU-München Fakultät Chemie, Biologie, Geowissenschaften, 1993.
9. R. Steinbrecher, Karbach I., Nitz S., Ziegler H., in: P.M. Borrell, P. Borrell, T. Cvitaš, W. Seiler (eds), *Proc. EUROTRAC Symp. '92*, SPB Academic Publishing bv, The Hague 1993, pp. 284–288.

10. R. Steinbrecher and Rabong, R., in: P.M. Borrell, P. Borrell, T. Cvitaš, W. Seiler (eds), *Proc. EUROTRAC Symp. '94*, SPB Academic Publishing bv, The Hague 1994, pp. 502–506.

11. G. Enders , Dlugi R., Steinbrecher R., Clement B., Daiber R., Eijk v. J., Gäb S., Haziza M., Helas G., Herrman U., Kessel M., Kesselmeier J., Kotzias D., Kourtidis K., Kurth H. - H., McMillen R. T., Roider G., Schürmann W., Torres L., *Atmos. Environ.* **26A** (1992) 171–189.

12. H. Ziegler, and Steinbrecher R., in: Rundgespräche der Komission für Ökologie Bd. 8 'Klimaforschung in Bayern', Verlag Dr. Friedrich Pfeil, München, 1994, pp. 95–102.

13. W. Schürmann, Ziegler H., Kotzias D., Schönwitz R., Steinbrecher, R., *Naturwissenschaften* **80** (1993) 276–278.

14. U. Fehsenfeld, Diplomarbeit, Lehrstuhl für Botanik, TU-München, 1992.

15. G. Eichstädter, Schürmann W., Steinbrecher R., Ziegler H., in: P.M. Borrell, P. Borrell, T. Cvitaš, W. Seiler (eds), *Proc. EUROTRAC Symp. '92*, SPB Academic Publishing bv, The Hague 1993, pp. 276–282.

16. A.B. Guenther, Zimmerman P.R., Harley P.C., Monson R.K., Fall R., *J. Geophys. Res.* **98D** (1993) 12609–12617.

17. A. Guenther A., Hewitt N.C., Erickson D., Fall R., Geron C., Graedel T., Harley P., Klinger L., Lerdau M., McKay W.A., Peirce T., Scholes B., Steinbrecher R., Tallamraju R., Taylor J., Zimmerman P., *J. Geophys. Res.* **100** DS (1995), 8873–8892.

18. R. Steinbrecher, Schürmann W., Schreiner A.-M., Ziegler H., in: J. Slanina *et al.* (eds), *Proc. Joint CEC/BIATEX Workshop, General Assessment of Biogenic Emissions and Deposition of Nitrogen Compounds, Sulfur Compounds and Oxidants in Europe.,* CEC Environmental Research Programme Report **47** (1993) 251–261.

19. R. Steinbrecher, in: Baars, H.P., *et al.* (eds), Proc. TNO/EURASAP Workshop on the Reliability of VOC Emission Databases, TNO Delft X, IMW-TNO Publication P93/040 1993, pp. 89–98.

20. R. Steinbrecher, in: in: P.M. Borrell, P. Borrell, T. Cvitaš, W. Seiler (eds), *Proc. EUROTRAC Symp. '94*, SPB Academic Publishing bv, The Hague 1994, pp. 448–455.

21. D. Simpson D., Guenther A., Hewitt C.N., Steinbrecher R., *J. Geophys. Res.* in press.

11.5 Emissions of VOCs from Stressed and Unstressed Vegetation

R.A. Street, S.C. Duckham, C. Boissard and C.N. Hewitt

Institute of Environmental and Biological Sciences, Lancaster University, LA1 4YQ, UK.

Summary

During 1989–1994 the emission rates of VOCs (primarily isoprene and monoterpenes) from various plant species were assessed by different sampling methods, including the use of excised plant material (leaf disc method) and a dynamic flowthrough branch enclosure, developed during the course of this work. Both stressed and unstressed plants have been studied.

Various tree (conifer and broadleaf), crop and herbaceous UK species were screened (leaf disc method), as were 19 typical Mediterranean species (enclosure method). All plants sampled with the enclosure method have been screened for their tendency to emit oxygenated compounds (BOVOCs). For example, maize emitted negligible amounts of isoprene and monoterpenes but various aldehydes and ketones.

The effects of both elevated ozone (O_3) and carbon dioxide (CO_2) have been investigated. Elevated ozone levels had an inconclusive effect on emission rates. On the other hand, exposure to elevated levels of CO_2 resulted in increased emissions of VOCs.

Relationships between environmental conditions and emission rates were studied. Of the UK species, emissions of isoprene from Sitka spruce, gorse and oak, and of monoterpenes from gorse, demonstrated a relationship with temperature. Relationships to the PAR regime were inferred, but not directly quantified, due to the uncontrolled nature of field measurements made under ambient conditions. Monoterpene emissions from the Mediterranean species *Quercus ilex* demonstrated a better relationship with light than temperature. Emissions from *Eucalyptus globulus* were exponentially related to temperature.

Comparisons between laboratory samples, field samples and different ages of vegetation, have revealed differences in chemical speciation and absolute emission rates, which may be a result of their different growth environments and physiology. It was concluded that extrapolations from a few samples, or those based on laboratory measurements, must be treated with caution.

Aims of the research

The principal aim of Lancaster's contribution to BIATEX was to study emission rates of VOCs from stressed and unstressed European plant species using field and laboratory experiments.

Principal scientific results

Controlled experiments

Ozone (O_3)

Eucalyptus leaves were exposed to ~ 120 ppb O_3 for 3–4 days on two different occasions, and demonstrated both higher and lower monoterpene emissions than the controls. In a longer term experiment, oak trees were exposed to higher than ambient ozone for 6 months, with no observable effects on VOC emission rates.

Carbon dioxide (CO_2)

Sitka spruce was exposed to 600 ppm for 9 months. Treated trees (sampled using incubation of needles) demonstrated a higher mean monoterpene emission rate. Oak trees were exposed to 600 ppm for 4 months. Isoprene emissions from treated trees (sampled with the branch enclosure) were elevated with respect to the controls (see Fig. 11.5.1).

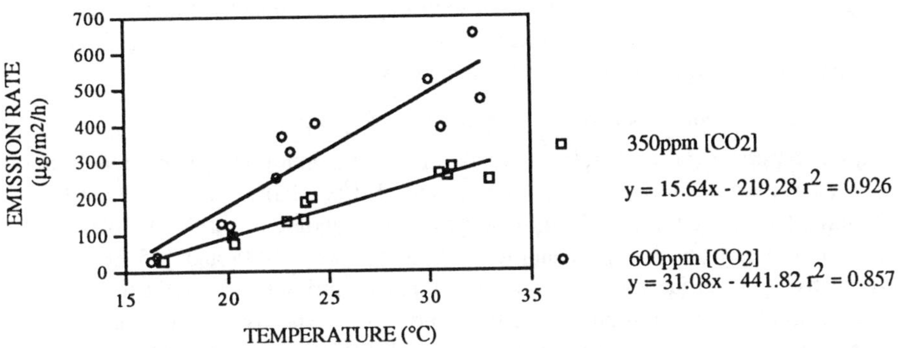

Fig. 11.5.1: Isoprene emissions from young oak (*Quercus petraea*) trees grown in a CO_2 enriched atmosphere.

Temperature - A Sitka spruce seedling was exposed to a range of temperatures between 21 and 35 °C in a controlled environment chamber. Isoprene demonstrated an exponential relationship with temperature which broke down at about 33 °C (Fig. 11.5.2).

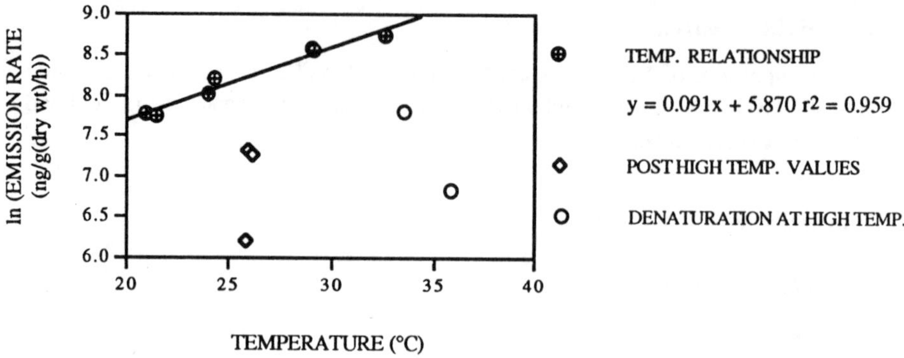

Fig. 11.5.2: Isoprene emissions from Sitka spruce seedlings (*Picea sitchensis*) subjected to different temperatures within a controlled environment chamber.

Field measurements

Northern European, Sitka spruce plantation, Grizedale forest, NW England.

Principal emissions from various aged Sitka spruce trees were identified as isoprene, α–pinene, β–pinene, myrcene, and a 5th peak, tentatively identified β–phellandrene. The relative proportion of isoprene and myrcene was changeable, but together they always represented at least 67 % of the emission rate. Total mean daily emission rates ranged between 42 and 2167 ng g^{-1} h^{-1}, and were more abundant from a young tree (7 years old) than an older tree (40 years old) regardless of the temperature and light regimes. It was concluded that in addition to environmental influences, physiological and seasonal variations between branches may be important in determining emission rates.

Northern European, Sitka spruce plantation, Rivox forest, SW Scotland

Emission rates from two heights in a Sitka spruce plantation were measured during one 24 h period in July 1993 as part of the NERC TIGER project. The VOCs identified were the same as from Sitka spruce at Grizedale forest. Mean total daily emission rates from the upper and lower branches were 1083 and 145 ng g^{-1} h^{-1} respectively. Isoprene was the dominant compound during the day (70 and 31 % of total emissions from the upper and lower branches respectively). All compounds demonstrated a diurnal cycle at the upper level. Emissions appeared to be related to both light and temperature, and were therefore used to derive emission algorithms. Based on these algorithms, a simple canopy model was used to predict daytime canopy scale emission fluxes of 660 ± 480, 90 ± 30, 180 ± 120, 70 ± 20 and 160 ± 80 μg m^{-2} h^{-1} for isoprene, α-pinene, β–pinene, myrcene and β-thujene/β-phellandrene respectively. Two other methods were also used to derive emission rates from the canopy. These were the aerodynamic gradient and relaxed eddy accumulation (or conditional sampling) methods. Data was sparse due to lack of sensitivity, but some agreement with the enclosure method was seen.

Southern European, Eucalyptus plantation, Portugal

The main compounds identified from *Eucalyptus globulus* (as part of the EC FIELDVOC project) were isoprene (92 % on average), and the monoterpenes α-pinene, sabinene, β-pinene, myrcene, limonene, linalool and α-terpineol (cineole was identified from the younger tree only). Three unknown peaks (the most abundantly emitted monoterpenes) were tentatively identified as *cis*- and *trans*-ocimene, and an oxygenated octatriene.

Isoprene emission rates ranged from 0.05–28 μg g^{-1} h^{-1}. Mean total monoterpene emission rates were 0.9 μg g^{-1} h^{-1} in the day, and 0.1 μg g^{-1} h^{-1} at night. Diurnal trends showed a minimum during the early hours of the morning and a maximum in the afternoon at about 16:00 h (local time) during the period of maximum temperature and light intensity. A 'burst' of monoterpenes in the early morning was associated with periods of very high humidity, and may partly explain the nocturnal maxima in air concentrations of some monoterpenes, assumed to be a result of low nocturnal emission rates of these compounds, nocturnal stability and the lack of reacting species. Emissions were exponentially related to the temperature regime.

Both chemical speciation and absolute emission rates varied between the young and older vegetation types. Young trees emitted more cineole, unknown 2 (*trans*-ocimene?), linalool and α-pinene. Absolute emission rates were 5 times greater from the young compared to the old tree (on average 70.5 and 13.8 μg g^{-1} h^{-1} respectively during the day). This was still the case after standardisation (Table 11.5.1). These data are further evidence that physiological factors can be as important as ambient environmental conditions in affecting emission rates.

An investigation of the applicability of the Günther model (based on environmental conditions alone) revealed that the model, coupled with the best estimate of a standard emission rate from the literature [1], overestimated isoprene emissions from old-type vegetation by about 300 %, but modelled the young tree emission rates with reasonable accuracy [2].

Central European maize plantation, Mankmoos, Germany

As part of the EC BOVOC POPCORN campaign emissions of isoprene, monoterpenes, aldehydes and ketones were determined from maize and other vegetation.

Both apparently healthy and necrotic leaves and cobs of maize emitted negligible amounts of isoprene and monoterpenes. Total mean oxygenated emissions (of formaldehyde, acetaldehyde, acetone and acrolein) from leaves were 6.8 μg g^{-1} h^{-1} during the night, 6.0 μg g^{-1} h^{-1} during the day, and from cobs 3.2 μg g^{-1} h^{-1}. Oak and poplar emitted substantial amounts of isoprene (87 and 53 μg g^{-1} h^{-1} respectively), whereas pine emitted a wide range of monoterpenes. Birch emitted significant amounts of compounds tentatively identified as *cis*- and *trans*-

Table 11.5.1 Summary of total summed, averaged emission rates measured for various plant species during 1992–1994.

Species / Type / Date		Mean Measured Emission Rates ng g(dry wt)$^{-1}$ h^{-1}	Standardised* Emission Rates ng g(dry wt)$^{-1}$ h^{-1}
Quercus petraea (n=4) (CO_2 expt.,L/cr1992)	control treated	2650 ± 160 5080 ± 1100	2359 4758
Quercus petraea (n=11) (O_3 expt., L/cr 1992)	control treated	829 ± 312 915 ± 322	9235 10193
Picea sitchensis (n=3) (Temp. expt., controlled env., L/cr 1993)	21.0 °C 32.6 °C	2391 6317	10380 6769
Picea sitchensis (n = 8–13) field samples, Rivox '93	Day (16 m) (13 m) Night (16 m) (13 m)	1083 ± 956 145 ± 120 135 ± 75 29 ± 20	10978 7611 – –
Picea sitchensis (n=4) (1 sample day, Grizedale 1993)	young old	1562 ± 418 218 ± 98	5506 1082
mean of 3 sample days June/July, 1992 & 1993	young (n=9) old (n=12)	1056 ± 1090 904 ± 559	7622 7396
Ulex europaeus (field samples)	flowering (April 1993) nonflowering (June 1993)	71–705† 1780	281–668† 1866
Quercus ilex (n=9) Mediterranean field samples, June 1993		29996 ± 20402	18297
Pinus pinea field samples Mediterranean,June '93	young(n=9) old (n=20)	2898 ± 1663 9301 ± 4505	2258 7385
Eucalyptus globulus field samples, S.Europe, June 1994	young (n=9) old (n=37)	70509 ± 46376 10435 ± 9491	61398 17981

NB. * values presented for standard conditions of 30 °C and 1000 μmol m^{-2} s^{-1} using current models (after, for example, Günther *et al.*, [3]). † isoprene exceeded detector range, therefore total estimated should be at least 1705 ng g^{-1} h^{-1}, and 1968 ng g^{-1} h^{-1} (standardised), all gorse values ± 25 % on average.

β-ocimene, and alloocimene (about 9 μg g^{-1} h^{-1} in total). Beech emitted mainly sabinene (6 μg g^{-1} h^{-1}).

Conclusions

Extensive measurements of emission rates were made from both stressed and unstressed plant material, and were compared after standardisation using existing models. However, there is a need for further measurements from different plant species. It has become clear that light and temperature may not affect emission rates from all species in the same way, and further studies of this nature are required, preferably under controlled environmental conditions. The assessment of field data with respect to the existing emission rate models has indicated that there is a serious need to reconsider the effects which other physiological parameters have on emission rates, both in the long and short term. It is also clear that BOVOC emissions may form a substantial proportion of the emissions, and need to be considered in future sampling programmes. The effect of elevated levels of ozone were inconclusive. Long term exposure to elevated carbon dioxide levels resulted in increased emission rates in the experiments reported here, whereas in the literature the effects of elevated CO_2 are ambiguous. Further study is required on the effects of these pollutants.

Acknowledgements

The funding provided by the EC and NERC (UK), for contribution to the BEMA, TIGER (Rivox), POPCORN, and FIELDVOC campaigns, is gratefully acknowledged.

References

1. Günther, A.B., P. Zimmerman, M. Wildermuth, *Atmos. Environ.* **28** (1994) 1197–1210.
2. Street, R.A., Emissions of non-methane hydrocarbons from three forest ecosystems, Ph.D. Thesis, Lancaster University, 1995.
3. Guenther, A.B., P.R. Zimmerman, P.C. Harley, R.K. Monson and R. Fall, *J. Geophys. Res.* **98**D (1993) 12609–12617.

11.6 Isoprenoïc Compounds: Emission and Atmospheric Concentration Measurements

L. Torres, B. Clement, J.L. Fugit, M. Haziza, V. Simon and M.L. Riba

Ecole Nationale Supérieure de Chimie de Toulouse, 118 Route de Narbonne, 31077 Toulouse Cedex France.

Summary

We have developed an automated analytical device, equipped with a pre-concentration/thermo-desorption module, in order to monitor natural hydrocarbons that are emitted into the atmosphere.

This on-line device was employed at various forested sites during the course of several measurement campaigns. Comparative tests revealed a satisfactory agreement between monoterpenes (MT) analyses carried out both in the field and in the laboratory by several groups.

Daily cycles of atmospheric concentrations in terpenes were found to exhibit minimum values at daytime and maximum ones at night. By contrast, the emission rates and fluxes exhibit maximum diurnal values and minimum nocturnal ones.

The emission rates and fluxes of terpenic compounds were correlated with the micrometeorological and physiological parameters of the vegetation.

Aims of the research

* Design and development of an analytical device for the analysis of the principal biogenic compounds present in the atmosphere.
* Adaptation of such a device to direct measurements at the experimental site in an automatic and continuous way over prolonged periods.
* Intercomparison between the results obtained in the field and in the laboratory by groups using different analytical methods.
* Measurement of the daily and seasonal cycles of atmospheric concentrations and of the emission rates and fluxes of the principal biogenic compounds.
* Search for correlations between the emission of biogenic compounds and micrometeorological and physiological parameters inherent in the type of vegetation.

Results

Sampling and analytical devices

During our participation in the BIATEX project (1989–1993), we have designed and developed a device allowing automatic sampling and analyses of biogenic VOCs. This system which has been previously described in detail [1,2], allows continuous atmospheric measurements 24 hours a day and samplings at different levels. The period of a complete sampling/analysis cycle is of the order of 45 minutes.

Terpenic concentrations

Fig. 11.6.1 shows an example of the daily cycles of atmospheric mixing ratio volumes measured for the principal terpenes released by *Picea abies* (L.) *Karst.* These data were recorded at a height of 2 m between September 19th and September 27th 1989 at the research station of Schachtenau, located in the Bavarian forest National Park [3]. The results displayed in Fig. 11.6.2 were recorded between June 12 and June 15, 1992 in a *Pinus pinaster Ait.* forest (Landes forest), located in south-western France, along the atlantic coast [4]. The profiles displayed in Figs. 11.6.1 and 11.6.2 show a variation between day and night with the highest terpenic concentrations being observed at night.

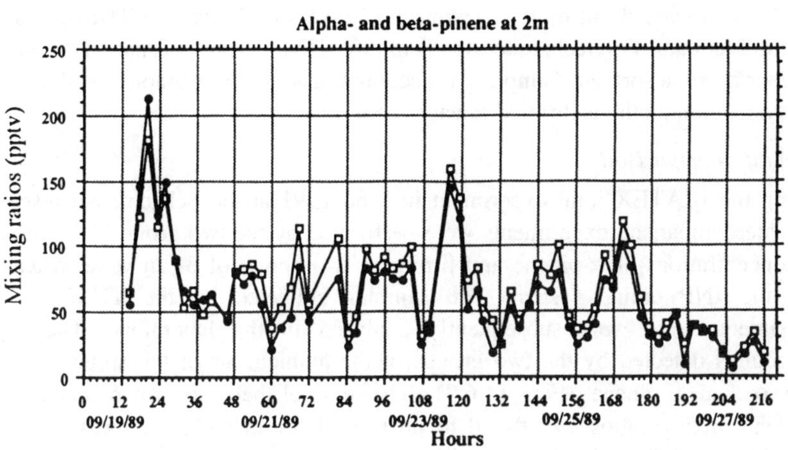

Fig. 11.6.1: Mixing ratio profiles of α-pinene (●) and β-pinene (•).

During the daytime, terpenes appear to be practically destroyed through photochemical reactions with ozone and $OH^{\bullet}$ radicals. At night, due to the absence of reaction with $OH^{\bullet}$ and the decrease of reactions with ozone, there is an important accumulation of terpenes in spite of their potential reaction with $NO_3^{\bullet}$ radicals. Immediately after sunrise, terpene concentrations are seen to decrease due to the predominance of photochemical process.

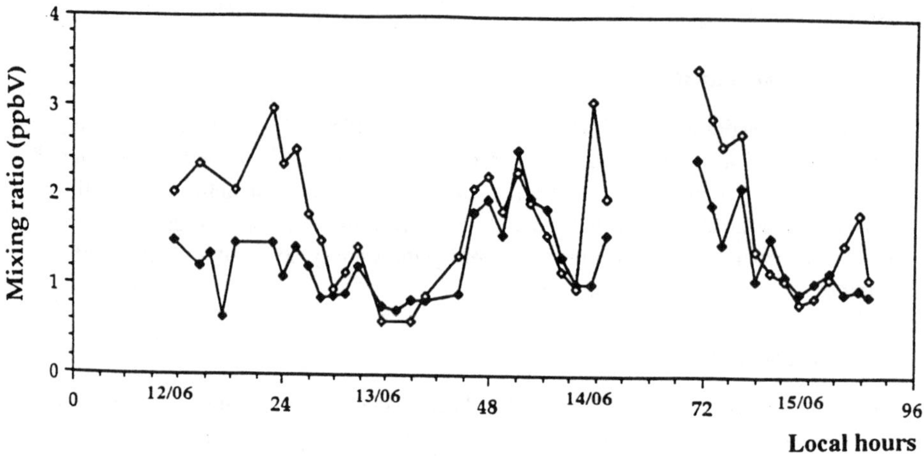

Fig. 11.6.2: Mixing ratio profiles of α-pinene (*) and β-pinene(o).

A number of research groups working in Bavaria [5, 6] and on other sites [2] have reported closely related terpene variations.

Intercomparison analysis

An intercomparison experiment on the preconcentration analysis of MT from air samples was carried out by the Technische Universität München [TUM], and the Institut National Polytechnique de Toulouse [INP], using Tenax TA porous polymer as an adsorbent. Sampling procedures, gas chromatographic analysis and standardisation methods differed in each laboratory.

Field intercomparison

During the BIATEX joint experiment in June 1991 at the Schachtenau research site, intercomparison experiments were performed by the two groups[7]. Ambient air concentrations of α-pinene and β-pinene at a height of 31 m were measured with the INP online system. The samples collected by the TUM on the experimental site were subsequently analysed in the laboratory. The main compounds detected by the two groups in the ambient air of the spruce canopy were α and β-pinene (Fig. 11.6.3). The diurnal behavior of α-pinene and β-pinene mixing ratios was found to be typical for that site and was recorded several times [3, 7].

In this intercomparison, in general, the values of INP and TUM were in excellent agreement. Only two values of β-pinene mixing ratios in the late evening differed by 20 % and 7 % from the average.

Prior to these field experiments, an intercomparison of laboratory experiments made by TUM and INP groups in 1990 led to encouraging results at that time [7].

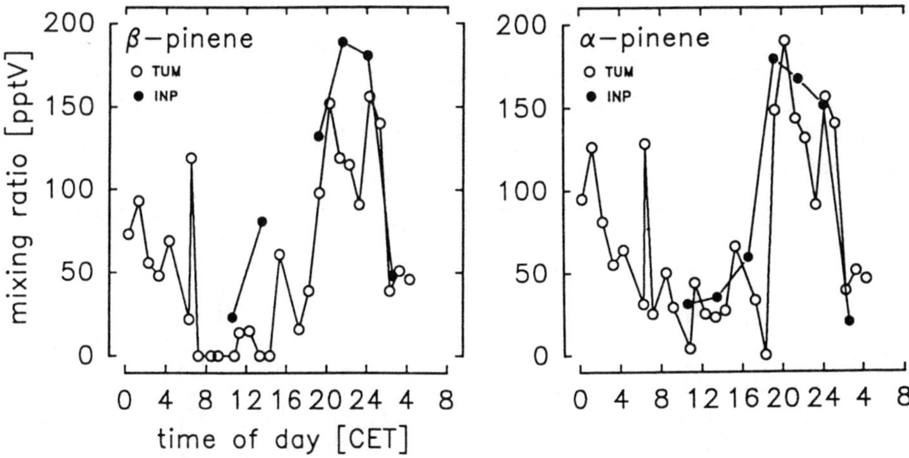

Fig. 11.6.3: Field intercomparison experiment of α-pinene and β-pinene mixing ratios.

Terpenic emission rates and fluxes

These measurements were made during the campaigns conducted in June 1991, June 1992 and June 1993 on the sites of Schachtenau, Germany, Landes, France and Aveiro, Portugal, respectively.

At Schachtenau, the emission fluxes were determined by the aerodynamic gradient method. Mixing ratios of α- and β-pinene were continously recorded at heights of 31 and 51 m. These levels were above the canopy, found to culminate at 28 m.

The wind velocity and temperature values recorded at different levels allowed to calculate the turbulent diffusivity coefficient for neutral atmospheric stability conditions.

α- and β-pinene fluxes were found to vary in a quite similar way all over the campaign (Fig. 11.6.4). Maximum values were currently observed at sunset and at the beginning of the night, though not systematically. The measured values range are of the same magnitude as those measured by Zimmerman [8] or Lamb *et al.* [9] in the same type of forest.

In the Landes forest, the values of emission rates for the six principal monoterpenes released by the Maritime pine *Pinus Pinaster* (Table 11.6.1) are in full agreement with recent measurements made on related species. In particular, Corchnoy *et al.* [10] measured average daytime emissions on 3 to 8 samples of various pine species. The average normalised values of daytime emission rates of α-pinene were 70 ng g^{-1} h^{-1} for Aleppo pine (*Pinus Halapensis*), 30 ng g^{-1} h^{-1} or Italian

Stone pine (P*inus Pinea*), and 370 ng g^{-1} h^{-1} for Monterey pine (*Pinus Radiata),* whereas the corresponding values of total daytime terpenic emissions for these three species were 150, 420, and 900 ng g^{-1} h^{-1} respectively. The average emission fluxes were calculated from the leaf biomass coefficient (0.9 kg m^{-2}) and from the corrected emissions rates. The biomass coefficient was determined by a statistical method based on logarithmic regressions (Lemoine *et al.* [11]), from allometric relationships established from a sample of 200 branches collected from 20 different trees.

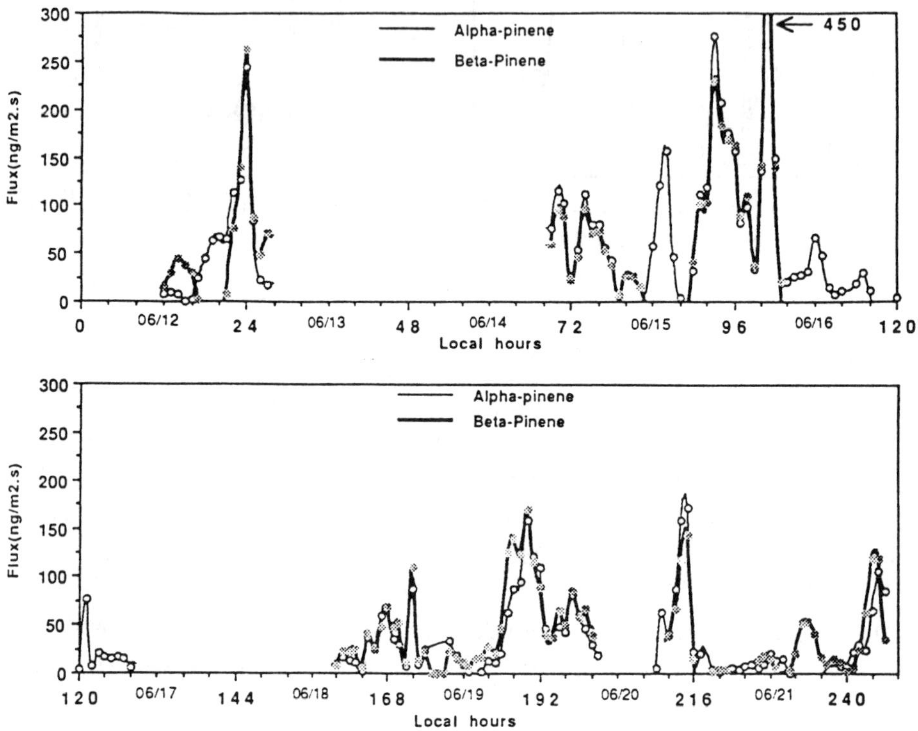

Fig. 11.6.4: α-pinene and β-pinene fluxes determined by the aerodynamic gradient method from the 12th to the 21st of June 1991.

In this work, the emission rates of terpenic compounds were correlated with temperature, light intensity, vapor pressure, and with the product between the relative monoterpene amount within the plant and the corresponding vapor pressure [4].

A study was conducted between June 26 and July 3, 1993 at Aveiro, Portugal in an artificial *Eucalytpus globulus* forest [12] in collaboration with Professor Pio's team from Aveiro University.

The determination of isoprene emissions was made via the enclosure method from an intact eucalyptus branch enclosed in a teflon (FEP) cuvette. The emission rates were determined with our on-line device by following the evolution of isoprenic concentrations within the cuvet as a function of time. The emission rates

Table 11.6.1 Terpenic emission rates and fluxes measured in the Landes forest (June 1992).

	Emission rates: ng g^{-1} h^{-1}					Fluxes: µg m^{-2} h^{-1}	
			MV (m)	MV (m)	MV (m)	MV (C)	MV
Compound	max (m)	min (m)	Day	Night	Day + Night	Day + Night	Day + Night
α-pinene	173.4	35.8	105.0	56.9	80.9	70.1	63.1
β-pinene	236.9	19.7	103.1	69.3	86.2	80.2	72.2
camphene	9.9	0.7	5.3	4.1	4.7		
myrcene	91.7	9.8	49.1	22.6	35.9	33.0	29.7
Δ_3-carene	155.2	9.8	57.2	25.7	41.4	34.9	31.4
limonene + 1-8 cineole	124.0	9.1	47.4	24.0	35.7	32.2	29.1
Σ terpenes			367.1	202.6	284.8	250.5	225.5

MV, mean value; m, measured; C, corrected.

determined on June 30 and July 1, 1993 are displayed in Figs. 11.6.5 and 11.6.6. The values appear to be maximum at about 2h p.m. (U.T.), whereas they are lower than the detection threshold of our analytical device during the night.

Thus, it clearly appears that isoprene is emitted for the most part between 8:00 a.m. and 8:00 p.m (U.T.). The values of emission rates were calculated by

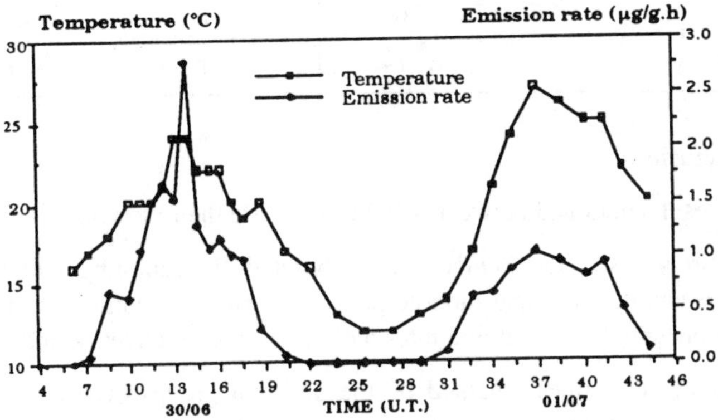

Fig. 11.6.5: Simultaneous profiles of isoprene emission rate and temperature (June 30–July 1, 1993).

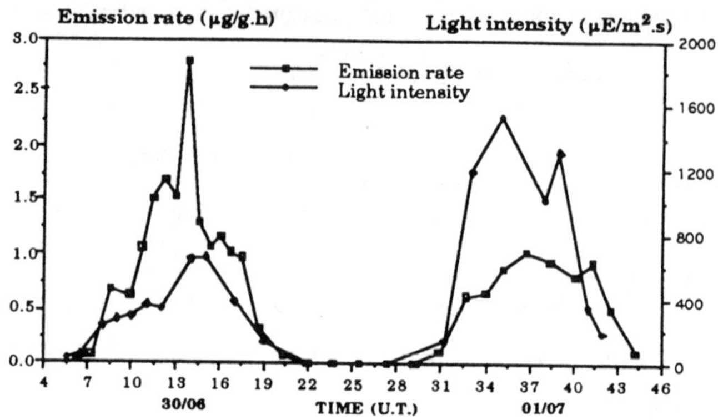

Fig. 11.6.6: Simultaneous profiles of isoprene emission rate and light intensity (June 30–July 1, 1993).

reference to the external temperature of the cuvet. The internal temperatures were generally higher. The results obtained indicate that the emission rate is correlated both with the temperature and with the intensity of the solar radiation.

Table 11.6.2 displays the maximum isoprene fluxes measured during the three days of the experiment. These fluxes were determined from the biomass coefficient of the eucalyptus forest. It is 0.28 kg m^{-2}.

Table 11.6.2: Isoprene emission fluxes measured during the daylight period.

Date	Minima		Maxima	
	ng m^{-2} s^{-1}	Time(UT)	ng m^{-2} s^{-1}	Time (UT)
June 30	6	07:10	217	13:45
July 1	8	07:00	80	12:50
July 2	5	07:15	261	13:35

Main conclusions

The main results obtained between 1989 to 1993 are the following :

* An entirely automatic device was elaborated for sampling and analysis of isoprene and the principal monoterpenes. It permits a continuous measurement of the concentration, emission rates and fluxes over a longer period.

* In general, the values obtained by the different groups involved in BIATEX were in good agreement. Intercomparisons made in 1993 and 1994 during the BEMA campaign (Biogenic Emissions in the Mediterranean Area) were in excellent agreement [13]. This tends to demonstrate that, even though the

analytical techniques may be different, they are now perfectly mastered and are becoming quite reliable.

* The diurnal cycles of the atmospheric concentration variations of the main monoterpenes released by trees generally show the alternance of nocturnal maxima and diurnal minima. In contrast, the daily variations of emission rates and fluxes exhibit diurnal maxima and nocturnal minima.

* Satisfactory correlations were established between the terpenic emissions and the micrometeorological, physical and physiological parameters [4].

Acknowledgements

We would like to thank the EUROTRAC French Commitee for their assistance in the financial support of this research. Thanks to Dr. G. Enders and his group for their technical support.

References

1. M.L. Riba, N. Tsiropoulos, B. Clement, B. Golfier, L. Torres, *J. Chromatogr.* **456** (1988) 165–173
2. B. Clement, M.L. Riba, V. Simon, L. Torres, *Int. J. Anal. Chem.* **50** (1993) 19–27
3. G. Enders, B. Clement, R. Dlugi, S. Gab, M. Haziza,U. Herrmann, M. Kessel, D. Kotzias, K. Kourtidis, H. Kurth, R. Mcmillen, G. Roider, R. Steinbrecher, U. Teichmann, L. Torres. *Atmos. Environ.* 26A (1992) 171–189
4. V. Simon, B. Clement, M.L. Riba, L. Torres *J. Geophys. Res.* **99D8** (1994) 16501–16510
5. R. Kreuzig, I. Gebefugi, F. Korte, *Forstw. Cbl.* **105** (1986) 435–441
6. R. Steinbrecher, R. Schonwitz, H. Ziegler, in : *19th Int. Symposium on Essential Oils and Natural Substrates.* Dubendorf L10 1988.
7. R. Steinbrecher, W. Schurmann, L. Torres, B. Clement, V. Simon, D. Kotzias, R. Daiber, J. Van Eijk, Inter. *J. Environ. Anal. Chem.* **54** (1994) 283–297
8. P.R. Zimmerman., EPA-904/9-77-028 (1979)
9. B. Lamb, H. Westberg, G. Allwine, T. Quarles, *J. Geophys. Res.* **90D1** (1985) 2380–2390
10. S.B. Corchnoy, J. Arey, R. Atkinson, *Atmos. Environ.* **26B** (1992) 339–348
11. B. Lemoine, J. Gelpe, J. Ranger, C.Nys, *Ann. Sci. For.* **43** (1986) 67–84
12. B. Clement, J.L. Fugit, M.L. Riba, L. Torres, T. Nunes, C. Pio, *Fresnius. Environ. Bull.* **3** (1994) 569-574
13. B. Larsen, I. Bartzis, T. Bomboi-Mingarro, P. Ciccioli, P. Foster, U. Hansen, T. Hoffmann, D. Kotzias, S. Owen, R. Perez-Pastor, G. Seufert, R. Steinbrecher, M. Staud, L. Torres, *Atmos. Environ.* submitted.

Chapter 12

Exchange of Particles

12.1 Vertical Flux Measurements of the Submicronic Aerosol Particles and Parametrisation of the Dry Deposition Velocity

J. Fontan, A. Lopez, E. Lamaud and A. Druilhet.

Laboratoire d'Aérologie, U.A. associée au CNRS No. 358
Université Paul Sabatier, Toulouse, France.

Summary

The dry deposition velocity of submicronic aerosol particles has been measured during three campaigns corresponding to different ground cover. The eddy correlation method cannot be computed blindly, because of low frequency fluctuations. After proper filtration, the dry deposition velocity can be obtained and parametrised. During unstable situations, the dry deposition velocity gives high values.

Introduction

Vertical flux measurements of atmospheric constituents in the surface layer are necessary to obtain source or sinks of a constituent. We have studied the flux of submicronic aerosol particles to obtain a parametrisation of their dry deposition velocity. Knowledge of these phenomena is necessary to calculate the aerosol residence time in the troposphere, to model the transport on different scales and to determine the possible impact of aggressive aerosol particles on the biosphere.

Previous measurements of vertical flux of aerosol particles are relatively scarce. With the gradient method, it is difficult to measure the difference between two altitudes during convective conditions, because the value is generally very low and during the night or neutral conditions the dry deposition is not very efficient to give a gradient in the surface layer. The eddy correlation method is also difficult to use because it is difficult to have a fast response and a large number of particles measured at the same time, to avoid a standard deviation due to poor statistics.

We have built a new fast analyser. The statistical fluctuations are eliminated by high flow rate sampling and it gives the turbulent atmospheric fluctuations. We have used it to measure turbulent vertical flux in the surface layer and the dry deposition velocity over different types of soil or vegetation. We have encountered difficulties in the flux measurements by the eddy correlation method due to important low frequency fluctuations. We present here the results we have obtained on the methodology of the vertical flux measurements and the parametrisation of the dry deposition velocity.

Measurement methods

Aerosol is charged by the corona effect and the electric charges carried by particles are collected and measured with an electrometer [1]. To avoid statistical fluctuations, a high flow sampling rate (3 L/s) is necessary. The time constant of the sensor is about 0.2 s. For a log-normal particle size distribution (geometrical mean 0.044 μm, standard deviation ≈ 2.5) representative of the experimental distributions obtained at our sites, the analyser has a response centred at 0.15 μm, with the same standard deviation. 80 % of the detected particles have a size between 0.05 and 0.5 μm. Vertical fluxes over the ground of aerosol, ozone, momentum and sensible heat are measured by the eddy correlation method [2, 3]. Wind is obtained with a 3-D sonic anemometer.

Particles smaller than 0.02 μm or larger than 1 μm have a negligible contribution to the current, similar to the surface distribution function. The analyser mainly measures the surface concentration of particles, which is located roughly on one decade size.

Experimental sites

The measurements have been made during 3 campaigns at different sites. In the Landes forest located in the southwest of France; the vegetation canopy consists of maritime pine (*Pinus pinaster Ait*) with a short (less than 1 m) gramineous under storey on a sandy soil. The canopy height is 15 m and the leaf area index of the trees is about 3. The displacement distance z_d = 11 m, the roughness length z_o = 1.4. It is equipped with a 25 m high tower. Fluxes were measured at the top [2]. The second site is the Ebre valley near Zaragoza in Spain and the campaign was called Stars II experiment. The ground is bare soil in the north westerly direction with a roughness length z_o = 0.25 cm, and bare soil also in the easterly direction with z_o = 1 cm and an almond tree plantation to the southwest with z_o = 10 cm. The altitude measurement is z = 7 m. The third site is in the Sahel near Tilabery in Niger, Africa. The campaign was called Stars I. The ground consists of bare clay soil with z_o = 0.20 cm.

Analysis of the aerosol particle fluctuations of concentration.

Fluctuations of low frequencies are much more important for aerosol than for other quantities like the vertical wind velocity, *W*, the temperature, *T*, and also ozone which are measured at the same time and at the same altitude. Fig. 12.1.1 gives the energy spectra for *W*, *T*, O_3 and aerosol (*C*) and we can observe the high contribution of the lower frequency range in the case of aerosol. These fluctuations are not produced by dry deposition but they give a contribution to the vertical flux. Figs. 12.1.2, a and b show mean co-spectra, *WT* and *WC*, obtained during the Stars II experiment on bare soil near Zaragoza in Spain. These co-spectra are represented with the classical adimensional frequency $f = nz/u$ (*u*: horizontal wind speed) for different values of the Monin Obukhov parameter *z/L*. In unstable situations, the low frequencies (between 5×10^{-3} and 5×10^{-2} for

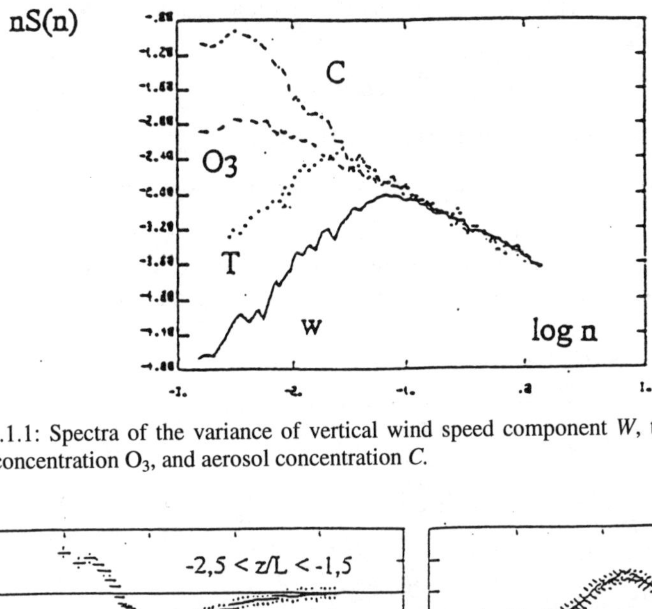

Fig. 12.1.1: Spectra of the variance of vertical wind speed component *W*, temperature *T*, ozone concentration O_3, and aerosol concentration *C*.

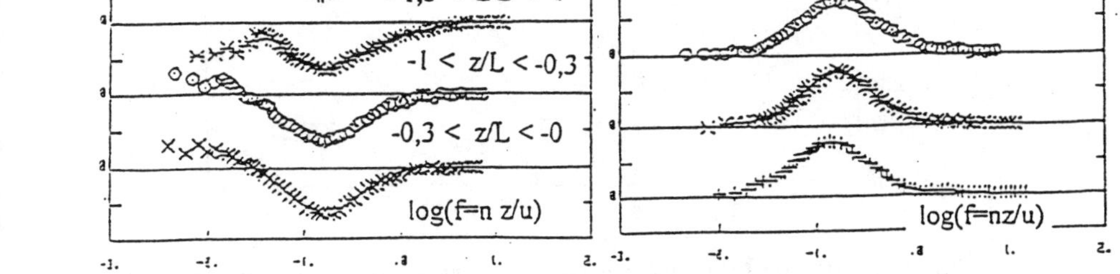

Fig. 12.1.2: Mean cospectra *WT* (Fig. a) and cospectra *WC* (Fig. b) with the adimensional frequency in the case of *z/L* (Starts II Experiment). The low frequency contribution is abundant in the case of aerosol.

adimensional frequencies and less than 10^{-2} Hz in the case of dimensional frequencies) are abundant in the case of aerosol. In the range $10^{-1}-1$ co-spectra *WT* and *WC* can be considered as universal with only a very weak shift of the maxima as a function of the instability.

The standard deviation S_c of a scalar linked to the local vertical flux Φ_{co} at ground level, is given by the similarity theory of the surface layer as a function of the Monin Obukhov stability parameter $-z/L$ [4-7].

The similarity theory representation is S_c/C^*

with $$C^* = F_c/u_* = V_d\, C / u_*$$

where F_c is the aerosol flux at ground level and V_d the dry deposition velocity.

$$S_c/C^* = a_1$$

when $z/L > x_1$, x_1 is a value corresponding to a low instability. Wesely *et al.* [5] give $x_1 = -0.31$

$$S_c/C^* = a_2(z/L)^{-2/3} \text{when } z/L < x_1$$

a_1 is about 2 and $a_2 = 1.25$ for a scalar like H_2O or CO_2

If V_d/u_* and S_c/C^* are functions of z/L, S_c/C must be also be a function of z/L.

Fig. 12.2.3 gives S_c/C in function of z/L before and after filtration of the lower frequency. Before there is no relationship between S_c/C and the stability parameter. After we have a good relationship. Fig. 12.2.4 represents, after

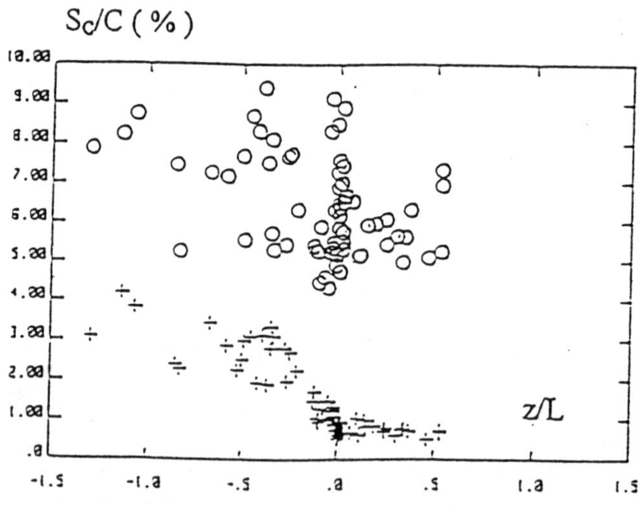

Fig. 12.1.3: Comparison between the normalized standard deviation S_c/C before (o) and after filtration(+) in function of the M.O. parameter. Before filtration there is no relation between S_c/C and the stability parameter.

filtration, S_c/C^* and S_t/T^*. The variations are very similar. The x_1 value given by Wesely *et al.* is indicated in the figure. In the case of aerosol x_1 is near zero. The dispersion is large in stable conditions but the accuracy in the measurements of flux is low. S_c/C is constant in neutral and stable conditions. It is an indication that V_d/u_* is constant. Fig. 12.2.5 gives the variation of S_t/T and S_c/C in the case of the Landes experiment. The variations are similar, but the dispersion is lower in stable and neutral conditions for aerosol. This good agreement with the similarity theory of the surface layer, based on the assumption that the turbulent characteristics, like

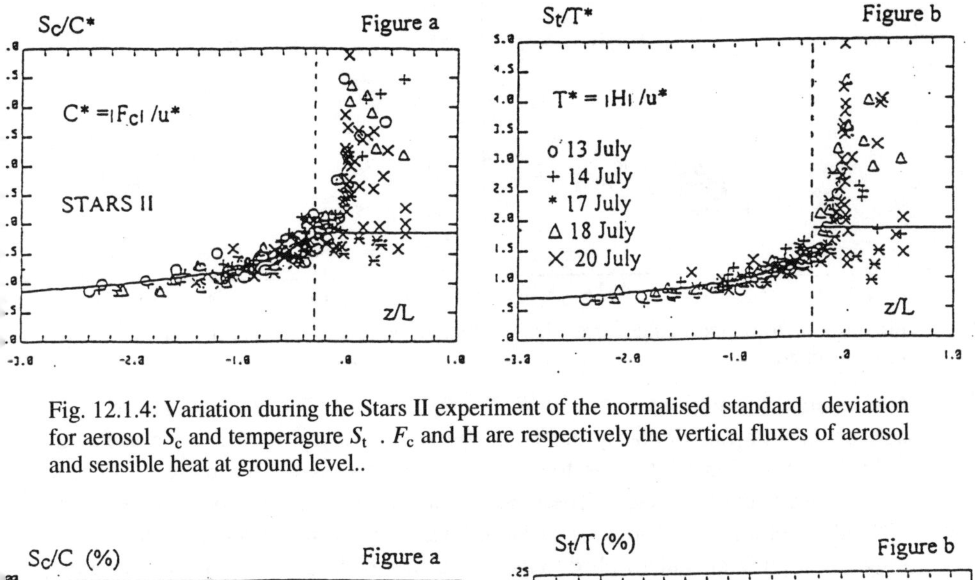

Fig. 12.1.4: Variation during the Stars II experiment of the normalised standard deviation for aerosol S_c and temperagure S_t . F_c and H are respectively the vertical fluxes of aerosol and sensible heat at ground level..

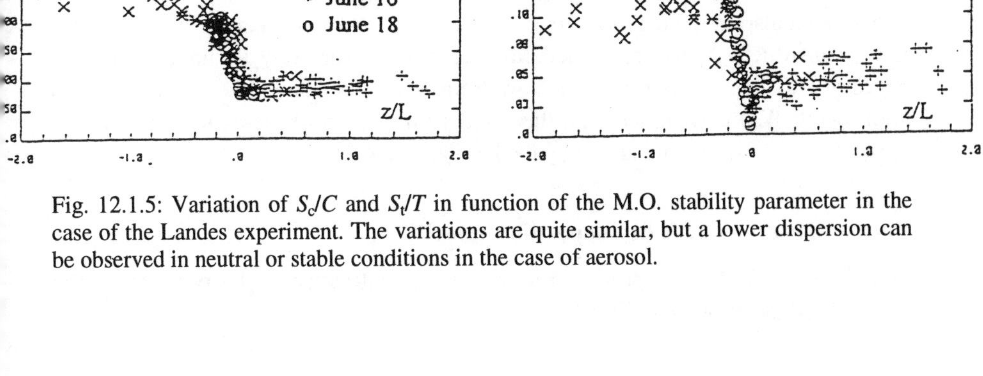

Fig. 12.1.5: Variation of S_c/C and S_t/T in function of the M.O. stability parameter in the case of the Landes experiment. The variations are quite similar, but a lower dispersion can be observed in neutral or stable conditions in the case of aerosol.

the standard deviation, are functions of F_c, u_*, H, the heat flux and β the buoyancy term, is a clear indication that the low frequency fluctuations have been well eliminated.

In the same way the correlation coefficients between vertical wind component and aerosol concentration, vertical wind and temperature have equivalent variations. During instability R_wT is between –0.4 and 0.7 and R_wC between –0.4 .and –0.6 (Fig. 12.2.6).

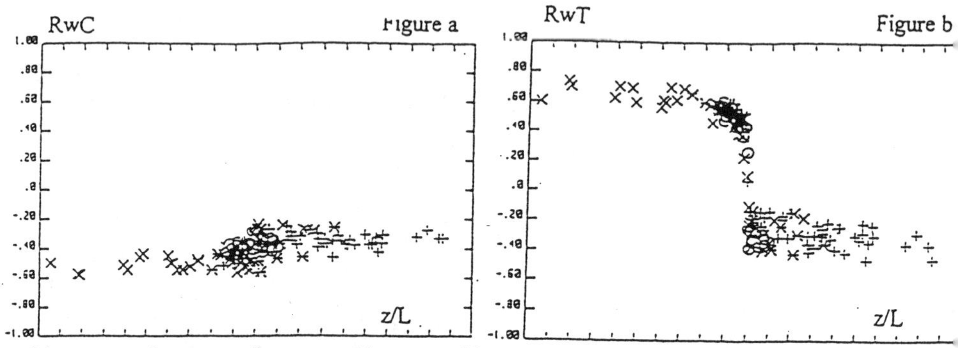

Fig. 12.1.6: Variation of the correlation coefficient between w and C, and W and T as a function of z/L.

During stable conditions, $z/L > 0$, $R_wT \approx R_wC \approx -0.3$. These results are in good agreement with the similarity theory of the surface layer in case of aerosol, but only after filtration of the low frequencies. The reasons for the presence of low frequencies is not quite clear. These non stationary events are important and more difficult to eliminate in the case of large instability, with low wind speeds. For a range of high frequencies linked to the interaction with ground, shifts are observed towards the low frequencies. It becomes obvious that turbulent fluxes must not be computed blindly. The simple observation of time changes in mean concentration is not sufficient for detecting detrimental low-frequency trends, and other methods such as flux gradient or eddy accumulation cannot be applied to the measurement of the dry deposition velocity of aerosol particles. We have observed this phenomena also in the case of ozone, but here it is less important. If we generalise for a constituent X, to apply methods like gradient or eddy accumulation, it is necessary to demonstrate that the low frequency contribution to the correlation coefficient, WX, is weak and that the standard deviation of the concentration is in agreement with the results given by the similarity theory.

Parametrisation of the dry deposition velocity

Wesely *et al* [8, 9] have proposed in the case of sulfate aerosol a parametrisation of the dry deposition velocity V_d

For stable or neutral situations $L \geq 0$: $V_d/u_* = 0.002$

In the case of unstable cases, $L < 0$

$$V_d / u_* = 0.002 \{ 1 + (-0.3 \, z_i/L)^{2/3} \} \quad (1)$$

with z_i as the height of the convective boundary layer.

Our experimental results show that when $L \geq 0$, V_d is function of the friction velocity and by dimensional analysis we can write $V_d/u_* = k_1 = Cte$. In the Landes experiment $k_1 = 0.004$ and $k_1 = 0.002$ for the Stars I and II campaigns. There is however a large variation of z_o as a function of wind direction during Stars. The main difference between Landes and Stars is that there was a substantial air flow through the forest canopy.

In convective situations, $L < 0$, the dry deposition velocity is a function of the friction velocity and also of the convective velocity scale W^*, and we can write dimensionally :

$$V_d / u_* = k_1 + k_2 \, W^{*2} / u^{*2} = k_1 + k_2 \, (\beta \, H \, z_i)^{2/3} / u_*^{\,2}$$
$$= k_1 + k_2 \, (-z_i/L)^{2/3} \quad (2)$$

H is the sensible heat flux, β the buoyancy parameter. In all our experiments, $k_2 = 0.0009$.

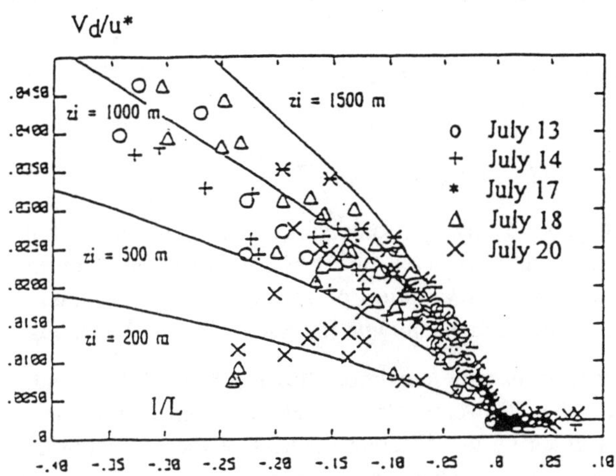

Fig. 12.1.7: Results of the Stars II experiment. Solid lines represent for different values of the z_i, relationship : $V_d/ u_* = 0.002 + 0.0009 \, (- z_i/ L)^{2/3}$

Fig. 12.2.7 gives the results of the Stars II experiment. Solid lines represent $V_d/u_* = 0.002 + 0.0009 \, (z_i/L)^{2/3}$ for different values of z_i. We have not measured the z_i values, but z_i increases during the morning with the development of the convection . If we represent the experimental results in function of the time of the day (Fig. 12.2.8) we have then a realistic distribution of the values with the curves corresponding to different z_i.

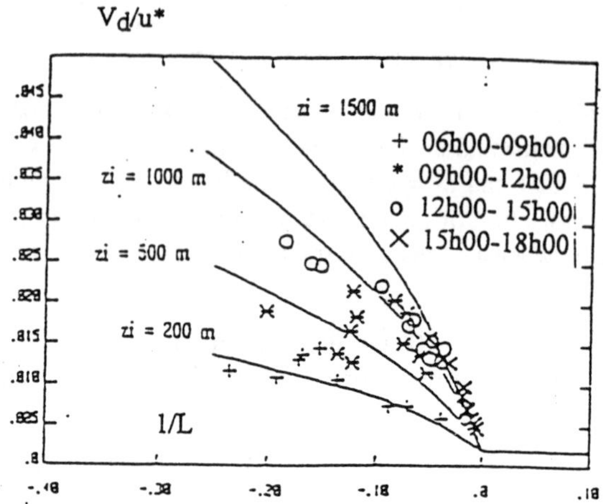

Fig. 12.1.8: Variation of V_d / u_* = f(1/*L*) for July 20 (Stars II) in unstable situations. Experimental results are represented by period of 3 hours and solid lines represent the same relationship as in Fig. 12.1.7 for different values of z_i.

Panofski *et al* [10] have proposed for the standard deviation of the horizontal wind velocity S_u

$$S_u/u_* = 12 + 0.5\,(-z_i/L)^{1/3} \tag{3}$$

and Wyngaard *et al* [11]:

$$S_u/u_* = \{\, 4 + 0.6\,(-z_i/L)^{2/3} \}^{1/2} \tag{4}$$

Relationships (3) and (4) give the same variation of z_i, changing in the relation (2) 0.6 by 0.52. An example is given in Fig. 12.2.9 for July 20 during the Stars II experiment. If we use this variation of z_i,, we obtain a variation of V_d/u_* as a function of z_i/L in good agreement with the relationhip in (2) (see Fig. 12.2.10)

During the day with unstable conditions V_d reaches 0.8 cm/s and is lower than the night-time value, 0.1 cm/s, in neutral or stable conditions. This type of variation is not given by the different theories of the dry deposition of aerosols particles. Values during unstable conditions are higher than those obtained for fine particles in wind tunnel measurements but confirm results previously obtained in *in-situ* experiments [5, 12, 13].

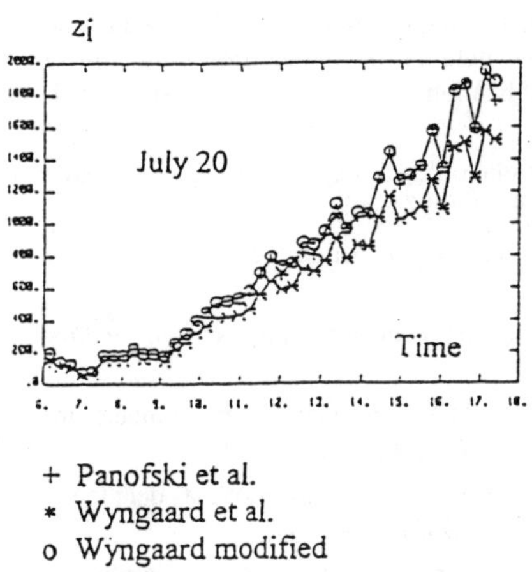

Fig. 12.1.9: Variations of z_i calculated from Panofski *et al.* [10]; Wyngaard *et al.* [11], relationships give S_u/u_* as a function of atmospheric stability.

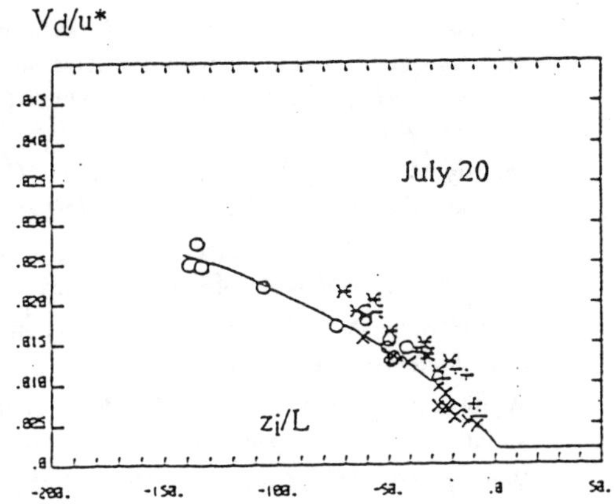

Fig. 12.1.10: Variation of $V_d/u_* = f(z_i/L)$ for July 20, z_i is deduced from the modified Wyngaard *et al.* [11] relationship.

Conclusions

In order to measure dry deposition velocity of submicronic aerosol, low frequency fluctuations of concentration need to be filtered out, and only the eddy correlation method can be applied. This dry deposition velocity can be parametrised as a function of the friction velocity and the atmospheric stability. The convection in the boundary layer influences the deposition at the ground; this phenomena is not only observed in the surface layer. The turbulent kinetic energy at ground level, partly produced by convective eddies in the ABL, seem to be the main parameter.

Acknowledgements.

This work has been supported by the scientific committee " Phase Atmosphérique des Cycles Biogéochimiques" of the French CNRS Environmental Programme .

References

1. El Bakkali Y., Etude et réalisation d'un analyseur pour la mesure rapide des fluctuations de l'aérosol atmosphèrique, Thesis, Université Paul Sabatier, Toulouse, France, 1991.
2. Lamaud E., Y.Brunet, A. Labatut, A. Lopez, J. Fontan, A. Druilhet, The Landes experiment. Biosphere-atmosphere exchanges of ozone and aerosol particles above a pine forest, *J. Geophys. Res.* **99 D8** (1994) 16511–16521.

3. Lamaud E., J. Fontan, A. Lopez., A. Druilhet, Parametrisation of the dry deposition velocity of submicronic aerosol particles, in: *Air pollution II, vol 2. Pollution control and monitoring*. Computational Mechanics publications Southampton, Boston 1994, pp. 433–440.

4. Wesely M.L., G.M. Thurtell, C.B. Tanner. Correlation measurements of sensible heat flux near the earth's surface. *J. Appl. Meteorol.* **9** (1970) 45–50.

5. Wesely M.L. Use of variance technique to measure dry air surface exchanges rates. *Boundary layer Meteoro.* **44** (1988) 13–31.

6. Högstrom U., A.S. Högstrom.. Turbulence mechanism at an agricultural site. *Boundary layer Meteoro*l. **7** (1974) 373–389.

7. Wyngaard J.C., Coté O.R. The budgets of turbulent kinetic energy and temperature variance in the atmospheric surface layer. *J. Atmos. Sci.* **28** (1971) 190–201.

8. Wesely M.N., D.R. Cook, R.L. Hart, Fluxes of gases and particles above a decideous forest in winter time, *Boundary layer Meteorol.* **27** (1983) 237–255.

9. Wesely M.N., D.R. Cook, R.L. Hart, Measurements and parametrisation of particulate sulfur dry deposition over grass, *J. Geophys. Res.* **90** (1985) 2131–2143.

10. Panofsky H.A., Tennekes H., Lenschow D.H., Wyngaard J.C., The characteristics of turbulent velocity component in the surface layer under convective conditions. *Boundary layer Meteorol.* **11** (1977) 355–361.

11. Wyngaard J.C., Cote O.R. The evolution of a convective planetary convective layer, a higher order-closure model study.. *Boundary layer Meteorol.* **7** (1974) 289–308.

12. Hicks, B.B., M.L. Wesely, J.L. Durham, and M.A. Brown, Some direct measurements of atmospheric sulfur fluxes over a pine plantation, *Atmos. Environ.* **17** (1982) 2899–2903.

13. Hicks, B.B., D.D., Baldocchi, T.P. Meyers, R.P. Hosker, and D.R. Matt, A preliminary multiple resistance routine for deriving dry deposition velocities from measured quantities, *Water, Air, and soil Pollut.* **36** (1987) 311–330.

12.2 Measurements of Fluxes of Particles, Ozone, Nitrogen Oxides and Carbon Dioxide to Coniferous and Deciduous Forests

Kim Pilegaard[1], P. Hummelshøj[2], N. O. Jensen[2] and L. Kristensen[2]

[1]Environmental Science and Technology Department,
[2]Meteorology and Wind Energy Department, Risoe National Laboratory,
P.O. Box 49, DK-4000 Roskilde, Denmark

Summary

Our contribution to the research covers field experiments and theoretical considerations. The fluxes of NO, NO_2, O_3, CO_2 and H_2O were determined by the eddy correlation method and the fluxes of particles by the gradient method. The details of materials and methods are given in the references [1–7]. The following gives an overview of the results.

Aims of the research

The aims of the contribution from Risoe to BIATEX are to quantify the fluxes of NO, NO_2, O_3, and particles and to parameterise the measured fluxes in order to provide input for modelling.

Principal scientific results

Dry deposition velocities for particles were obtained by gradient measurements with an optical particle counter, for three different particle sizes. The measurement site was a field covered with 5 cm tall green winter barley, giving a surface roughness below 1 cm. It was found that the mean dry deposition velocities for three different particle sizes were: V_d(0.065–0.15 μm) = 0.16 cm $s^{-1,}$ V_d(0.15–0.4 μm) = 0.20 cm s^{-1} and V_d(0.4–0.9 μm) = 0.28 cm s^{-1} , during stable stratification. The dry deposition velocity increased with increasing particle size in the particle diameter interval of 0.1 to 1.0 μm, indicating that the minimum dry deposition velocity should be found for particles below 0.1 μm.

A lot of uncertainties are still connected to determination of dry deposition of particles. It seems that some processes are missing to explain what is controlling the deposition velocities as a function of the particle size. During stable stratification the dry deposition velocity was found to decrease with increasing stability. On the other hand the dry deposition velocity was found to increase with increasing friction velocity u_*. Local emission plumes containing particles were shown to be a problem at the field site during the measurement of fluxes of particles. This problem can be solved by a careful check of the gradients before deposition velocities are calculated.

Fluxes of O_3, NO_2, CO_2 and H_2O were measured over agricultural fields and coniferous and deciduous forests. Aerodynamic and viscous sublayer resistances were calculated from meteorological parameters, and surface resistances as the residual resistance when the inverse deposition velocity was taken as the total resistance.

Over a harvested wheat field at Scherzheim in the Rhine valley (TRACT experiment) a clear diurnal variation was found for fluxes of O_3 and NO_2 as well as the deposition velocities and resistances which indicated stomatal activity as a regulating factor. The fluxes were highest during the early afternoon with maximum fluxes of –0.6 μg O_3 m^{-2} s^{-1} and –0.2 μg NO_2 m^{-2} s^{-1} . The average daily maximum deposition velocity of O_3 was 5 mm s^{-1} and of NO_2 3.5 mm s^{-1}. Surface resistances (r_c) to O_3 uptake were up to 600 s m^{-1} during night-time and 200 s m^{-1} during daytime. r_c values to NO_2 were even higher; 800–1000 s m^{-1} during the night and 200–400 s m^{-1} during the day.

The diurnal variation of the concentration, flux and deposition velocity of O_3 over a coniferous forest is shown in Fig. 12.2.1. The concentrations show a diurnal

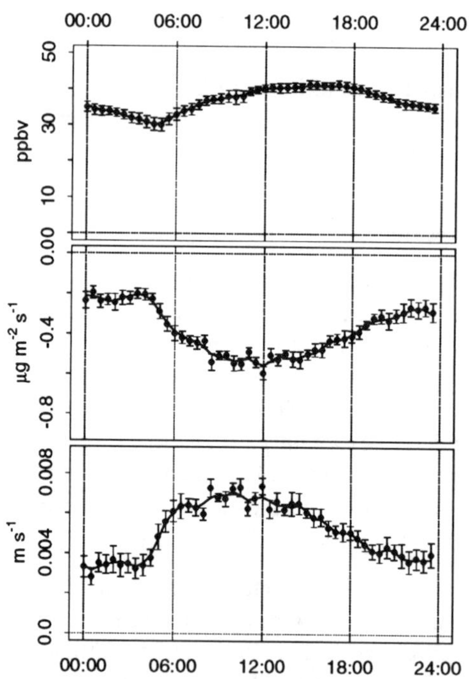

Fig. 12.2.1: Diurnal variation of ozone concentration, flux and deposition velocity over a Norway spruce plantation in Ulborg, Denmark, 7–17 June, 1994.

pattern with a minimum in the early morning and a maximum in the late afternoon. The difference between the minimum and maximum is about 15 ppb, indicating that the air is only moderately polluted. The dominating wind direction during the experiment was NW; this wind brings relatively clean air from the North Sea.

The fluxes of ozone ranged from about –0.25 μg m^{-2} s^{-1} to –0.6 μg m^{-2} s^{-1} with a maximum around noon. The deposition velocities ranged from 3.5 mm s^{-1} to 7 mm s^{-1} with a sharp rise at dawn and a maximum already in the morning.

This diurnal pattern found for O_3 indicates that stomatal processes are important for the deposition. However, the rather high deposition velocities during the night indicate that other processes for removal of O_3 also play a significant role. The results of the calculation of resistances are given in Table 12.2.1, where data from earlier experiments in the spruce forest [7] and open land [8] are included for comparison. The resistances show a diurnal pattern with the lowest values during daytime. The surface resistance is highly dominating during daytime and the influence of meteorology low. During night-time the surface resistance to O_3 was substantially lower than the canopy stomatal resistance, which again shows that stomatal uptake is not the only sink for O_3. Table 12.2.1 shows that low surface resistances to O_3 also can be found during winter time, when the activity of the trees is low.

Table 12.2.1: Average day (06:00–18:00) and night (18:00–06:00) values of resistances (s m^{-1}) from Scherzheim (re-grown wheat), Ulborg (Norway spruce) and Corselitze (beech).

	re-grown wheat		Norway spruce						beech			
	September 92		June 94		July 92		February 93		April 94		May 94	
	day	night	day	night	day	night	day	night	day	night	day	night
r_a	74	120	8	31	6	68	16	27	17	69	7	11
r_b	26	32	9	14	9	20	16	17	14	21	7	8
r_c (O_3)	182	396	*151*	288	368	800	208	179	330	692	269	602
r_c (NO_2)	302	771	347	1023	-	-	1153	428	-	-	-	-

Fig. 12.2.2 shows that the flux of NO_2 over a beech forest was mostly upwards. Diurnal patterns with the largest fluxes during daytime was found both before and after bud burst. There seems to be a larger NO_2 flux in the first period than in the second. This might be either a result of lower bacterial activity in the second period due to the shadowing effect of the beech leaves, which leads to lower temperatures, or the result of uptake in the canopy. Experiments in coniferous forests (Ulborg and Speulderbos, NL [8]) have also shown that the flux of NO_2 can be both downwards and upwards. The upward flux of NO_2 is most likely explained by the reaction of O_3 with NO, emitted as a result of bacterial activity in the forest soil. The amount of O_3 removed in this way is, however, normally only a small part of the total O_3 flux.

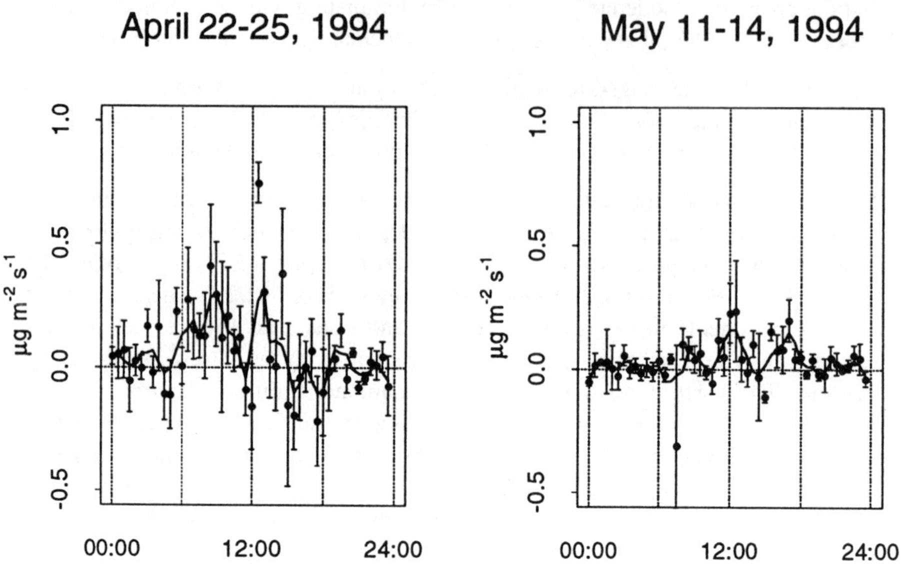

Fig. 12.2.2: Diurnal variation of NO_2 fluxes over a beech forest in Corselitze, Denmark before and after bud burst.

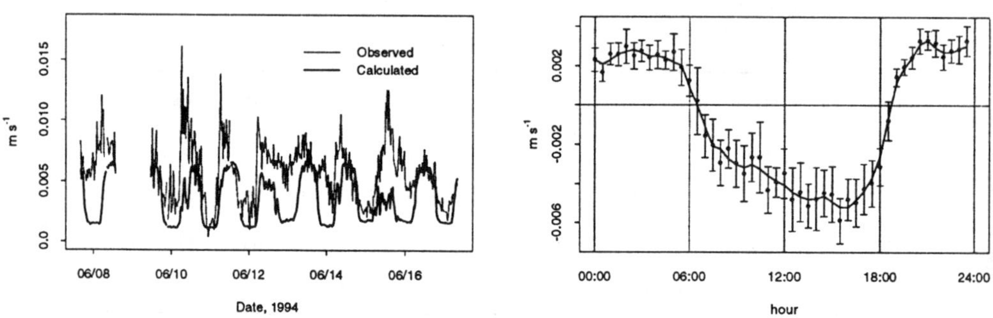

a) Time series of observed and calculated conductivity using $r_i = 130$ s m^{-1}

b) Diurnal variation of the difference between observed and calculated conductivity with $r_i = 50$ s m^{-1}

Fig. 12.2.3: Comparison of observed and modelled surface conductivity for ozone to a Norway spruce plantation.

The parameterisation for surface resistance proposed by Erisman, van Pul and Wyers [9] was evaluated using the previously mentioned data-set from Ulborg 7–17 June 1994. The proposed parameterisation generally underestimates the conductivity with 0.001–0.004 m s^{-1} (Fig. 12.2.3). However, when using the measured minimum canopy resistance (r_i) for water vapour (50 s m^{-1}) instead of the one tabulated by Wesely [10] (130 s m^{-1}), the average calculated conductivity fits the average observed, but the residuals show a clear diurnal variation with higher observed conductivity during the night and lower observed conductivity during the day than predicted by the parameterization. The diurnal variation of the residuals closely resembles the variation of relative humidity, suggesting that the parameterization should be further adjusted with respect to humidity.

Theoretical studies of the mean concentration and flux profiles of NO_2, NO, and O_3 [6] showed that it is possible to describe the profiles by an ordinary differential equation under stationary, horizontally homogeneous conditions, provided that the only processes we need to consider are the reaction of NO with O_3 and the photolytic conversion of NO_2, that there are only sources and sinks at the surface, and that there is a simple flux-gradient relationship. If it is assumed that the mean photostationary ratio (R) is constant with height, the differential equation yields analytic solutions in a neutral and an extreme convective case. The concentrations and fluxes are those of passive scalars when the production of NO_2 balances its destruction (R = 1). If the photolytic conversion of NO_2 is greater than the production, the flux changes sign at some height above the surface. It seems convenient to use the photostationary ratio as a parameter describing the behavioural difference between a passive scalar and the chemical active scalars NO_x and O_3. The differential equation can be solved numerically even in cases where R is not constant with height, provided that it is specified as a function of height.

Conclusions

We have found that the flux of O_3 was always downwards, whereas the flux of NO_2 could be both downwards and upwards depending on the concentration in the atmosphere and the underlying surface. Emission of NO from the soil reacting with O_3 is believed to be the local source of NO_2. The flux of O_3 is thus governed by several processes such as stomatal uptake, reaction with NO and possibly also destruction at surfaces.

Acknowledgements

The work has been funded by the National Agency of Industry and Trade, the Danish Environmental Research Programme and the Commission of the European Communities (contracts no. EV5V-CT92-0060 and EV5V-CT93-0276). The RIMI project (funded directly by Risø National Laboratory) provides the field station at Ll. Valby and meteorological and chemical equipment. The field station at Ulborg is provided by the Ion Balance Project through a contract with the Danish Forest and Landscape Research Institute and the National Environmental Research Institute.

References

1. P. Hummelshøj, Dry deposition of particles and gases. Technical Report Risoe-R-658, Risoe National Laboratory, Roskilde, Denmark, 1994.
2. P. Hummelshøj, K. Pilegaard, N. O. Jensen, Flux measurements of O_3 and NO_2 over a regrown wheat field. in: P.M. Borrell, P. Borrell, T. Cvitaš, W. Seiler (eds), *Proc. EUROTRAC Symp. '94*, SPB Academic Publishing bv, The Hague 1994, pp. 544–548.
3. N. O. Jensen, P. Hummelshøj, Derivation of canopy resistance for water vapour fluxes over a spruce forest, using a new technique for the viscous sublayer resistance. *Agricultural and Forest Meteorology* **73** (1995) 339–352.
4. L. Kristensen, The NO_x–O_3 triad in the turbulent atmospheric surface layer, in: S. Beilke, J. Slanina, G. Angeletti (eds) *Field Measurements and Interpretation of Species related to Photooxidants and Acid Deposition, Air Pollution Research Report* **39,** CEC 1992, pp. 37–49.
5. K. Pilegaard, P. Hummelshøj, N. O. Jensen, Deposition of ozone and nitrogen dioxide to open land and forest, in: J. Slanina, G. Angeletti, S. Beilke (eds) *General assessment of biogenic emissions and deposition of nitrogen compounds, sulphur compounds and oxidants in Europe. Air Pollution Research Report* **47**, CEC 1993, pp. 157–164.
6. K. Pilegaard, P. Hummelshøj, N. O. Jensen,. Fluxes of ozone and nitrogen dioxide measured by eddy correlation over a harvested wheat field, *Atmos. Environ.* in press.
7. K. Pilegaard, N. O. Jensen, P. Hummelshøj, Deposition of nitrogen oxides and ozone to Danish forest sites. in: T. Schneider (ed) *Proc. Specialty Conf. Acid Rain Research. Do we have enough Answers?* Elsevier 1994.
8. T.W. Choularton, H. Coe, S. Walton, M.W. Gallagher, K.M. Beswick, C. Dore, J.K. Duyzer, H. Westrate, K. Pilegaard, N.O. Jensen, P. Hummelshøj, Photochemical modification of ozone and NO_x deposition to forests: Results from the Speulderbos and Rivox experiments. in: P.M. Borrell, P. Borrell, T. Cvitaš, W. Seiler (eds), *Proc. EUROTRAC Symp. '94*, SPB Academic Publishing bv, The Hague 1994, pp. 645–650.
9. J. W. Erisman, A. van Pul, P. Wyers, Parameterization of surface resistance for the quantification of atmospheric deposition of acidifying pollutants and ozone, *Atmos. Environ.* **28** (1994) 2595–2607.
10. M. L. Wesely, Parameterization of surface resistances to gaseous dry deposition regional scale numerical models. *Atmos. Environ.* **23** (1989) 1293–1304.

12.3 Sulfate Dry Deposition Measurements over Low Vegetation and in a Forested Area

M.L. Sánchez, M.J. Domínguez and R. Rodríguez

Dept. of Applied Physics I. University of Valladolid. Spain.

Summary

Several sampling campaigns over flat terrain and forested areas were carried out from 1992 to 1994. The campaigns were aimed at obtaining preliminary information on the ambient concentrations of acidic aerosols and particle size and at analysing the daily cycles in order to examine their possible reactivity.

Using filter packs, sulfate dry deposition and velocities were determined over flat vegetation and in a forested area by means of the gradient method. Micrometeorological parameters were evaluated using both the gradient method and the eddy correlation technique. The average values of the dry deposition velocities were 0.95 and 0.17 cm s^{-1} over low vegetation and in the forested area respectively. A strong influence of the Monin Obukhov length on the deposition velocity was observed at both sites and a satisfactory dependence on $(-1/L)^{2/3}$ was found.

Aims of the research

The goal of this project is to measure and parametrise dry deposition fluxes and velocities in different land uses and to incorporate these results into a mesoscale model in order to contribute to an improvement in the current state of the art on the critical load values of acidifying compounds on a regional scale in Castile and Leon.

From 1992 to 1994 we accomplished the following objectives

- A setting up of the analytical technique to measure dry deposition of acidic compounds using filter packs. We have also characterised the particle size distribution.
- Six sampling campaigns over an Mediterranean oak forest and flat terrain were conducted. The campaigns were aimed at obtaining preliminary information on the ambient concentrations of aerosols, seasonal variation, particle size and daily patterns, at analysing the possible contribution of biogenic emissions from vegetation and trees. NO, NO_2, O_3, SO_2, HNO_3, VOC and CH_4 as well as SO_4^{2-}, NO_3^- and particle size distribution from 0.1 to 7.5 μm were also measured.

- A mobile mast equipped with slow and fast response meteorological instrumentation was installed as was the experimental device to collect aerosols using filter packs at 4 levels. Intercomparison exercises to check the accuracy of the collecting system were performed. The analytical error of the system was always lower than 5 %.

- A comparison between the micrometeorological parameters obtained using the eddy correlation technique and the gradient approaches and their influence on dry deposition fluxes was analysed. Also, two parametrisations were applied to determine sulfate dry deposition fluxes. The comparison of the results led to minor differences in the fluxes.

- The dry deposition fluxes initially proposed in a forested area were extended to a flat terrain and a comparison between the results of fluxes and deposition velocities in both sites was carried out.

- Measurements of sulfate aerosol fluxes were made over flat terrain in Castile and Leon during two sampling campaigns carried out in July, 1993 and in May, 1994. By June 1994, a second sampling campaign was performed in Les Landes forest in cooperation with the Laboratory of Aeorology at the University of Toulouse.

- Measurements of ozone dry deposition fluxes were also included using the gradient approach and a comparison between the results obtained in flat terrain and in a forest was performed.

- Great progress has also been made in the implementation of a mesoscale model, with a 20 × 20 km spatial resolution, to determine dry deposition of acidifying compounds. The following input parameters of the model have already been evaluated: inventory of emissions, wind fields, mixing height, land uses and turbulent parameters. Preliminary maps of dry deposition sulfates were also obtained using the parametrisations found during the development of BIATEX.

Scientific results

Data and procedure

Sulfate vertical profiles were measured in both sites using a four-lane piping system by equipping each pipe with an individual membrane pump. The sampling time was 2 h. In the flat terrain, the filter packs were located at 1.8, 4.0, 6.9 and 11.5 m from the ground and 54 vertical profiles were measured in July, 1993 and May, 1994. In Les Landes forest, the filter packs were situated at 6, 10, 17.5 and 25 m and 30 vertical profiles were measured from 18 to 22 of June, 1994. Most of the data were obtained during the daytime, from 6:00 a.m. to 8:00 p.m. and only one nocturnal profile per day was measured at around 2:00 a.m.

In order to determine the sulfate particle size, concurrent measurements of the aerosol spectra were performed with a laser spectrometer, PMS, model LAS-X, provided with 15 channels covering the dynamic range 0.10–7.50 μm. A linear fit between the concentrations and the number of particles recorded at each range was then performed. The examination of the variance of each linear fit revealed that most of the aerosol loading mass is associated with the spectra covering the 0.2–1 μm range with a maxima from 0.25 to 0.75 μm [1].

Temperature and wind profiles were systematically recorded using slow response temperature sondes. In addition, fast response instrumentation consisting of a three propeller (10 Hz) and a couple thermometer (25 Hz) located at 11.5 m and 24 m in each site was installed. Other common meteorological measurements such as relative humidity, solar radiation and wind speed were also systematically taken.

Sulfate fluxes were calculated following two methods. The first was based on the assumption that the eddy diffusivity of trace components may be replaced by the diffusivity of heat [2]. The second method was based on c^*

$$F = u_* c^*$$

which was calculated using the same procedure as described by Erisman [3]. The comparison between both methods led to minor differences. Micrometeorological parameters were calculated on a half hourly basis and each 2 h sulfate sampling period was covered by four half hour intervals of micrometeorological data by averaging.

In the low vegetation site, the measurements at four levels were used to determine the vertical profiles. In order to ensure quality of the vertical profile results, the following rejection criteria was used. Initially, linear fits between the logarithm of height versus the sulfate concentrations were carried out using the data recorded at the four levels. Those gradients with an error in the slope greater than 35 % were rejected. From the rejected gradients, the data which showed greatest uncertainty was removed and the same rejection criteria was again applied to the linear fits obtained with the three levels retained. When these last profiles presented an error in the slope greater than 35 %, they were definitively rejected. A total of 38 gradients covering unstable, stable and neutral atmospheric conditions were finally considered. In the forest, the displacement height concept was applied and the calculation of this parameter as well as the roughness length yielded an average value of 16.5 m and 1.9 m respectively. The deposition fluxes were then only evaluated above the canopy using the measurements provided by the two highest levels.

Main results

Low vegetation

The examination of data-to-data showed a strong influence of atmospheric stability on the deposition velocities and fluxes. Upward fluxes were generally obtained in the central period of the day and downward or slight upward fluxes were observed in the early morning and in late afternoon. However, deviations from this general behaviour were also obtained and a great variability of data-to-data was the most common feature. Based on these general considerations, an average daily pattern was calculated. The results obtained are shown in Fig. 12.3.1. As the figure reveals, maxima deposition velocities were obtained from 8:00 a.m. GMT to 2:00 p.m. GMT and minima values or emissions prevailed during the remaining period of time. Taking the Monin Obukhov length as a key to classify atmospheric stability, the average values were 1.29 and 0.012 cm s^{-1} for unstable and stable conditions respectively and the overall average value was 0.95 cm s^{-1}.

The analogy resistance model is not applied for particles and very few parametrisations are described in the literature [4, 5]. For unstable conditions Erisman [6] suggest a parameterisation on u_* and $1/L$:

$$V_d = u_*/100 \times (a + [\, b \times 300 / L]^{2/3})$$

where a=1 and b=1 in a forested area in the Netherlands. In order to find out the best fit for our data, the following procedure was applied. Firstly, a normalised deposition velocity was calculated for each individual measurement by dividing the dry deposition velocity by the friction velocity, V_d/u_*. Secondly, the data obtained were sorted on $1/L$ and for each $1/L$ range an average normalised dry deposition velocity was determined. Finally, as a strong influence of this parameter was found (see Fig. 12.3.2) and after taking into consideration the parametrisations described in the literature, a linear fit between V_d/u_* and $(-1/L)^{2/3}$ was performed, and the following equation was obtained:

$$V_d = u_*/100 \times (0.65 + [\, 0.54 \times 300 / L]^{2/3}) \qquad r = 0.94$$

Forested area

The average daily pattern shown in Fig. 12.3.3 was rather similar to that obtained over low vegetation. The data-to-data profiles presented less variability than in the low vegetation site and the average values for unstable and stable conditions, 0.2, and 0.04 cm s^{-1} for the overall data yield lower deposition velocities.

Fig. 12.3.4 shows the influence of $1/L$ on V_d/u_*. Following the same procedure as described above, the linear fit between V_d/u_* and $(-1/L)^{2/3}$ yields:

$$V_d = u_*/100 \times (-0.625 + [\, 0.21 \times 300 / L]^{2/3}) \qquad r = 0.998$$

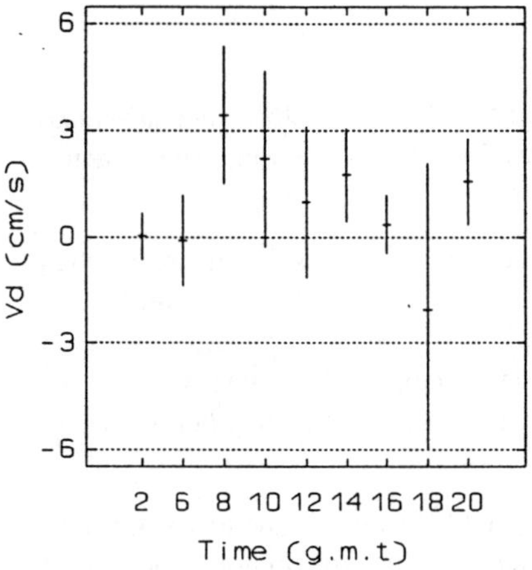

Fig. 12.3.1: Average dry deposition velocity daily pattern over low vegetation.

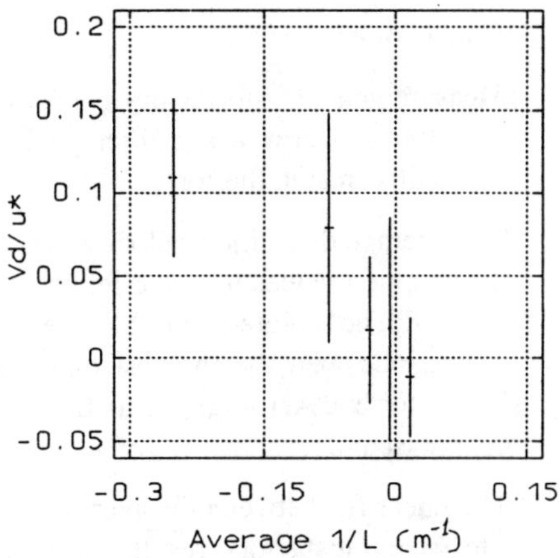

Fig. 12.3.2: Normalised deposition velocity versus L^{-1} over low vegetation.

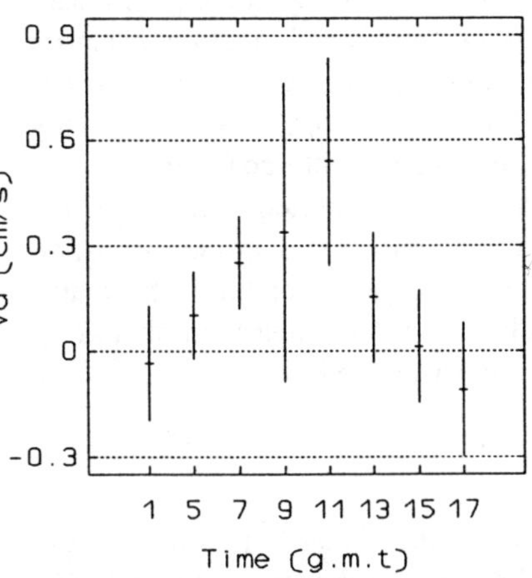

Fig. 12.3.3: Average dry deposition velocity daily pattern in a forested area.

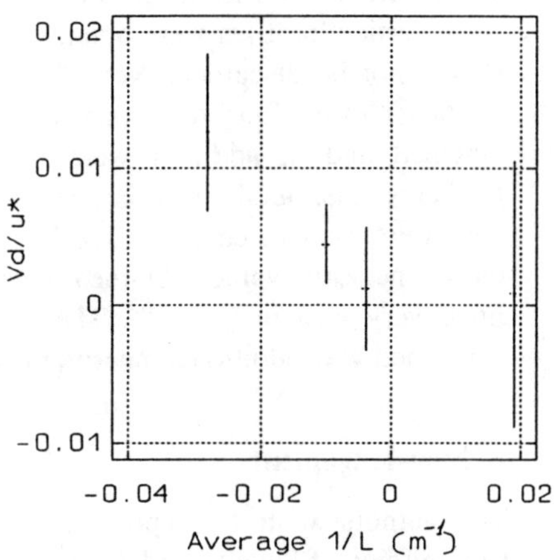

Fig. 12.3.4: Normalised deposition velocity versus L^{-1} in a forested area.

Conclusions

Measurements of sulfate dry deposition were carried out over low vegetation and in a forested area using filter packs. The results over flat terrain show greater variability than in the forest.

The average deposition velocity over low vegetation, 0.95 cm s^{-1}, was greater than the common values reported in the literature for submicron particles. Nevertheless, it should be pointed out that the highest sulfate deposition velocities were in accordance with the greatest ozone deposition velocity values measured by the Laboratoire d'Aerologie with the eddy correlation technique in other plains in central Spain.

The linear fits between the average V_d/u_* and $(-1/L)^{2/3}$ values in the low vegetation site gave satisfactory results. However, the slope value, the major contributing factor in V_d/u_*, was intermediate to those found for low vegetation and forested areas. The limited data available for stable conditions did not allow us to reach any definite conclusions.

The results related to the forested area should only be considered as preliminary since vertical profiles were obtained using only two levels and great uncertainties should not be discarded. Nevertheless, it should be mentioned that satisfactory gradients using four filter packs - in and above the canopy - were generally obtained and in addition were in general agreement with concurrent ozone gradients measured in this site during the daytime period. Although the measurements carried out in the Les Landes campaign are still in the evaluation phase, the V_d/u_* values obtained under unstable atmospheric conditions have also fitted very well in $(-1/L)^{2/3}$. The unexpectedly low values measured should be confirmed with additional measurements, using at least three levels.

Acknowledgements

The authors wish to express their gratitude to the CICYT (Interministerial Commission of Science and Technology), the Regional Government of Castile and Leon and the DG XII in the European Community for their financial support in the development of this project. We would also like to thank the Laboratoire d'Aerologie for sending us the meteorological data from the slow response instrumentation installed at Les Landes.

References

1. Sánchez M.L., Domínguez J., Rodríguez R., Experimental results of acidic aerosols and particle size distribution in a forested area, *Report EUR 15609/2 EN,* 1993, p. 1005.
2. Meixner F.X., Franken H.H., Duijzer J.H., van Aalst R.M., Dry deposition of HNO_3 to a pine forest, *Air Pollution Modeling and its Application VI*, Plenum Press, London 1988, p. 23.
3. Erisman J.W., Atmospheric deposition of acidifying compounds in the Netherlands. PhD Thesis, University of Utrecht, Utrecht, the Netherlands, 1992.
4. Wesely M.L., Cook D.R., Hart R.L., Speer R.E.. Measurements and parametrization of particulate sulfur dry deposition over grass. *J. Geophys. Res.* 90 (1985) 2131.
5. Wyres P., Geusebroek, M. Wayers, J.J Möls, Velkam A.C.. Dry deposition of submicron aerosol on a coniferous forest, in: P.M. Borrell, P. Borrell, T. Cvitaš, W. Seiler (eds), *Proc. EUROTRAC Symp. '94*, SPB Academic Publishing bv, The Hague 1994, p.712.
6. Erisman J.W., van Paul, A., Wyers P. Parametrisation of dry deposition mechanisms for the quantification of atmospheric input to ecossytems. *Air Pollution Research, Report 47.* (1993) 215.

Chapter 13

Mechanistic Studies

13.1 Emission of Gaseous Hydrocarbons and NH_3 out of Soils

Pascal Boeckx, A. Vermoesen and O. Van Cleemput

University of Ghent Faculty Agricultural and Applied Biological Sciences, Coupure links 653, B-9000 Ghent, Belgium

Summary

The biosphere, and soils in particular, interact importantly with the atmosphere. It was found that fertilisation with inorganic nitrogen (NH_4^+, NO_3^- and urea) contributes to the emission of NH_3, NO and N_2O. Field measurements of NO emission showed a clear diurnal fluctuation. The $CaCO_3$ content and pH of the soil are important soil parameters controlling NH_3, as well as NO and N_2O emission. However, incorporation of the fertilisers into the soil instead of surface application can substantially reduce the NH_3 emission.

If upland soils are water saturated and contain enough organic matter they can emit CH_4. On the other hand, they can be a potential sink for atmospheric CH_4 if the methane diffusion into the soil is sufficient. The sorption is substantially reduced by the N turnover of NH_4^+ or organic matter residues with low C/N ratios.

On a global scale wetlands are the largest source of atmospheric methane. From our study it was clearly shown that wetlands have a large spatial and seasonal emission variation, mainly controlled by water status and temperature interactions. These variations are important in assessing flux estimates. Landfills, on the other hand, are an important anthropogenic source of methane. However, a good management of the landfills can reduce the emission substantially. A good and cheap managing practice is covering the landfill with an effective methane oxidising soil layer.

Aims of the research

In general, we studied, both under field and laboratory conditions, the parameters influencing the emission and/or sorption of ammonia (NH_3), nitric oxide (NO), nitrous oxide (N_2O) and methane (CH_4) from soils. Ammonia volatilisation importantly contributes to acid deposition and causes a significant loss of nitrogen from agricultural soils. It was the aim to study the parameters influencing NH_3 volatilisation upon application of urea and NH_4^+ containing fertilisers on different soil types under different conditions. Next to the fact that N_2O has an important impact on atmospheric chemistry, it is a greenhouse gas contributing to global warming. NO emission contributes to acid deposition and atmospheric chemistry as well. We studied the influence of different soil properties and fertiliser amendments on NO and N_2O emission and the diurnal emission fluctuation of NO. After CO_2, CH_4 is the most important greenhouse gas. We looked at several factors controlling CH_4 production in different soil types, methane emission out of wetlands and landfills, as well as at the influencing parameters determining CH_4 sorption in soils.

Principal scientific results

NH_3 *volatilisation*

The global annual NH_3 emission is estimated at 75 (50–128) Tg yr^{-1}. The NH_3 emission from fertiliser use accounts for between 6.7 and 13.4 %. Urea is responsible for about 80 % of the NH_3 loss from fertilisers [1].

The influence of temperature, $CaCO_3$, moisture content, as well as fertiliser dose, physical form and depth of application on the NH_3 volatilisation from different soil types was studied under laboratory conditions. Four different types of fertiliser were used: ammonium sulphate (AS), ammonium nitrate (AN), urea (U) and liquid fertiliser (LF) (50 % U and 50 % AN). These data were compared with field measurements.

It was found that NH_3 emission increased with increasing soil temperatures. An increasing $CaCO_3$ content caused an increase of the ammonia emission (Equations 1 a–d). The emission was enhanced when AS was applied, because of the formation of $CaSO_4$. In case of U the influence of $CaCO_3$ was limited, because U hydrolysis causes a pH increase and a subsequently high emission from the beginning on. In general, application of AS on a $CaCO_3$ rich soil with a high pH enhanced NH_3 volatilisation , while a soil with a neutral to acid pH and a low buffering capacity promoted emission upon U application.

$$\text{AS:} \quad \%\ \text{N–loss} = 19.1\ (CaCO_3)^{0.22} \qquad R^2 = 0.93 \qquad (1a)$$

$$\text{AN:} \quad \%\ \text{N–loss} = 9.1 + 0.48\ (CaCO_3) \qquad R^2 = 0.96 \qquad (1b)$$

$$\text{LF:} \quad \%\ \text{N–loss} = 17.6 + 0.68\ (CaCO_3) \qquad R^2 = 0.92 \qquad (1c)$$

$$\text{U:} \quad \%\ \text{N–loss} = 24.2 + 0.30\ (CaCO_3) \qquad R^2 = 0.52 \qquad (1d)$$

The maximum NH_3 losses decreased linearly with increasing moisture content of the soil ($R^2 > 0.92$). At high moisture contents (field capacity) less NH_4^+ is available due to rapid nitrification and the exchange of NH_3 with the atmosphere is being hindered. The physical form and dose of the fertiliser had only a minor effect on the NH_3 emission. On the other hand, incorporation of the fertiliser at 2 cm depth drastically reduced the ammonia emission. For AS and AN the volatilisation was reduced by a factor of 10, while for U and LF the emission was reduced by a factor of 5. In general, it was observed that the sequence of ammonia volatilisation was the following: AS > U > LF > AN.

Field measurements were conducted both with an enclosure method [2] and the micrometeorological method [3]. Statistical analyses of the results showed no significant difference between the two methods (P = 0.568). From field experiments, on a clayey soil with 7 % $CaCO_3$ and pH 8, it was confirmed that the ammonia volatilisation from AS is higher than from U, LF and AN. It could be stated that high temperatures and a high soil moisture content followed by dry weather conditions favoured NH_3 volatilisation. Total losses between 10 and 20 % were very common.

The following measures could be taken to reduce NH_3 volatilisation from AS, AN and U. The application of AS should be avoided on soils with a pH > 7 and free $CaCO_3$. NH_3 emission from AN is low and can be further reduced using calcium ammonium nitrate. U application on a neutral to acid soil with a low buffer capacity is not recommended. Many problems with ammonia volatilisation can be avoided by incorporation of the fertilizer into the soil directly after spreading.

NO/N_2O *Emission*

The emission of NO and N_2O out of four different soils (Table 13.1.1) was found to increase both upon the application of NH_4^+ and NO_3^-, indicating that both nitrification and denitrification linked production occurred.

Table 13.1.1: Characteristics of the soils studied in the NO/N_2O - fertiliser experiment.

	Soil 1	Soil 2	Soil 3	Soil 4
Soil type	sand	sandy-loam	clay	heavy clay
pH	4.8	5.9	7.1	7.8
$CaCO_3$	0.30	0.30	0.39	16.8
NH_4^+ (t=0) mg kg^{-1}	11.2	25.2	16.5	8.0
NO_3^- (t=0) mg kg^{-1}	0.0	3.0	12.0	2.9

At low pH (pH = 4.8), denitrification was the dominant process producing N_2O and NO. The low N_2O/NO ratio observed in the NO_3^- treatment may indicate the occurrence of chemo-denitrification. At high pH (pH = 7.7) nitrification was dominant. A stimulating effect of nitrification on denitrification was observed in all soils except soil 1 (pH = 4.8). This seems to be caused by nitrification linked O_2 consumption creating conditions more favourable to denitrification. ANOVA showed that the variation in total NO (96.63 %) and N_2O emission (100 %) was almost entirely explained by the factor Soil, the factor Fertiliser explained the remaining 3.37 % of variation in total NO emission. Stepwise regression analysis was used to determine the soil characteristics explaining the highly explicative value of the factor soil in the ANOVA. For total NO emission the best description was obtained with the variables $CaCO_3$ content and $NH_4^+{}_{t=0}$ (R^2 adj.= 0.7328), whereas for total N_2O a model was selected containing pH, sand content and $NH_4^+{}_{t=0}$ (R^2 adj.= 0.6750), which seemed to give a more realistic description than a model with $CaCO_3$ content, water soluble carbon (WSC) and $NH_4^+{}_{t=0}$ (R^2 adj.= 0.6915). For six of the eight soil-fertiliser combinations a good to excellent correlation was obtained between N_2O emission rates and NO emission rates. The N_2O/NO ratio was found to be highly variable, indicating that the processes responsible for the emission do not depend in the same way on different conditions. The lack of correlation between the N_2O emission rates and NO emission rates in two soil-fertiliser combinations and the fact that the total NO emission did not respond in the same way to increasing doses of fertiliser as did the N_2O emission, indicate that NO production and N_2O production are not necessarily linked.

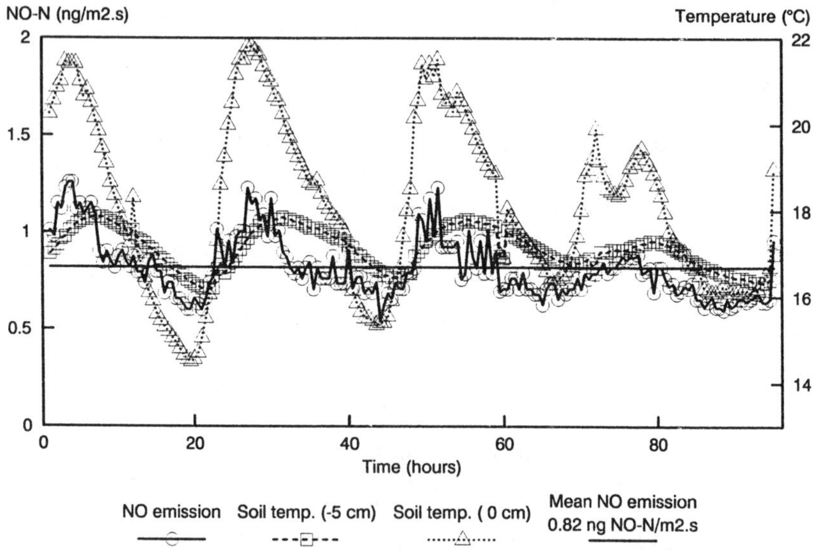

Fig. 13.1.1: Diurnal fluctuation of NO emission out of an acid forest soil

NO emissions were also measured from an acid forest soil (pH 3.6), using a dynamic chamber. In this forest soil a humic layer (± 5 cm) covered the sandy soil. From Fig. 13.1.1 it is clear that the NO emission followed a clear diurnal fluctuation, which was very well correlated with the soil temperature. This suggests that, although the acid pH, the NO production is of microbial origin. During two measuring periods of 4 consecutive days an average emission of 0.82 and 0.71 ng NO–N m^{-2} s^{-1} was found. The mean soil temperatures were 17.2 and 15.1 °C respectively. When the humic layer was removed, an emission of only 0.30 ng NO–N m^{-2} s^{-1} was found. These lower emissions are explained by the lower NH_4^+ (factor 5) and NO_3^- (factor 10) content of the sandy layer.

CH_4 *emission and sorption*

The global annual CH_4 emission is estimated at 500 Tg yr^{-1}, with an uncertainty of 10–20 % [3]. Both wetlands (115 Tg yr^{-1}) and landfills (30 Tg yr^{-1}) are important contributors to global CH_4 fluxes [3]. Projections for the year 2100 suggest that due to an increased population and waste generation, landfills will become a major source of atmospheric methane [4]. The methane emission from natural wetlands is estimated to remain constant during the next 100 years [4].

Methane emission out of soils is influenced by a large set of parameters. From 18 Belgian soils, with a sandy to heavy clayey texture, a pH range of 4–8, a $CaCO_3$ content of 0–17 % and an organic matter content of 1.6–7.5 %, the CH_4 production varied between 5 and 5.10^5 μg kg^{-1} soil after 31 days of field capacity conditions. Thus, upland soils can be a source of atmospheric methane. Although, no clear relation could be drawn between the soil texture and its methane production, a good correlation was found between the WSC content of different soils and their methane production ($R^2 = 0.91$, $CH_4 = 344.2$ WSC - 314.5). The incorporation of harvest residues increased the methane production, but the "quality" of the residues importantly determined the increase. One type of residue gives also other results in different soil types. The pH of the soil seems to control this effect. Grassland soils tend to produce more methane than arable land, because of the higher organic matter content. A good correlation was found between the organic matter content (OM, %) of different soils and the methane production ($R^2 = 0.97$, $CH_4 = 43.8$ OM - 82.4). The key parameter controlling the CH_4 emission from soils is the redox potential (*Eh*). A typical curve is shown in Fig. 13.1.2. Decreasing *Eh* levels caused an increase in methane production.

Methane emission out of a managed natural freshwater wetland was measured with the closed box method. The wetland is yearly flooded from about November until February. From March onwards, the remaining water is drained.

In this natural wetland, three fields were chosen with different drainage classes and neutral pH. The CH_4 emission was measured in July 1994. The wetland with a 'bad natural drainage' showed the highest CH_4 flux (5408 μg CH_4 m^{-2} d^{-1}). The methane flux out of the field with a 'fairly bad natural drainage' was 2045 μg CH_4 m^{-2} d^{-1}, while the soil with an 'insufficient natural drainage' had an average CH_4 flux of 1440 μg CH_4 m^{-2} d^{-1}. So, a clear relation between the drainage capacity of the soil and its methane emission was found.

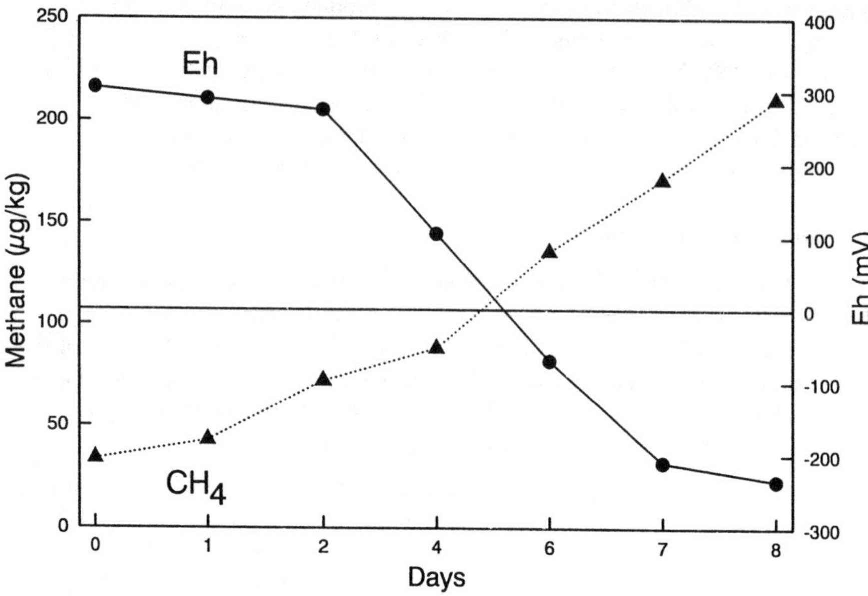

Fig. 13.1.2: Evolution of Eh (mV) and methane emission (µg kg^{-1})

The methane emission out of the wetland with the 'fairly bad natural drainage' was followed from March until December 1994 (Fig. 13.1.3). The methane flux varied between 2 and 136 mg CH_4 m^{-2} d^{-1} and the calculated methane emission during the measured period was 60 kg CH_4 ha^{-1}. From wetlands in northern Germany a similar emission was found [5]. The methane emission intensity could be subdivided in several periodic events, meanly controlled by the water status and temperature of the soil. From March until the end of May the emission increased gradually from 1.8 to 44 mg CH_4 m^{-2} d^{-1}, parallel with increasing soil temperatures. In June, high emissions were found (up to 136 mg CH_4 m^{-2} d^{-1}), due to a relatively high temperature and water saturation of the soil. During July and August the emission dropped to ± 8.6 mg CH_4 m^{-2} d^{-1}, while during September and October a second, smaller peak of methane emission was observed (max. 30 mg CH_4 m^{-2} d^{-1}). Although, the soil was flooded during November and December hardly any CH_4 emission could be detected (< 3 mg CH_4 m^{-2} d^{-1}).

To determine the methane emission at different depths, boxes were placed on the surface, at 15 cm depth and at 30 cm depth. From February until June 1994 a total methane emission of 40 kg CH_4 ha^{-1} was calculated from the boxes placed on the surface. For the boxes placed at 15, respectively 30 cm depth, an emission of 11.3 kg CH_4 ha^{-1} and 3.3 kg CH_4 ha^{-1} was determined. Obviously most of the emitted methane is produced in the upper humic layer of 15 cm.

In developed countries, landfills contribute importantly to the annual CH_4 emission. We measured methane emissions (measured on the surface) from two different soil-covered landfills ranging from 0 until 2,650 mg CH_4 m^{-2} d^{-1}. The methane productions (measured directly on the buried waste) varied between 760 to 24,600 mg CH_4 m^{-2} d^{-1}. Thus, between 100 and 90 % of the methane produced could be oxidised. From these field campaigns it was observed that a cover soil can be an important mitigating option to reduce methane emission from landfills.

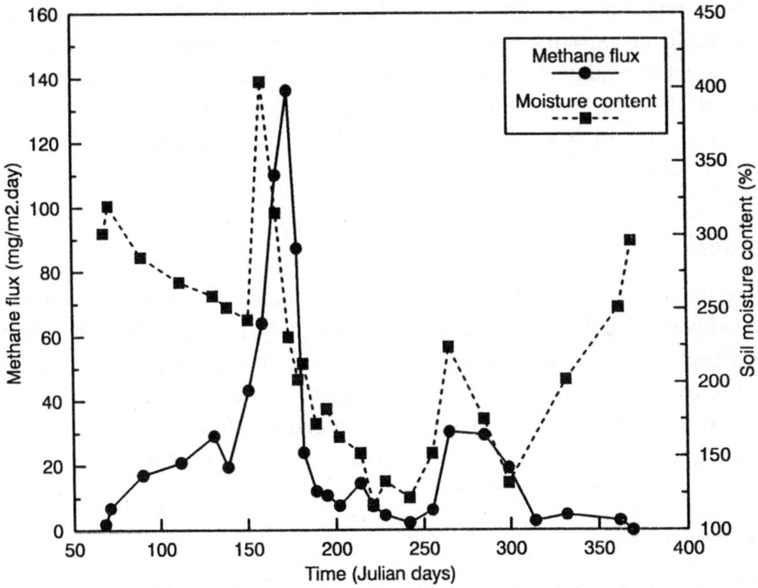

Fig. 13.1.3: The evolution of the methane emission intensity from a natural wetland

In order to study the parameters influencing the methane oxidation, laboratory experiments with a landfill cover soil (pH 7.2, sandy loam texture) were carried out. It was found that methane oxidation was influenced both by physical – microbial and chemical – microbial interactions. A moisture content of 15 % gave the highest methane oxidation rate (2.36 ng CH_4 h^{-1} g^{-1} soil). This is the result of an enhanced gas transport together with a sufficient microbial activity. Higher soil moisture contents reduced methane oxidation substantially due to a limited gas diffusion to the cell surface of the methane oxidising bacteria (30 %, 1.60 ng CH_4 h^{-1} g^{-1} soil). On the other hand, methane oxidation was also decreased at low moisture contents because of a reduced microbial activity. At 5 % moisture 0.84 ng CH_4 h^{-1} g^{-1} soil was removed. As well as for the moisture content an optimum temperature was found. The maximum methane oxidation rate is situated between 25 and 30 °C. An activation energy of 56.5 kJ K^{-1} mol^{-1} was calculated. It was also found that methane oxidation is controlled by nitrogen turnover, rather than by the actual mineral N content of the soil. Increased NH_4^+ additions to the soil resulted in an increased N_2O emission and restricted methane consumption. A negative linear correlation was found between the CH_4 oxidation rate (Y) and the nitrous oxide production rate (X) ($R^2 = 0.96$, $Y = 2.7 - 0.44\,X$). Nitrate additions, resulting in low N_2O production rates, had no significant effect on the methane oxidation. Methane oxidation was also strongly influenced by the C/N ratio of crop residue amendments. When the soil was amended with crop residues with low C/N ratios (13.5 and 11.3) high ammonium and nitrite contents were found (mineralisation), inhibiting the methane oxidation almost entirely [6]. When crop residues with higher C/N ratios (97.2 and 24.8) were applied, minor amounts of ammonium and nitrite were detected (immobilisation) and no effect on the methane oxidising capacity was observed. These results indicate that N turnover of organic matter amendments may have significant effects on the methane oxidising capacity of the soil. However, further experiments need to be done to check whether the effect of N turnover on the methane consumption is a temporary or long term effect.

Conclusions

With emphasis on the factor soil we studied the interaction of the biosphere with the atmosphere. We determined exchange rates of NH_3, CH_4, NO and N_2O from different ecosystems, as well as in laboratory experiments. In all our experiments we tried to explain the underlying controlling mechanisms. All these achievements fit fairly well to our original aims of emission and sorption of NH_3 and CH_4. To be complete we included NO and N_2O emission/sorption experiments.

Acknowledgements

We thank the Ministry of Science Policy of the Belgian Government (Project: EUREKA EUROTRAC BIATEX E7/02/03P) for their financial support. In particular, the smooth interaction with Mrs. M. Vanderstraeten is highly appreciated.

References

1. Schlesinger, W.H. and Hartley, A.E., A global budget for atmospheric NH_3, *Biogeochem.* **15** (1992) 191–211.
2. Van Den Abbeel, R., Claes, A. and Vlassak, K., Gaseous nitrogen losses from slurry manured land, in: Harmsen, J. and Hendriksen, K. (eds.), *Nitrogen in organic wastes applied to soils.* A.P. Harcourt Brace Jovanovich, London 1991, pp. 213–224.
3. IAEA (International Atomic Energy Agency), Manual on measurement of methane and nitrous oxide emissions from agriculture, *IAEA-TECDOC-674.* (1992) 91 p.
4. Kreileman, G.J.J. and Bouwman, A.F., Computing land use emissions of greenhouse gases, *Water, Air, Soil Pollut.* **76** (1994) 231–258.
5. Pfeiffer, E.M., Methane fluxes in natural wetlands (marsh and moors) in Northern Germany, *Current Topics in Wetland Biogeochemistry.* **1** (1994) 36–47.
6. Schnell, S. and King, G.M., Mechanistic analysis of ammonium inhibition of atmospheric methane consumption in forest soils, *Appl. Environ. Microbiol.* **60** (1994) 3514–3535.

13.2 Uptake and Distribution of NO_2 in Norway Spruce (*Picea Abies* [L.] *Karst.*)

Christian Brunold[1], Peter von Ballmoos[1], Stephan Nussbaum[1], Markus Ammann[1], Michael Stalder[1], Urs Peter Schlunegger[2] and Hans Gfeller[2]

[1] Pflanzenphysiologisches Institut der Universität Bern, Altenbergrain 21, 3013 Bern, Switzerland
[2] Institut für organische Chemie der Universität Bern, Freiestrasse 3, 3012 Bern, Switzerland

Summary

Potted spruce trees (*Picea abies* [L.] *Karst.*) were fumigated with NO_2 or $^{15}NO_2$ under controlled conditions in constant light. Current and previous flush needles, the bark and the fine roots were analysed for total ^{15}N content and incorporation of ^{15}N into the α-amino nitrogen of free amino acids. In addition, *in vitro* nitrate reductase activity and stomatal conductance of the needles were measured. In fumigated trees, nitrate reductase activity increased after a lag period of about four hours to levels which were five times higher than in controls exposed to filtered air. After a fumigation period of 24 hours nitrate reductase activity was linearly dependant on the NO_2 concentration used for fumigation both in previous and current flush. With an average of 9.1 % ^{15}N, free glutamate was the pool with the most label. Taking into account the time-course of the labelling of this pool, this figure can be taken as an estimate of the minimum contribution of NO_2 to the N nutrition of the needles. ^{15}N labelled amino acids were also detected in the bark and the roots, indicating export from the needles. There were linear correlations between the stomatal conductance and the ^{15}N uptake and between the ^{15}N uptake and nitrate reductase activity.

Aims of the research

Quantitative description of NO_2 deposition to spruce forests requires detailed knowledge of the mechanisms involved at the plant level. The aim of the present contribution was to study uptake of NO_2 in relation to gas exchange, induction of the nitrate reducing enzyme forming the basis for assimilation of NO_2–N and incorporation of NO_2–N into amino acids. In addition, a quantitative assessment of the contribution of NO_2 to the N nutrition of spruce trees was of great interest.

Principal scientific results

Fumigation of potted spruce trees with 100 μg NO_2 m^{-3} resulted in an increase of *in vitro* nitrate reductase activity in needles after a lag period of about four hours

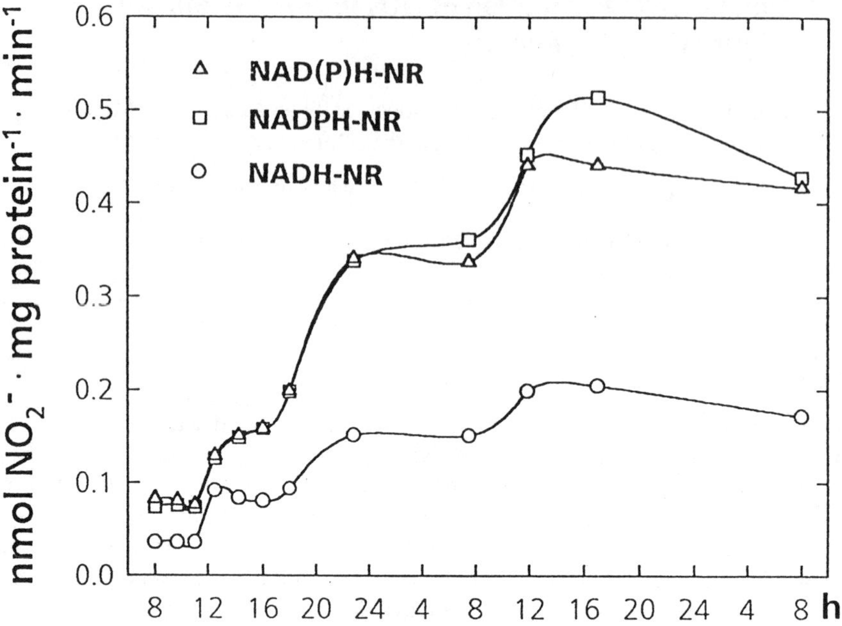

Fig. 13.2.1: Time course of nitrate reductase activity (nmol NO_2^- mg protein $^{-1}$ min^{-1}) needles of 5 year-old spruce trees induced by 100 µg NO_2 m^{-3} fumigation started at 8:00 on the first day and was stopped after 33 h. Nitrate reductase activity was measured with NADH, NADPH or a combination of NADPH and NADH (NAD(P)H) as indicated. The values presented are means of two measurements.

(Fig. 13.2.1), whereas the enzyme activity of control trees remained at the initial level. When NADH was used as a reductant the final level of this enzyme activity was 100 % higher than that of controls kept in charcoal filtered air containing 3 ± 1 µg NO m^{-3} and 4 ± 2 µg NO_2 m^{-3}. When NADPH or a combination of NADH and NADPH were used as reductant, the final level of nitrate reductase activity reached after 24 hours was 400 % higher than that of the controls.Varying NO_2 levels under otherwise constant conditions induced changes in nitrate reductase activity of previous and current spruce needles (Fig. 13.2.2). The activity was proportional to the NO_2 concentrations applied (Fig. 13.2.3).

Combining results from $^{15}NO_2$ fumigations performed in spring and autumn, 23 ± 10 % (N = 6) and 15 ± 6 % (N = 6) of the total label in needles was detected in free amino acids in current and previous flush, respectively, after fumigation for four days (Fig. 13.2.4 A and B).

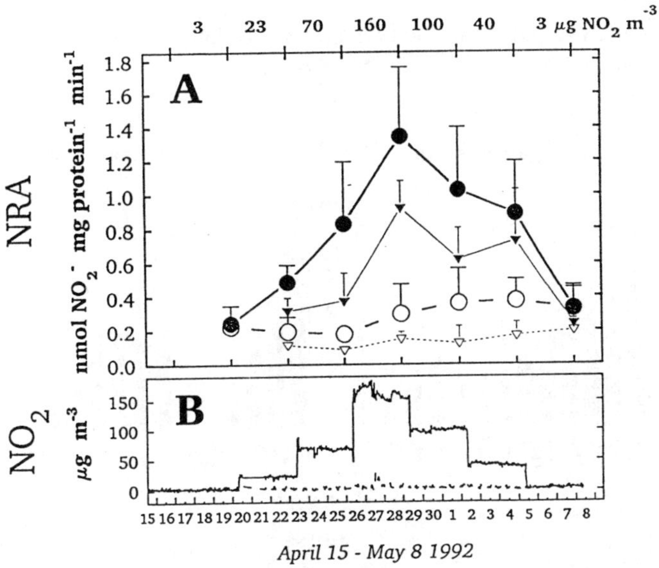

Fig. 13.2.2 A, B: Nitrate reductase activity (NRA, A) of previous (circles) and current flush needles (triangles) of 5-year-old *Picea abies* exposed to different levels of NO_2 (filled symbols) or Purafil and acitvated charcoal filtered air (open symbols) in spring 1992. Mean values ± SD of three trees are shown. At the beginning of the measurements current year's needles were about 2 weeks old. NO_2 concentrations (B) were continously measured.

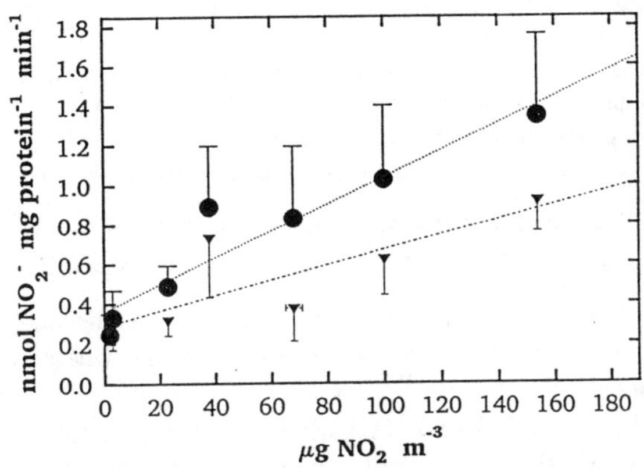

Fig. 13.2.3: Correlation between nitrate reductase activity (nmol NO_2^- mg protein^{-1} min^{-1}) of previous (circles, r = 0.9474, $p \leq 0.005$) and current flush (triangles, r = 0.7970, $p \leq 0.1$) spruce needles and NO_2 concentration used for fumigation. The results of Fig. 13.2.2 were used for the calculation.

At the same time 13 out of 18 detectable amino acids of the needles were labelled in the α-amino N with ^{15}N from $^{15}NO_2$. The glutamate pool was the most heavily labelled and contained 9.1 ± 1.2 % ^{15}N (n = 12). In samples of the spring experiment analysed after one and two days, the α-amino N of glutamate contained 5.3 ($n = 1$) and 5.3 ± 1.3 % ($n = 4$) ^{15}N, respectively.

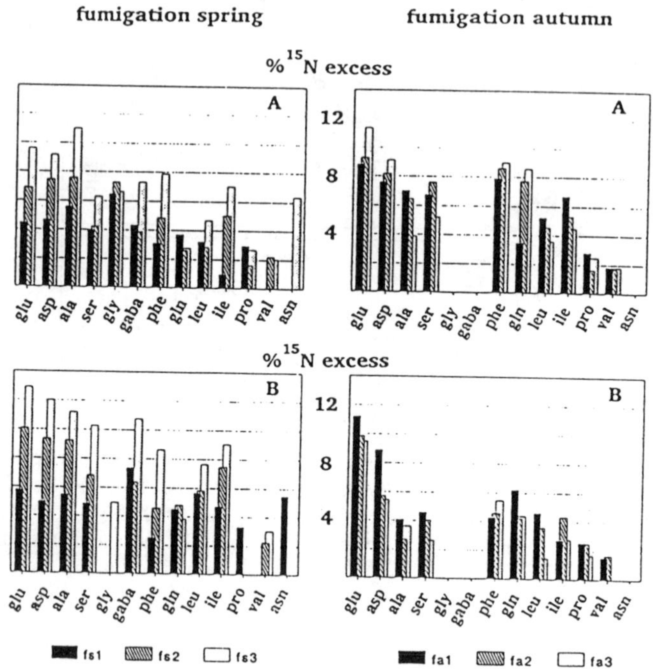

Fig. 13.2.4 A and B: Percent ^{15}N excess in the α-amino nitrogen of the free amino acids in A: current and B: previous flush needles of 6-year-old *Picea abies* trees fumigated with 60 ppb $^{15}NO_2$ in spring (*fs*) and autumn (*fa*). Numbers and shadings indicate individual trees: *fs* 1, 2, 3 and fa 1, 2, 3 are solid, shaded and stippled columns, respectively.

The $^{15}NO_2$ labelling experiment also demonstrate that ^{15}N label was transported out of the needles into the bark and the roots (Fig. 13.2.5 A and B).

In the bark the highest ^{15}N label was detected in glutamine and glutamic acid. These amino acids were also ^{15}N labelled in the roots where alanin was also substantially labelled. Total ^{15}N content was 43–156, 9.0–15.4 and 5.3–12.8 μg g $^{-1}$ dry weight in needle, bark and root tissue, respectively. With the aim of obtaining an estimate of possible uptake of ^{15}N from the soil, the top two centimeters were analysed with respect to ^{15}N enrichment. The resulting $\delta^{15}N$ was 30–40 ‰ relative to air.

The correlation between external NO_2 concentrations and nitrate reductase activity indicated that this enzyme activity is dependant on NO_2 uptake. To test this hypothesis, the rate of uptake of $^{15}NO_2$–N was analysed as function of the stomatal conductance measured on the day before harvest. A linear relationship between stomatal conductance and ^{15}N uptake rate is evident from Fig. 13.2.6. An internal resistance is negligeable at concentrations used in this study. It should be stressed, however, that our calculation is based on single point measurements at the given time with one-year-old needles. With two-year-old needles no comparable relationship was obtained.

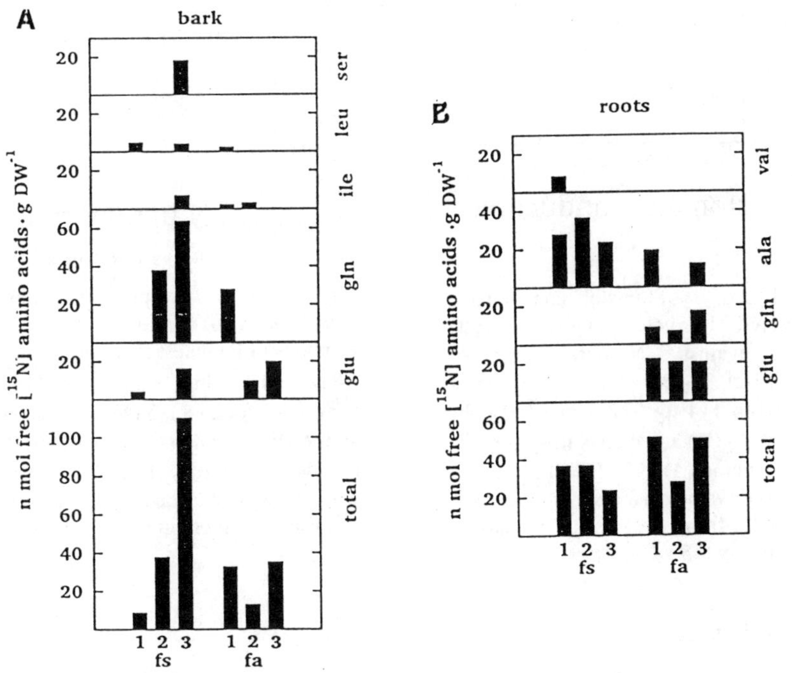

Fig. 13.2.5 A and B: ^{15}N label in total free amino acids and in selected amino acids of 4 year-old bark and of the fine root of 6 year-old spruce trees fumigated for 4 days with 60 ppb $^{15}NO_2$ in spring (*fs*) and autum (*fa*). Numbers indicate individual trees.

Fig. 13.2.7 shows a good correlation between nitrate reductase activity and the ^{15}N-uptake rate, calculated by dividing total ^{15}N uptake into the needles by the minutes of treatment. This result is in agreement with that presented in Fig. 13.2.6. The data in Figs. 13.2.6 and 13.2.7 indicate that almost the whole amount of NO_2 deposited on the needles was assimilated and that the efflux of label from the needles to the bark and the roots was low. This corresponds to the very low levels of ^{15}N in the bark and the roots of trees fumigated with $^{15}NO_2$ for four days (Fig. 13.2.5 A and B).

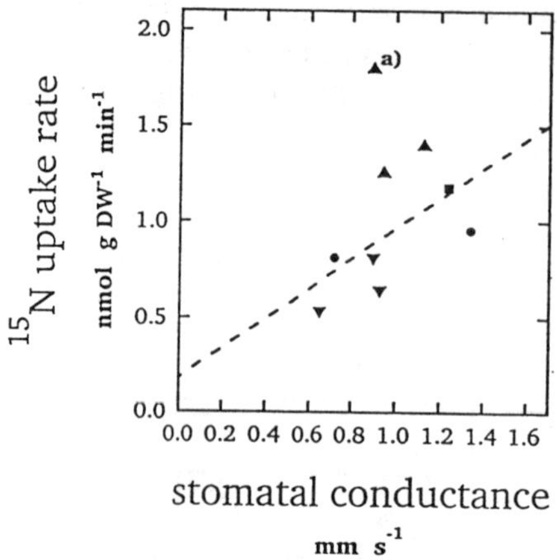

Fig. 13.2.6: Stomatal conductance (measured 16 hours before harvest) plotted against ^{15}N uptake rate of needles of 6-year old *Picea abies* exposed to 114 µg $^{15}NO_2$ m^{-3} (60 ppb) for 2 (●), 3 (■) and 4 days in spring (∇) and in autumn (Δ) 1991 ($r = 0.6173$, $p \leq 0.1$). The value marked with a) was excluded from regression analysis because $y - y > 3\ s_y$.

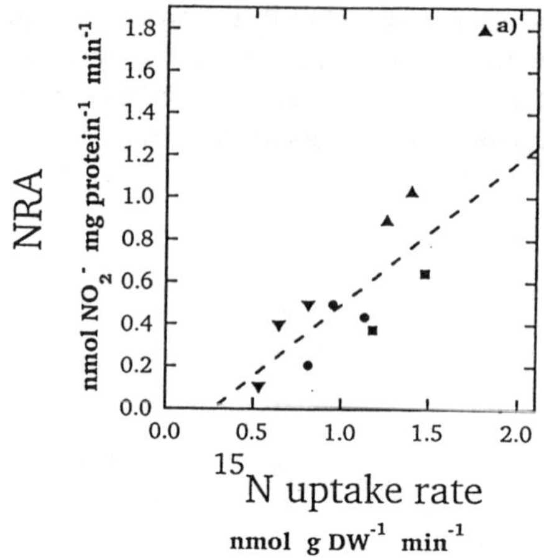

Fig. 13.2.7: ^{15}N uptake rate plotted versus *in vitro* nitrate reductase NRA activities of the latest flush of needles of 6-year-old P. abies exposed to 114 µg $^{15}NO_2$ m^{-3} for 2 (●), 3 (■) and 4 days in spring (∇) and in autumn (Δ) 1991 ($r = 0.7664$, $p \leq 0.01$). The value marked with a) was excluded from regression analysis corresponding to Fig. 13.2.6.

Conclusions

The data presented here provide evidence in support of the hypothesis, that in spruce needles NO_2 is taken up via the stomata and that elevated NO_2 results in increased incorporation of NO_2–N into amino acids. The amount of NO_2 taken up through the needles can be comparatively high in polluted environments such as in cities and along traffic rich roads. It is linearly correlated over a wide range to the NO_2 concentration in the atmosphere. As calculated by Schulze [1] the respective contribution of NO_x to the N nutrition of a healthy spruce stand exposed to an average of 15–20 ppb amounts to 1.7 %. Taking into account the linear relationship between NO_2 concentration and its absorption by *Picea abies* trees in this range [2] and assuming that the proportion of NO_2 incorporated into organic compounds remains constant as shown for sunflower [3] our estimated contribution from NO_2–N is compatible with the value calculated by Schulze [1].

^{15}N enrichment in free amino acids was also detected in bark and roots. Since both organs contain appreciable nitrate reductase activity [4], import of amino acids from the needles as well as direct uptake and assimilation *in situ* are both possible explanations for the presence of the label. In the bark none of the two possible ways of labelling can be excluded. In roots exclusive or almost exclusive import of the tracer from the shoot is the more likely explanation for the label in the amino acids since the bulk of the mass of the roots found in the bottom half of the pots and the total ^{15}N enrichment of the fine roots was distinctly higher than that of the top soil layer.

Our results show that NO_2 at moderate concentrations can act as nitrogen fertilizer for spruce. Whether this input is relevant with respect to possible adverse effects of increased nitrogen deposition postulated by Schulze [1] remains to be elucidated.

Acknowledgements

The project was financed by the Bundesamt für Bildung und Wissenschaft, Berne, Switzerland.

References

1. Schulze E-D., Air pollution and forest decline in a spruce (*Picea abies*) forest. *Science* **244** (1989) 776–783
2. Thoene B, Schröder P, Papen H, Egger A and Rennenberg H., Absorption of atmospheric NO_2 by spruce (*Picea abies* [L.] Karst.) trees. I. NO_2 influx and its correlation with nitrate reduction. *New Phytol.* **117** (1991) 575–585
3. Latus C, Förstel H, Führ F, Neubert A, Kley D., Quantitative measurements of NO_2 uptake and metabolism by sunflower plants. *Naturwissenschaften* **77** (1990) 283–285
4. Vogel J., Einfluss von Bodendüngung und Jahreszeit auf die Verteilung und die Induzierbarkeit der Nitratreduktaseaktivität bei Fichte (*Picea abies* [L.] Karst). Diplomarbeit, Universität Bern (1992).
5. Nussbaum S, von Ballmoos P, Gfeller H, Schlunegger U-P, Fuhrer J, Rhodes D, Brunold C., Incorporation of atmospheric $^{15}NO_2$ nitrogen into free amino acids by Norway spruce *Picea abies* [L.] Karst. *Oecologia* **94** (1993) 408–414
6. von Ballmoos P, Nussbaum S, and Brunold C The relationship of nitrate reductase activity to uptake and assimilation of atmospheric $^{15}NO_2$ nitrogen in needles of norway spruce (*Picea abies* [L.] *Karst.*) **29** (1993) 59–70.

13.3 Mechanisms of Release of NO_x and COS from Soil and Soil Micro-organisms

R. Conrad

Max-Planck-Institut für Terrestrische Mikrobiologie, Karl-von-Frisch-Str., D-35043 Marburg, Germany.

Summary

The flux of NO between soil and atmosphere was demonstrated to be a function of simultaneous production and consumption reactions which are both due to micro-organisms. Measurement of production and consumption rates in the laboratory allowed us to predict the NO flux *in-situ* by using a simple diffusion model. Both production and consumption of NO are controlled by soil variables such as temperature, pH, moisture and nutrients. NO is produced by many different micro-organisms of which nitrifiers and denitrifiers seem to dominate. Their relative contribution to NO production is affected by soil conditions, *e.g.* moisture. NO is reduced to N_2O by denitrifiers, but is also oxidised to nitrate by aerobic bacteria. The flux of COS between soil and atmosphere is also due to simultaneous production and consumption reactions which are at least partially due to biological processes. NO_2 turnover in soil is a purely chemical process.

Aims during the research

- parameterisation of the flux of NO_x and COS between soil and atmosphere as a function of production and consumption.
- Characterisation of the NO metabolism in different micro-organisms.
- Characterisation of the processes contributing to the flux of NO_x, in particular the role of nitrification and denitrification.

Principal scientific results

Processes involved in NO_2 *turnover*

Release of NO_2 from soils was only rarely observed. In contrast to the turnover of NO, only chemical processes seem to be involved in the turnover of NO_2 [2]. Production of NO_2 occurs at relatively high temperatures (45–65 °C) when the soil is drying up. In fertilized soils release of NO_2 then reaches rates that are comparable to those of NO release. We also studied uptake of atmospheric NO_2. The NO_2 uptake is apparently also due to non-biological processes. Uptake of NO_2 is stimulated by high humus contents. In dry and acidic soils, NO_2 is reduced to

NO, apparently by involving ferrous iron present in soil. It was possible to regenerate ferrous iron from ferric iron by exposure of the soil to SO_2.

Parameterisation of the flux of NO *and* COS

The flux of NO between soil and atmosphere was demonstrated to be a function of simultaneous production and consumption reactions according to the following equation [21]:

$$J = P - k\,m$$

$$m_c = P/k$$

with J = flux, P = gross production rate, k = uptake rate constant, m = NO mixing ratio in the atmosphere, and m_c = compensation mixing ratio at which NO production is balanced by NO consumption resulting in zero flux. This finding corroborated and extended the results of Johansson and Galbally [9]. The same relationship was also shown for the flux of carbonyl sulfide (COS) between soil and atmosphere (Lehmann and Conrad, submitted).

By measuring the flux of NO (or COS) as function of the NO mixing ratio in the atmosphere, it was now possibly to distinguish between the production and consumption processes, and by sterilisation and inhibition experiments it was possible to show that both production and uptake of NO are due to microbial activities. Chemical decomposition of nitrite to NO is also possible, but is usually limited by low microbial production and release of nitrite. In the case of COS, microbial processes contribute to production and uptake but, depending on soil type, also involve chemical processes to different extent.

Furthermore, it was now possibly to study the effects of environmental variables on production and consumption processes of NO separately. We studied the effects of soil pH [15], O_2 partial pressure [12], soil moisture [24], nitrate concentration [3], and soil temperature [22]. It became evident, that the NO production and consumption processes usually react completely different upon changes in the soil environment, with the consequence that the compensation mixing ratio (m_c) changes with changing soil conditions. This result implies that at a given NO mixing ratio in the atmosphere, the soil may change between acting as a source and a sink.

Determination of production rates (P) and uptake rate constants (k) in soil samples also allowed to predict the flux of NO in the field using the simple diffusion model described by Galbally and Johansson [7]. Comparison of the predicted fluxes with those that were actually measured in the field resulted in a relatively good agreement by a factor of 2–4 [20]. Experiments using intact soil cores under closely controlled conditions sometimes resulted an agreement of as good as ± 20 % (Rudolph and Conrad, submitted). These results show that fluxes in the field can possibly be predicted with a relatively good accuracy by using simple laboratory experiments.

Characterisation of the NO *metabolism in different micro-organisms*

At the beginning of the project, it was known from the literature that denitrifying bacteria produce NO although much of the metabolism was still unclear [4]. In the meantime, it is well accepted that NO is an obligatory intermediate in the sequential reduction of nitrate via nitrite, NO, N_2O to N_2 [26]. We observed that denitrifying bacteria indeed produce relatively large amounts of NO [17]. However, bacteria that reduce nitrate only to nitrite (nitrate respirers) also produce NO in similar amounts [10].

We have shown that denitrifiers reduce NO to N_2O with an extremely high affinity (K_m values < 6 nM) and that similar kinetics of NO consumption can also be observed in soil samples [17]. Although denitrification is a process that operates only under anoxic conditions, soils contain sufficient microenvironments where active denitrification is possible. In particular, the ability of denitrifiers to reduce NO was found to be relatively insensitive to increased O_2, thus explaining much of the NO consumption reactions that are observed in well aerated soils [19, 23].

However, NO consumption in soil can also be caused by aerobic bacteria. An interesting result was the observation that methanotrophic bacteria also consume NO and thus may contribute to the NO consumption process in soil, especially in those soils which have a high CH_4 oxidation activity [11]. It is unclear why the methanotrophs are consuming NO, as it is not used for any obvious purpose. We also isolated a *Pseudomonas* species which is able to oxidise NO to nitrate (Koschorreck *et al.*, in preparation). This oxidation reaction is also of no obvious use to the organism which grows aerobically on organic substrates.

It was known from the literature that nitrifying bacteria produce NO [4]. Our experiments demonstrated that nitrifiers oxidise ammonium and then partially reduce the product nitrite to NO and N_2O [16]. Thus, NO production by nitrifiers in principle is also a denitrification process. The difference of NO production between nitrifiers and denitrifiers is that the ultimate precursors are ammonium and nitrate, respectively. This is also the basis to distinguish between the two processes when analysing the contribution of the two groups of organisms to NO production in soil (see below).

Characterisation of the processes contributing to the flux of NO

It is possible to distinguish between the contribution of nitrification and of other processes (mainly denitrification) to the production of NO. This distinction can either be based on the use of compounds that specifically inhibit the oxidation of ammonium (nitrapyrin, acetylene) or on ^{15}N labelling experiments [18]. Both techniques do not discern between heterotrophic and autotrophic nitrification, and also not between the different types of nitrate dissimulation, i.e. denitrification, nitrate respiration, and nitrate reduction to ammonia. However, they allow the distinction between the pathways based on oxidation of ammonia (summarily nitrification) versus those based on reduction of nitrate (summarily denitrification).

Both nitrification and denitrification contribute to NO production in soil, but the percentage contribution changes with changing soil conditions. In this respect, the water content is an important soil variable as it affects the gas-filled porosity and thus the aeration of the soil. Basically, denitrification is favoured by high soil moisture contents, nitrification by low ones. The data base is presently still poor, but the picture that is emerging is usually consistent with the theory [5] (Bollmann and Conrad, in preparation). However, several observations indicate that nitrification may be the dominant NO production process even under wettest soil conditions [8, 24, 25]. Furthermore, it is unknown to what extent other soil variables actually influence the relative contribution of nitrification and denitrification to NO production. From most of the studies it is also obvious that the relative contribution to NO production may be significantly different from that to N_2O production which is taking place simultaneously.

Conclusions

The principal aims during the project were all achieved. At the beginning of the project, little explicit knowledge existed about the processes involved in the release of NO from the soil into the atmosphere, *e.g.* [1, 6, 9, 13, 14]. Much was based on assumptions in analogy to that what was known about the processes of N_2O production. Production processes of N_2O had been studied much earlier, since the instrumentation for the analysis of atmospheric N_2O (electron capture detector) was available much earlier and cheaper than that for analysis of atmospheric NO_x (chemiluminescence detectors). The results of our studies demonstrated that despite some principal similarities to N_2O production, the processes involved in NO turnover exhibit many unique features.

The most striking feature, and also that which is most relevant for NO fluxes in the field, is the fact that NO is as rapidly consumed in soil as it is produced. We showed a relatively easy way to parametrise production and consumption in laboratory experiments and use the data to model the NO fluxes in the field. However, the applicability of this procedure has still to be tested by more studies comparing field and laboratory data.

The next logical step in prediction of NO fluxes in the field would now be the modelling of NO production rates and NO uptake rate constants from simpler soil variables. An explicit mechanistic modelling will not be easy because of the following reasons:

(1) Empirical experiments show that both production and consumption of NO react independently on changes of all kinds of soil variables, *e.g.* soil texture, soil moisture, pH, temperature, concentration of nitrogen and organic carbon *etc.*

(2) NO production is dominated by at least two completely different processes (nitrification and denitrification) which a priori should be regulated differently.

(3) NO consumption is probably not only dominated by reduction through denitrification but in addition by oxidation through still enigmatic processes. We have to expect that these NO consumption processes are also differently regulated. Therefore, it is evident that more process studies are required to help further in the development of mechanistic models. These models should be as simple as possible but should not neglect the relevant controls.

Acknowledgements

I thank my coworkers who did all the experimental work, *i.e.* M. Baumgärtner, M. Blösl, A. Bollmann, I. Kalkowski, M. Koschorreck, M. Krämer, S. Lehmann, W. Nägele, A. Remde, J. Rudolph, O. Saad, F. Schäfer, and M. Schuster. I thank the members of BIATEX for helpful cooperation and discussion, and the German Bundesministerium für Forschung und Technologie for financial support during 1989–1991.

References

1. Anderson, I. C., J. S. Levine, Relative rates of nitric oxide and nitrous oxide production by nitrifiers, denitrifiers, and nitrate respirers. *Appl. Environ. Microbiol.* **51** (1986) 938–945.
2. Baumgärtner, M., E. Bock, R. Conrad, Processes involved in uptake and release of nitrogen dioxide from soil and building stones into the atmosphere. *Chemosphere* **24** (1992) 1943–1960.
3. Baumgärtner, M., R. Conrad, Role of nitrate and nitrite for production and consumption of nitric oxide during denitrification in soil. *FEMS Microbiol. Ecol.* **101** (1992) 59–65.
4. Conrad, R., Flux of NO_x between soil and atmosphere: Importance and soil microbial metabolism, in: N. P. Revsbech, and J. Soerensen (eds), *Denitrification in Soil and Sediment.* Plenum, New York 1990, pp. 105–128.
5. Davidson, E.A., Soil water content and the ratio of nitrous oxide to nitric oxide emitted from soil, in: R.S. Oremland (ed), *Biogeochemistry of Global Change.* Chapman & Hall, New York 1993, pp. 369–386.
6. Firestone, M.K., R.B. Firestone, J.M. Tiedje, Nitric oxide as intermediate in denitrification: evidence from nitrogen-13 isotope exchange. *Biochem. Biophys. Res. Comm.* **91** (1993) 10–16.
7. Galbally, I. E., C. Johansson, A model relating laboratory measurements of rates of nitric oxide production and field measurements of nitric oxide emission from soils. *J. Geophys. Res.* **94** (1989) 6473–6480.
8. Hutchinson, G. L., W. D. Guenzi, G. P. Livingston, Soil water controls on aerobic soil emission of gaseous nitrogen oxides. *Soil Biol. Biochem.* **25** (1993) 1–9.
9. Johansson, C., I.E. Galbally, Production of nitric oxide in loam under aerobic and anaerobic conditions. *Appl. Environ. Microbiol.* **47** (1984) 1284–1289.
10. Kalkowski, I., R. Conrad, Metabolism of nitric oxide in denitrifying *Pseudomonas aeruginosa* and nitrate–respiring *Bacillus cereus. FEMS Microbiol. Lett.* **82** (1991) 107–112.

11. Krämer, M., M. Baumgärtner, M. Bender, R. Conrad, Consumption of NO by methanotrophic bacteria in pure culture and in soil. *FEMS Microbiol. Ecol.* **73** (1990) 345–350.

12. Krämer, M., R. Conrad, Influence of oxygen on production and consumption of nitric oxide in soil. *Biol. Fertil. Soils* **11** (1991) 38–42.

13. Lipschultz, F., O.C. Zafiriou, S.C. Wofsy, M.B. McElroy, F.W. Valois, S.W. Watson, Production of NO and N_2O by soil nitrifying bacteria. *Nature* **294** (1981) 641–643.

14. McKenney, D. J., G. P. Johnson, W. I. Findlay, Effect of temperature on consecutive denitrification reactions in Brookston clay and Fox sandy loam. *Appl. Environ. Microbiol.* **47** (1984) 919–926.

15. Nägele, W., R. Conrad, Influence of pH on the release of NO and N_2O from fertilized and unfertilized soil. *Biol. Fertil. Soils* **10** (1990) 139–144.

16. Remde, A., R. Conrad, Production of nitric oxide in *Nitrosomonas europaea* by reduction of nitrite. *Arch. Microbiol.* **154** (1990) 187–191.

17. Remde, A., R. Conrad, Metabolism of nitric oxide in soil and denitrifying bacteria. *FEMS Microbiol. Ecol.* **85** (1991) 81–93.

18. Remde, A., R. Conrad, Role of nitrification and denitrification for NO metabolism in soil. *Biogeochem.* **12** (1991) 189–205.

19. Remde, A., R. Conrad, Production and consumption of nitric oxide by denitrifying bacteria under anaerobic and aerobic conditions. *FEMS Microbiol. Lett.* **80** (1991) 329–332.

20. Remde, A., J. Ludwig, F. X. Meixner, R. Conrad, A study to explain the emission of nitric oxide from a marsh soil. *J. Atm. Chem.* **17** (1993) 249–275.

21. Remde, A., F. Slemr, and R. Conrad, Microbial production and uptake of nitric oxide in soil. *FEMS Microbiol. Ecol.* **62** (1989) 221–230.

22. Saad, O.A.L.O., R. Conrad, Temperature dependence of nitrification, denitrification, and turnover of nitric oxide in different soils. *Biol. Fertil. Soils* **15** (1993) 21–27.

23. Schäfer, F., R. Conrad, Metabolism of nitric oxide by *Pseudomonas stutzeri* in culture and in soil. *FEMS Microbiol. Ecol.* **102** (1993) 119–127.

24. Schuster, M., R. Conrad, Metabolism of nitric oxide and nitrous oxide during nitrification and denitrification in soil at different incubation conditions. *FEMS Microbiol. Ecol.* **101** (1992) 133–143.

25. Skiba, U., K. A. Smith, D. Fowler, Nitrification and denitrification as sources of nitric oxide and nitrous oxide in a sandy loam soil. *Soil Biol. Biochem.* **25** (1993) 1527–1536.

26. Zumft, W.G., The biological role of nitric oxide in bacteria. *Arch. Microbiol.* **160** (1993) 253–264.

13.4 The Chemical Behaviour and Deposition of Ammonium Nitrate, Nitric Acid and Ammonia

Ralph Dlugi[1], Lucia Kins[1], Thomas Seiler[1], Winfried Seidl[1], Peter Seifert[1], Anne Kasper[2], Herman Sievering[3], Michael Zelger[1], Eva Köhler[1] and Karin Reußwig[1]

[1]Meteorologisches.Institut, Universität (MIM), Theresienstraße 37, D-80333 München, Germany,
[2]Institut für Analytische Chemie, Technische Universität, Wien, Austria,
[3]Center for Environmental Sciences, University of Colorado at Denver, CO, USA

Introduction

The TRACT experiment and the BIVOTEX studies at the Bavarian Forest site offered the possibility of investigating the composition and deposition behaviour of aerosol particles and gaseous compounds above and within tall vegetation (spruce forest with 14 % beech) and in an agricultural area in the Rhine valley. During the intensive field campaign of TRACT a study was also performed on the thermodynamic behaviour of the chemically coupled system HNO_3–NH_3–NH_4NO_3 with respect to the phase transition and dry deposition or/and emission.

The interaction of aerosol particles with the gas phase has already been investigated in a number of studies. Experimental investigations were performed *e.g.* by Harrison and Pio, and Clarke *et al.* in England, Wall *et al.* in Los Angeles, Dlugi *et al.*, and Seidl *et al.* and Brunnemann *et al.* near Leipzig, Germany [1–6]. All these measurement sites were located in regions of both rural (maritime) and anthropogenic influences on the chemical composition of the atmosphere.

A number of theoretical studies described the aerosol particles in equilibrium with the gas phase [7–8], the uptake of water by hygroscopic particles [9] and considered also the transport limitation of the exchange between the gaseous and particulate phase [10, 11]. Thus $NH_4NO_3^-$ particles with radii larger than about 1 µm should not be in equilibrium with the surrounding air. But it can be also assumed that the equilibrium in the chemical system under consideration for particles with $r < 1$ µm can be kinetically limited. Kramm and Dlugi [12] and Huebert *et al.* [13] discussed influences of the equilibrium and non-equilibrium conditions on the determination of trace gas and particulate dry deposition and emission fluxes.

Measurement procedure

The aerosol particles were sampled by filter packs (sampling time ≤ 1 h) and by five stage impactors (4 hour samples) and were analysed for NH_4^+, Na^+, K^+, SO_4^{2-}, NO_3^-, Cl^- NO_2^-, SO_3^{2-} and HPO_4^{2-}. The aerodynamic cut-off diameters of the impactor stages 1 to 5 were 0.04 μm, 0.14 μm, 0.49 μm, 1.72 μm, and 6.0 μm. NH_3 and HNO_3 have been sampled on filter packs. In general 3 impactor samples and 7 to 8 filter samples have been taken per day. The detailed sampling and analytical procedures are described by Brunnemann *et al.* [6] and Seidl *et al.* [5].

Size dependence of the particle properties

With respect to the particle origin and the different relaxation times for the establishment of the thermodynamic equilibrium between the gas and the particulate phase for different particle sizes, one has to divide the aerosol size distribution into the three modes, nucleation mode, accumulation mode and coarse mode, as defined by Whitby [14]. The accumulation mode particles are either emitted by anthropogenic sources, or they have grown from the nucleation mode by coagulation, condensation and cloud liquid phase reactions. Most measurements showed a distinct accumulation mode for sulfate. If a maximum radius of 1.0 μm is assumed as an upper limit for the accumulation mode, then these particles can reach thermodynamic equilibrium, *e.g.* with gas phase NH_3 within the one-hour filter sampling time and the four-hour impactor sampling period even at temperatures below 280 K [11]. Thermodynamic equilibrium in a shorter time scale than the collection period cannot be observed by particle sampling with an impactor, filter packs or denuders. Therefore actually the available sampling techniques limit the possibilities for tests of the equilibrium/non-equilibrium conditions in the atmosphere. But if the conditions of stationarity and homogeneity exist the influences of reactions on the mean flux profile scan be examined [12].

Results

The prediction of the equilibrium constant for the NH_4NO_3–HNO_3–NH_3 system compared to the equilibrium constant as empirically determined is given in Fig. 13.4.1 for different experiments being performed with the same experimental techniques (*e.g.* filter packs, impactors; profile technique and cuvette measurements for deposition and emission).

Sulfate, nitrate and ammonium have been the dominating ions in the accumulation mode particles at the measurement sites. If the equivalent concentrations of NH_4^+, K^+ and Na^+ are plotted against the sum of the equivalent concentrations of Cl^-, SO_4^{2-} and NO_3^-, one can see that the particles are not neutralised by ammonia emissions from agricultural activity in the nested area of the TRACT experiment. This behaviour was also found in urban areas like Munich or Leipzig [4, 5]

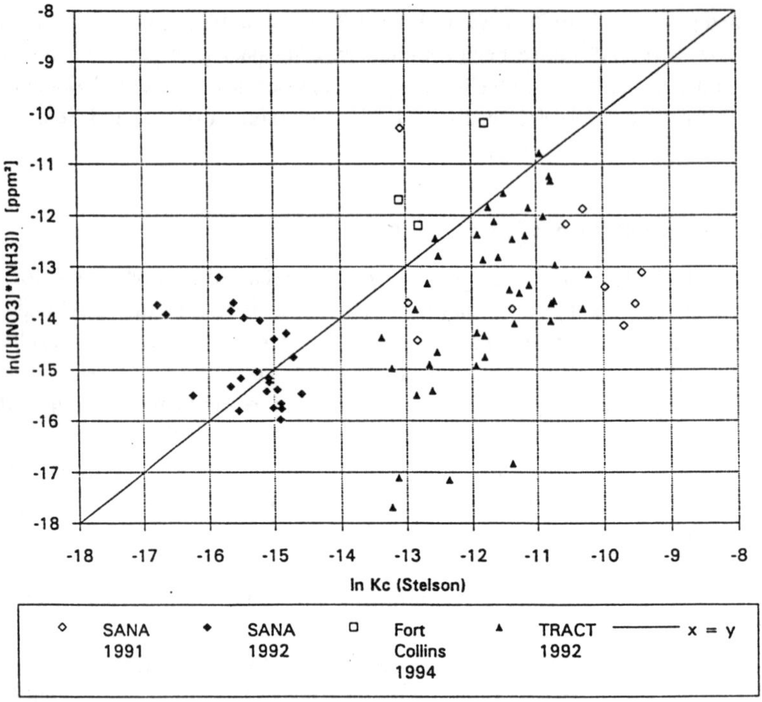

Fig. 13.4.1: The comparison of the predicted and the experimentally determined equilibrium constant *K*p for the system HNO_3–NH_3–NH_4NO_3.

Nitrate gas - particle phase partition

For nitrate, a distinct peak in the accumulation mode has been found in the morning aerosol samples in the TRACT area but also during other field studies [5, 6]. As nitrate, sulfate and ammonium were the only major ions in the accumulation mode, nitrate must have existed as ammonium nitrate.

However, ammonium nitrate is a rather thermodynamically unstable compound which dissociates as a function of temperature [15, 16]. Fig. 13.4.2 shows the ratio of the nitrate in the gas phase to the nitrate in the accumulation mode particles in dependence of the temperature for conditions with nitrate being present as NH_4NO_3 in different areas of Germany [5]. Although this ratio is no specific quantity like the dissociation constant of ammonium nitrate but also depends on the total nitrate concentration, one can clearly see a temperature dependence. At temperatures below approximately 15 °C, most nitrate is in the accumulation mode particles, and at temperatures above 20 °C, most nitrate is found in the gas phase as HNO_3.

The coarse mode nitrate was excluded from these considerations mainly because it was present as other salts than ammonium nitrate. These salts are thermodynamically more stable and can exist at higher temperatures. In addition, the coarse mode particles are assumed to be not in equilibrium with the gas phase. The coarse mode nitrate originates from diffusion and subsequent absorption of HNO_3 or its precursor gases to the coarse mode particles. This process at the Schachtenau site in the Bavarian Forest appears to be important for the removal of nitrate from the atmosphere because the larger particles have large sedimentation velocities [17].

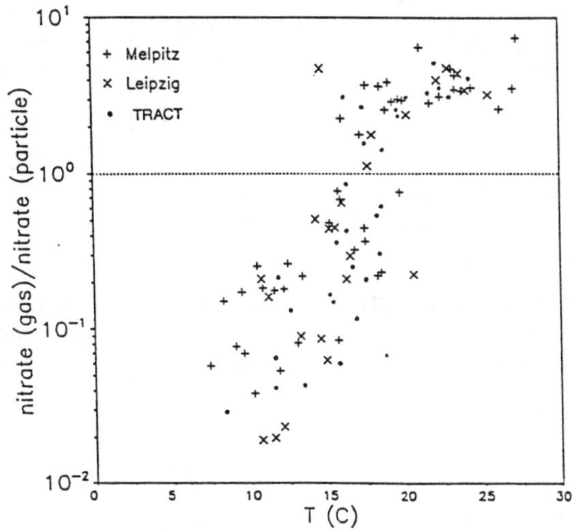

Fig. 13.4.2: The temperature dependence of the relation nitrate (gas)/nitrate (particle) for conditions with particulate nitrate being mainly NH_4NO_3.

Deposition and emission

The deposition and emission of trace substances like HNO_3, NH_3 and NH_4NO_3 was studied by cuvette and profile techniques. The NH_3 concentrations as function of surface temperature from the Bavarian forest site and the TRACT site (meadow) are highly variable and can be above or below the possible ranges of compensation points (Fig. 13.4.3).

The determination of resulting dry deposition velocities of HNO_3 shows that for some conditions the reaction between HNO_3 and NH_3 can enhance the HNO_3 flux or V_d (HNO_3) in a way that the relation $V_d < (HNO_3)_{reac} > (r_t + r_{mt})^{-1}$ is fulfilled. Then V_d $(HNO_3)_{reac}$ is larger than the stability dependent theoretical upper values V_d (max) for $r_c = 0$.

Thus, under these conditions, the influence of chemical reactions on the concentration profile and the fluxes of these nitrogen compounds can be detected experimentally (Fig. 13.4.4).

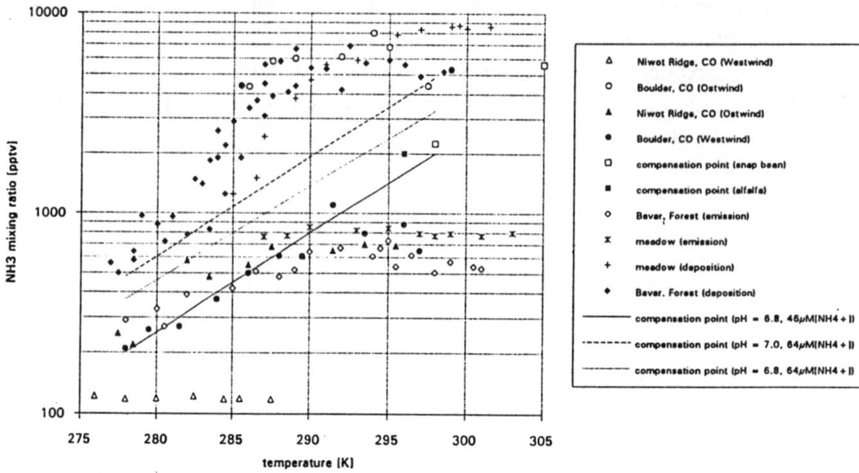

Fig. 13.4.3: Temperature and pH-dependence of NH_3 mixing ratios at the TRACT measuring site (meadow) and the BIVOTEX site (Bavarian forest) compared to data of Langford and Fehsenfeld [17] and to different estimations of compensation points [18]

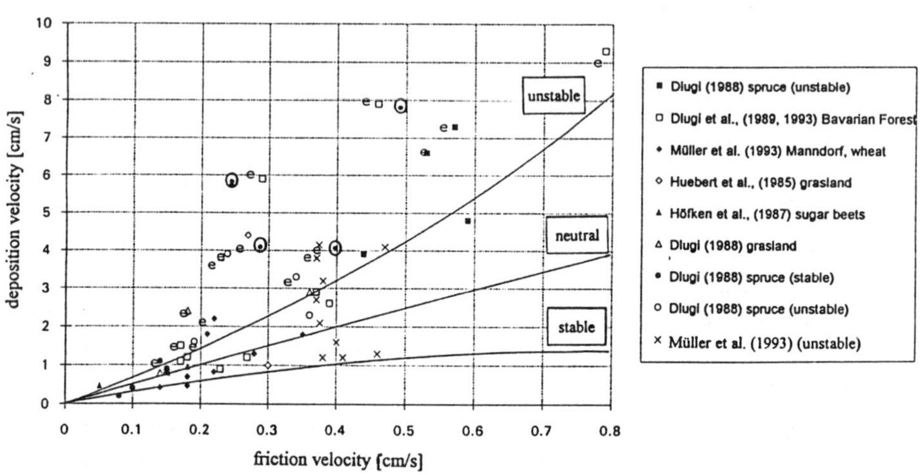

Fig. 13.4.4: The deposition velocity of HNO_3 as function of u* and for different ranges of stability indicating that values above a certain range are influenced by chemical reactions in the systems HNO_3–NH_3–NH_4NO_3. Some first results from model calculations (o) indicate that the measured and calculated enhancement of V_d (HNO_3) agrees within the errors of experimental data and physico-chemical parameter. Enhanced values are marked by 'e'.

But as shown by Kramm and Dlugi [12] any kinetic limitations (or deviation from equilibrium) can only be detected by the sampling methods currently used if the concentrations are nearly constant over the sampling time (≤ 1 h). Consequently as shown recently by Seidl *et al.* [5] any intercomparison with models like MARS [7] can only be successful if the thermodynamic and chemical conditions for filter and impactor samples are at least approximately constant.

The intercomparison between measured and predicted fluxes (or V_d) for HNO_3 indicates that during the field studies being performed the time constant to achieve equilibrium conditions should be in the range larger than about 40–60 s. This is still below the sampling times and can only be estimated from flux determination studies.

Thus the particle samples itself appear to be approximately in equilibrium with the gaseous compounds NH_3 and NHO_3 and therefore results from MARS mostly agree with measured results [5] during the studies being performed. During the TRACT campaign in rural parts of the upper Rhine valley the aerosol particles were more acidic like in urban environments. Although NH_3 was present in considerable amounts during daytime (2–10 ppb or more) the particulate acidic compounds were not completely neutralised. The analysis yields that both inhomogeneous mixing or/and advection of NH_3 as well as kinetically limitations would yield such results. Therefore these findings are also further analysed and compared with model calculations for various influences of transport processes, emission and chemical reactions [12].

References

1. Harrison, R.M., C.A. Pio, *Atmos. Environ.* **17** (1983) 1733–1738.
2. Clarke, A.G., M.J. Willison, E.M. Zeki, *Atmos. Environ.*, **18** (1984) 1767–1775.
3. Wall, S.M., W. John, J.L. Ondo, *Atmos. Environ.* **22** (1988) 1649–1656.
4. Dlugi, R. *et al.*, Charakterisierung der Chemie und Physik des anthropogenen Aerosols in der belasteten Atmosphäre. *BMFT-Bericht*, MIM-FB93, Meteorologisches Institut, Universität, München, 1993.
5. Seidl, W., G. Brunnemann, L. Kins, E. Köhler, K. Reußwig, R. Dlugi, On the composition of aerosol particles and related gas phase species at two sites during the SANA II campaign. *Meteorol. Z.*, submitted
6. Brunnemann, G., L. Kins, R. Dlugi, Physical and chemical characterization of the atmospheric aerosol: An overview of the measurements during the SANA II campaign at Melpitz; *Meteorol. Z.*, submitted.
7. Saxena, P., A.B. Hudischewskyi, C. Seigneur, J.H. Seinfeld, *Atmos. Environ.* **20** (1986) 1471–1483
8. Pilinis, C., J.H. Seinfeld, *Atmos. Environ.* **21** (1987) 2453–2466.
9. Hänel, G., *Adv. Geophys.* **19** (1976) 73–188
10. Harrison, R.M., A.R. MacKenzie, *Atmos. Environ.* **24A**, (1990) 91–102.
11. Wexler, A.S., J.H. Seinfeld, *Atmos. Environ.* **24** (1990) 1231–1246.
12. Kramm, G., R. Dlugi, *J. Atmos. Chem.* **18** (1994) 319–357.
13. Huebert, B.J. et al., *J. Geophys. Res.* **93** (1988) 7127–7136.
14. Whitby, K. T., *Atmos. Environ.* **12** (1978) 609–633

15. Stelson, A.W., J.H. Seinfeld, *Atmos. Environ.* **16** (1982) 983–992.

16. Mozurkewich, M., *Atmos. Environ.* **27A** (1993) 261–270.

17. Langford, A.O., F.C. Fehsenfeld, *Science* **225** (1992) reports, 1–3.

18. Dlugi, R. *et al.*, Entstehung, Verhalten und Ablagerung von gas-, partikel- und tropfenförmigen Reaktionsprodukten. *BMBF-Abschlußbericht*, Vorh 07 VND 098, Meteorol. Inst., Univ. München, 1995.

19. Harrison, R.M., W.T. Sturges, A.-M.N. Kitto, Y. Li *Atmos. Environ.* **24A** (1990) 1883–1888.

20. Sinclair, J.D., L.A. Psota-Kelty, C.J. Weschler, *Atmos. Environ.* **22** (1988) 461–469.

13.5 Studies on Deposition, Emission and Chemical Transformation above and within Tall Vegetation

Ralph Dlugi, Lucia Kins, Dieter Köhler, Eva Köhler, Karin Reußwig, Jochen Reuder, Gunter Roider, Kurt Ruoss, Jan Schween and Michael Zelger

Meteorologisches Institut, Universität (MIM) Theresienstraße 37, D-80333 München, Germany

Introduction

The exchange of inorganic and organic trace compounds in the soil-vegetation-atmosphere system is influenced by plant physiological, and physical and chemical processes in the soil and the atmosphere. A number of studies (called BIVOTEX I-V) to understand these complex interactions were performed at the Schachtenau experimental research site in the Bavarian Forest National Park, Germany [1–3]. The forest (about 86 % spruce (*Picea abies* [L.] *Karst*), 14 % beech (*Fagus sylvatica* [L.]) has an average height of trees ≈ 28 m with LAI ≈ 5.5, crown between 14 m to 28 m, age 80–100 years, and a stand density of 721 trees/ha [1, 2].

The general objectives of this contribution by the MIM group within the BIVOTEX experiments were:

- to determine quantitatively the emission and deposition fluxes of inorganic compounds like SO_2, O_3, H_2O_2, NH_3 and particles,
- to perform studies on the co-deposition of SO_2 with O_3, H_2O_2 and NH_3,
- to study the influences of transport processes on the formation of reaction products from the NO_x–O_3–VOC–SO_2 system within and above the canopy,
- to understand the influences of the hydrological cycle, radiation and different exchange processes for energy on the budgets of different compounds.

Chemical conditions at the experimental site

A considerable number of chemical measurements were performed at this site during 1986–1993 [1–8]. The range of mixing ratios for different compounds is summarized in Table 13.5.1. From these data, relationships, *e.g.* $0.55\ (0.1) \leq NO_x/NO_y \leq 0.84\ (0.96)$ and $0.5\ (0.1) \leq VOCs/NO_x \leq 5–18$, can be calculated [8]. This shows that the reactivity of the gaseous mixtures *e.g.* for the formation of radicals, O_3 and compounds like HNO_3, PAN, H_2O_2 organic peroxides or carbonyl compounds inside and above the canopy, can vary considerably by orders of magnitude.

Table 13.5.1: Range of mixing ratios for different chemical compounds. Extreme values were occasionally determined, shown here in brackets.

Compound	Concentration (ppb)	
NO	0.2–3.8	
NO_2	0.8–15.0	
HNO_3	0.1–1.4	
NO_3^-	1.0–28.0	
SO_2	0.5–5.5	(15)
SO_4^{2-}	1.8–35	(47)
$H_2SO_4 + NH_4HSO_4$	0.5–7.8	(11.9)
HCHO	0.0–2.4	
PAN	0.0–1.9	
NH_3	0.1–1.2	
NH_4^+	1.9–37.6	
H_2O_2	0.08–1.4	(2.5)

A further aspect is related to the photodissociation of molecules such as O_3, NO_2, NO_3 and aldehydes inside and above the canopy. The measurements by MIM related to this topic showed that the photolysis rate for NO_2, $J(NO_2)$ was reduced proportionally to the reduction of global radiation G inside the canopy, where the relationship G_{in} (z = 2 m) ≈ 0.15 G (z = 50 m) holds. Thus for $J(NO_2)$ we found experimentally $J(NO_2)$ (z = 2 m) ≈ 1.70×10^{-6} G (z = 50 m) while above the forest $J(NO_2) \approx 1.38 \times10^{-5}$ holds. $J(O_3)$ (z = 2 m) was reduced by a factor of about 9–12 by the canopy depending on the amount of turbidity and cloud cover. Also the lifetime of NO_3 increased strongly inside the forest with $J(NO_3)(z = 2\text{ m}) \approx 10^{-2}\ s^{-1}$ during noon. As pointed out by [8, 11] the chemical lifetimes τ_c of many species therefore are considerably larger inside the forest than above. It is obvious that these findings strongly depend on local biometric parameters like stand density, LAI, h and crown structure which control the light regime and therefore also strongly influences the exchange processes in the soil-vegetation-atmosphere system. Therefore any reliable up-scaling procedure for the determination of deposition and emission fluxes needs such information. Currently therefore studies are performed [14] to determine the influences of biometric parameter and hydrological conditions on the emission and deposition fluxes for tall canopies.

Transport in plant canopies

Directly above and inside tall vegetation the traditional framework of 'K-theory' fails to describe quantitatively the fluxes of momentum, heat, water vapour, CO_2 and other trace substances [9]. One reason for this behaviour is that inside plant canopies often coherent eddies control the mass and energy exchange in the low frequency spectral range. The experiments being performed with respect to this topic showed that, depending on the water status of the soil and the boundary layer flow field characteristics, the characteristic transport times τ_t for the compartments 'upper crown', 'lower crown', and 'trunk space' varied considerably in relation to τ_c Thus the influence of chemical reactions on the chemical transformation of compounds varies within the diurnal energy cycle of the forest [2, 8, 14]. Recently the model FLAME (Forest Land Atmosphere Model) [10] was successfully applied to analyse some of the main features of the exchange processes between the forest and the boundary layer. Both the experiments and the model results showed that if the forest soil is humid, time scales of $\tau_t \approx 5 \times 10^2$–$1.5 \times 10^3$ s are determined, while if the forest soil dries out the exchange times decrease $\tau_t \approx 70$–10^2 s around noon with cloud free sky in the summer [14]. The empirical minimum value for the time scale of the exchange of heat between the crown and the overlying atmosphere for different types of tall canopies and conditions with dry soils and high solar insolation is $\tau_{t\,min} \approx 40$–60 s [8, 11, 14].

Deposition of sulfur and nitrogen compounds

The deposition fluxes at the measuring site are summarised in Table 13.5.2 as yearly minimum values, because all parameters controlling the co-deposition were not known for all times of the years of measurements.

The mean total N flux from the atmosphere is about 9.5×10^{-8} g (N) m^{-2} s^{-1} with a contribution from NH_3 by about 2.4×10^{-8} g (N) m^{-2} s^{-1}. The mean emission fluxes from the needles in the crown were about (3–8) $\times 10^{-9}$ g (N) m^{-2} s^{-1}. If the strong variable emission by the dissociation of NH_4NO_3 from needles surfaces is considered this flux can reach values of (0.9–1.4) $\times 10^{-8}$ g (N) m^{-2} s^{-1}. The total nitrate flux (NO_3^- + HNO_3) therefore is strongly variable (2.5–19 kg (N) ha^{-1} yr^{-1}) [11, 12] and it's 'dry fraction' often has a significant contribution by the deposition of large NO_3^- particles with mean diameters > 2 μm. The total flux of ammonium (NH_3 + NH_4^+) appears to be larger than the total flux of nitrate for the period 1988–1993. If one estimates the contribution of emissions of NH_3 (and HNO_3 ?) by the dissociation of NH_4NO_3 to the N budget the fluxes of total nitrate and total ammonium are reduced by 20–35 %. The results suggest that NH_3 can be emitted by the canopy in this way but considerable amounts of HNO_3 can be deposited in the crown region again. This process strongly depends on the character of the turbulent exchange processes in the canopy.

Table 13.5.2: The yearly minimum fluxes (kg ha^{-1} yr^{-1}) by wet and dry deposition with the effect of co-deposition [11] at the Schachtenau site.

Ion	wet deposition	dry deposition
H^+	0.48	0.1
		0.6 co-dep. crown
		0.8
SO_4^{2-}	23.9	2.5 aerosol
co-dep. from		29.0 co-dep. crown
$SO_2 + O_3 + NH_3^-$		11.0 co-dep. soil
and $SO_2 + H_2O_2$		
HNO_3–N		1.7
NO_3^-–N	5.9	2.1
-	(9.0: precip. - leaching)	
NH_3^-–N		7.5
NH_4^+–N	6.9	2.0
-	(9.3: precip. + leaching)	
NO_2–N	-	5.0

The dry deposition of SO_2, H_2O_2 and O_3 was often controlled by the processes of co-deposition. SO_2 was deposited without any reactions with $V_d \leq 0.7$ cm s^{-1}. The influence of surface reactions, the co-deposition, leads to $0.9 \leq V_d \leq 1.6$ cm s^{-1}. If in addition a strong dry particle deposition was observed these values increased over 1–2 weeks with no significant rainfall to maximum values being observed in the range of $2.0 \leq V_d \leq 2.6$ cm s^{-1}, while model calculations according to the inferential method [*e.g.* 13] would yield V_d (SO_2) ≤ 0.9 cm s^{-1}. For SO_2 mixing ratios $1.5 \leq [SO_2] \leq 10$ ppb the deposition flux for O_3 was enhanced by a factor of 1.1–1.5 compared to situations where no surface reaction (SO_2 + O_3) can be detected. The maximum deposition velocity for H_2O_2 was V_d = 3–4 cm s^{-1} for $u_* = 0.4$–0.5 m s^{-1}. This is comparable to $r_c \approx 0$ s m^{-1} and $r_{mt} + r_t \approx 25$ s m^{-1} . But larger values were obtained by the profile method (V_d = 5–10 cm s^{-1}) conditions with high relative humidities and aerosol (dry) mass concentrations of 28–53 μg m^{-3} together with $[SO_2] > 3$ ppb above the forest. For these conditions the data analysis suggests a reactive uptake of H_2O_2 into wet particles with a corresponding increase of the concentration gradient above the canopy which results in fluxes being above the theoretical limit $V_d(\max) = (r_{mt} + r_t)^{-1}$.

If the condition $[NH_3] > 1$ ppb was fulfilled a reactive uptake of SO_2 and the formation of compounds like H_2SO_4, $(NH_4)_xH_y(SO_4)_z$ and in addition HN_4NO_3 or related mixed salts was possible. If these processes occur the limitation for V_d with $r_c > 100\ s\ m^{-1}$ for the uptake of SO_2 (but also O_3) into spruce needles is no longer valid and the influence of the molecular turbulent resistance r_{mt} on the magnitude of the SO_2 and O_3 flux becomes more important. Therefore studies are also performed for a better quantification of r_{mt} for heat and matter over different surfaces [14].

References

1. Enders, G., Micrometeorological-chemical interactions in a forest atmosphere. in: *EUROTRAC Ann. Rep. 1991, Part 4*, EUROTRAC ISS, Garmisch-Partenkirchen 1992, p. 183.
2. Enders, G., R. Dlugi, R. Steinbrecher *et al.*, *Atmos. Environ.* **26A** (1992) 171.
3. Ziegler, H., G. Eichstädter, W. Schürmann, R. Steinbrecher, Volatile monoterpenes in a spruce forest (*Picea abies* [L.] *Karst*): Sources and mechanisms of emissions and behaviour in the atmosphere; in: *EUROTRAC Ann. Rep. 1991, Part 4*, EUROTRAC ISS, Garmisch-Partenkirchen 1992, p. 224
4. Dlugi, R., U. Meier, M. Paffrath, H. Quenzel, Entstehung, Verhalten und Ablagerung von gas-, partikel- undtropfenförmigen Reaktionsprodukten, *PBWU-Bericht,* GSF-Neuherberg, 1989.
5. Steinbrecher, R., Uncertainties of biogenic VOC-emission databases, in: Baars H.P. *et al.* (eds) *Proc. TNO/EURASAP workshop on the reliability of VOC emission databases*, TNO Delft (NL), IMW-TNO Publications P93/040, 1993, pp. 89–98
6. Kesselmeier, J., L. Merk, M. Bliefernicht, G.Helas, Trace gas exchange between terrestrial plants and atmosphere: Carbondioxide, carbonyl sulfid and ammonia under the rule of compensation points. in: J. Slanina, G. Angeletti, S. Beilke (eds) *CEC-Air Pollution Research Reports*, Joint workshop CEC/BIATEX (ISBN2-87236-095-3), 1993, pp. 71-80
7. Enders, G., *Environ. Pollut.* **75** (1991)
8. Dlugi, R., 1993 Interaction of NOx and VOCs with vegetation. in: P.M. Borrell, P. Borrell, T. Cvitaš, W. Seiler (eds), *Proc. EUROTRAC Symp. '92*, SPB Academic Publishing bv, The Hague 1993, pp. 682–688
9. Denmead, O.T., Bradley, E.F., Flux-gradient relationships in a forest canopy. in: Hutchison, B.A., Hicks, B.B. (eds), *The Forest - Atmosphere Interaction,* Reidel Publ., Dordrecht 1985
10. Inclan, M.G., R. Forkel, R. Dlugi, R. Stull, 1995 Application of the transilient turbulent theory to study the interactions between the atmospheric boundary layer and forest canopies, *Boundary Layer Meteorol.*: submitted
11. Dlugi, R. *et al.*, Entstehung, Verhalten und Ablagerung von gas-, partikel- und tropfenf°rmigen Reaktionsprodukten, *BMBF-Bericht,* Vorhaben 07VND098, Meterorologisches Institut, Universität, D-80333 München, MIM-FB 95 S-1, 1995.

12. Sievering, H., G. Enders, L. Kins, G. Kramm, K. Ruoss, G. Roider, M. Zelger, L. Anderson, R. Dlugi, Nitric acid, particulate nitrate, and ammonium profiles at the Bayerischer Wald: Evidence for large deposition rates of total nitrate. *Atmos. Environ.* (1993) 311-315

13. Wesely, M.L., Parameterization of surface resistances to gaseous dry deposition in regional-scale numerical models. *Atmos. Environ.* (1989) 1293-1304.

14. Dlugi, R., M. Zelger, J. Schween, Th. Seiler, Th. Foken, G. Inclan, R. Forkel, Studies on exchange processes of momentum, energy and matter above and within tall canopies. *Report on a contribution to subproject BIATEX*, Meteorological Institute, D-80333 Munich, Theresienstraße 37, Germany, 1995.

13.6 Studies on Exchange Processes of Momentum, Energy and Matter above and within Tall Vegetation

R. Dlugi[1], M. Zelger[1], J. Schween[1], Th. Seiler[1], Th. Foken[2], G. Inclan[1] and R. Forkel[1]

[1]Meteorologisches Institut, Universität (MIM), Theresienstraße 37, D-80333 München, Germany,
[2]Meteorologisches Observatorium Lindenberg, D-15884 Lindenberg, Germany

Introduction

The exchange of momentum, heat and matter (*e.g.* H_2O, CO_2, SO_2, NO, NO_2, O_3, NH_3, HNO_3, VOC) in the soil-vegetation-atmosphere system over tall canopies is influenced by non-local transfer processes and the presence of distributed sources and sinks [1–4]. Depending on their water status tall plant canopies therefore have a microweather and a microclimate which significantly differ from those above neighbouring areas covered with bare soil or low vegetation. Thus also their nutrient budgets and especially the fluxes of trace substances between the atmosphere and the plant-soil system can be altered with respect to time and space. A number of studies suggest that chemical cycles inside the canopy can have a different behaviour compared to those from outside and can be coupled or decoupled from cycles in the overlying atmosphere by the various influences of dynamic and thermodynamic processes (*e.g.* [5–7]). Therefore a study was performed by the joint MIMMOL group to understand better the interaction between transport processes, trace gas emissions and deposition and chemical reactions within and above tall vegetation.

Experiments and model developments

Different experimental studies within the topic mentioned above were performed within the BIATEX experiments BIVOTEX at the Schachtenau site in the Bavarian Forest, Germany (about 86 % spruce, 14 % beech) [5–7] and during the BEMA experiments 1993 and 1994 at Castelporziano, Italy, near Rome in a mixed forest of Mediterranean oaks (*Q. ilex*) and pines (*P. pinea*).

According to the BEMA objectives the groups at the forest site provided measurements to determine fluxes of energy including water vapour, CO_2, non-methane hydrocarbons, peroxides, aldehydes and O_3. A tracer experiment by JRC, Ispra, Italy was performed to help to understand the mixing process in the canopy. The experiments also were designed to provide data to compare the fluxes being determined with micrometeorological methods with fluxes as determined with cuvettes.

The integrated BEMA concept for the *Q. ilex/P. pinea* forest includes the following experimental activities by the joint group from MIM and DWD (MOL). Micrometeorological profile (21 m) with 12 levels of wind, temperature and humidity; wind direction at 21 m, radiation at 14 m, radiation under the canopy (2 m) and soil heat flux and soil water content at different sites in the forest. Turbulent fluxes of energy, temperature, and water vapour in 5 heights (19 m, 16 m, 13 m, 8.5 m, and 4 m) and CO_2 fluxes at 16 m and 6 m were determined. In addition fluxes of water vapour, heat and CO_2 at the soil surface were determined with use of a cuvette technique.

Different concepts for the determination of turbulent fluxes of momentum, heat and matter (*e.g.* H_2O, CO_2, trace gases) were compared. The structure of turbulence above and within the forest was studied

- to determine complete data sets for intercomparisons with vertically resolved soil vegetation-atmosphere models [8]
- to analyse the characteristics of exchange processes for momentum, heat, H_2O and CO_2 and other trace substances
- to understand better the role of viscous sublayer resistance on transport characteristics
- to understand better the influences of mixing processes on chemical reactions inside and directly above tall vegetation
- to quantify fluxes of reactive organic hydrocarbons from the soil and the trees.

A new Forest Land Atmosphere ModeEL (FLAME) was developed [8] and is applied to study the influences of mixing processes on trace gas transfer and chemical transformations. The prognostic model can describe vertically resolved the transport of radiation, momentum, heat and matter (*e.g.* H_2O, CO_2, trace substances) throughout the atmospheric boundary layer (ABL) and the vegetation down to the soil surface by use of a modified transient turbulence theory [9–8].

This especially allows us to consider the effects of non-local transport by larger eddies on fluxes of momentum, heat and matter.

Results

Directly above and inside tall vegetation the influences of different sinks and sources of heat, momentum and matter on the fluxes considerably alters the transport characteristics of the flow field. As discussed recently [7] the experiments being performed and results from literature [1–6, 10, 11] showed that the characteristic transport times τ_T for momentum, heat and matter for the compartments 'upper crown', 'lower crown', and 'trunk space' vary considerably by the influences of soil and plant water status, driving meteorological parameters and biometric stand characteristics. Also the characteristic reaction time scale τ_c of various compounds can vary up to orders of magnitude. Therefore all three possibilities

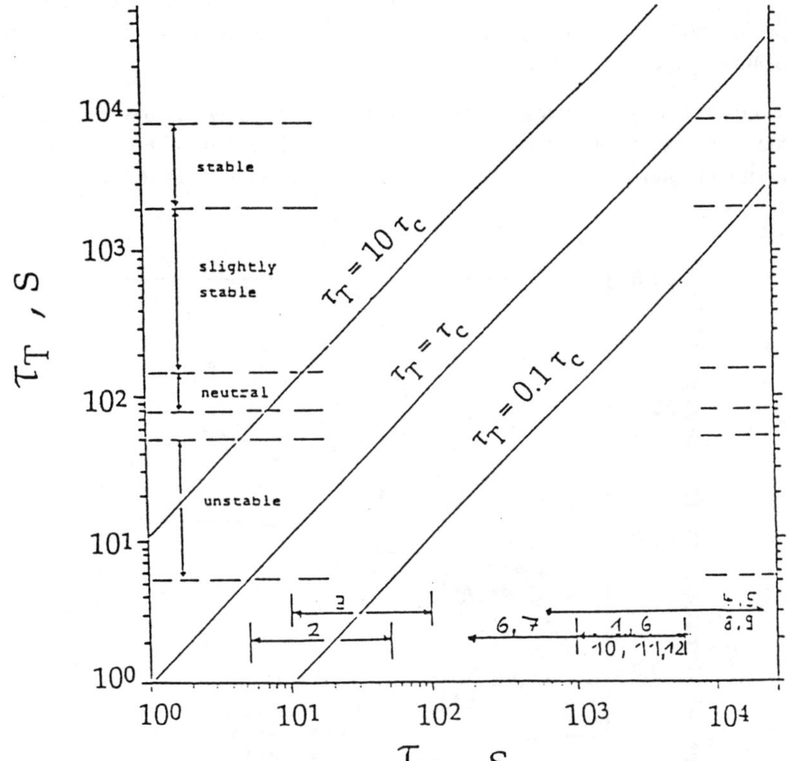

1. HO_2	+ HO_2	⇒	H_2O_2		10. O_3	+ olefine	⇒	NO_3	
2. NH_3	+ HNO_3	–	NH_4NO_3		11. O_3	+ RO_2	⇒	RO	+ O_2
3. O_3	+ NO	⇒	$NO_2 + O_2$		12. O_3	+ NO_2	⇒	NO_3	+ O_2
4. O_3	+ isoprene	⇒	products(p)		13. N_2O_5		⇒	NO_3	+ O_2
5. O_3	+ monoterp.	⇒	p		14. NO_3	+ $h\nu$ (O_2)	⇒	NO_2	+ O_3
6. NO_3	+ monoterp.	⇒	p		15. NO_3	+ NO	⇒	NO_2	
7. NO_3	+ isoprene	⇒	p		16. NO_3	+ NO_2	⇒	N_2O_5	
8. OH	+ isoprene	⇒	p		17. N_2O_5	+ H_2O (fl)	⇒	2 HNO_3	
9. OH	+ monoterp.	⇒	p		18. NO_2	+ $h\nu$	⇒	O	+ NO

(for: O_3 = 10ppb; NMHC = 1ppb; NO = 10ppt (night); OH = 0.02ppt (day)).

Fig. 13.6.1: The relation between τ_t and τ_c for different reactions, a layer thickness of 10 m and different stability regimes (see [7]).

$$\tau_t << \tau_c \quad (1)$$

$$\tau_t \approx \tau_c \quad (2)$$

$$\tau_t >> \tau_c \qquad (3)$$

may happen. If (1) holds, a chemical compound may be considered as a tracer, for (2) a considerable influence of transport processes on chemical reactions can be expected and for (3) a chemical substance is in a *quasi*-equilibrium with the surrounding air (Fig. 13.6.1).

The observations of characteristic exchange times τ_t were first evaluated with respect to the 'upper crown' region and are presented within the data from experiments above and within tall vegetation by PawU *et al.* [12].

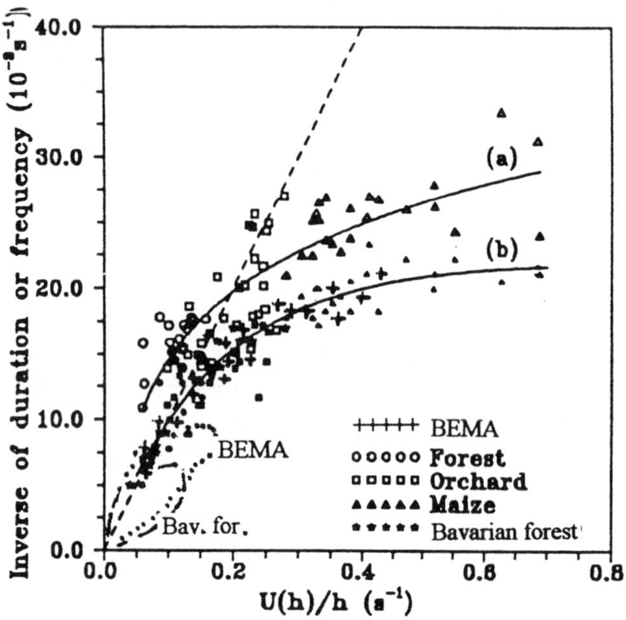

Fig. 13.6.2: Inverse of mean structure duration (open symbols) and frequency of occurrence of coherent structures (closed symbols, [12]) versus bulk mean wind shear at canopy top. Circles represent forest, squares represent orchard, triangles represent maize, stars represent data from Bavarian forest and crosses represent data from the forest at the Castelporziano site. Also the range for the 'lower crown to upper trunk space' in the Bavarian Forest and the 'BEMA forests are given. (solid line (a): power law fit of all open symbols; solid line (b): power law fit of all closed symbols; dashed line: results of linear theory with inverse Strouthal number of 10) (modified Figure 2 from [13]).

If the structures of so-called temperature ramps for dry conditions are analysed with the structure duration L and the separation S between structures, a measure of intermittency is given by $U(L+S)$. Data for forests are spread over a large range but seem to be independent of the bulk mean wind shear for $U(\mathrm{h})/\mathrm{h} > 0.35\ \mathrm{s}^{-1}$ with an average value between 0.75–0.8.

For $U(\mathrm{h})/\mathrm{h} < 0.3\ \mathrm{s}^{-1}$ the range of intermittency is $0.23 < U(L+S) < 0.98$. This result can be caused by the influences of convective processes for $w^* > u^*$, but also by the influence of larger inhomogeneities of natural vegetation. Compared to [12] and [13] the results for forests with a well watered soil and also dew fall in the crown region are considerably different. This is represented in Fig. 13.6.2 with data from the Bavarian forest and $U(\mathrm{h})/\mathrm{h} < 0.1\ \mathrm{s}^{-1}$.

Recently, the first data become available to study if the intermittency is also a function of the Bowen ratio ($H/\lambda E$) or the evaporation fraction ($\lambda E/(Rn-G)$). For a range of data with $0.18 < U(\mathrm{h})/\mathrm{h} < 0.27\ \mathrm{s}^{-1}$ the intermittency in the upper crown increased with decreasing evaporation fraction from $0.30 < U(L+S) \sim 0.38$ for $\lambda E/(Rn-G) > 0.65$ to $0.7 < U(L+S) \sim 0.92$ for $\lambda E/(Rn-G) < 0.5$ ($Bo > 0.9$).

If results from the lower crown region or the upper trunk space are considered also biometric parameters like stand density should have an influence on the magnitude and range of τ_t (*e.g.*: [11]). For the Bavarian forest and the BEMA forest at Castelporziano τ_t in the height interval 'lower crown to upper trunk space' tends to larger values compared to τ_t for the upper crown. Thus the intensity and frequency of momentum and heat transfer (as a measure for transfer of scalars) will considerably differ with height and time. For example, if the time scale τ_t in the unstable air of the upper crown is around 100 s the corresponding values in the lower layers can be still $250 \leq \tau_t \leq 500$s (*e.g.* [6, 7]) and the influences of exchange processes on chemical reactions therefore strongly depend on height (*e.g.* [6, 7, 14]. Therefore inside forests chemical compounds can develop and react which have a considerably lower lifetime ~G in upper layers.

This study will be extended to use all available data sets of the BIVOTEX and the BEMA experiments with a more extended range of dynamic and chemical situations.

References

1. Raupach, M.R., Canopy transport processes; in: W.L. Steffen, O.T. Denmead (eds) *Flow and Transport in the Natural Environment: Advances and Applications*, Springer, Berlin 1988, pp. 95–127.
2. Finnigan, J.J., M.R. Raupach, in: E. Zeiger, G.D. Farquhar, I.R. Cowan (eds), *Stomatal Function,* Stanford Univ. Press 1987. pp. 385–429.
3. Denmead, O.T., E.F. Bradley, in: B.A. Hutchison, B.B. Hicks (eds), *The Forest Atmosphere Interaction*, Reidel, Dordrecht 1985, pp. 421–442.
4. Denmead, O.T., E.F. Bradley, *Irrig. Sci.* **8** (1987) 131–149.
5. Enders, G., R. Dlugi, R. Steinbrecher *et al.*, *Atmos. Environ.* **26A** (1992) 171–189.
6. Dlugi, R. *et al.*, *Entstehung, Verhalten und Ablagerung von gas-, partikel und tropfenförmigen Reaktionsprodukten*, BMBF-Bericht, Vorhaben 07VND098, Meteorologisches Institut, Universität, D-80333 München, MIM-FB95 S-1, 1995.

7. Dlugi, R., Interaction of NOx and VOCs with vegetation. in: P.M. Borrell, P. Borrell, T. Cvitaš, W. Seiler (eds), *Proc. EUROTRAC Symp. '92*, SPB Academic Publishing bv, The Hague 1993, pp. 682–688.

8. Inclan, M.G., R. Forkel, R. Dlugi, and R. Stull, Application of the transilient turbulent theory to study the interactions between the atmospheric boundary layer and forest canopies, *Boundary Layer Meteorol.* submitted.

9. Stull, R.B., *Boundary Layer Meteorol.* **62** (1993) 21–96.

10. Stull, R.B., Parameterization of asymmetric transilient matrices for convective PBLs. *Proc. 11th Symp. Boundary Layers and Turbulence,* AMS, Charlotte, N.C. 1995, pp. 57–58.

11. Raupach, M.R. *et al.*, *Appl. Mech. Rev.* **44** (1991) 1–25.

12. PawU, K.T. *et al.*, *Agric. Forest Meteorol.* **61** (1992) 55–68.

13. Qiu, J., K.T. PawU, R.H. Shaw, Pseudo–wavelet analysis of turbulence pattems in three vegetation layers. *Proc. 21st Conf. on Agricultural and Forest Meteorology*, AMS 1994, pp. 106–109.

14. Dlugi, R., *et al.*, Studies on deposition, emission and chemical transformation above and within tall vegetation. *this volume*.

13.7 Ammonia Exchange between Terrestrial Plants and the Atmosphere Controlled by Plant Physiology: Compensation Point and CO_2 Exchange.

F.X. Meixner[1], M. Bliefernicht[1], G. Helas[1], J. Kesselmeier[1], P.G. Wyers[2] and M.O. Andreae[1]

[1]Max Planck Institute for Chemistry, Biogeochemistry Dept.,P.O.Box 3060, D-55020 Mainz,
[2]Netherlands Energy Research Foundation (ECN), P.O.Box 1, NL-1755 ZG Petten

Summary

Based on forest experiments, we found a compensation point of 0.4 ppb for the exchange of ammonia with spruces at the Schachtenau site in June 1991. Thus, emission of ammonia from spruces was observed, when the atmospheric concentration was below 0.4 ppb, and uptake was found under conditions with higher mixing ratios. In general, a spruce site like the Schachtenau site acts as a sink for ammonia emitted from other ecosystems (*e.g.* agricultural sites). This is further supported by gradient measurements in correlation with cuvette data under higher atmospheric mixing ratios in September 1990. For a wheat field site at Manndorf we report a strong diurnal variation of net NH_3 fluxes showing emission over the day and deposition during the night in close correlation to CO_2 fluxes induced by the assimilation and respiration of the wheat field. This behaviour points to the plant physiological control of the NH_3 exchange.

Aims of the research

Exchange of ammonia between the biosphere and the atmosphere.

Introduction

A lot of climatically relevant trace gases are exchanged between the biosphere and the atmosphere. Therefore, it is necessary to know the amount of biogenically produced and consumed trace gases. However, global and even regional budget calculations exhibit enormous uncertainties, presumably because of insufficient technical development and perhaps more important fundamental gaps in our understanding of the exchange processes [1] including chemical as well as physical and biological approaches. Exchange of trace gases between higher plants and the atmosphere includes emission and production as well as deposition and uptake, terms which are not exchangeable in all cases [2]. Understanding the basic plant physiological pathways for the production or consumption, as well as the physiological meaning of compounds which are of importance for atmospheric chemistry, is necessary for the discussion of exchange processes. One important parameter for processes of consumption and production is the so called compensation point. This term points to a general behaviour of plants to adapt to the actual environment.

Principal scientific results

Measurements of the exchange of ammonia: forest site

In June 1991 we were able to measure the exchange of ammonia between spruce twigs and the atmosphere at the Schachtenau site (Tower of the University of Munich, Bavarian forest [3]) using a dynamic cuvette system at a height of 23 m [4]. Gas phase ammonia was sampled with a mist scrubber and analysed by ion chromatography [5]. In this early summer of 1991 we measured the exchange with spruce twigs just after the development of the new needle generation. At this time of the year the mixing ratios of gas phase ammonia around the cuvette at 23 m height ranged between zero and 1.5 ppb. Above the forest we observed values up to 4 ppb. We found emission as well as deposition of NH_3 depending on the ambient concentration. The correlation between ammonia exchange and ambient mixing ratios present evidence that spruce twigs at this time of the year may be both sources and sinks for ammonia depending on the ammonia concentration in the surrounding air (Fig. 13.7.1) and thus exhibiting a NH_3 compensation point.

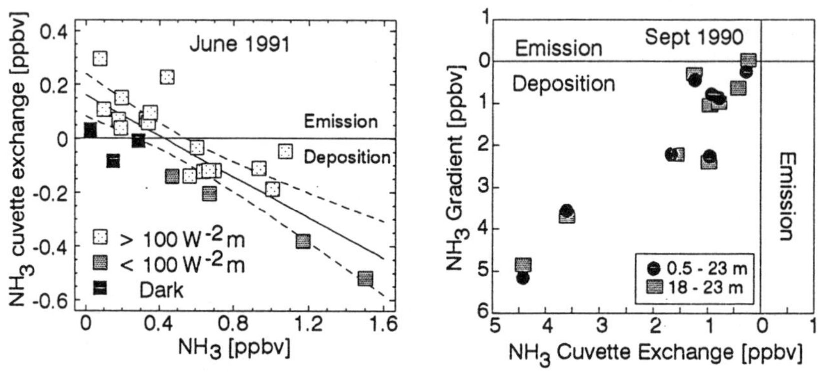

Fig. 13.7.1: June 1991: NH_3 exchange measured with spruce twigs at 23 m height at the Schachtenau site in relation to the ambient NH_3 concentration. Solid line, calculated regression; dotted lines give the 95 % confidence level. NH_3 compensation point calculated was 0.43 ppb (0.28–0.62 ppb), given within the 95 % confidence level. Note the slight shift of the compensation point from a lower value under dark or low light conditions to a higher value under higher light intensities. September 1990: Concentration gradients of gas phase ammonia within a spruce stand at the Schachtenau site in September 1990 in close correlation with the exchange data obtained by cuvette measurements enclosing an intact branch at the height of 23 m.

Calculations by linear regression gave a value around 0.43 ppb (0.28–0.62 ppb; 95 % confidence level). In this context it is of interest that during night and under low light conditions (< 100 W m^{-2}) the exchange data move towards a zero emission or a higher deposition (Fig. 13.7.1), indicating a decrease of the compensation point. This behaviour is understood as a decrease or a standstill of nitrate reduction, which is a light driven process. Under autumn conditions in September 1990 with atmospheric NH_3 mixing ratios between 0.2 and 6 ppb we always observed an uptake of ammonia by the forest measured with our cuvette system, supported by gradient observations which always pointed to a deposition flux (Fig 13.7.1).

Measurements of the exchange of ammonia over an agricultural site

Ammonia fluxes were measured over a wheat field at Manndorf (eastern Bavaria) in July 1990 comparing different micrometeorological methods, such as the aerodynamic method and the modified Bowen ratio method. During these measurements over wheat we observed bi-directional fluxes, ranging from –35 ng m^{-2} s^{-1} during the night and early morning to +40 ng m^{-2} s^{-1} during late morning to late afternoon. These data showed a significant correlation of the observed NH_3 flux to the net CO_2 flux (Fig. 13.7.2.) measured by eddy correlation. We found an emission over the day and deposition during night in close correlation to CO_2 fluxes induced by the assimilation and respiration of the wheat field. The coupling of the ammonia emission to the photosynthetically driven CO_2 assimilation points to the plant physiological control of the NH_3 exchange and is on accordance with our knowledge on nitrogen fixation by the vegetation (see conclusions). Correlations between ambient NH_3 and NH_3 flux in mid July 1990 over the wheat field indicated a canopy compensation point of 0.5–3.7 ppb under the actual senescing conditions during this time of the year with ripening seeds and already mainly discoloured leaves.

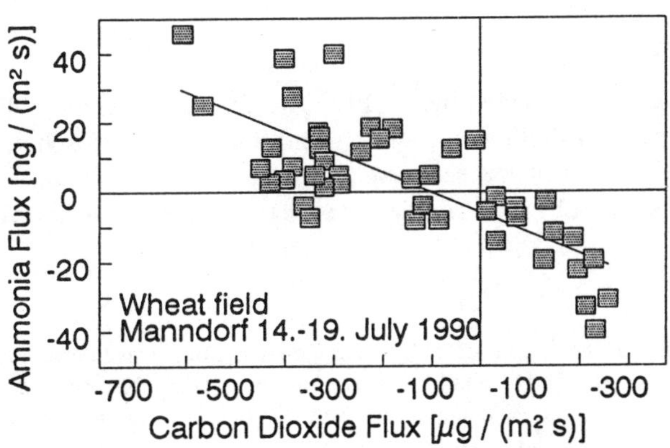

Fig. 13.7.2: Net vertical NH_3 flux (average of 6 different flux estimates) versus net vertical CO_2 flux (measured by eddy correlation). n = 46; $r^2 = 0.6$; $P < 0.0001$.

Conclusions

The low value for the observed compensation point gives reason to believe that forests like the spruce stand at Schachtenau at this time of the year act as a net sink for ammonia emitted by agricultural ecosystems (the Manndorf site, for example), where we found a strong diurnal variation of net NH_3 fluxes showing emission over the day and deposition during night in close correlation to CO_2 fluxes induced by the assimilation and respiration of the wheat field. This behaviour highlights the plant physiological control of the NH_3 exchange, as the nitrogen fixation delivering NH_3 by plants is light driven. The general existence of an NH_3 compensation point in plants was established by Farquhar *et al.* [6]. According to Schjerring [7] the existence of an NH_3 compensation point fairly close to natural ambient levels in the case of agricultural plants (1.5–6 ppb) explains why NH_3 may either be adsorbed or emitted by the plant. However, NH_3 compensation points may vary during the life cycle and reach values up to 20–35 ppb as has been shown for wheat [8, 9]. Observations for forest trees are rare, however in accordance with our measurements exhibiting very low compensation points [10–12]. Here again we have to check for possible physiological pathways exhibiting production and consumption. In the case of ammonia, this is quite easy to do, as the nitrogen metabolism as well as emission and deposition has been intensively studied during the last years [7, 12]. For the production and emission mainly three processes are responsible: light driven nitrate reduction, senescence (ripening) or protein degradation and photorespiration. The deposition, in this case an uptake, may be explained by a metabolisation by the enzymes glutamine synthetase and glutamate synthase. These processes fulfil the premise of independent and counteracting pathways, thus are sufficient to explain the existence of a compensation point. Being aware of compensation points for the exchange of trace gases between plants and the atmosphere we will be able to understand and discuss questions of emission and deposition under environmental conditions in a more reliable way.

Acknowledgements

This work was partly funded by the EEC and the Bundesminister für Forschung und Technologie (BMFT) as a project within the section Biosphere Atmosphere Exchange (BIATEX) of the European Environmental Programme EUROTRAC. Fundamental support by the Max Planck Society is gratefully acknowledged.

References

1. Kesselmeier, J. Emission of sulfur compounds from vegetation and global-scale extrapolation. in: Th.D. Sharkey, E.A. Holland, H.A. Mooney (eds.),*Trace Gas Emissions from plants*, Academic Press, San Diego, USA 1991 pp. 261–265.
2. Kesselmeier, J., Plant physiology and the exchange of trace gases between vegetation and the atmosphere, in: S.E.Schwartz and W.G.N.Slinn (eds), *Precipitation Scavenging*

and Atmosphere-Surface Exchange. Volume 2-The Semonin Volume: Atmosphere-Surface Exchange Processes (Hemisphere, Washington DC 1992, pp. 949–966.

3. Enders, G., Dlugi, R., Steinbrecher, R., Clement, B., Daiber, R., Eijk, J.v., Gäb, S., Haziza, M., Helas, G., Herrmann, U., Kessel, M., Kesselmeier, J., Kotzias, D., Kourtidis, K., Kurth, H.H., McMillen, R.T., Roider, G., Schürmann, W., Teichmann, U., Torres, L. Biosphere/Atmosphere interactions: Integrated research in a European coniferous forest ecosystem. *Atmos. Environ.* **26A** (1992) 171–189.
4. Schäfer, L., Kesselmeier, J., Helas, G., Formic and acetic acid emission from conifers measured with a "cuvette" technique. in: G. Angeletti, S. Beilke, J. Slanina (eds), *Field Measurements and Interpretation of Species related to Photo-oxidants and Acid Deposition,* CEC Air Pollution Research 39, E. Guyot SA, Brussels 1992, pp. 319–323.
5. Hartmann, W. R., M. O. Andreae, G. Helas, Measurements of organic acids over central Germany,. *Atmos. Environ.* **23** (1989) 1531–1533.
6. Farquhar, G.D., Firth, P.M., Wetselaar, R., Weir, B., On the gaseous exchange of ammonia between leaves and the environment: Determination of ammonia compensation point, *Plant Physiol.* **66** (1980) 710–714.
7. Schjørring, J.K., Ammonia emission from the foliage of growing plants. in: Sharkey, Th.D., Holland, E.A., Mooney, H.A. (eds), *Trace Gas Emissions from Plants,* Academic Press, San Diego, USA 1991, pp. 267–292.
8. Harper, L.A., Sharpe, R.R., Langdale, G.W., Giddens, J.E., Nitrogen cycling in a wheat crop: Soil, plant, and aerial nitrogen transport, *Agron. J.* **79** (1987) 965–973.
9. Morgan, J.A, Parton, W.J., Characteristics of ammonia volatilisation from spring wheat, *Crop Sci.* **29** (1989) 726–731.
10. Breiding, H., Gravenhorst, G., Der NH_3-Transfer zwischen Atmosphäre und Nadelbäumen. In: Ammoniak in der Umwelt, *Kuratorium für Technik und Bauwesen in der Landwirtschaft e.V. (KTBL) und Verein Deutscher Ingenieure (VDI),* KTBL-Schriften Vertrieb im Landwirtschaftsverlag GmbH, Münster, Germany 1990.
11. Langford, A.O., F.C. Fehsenseld, Natural vegetation as a source or sink for atmospheric ammonia: A case study, *Science*, **255** (1992) 581–583.
12. Sutton, M.A., Pitcairn, C.E.R., Fowler, D., The exchange of ammonia between the atmosphere and plant communities. *Adv. Ecol. Res.* **24** (1993) 301–393
13. Hoagland, D.R., Arnon, D.J.,. The water culture method of growing plants without soil. *Berkeley Coll. Agr. Circ.* Univ. California 1950, p. 347.
14. Larcher, *Ökologie der Pflanzen*, 4. Auflage. Eugen Ulmer Verlag, Stuttgart, Germany 1984.
15. Mohr, H., Schopfer, P., *Lehrbuch der Pflanzenphysiologie,* 3.Auflage. Springer Verlag, Berlin, Germany 1985.
16. Strasburger, E., Sitte, P., Ziegler, H., Ehrendorfer, F., Bresinsky, A. (eds), *Lehrbuch der Botanik,* Fischer Verlag, Stuttgart, Germany 1991.

13.8 Photolysis Frequencies of Nitrogen Dioxide and Ozone: Measurements and Model Calculations

Joachim Reuder, Tullio Gori, Ansgar Ruggaber, Lucia Kins and Ralph Dlugi

Meteorologisches Institut, Universität München, Theresienstraße 37, D-80333 München, Germany

Introduction

Photolysis of ozone at wavelengths below 320 nm is the main source of tropospheric OH radicals:

$$O_3 + h\nu \rightarrow O(^1D) + O_2 \tag{1}$$

$$O(^1D) + H_2O \rightarrow 2\,OH \tag{2}$$

Photolysis of nitrogen dioxide occurs in the wavelength range below 420 nm and is the basic supposition of tropospheric ozone production:

$$NO_2 + h\nu \rightarrow NO + O(^3P) \tag{3}$$

$$O_2 + O(^3P) \rightarrow O_3 \tag{4}$$

The reaction rate constants of the photolytic processes (1) and (3) are given by the photolysis frequencies $J(O_3)$ and $J(NO_2)$. An accurate knowledge of these quantities is the basis for numerical modelling of photochemical reaction cycles.

Measurements of $J(O_3)$ and $J(NO_2)$ were performed during TRACT 1992 Lichtenau (Rhine valley near Bühl), BIATEX 1990/1991 Schachtenau (Bavarian Forest), and BayFORKLIM 1992/1993 Hohenpeißenberg. These data were compared with photolysis calculations carried out by STAR (System for Transfer of Atmospheric Radiation) developed in the subproject EUMAC [1].

Theory and instrument design

The photolysis frequency J of any absorbing gaseous component is given by:

$$J = \int_{\lambda_1}^{\lambda_2} \sigma(\lambda) \cdot \phi(\lambda) \cdot I(\lambda)\, d\lambda \tag{5}$$

λ_1, λ_2: wavelength thresholds of the interesting photolytic process

$\sigma(\lambda)$: absorption cross section of the gas molecule

$\phi(\lambda)$: quantum yield of the photolytic process

$I(\lambda)$: actinic flux (number of photons per volume and time)

Photoelectric detectors for the direct measurement of photolysis frequencies have to fulfill two main requirements. First, an appropriate combination of filter and photodetector must be found, which matches the product of $\sigma(\lambda) \times \phi(\lambda)$ as close as possible. Furthermore a light sampling device is needed, which estimates incoming photons equally, independent of the angle of incidence.

The instrument design and technical details concerning filters and photodetectors of our photoelectric detectors are depicted in Fig. 13.8.1. Further information on the measuring method are given by Junkermann *et al.* [2].

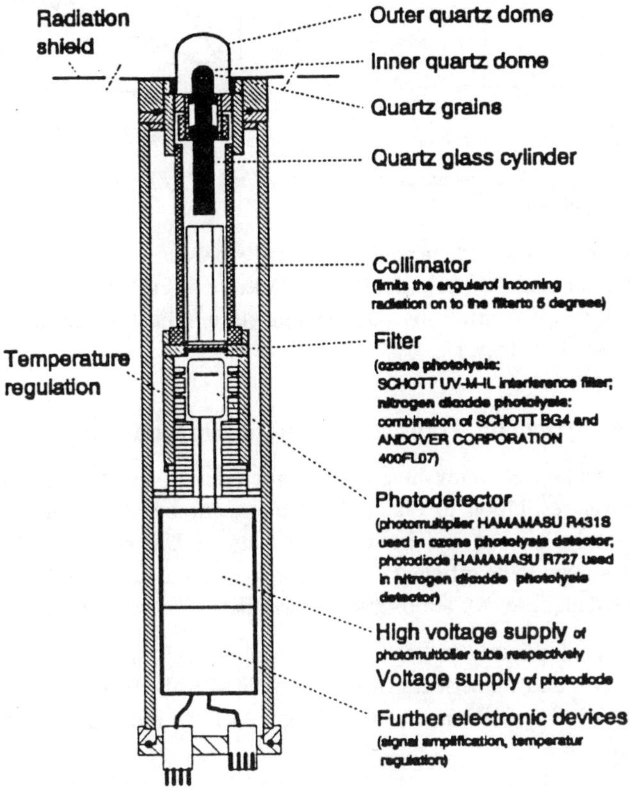

Fig. 13.8.1: Schematic cross section of the photoelectric detectors used for photolysis frequency measurements.

A single detector can only receive radiation from one hemisphere (2π sr). For the measurement of all incoming photolytic active radiation two detectors are used, one facing upwards, the other facing to the ground.

Fig. 13.8.2 shows typical deviations of the theoretical values of $\sigma(\lambda) \times \phi(\lambda)$ used for STAR calculations (thick solid line) from the spectral sensitivity of the detectors measuring $J(O_3)$ (left) and $J(NO_2)$ (right). These sensitivity values, depending on the chosen combinations of filter and photodetector, were measured at our laboratory with a wavelength resolution of 0.5 nm.

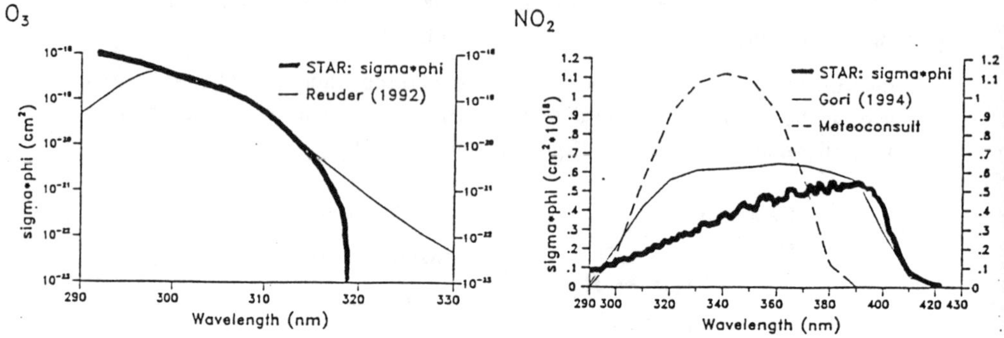

Fig. 13.8.2: Comparison between theoretical values of $\sigma(\lambda) \times \phi(\lambda)$ and the spectral response of the photoelectric detectors.

The diagram for $J(NO_2)$ depicts the difference between a commercial detector (Meteorologie Consult GmbH, denoted as Meteoconsult, dashed line) and the instruments developed further by our institute (thin solid line) using a new filter combination (see Fig. 13.8.1).

Model STAR (System for Transfer of Atmospheric Radiation)

STAR [1] considers all relevant physical processes in the troposphere with an optimal spectral resolution to obtain adequate accuracy for calculations of UV radiation, photolysis rates and biological effects of UV radiation. Furthermore STAR was designed to be efficient enough for routine calculations. STAR includes representative data bases for atmospheric constituents, especially aerosol particles.

Total accuracy of photolysis calculations using STAR due to uncertainties in model input data and numerical errors is estimated to be ± 14% for $J(NO_2)$ and ± 15% for $J(O_3)$. The influence of the accuracy of the input parameters alone is shown in Table 13.8.1.

STAR provides the possibility to calculate photolysis frequencies by using the spectral response measurements of the different detectors. Therefore it can be used for the derivation of correction functions considering individual instrumental characteristics. Results of these simulations are presented in Fig. 13.8.3 for nitrogen dioxide photolysis detectors. Furthermore a comparison between calculated (STAR) and measured (solid lines) values of ozone and nitrogen dioxide photolysis frequencies is given for two selected days during the TRACT and BayFORKLIM campaigns.

Table 13.8.1.

Input parameter	Uncertainty:	$J(O_3)$	$J(NO_2)$
solar zenith angle:	± 0.3°	2 %	2 %
surface albedo:	± 0.05	3 %	3 %
ozone content:	± 3 %	5 %	–
ozone profile:	–	4 %	–
aerosol optical depth:	± 0.02	1 %	3 %
aerosol profile:	–	1 %	3 %
aerosol: ω_0	± 0.05	3 %	4%
aerosol model:	–	< 1 %	< 1 %
pressure:	± 1h Pa	–	–
NO_2 content:	± 100 %	1 %	3%

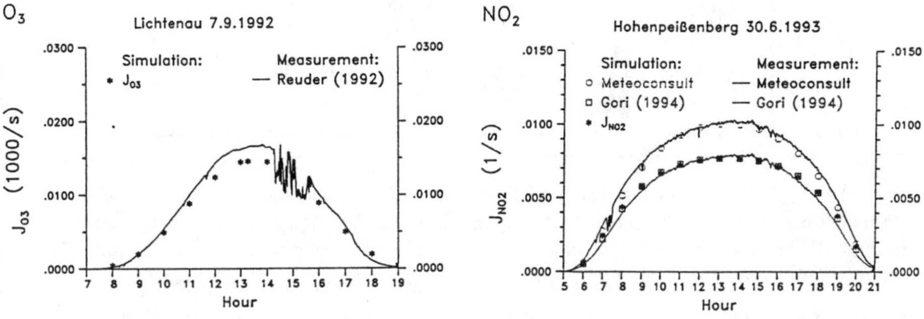

Fig. 13.8.3: Comparison between measured (solid lines) and calculated values (symbols) of ozone and nitrogen dioxide photolysis frequencies.

Measurements

The accuracy of measured photolysis frequencies by photoelectric detectors is assumed to be 15 % for $J(NO_2)$ and 25 % for $J(O_3)$. The main uncertainty factors are:

- uncertainty in the photolysis calibration method using chemical actinometry,
- deviation between theoretical values of $\sigma(\lambda) \times \phi(\lambda)$ and spectral response of the detector,

- deviation of the light sampling and scattering device of the detector from uniform angular response respective to incoming radiation,
- ageing of filters and photodetectors between different calibration campaigns

Fig. 13.8.4 depicts measured $J(O_3)$ and $J(NO_2)$ during a campaign at Hohenpeißenberg in June 1993. Maximum values at cloudless days range up to 3.9 $\times 10^{-5}$ s^{-1} for $J(O_3)$ and 1.05 $\times 10^{-2}$ s^{-1} for $J(NO_2)$. Broken cumulus clouds (June 25 and afternoon of June 28) cause an increase in maximum values of $J(O_3)$ and $J(NO_2)$ due to additional reflected and scattered radiation compared to cloudless conditions (June 29 and 30). An interrupt of the direct solar radiation by clouds causes a loss of about 50 % in $J(NO_2)$, while the decrease of $J(O_3)$ is only about 30 %. This is an effect of a higher percentage of diffuse radiation at shorter wavelength, where ozone photolysis occurs.

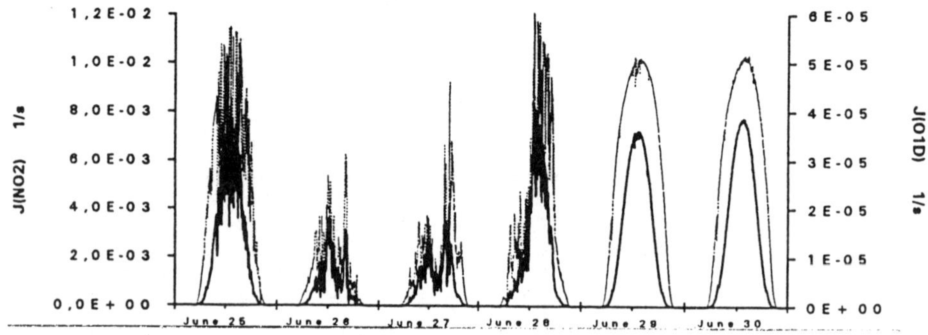

Fig. 13.8.4: Selected time series (CEST) of ozone(thick solid line) and nitrogen dioxide (dotted line) photolysis frequencies during BayFORKLIM campaign at Hohenpeißenberg in June 1993. The time interval includes almost cloudless conditions (June 29 and 30), broken cloud cover (June 25 and 28) as well as almost overcast sky (June 26 and 27).

During the BEMA campaign at Castelporziano near Rome in May 1994 additional photolysis profile measurements were performed to investigate the height dependence of chemically relevant radiation within a forest site. Two photoelectric detectors, measuring $J(O_3)$ and $J(NO_2)$ from the upper hemisphere were mounted on a lift at a 30 m pole, moving at a vertical speed of about 0.2 m/s. The measurement site was located in the middle of a clearing, 10 m in diameter, within a typical Mediterranean forest consisting of *Quercus Ilex*, *Quercus Suber* and *Pinus Pinea*. The top of the canopy was about 10 m above ground. An additional set of photolysis frequency detectors was mounted at 14 m elevation to provide reference data, which are undisturbed by canopy effects.

Fig. 13.8.5 depicts exemplary the height dependence of the ratio of measured $J(O_3)$ related to the reference measurements at 14 m above ground. For overcast sky (12.05.94) about 30 % of the incoming photolytic active radiation at the top of

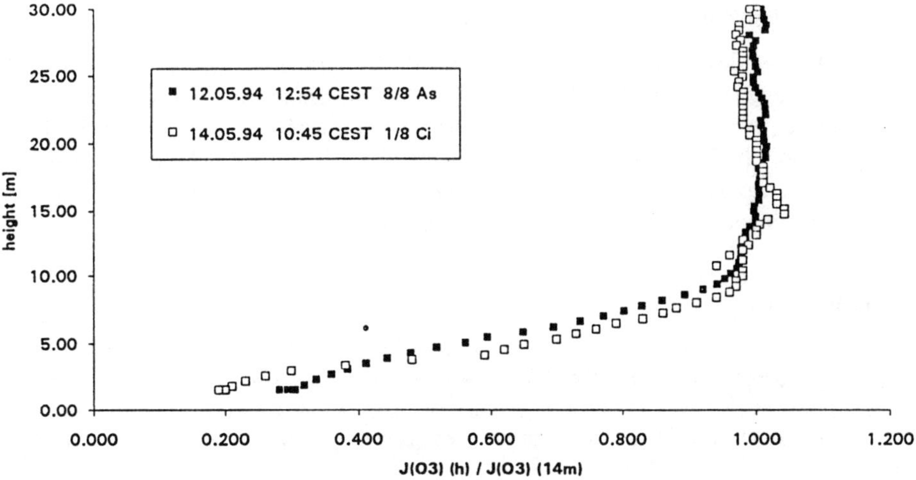

Fig. 13.8.5: Profiles of measured ozone photolysis frequency ratios during the BEMA campaign in May 1994 for clear (14.05.94) and overcast (12.05.94) situations.

sky (12.05.94) about 30 % of the incoming photolytic active radiation at the top of the canopy reaches the ground. This ratio decreases to approximately 20 % in the presence of direct solar radiation above the trees (14.05.94). In both situations the measurements start at 1.5 m while the detector is not exposed to direct sunlight. The sharp increase in the ozone photolysis ratio of May 14 at a height of about 4 m is caused by the influence of direct solar radiation that hits the detector on its upward move at that moment. The slight increase of about 5 % around 15 m at that profile could possibly be caused by radiation reflected at the top of the canopy. The detailed analysis of all measured (about 30) profiles is the topic of current investigations.

Conclusions

Photolysis frequencies of ozone and nitrogen dioxide were continuously measured during TRACT, BIATEX, and BayFORKLIM field campaigns. The measurements were performed by photoelectric detectors. The model STAR was developed within EUMAC to calculate accurately and efficiently photolysis frequencies in the troposphere. A comparison of measured and simulated photolysis values shows good agreement within the discussed uncertainty intervals. The agreement can be improved, if all necessary atmospheric parameters for the simulation are measured simultaneously and if the spectral response and the angular dependence of the detectors are considered in the simulation.

Acknowledgements

STAR was developed with support from the Federal Ministry of Research and Technology (BMFT) (521-4007-07EU738/6). The photolysis measurements were financed by the BMFT under contract no. 07EU818 and 07VND09/8. The Bavarian Climate Research Program (BayFORKLIM) supported the construction of the measuring instruments. The authors thank the staff of the Met. Obs. Hohenpeißenberg for their generous help during measurement campaigns.

References

1. Ruggaber, A., R Dlugi, T. Nakajima, Modelling radiation quantities and photolysis frequencies in the troposphere, *J. Atmos. Chem.* **18** (1994) 171–210.
2. Junkermann, W., U. Platt, A. Volz-Thomas, A photoelectric detector for the measurement of photolysis frequencies of ozone and other atmospheric molecules. *J. Atmos. Chem.* **8** (1989) 203–227
3. Reuder, J., Experimentelle Untersuchungen zum Einfluß atmosphärischer Parameter auf die Photolyse von Ozon in der bodennahen Troposphäre. Diplomarbeit, Ludwig-Maximilians-Universität München, 1991
4. Gori, T., Bestimmung der NO_2-Photolysefrequenz in der bodennahen Troposphäre mit Hilfe photoelektrischer Detektoren. Diplomarbeit, Ludwig-Maximilians-Universität München, 1995

Chapter 14

Modelling and Validation

14.1 Dry Deposition Inferential Measurement in a Rural Region of Northern Italy

C. Cavicchioli[1], G. Manzi[2], G. Catenacci [1], G. Brusasca[2] and M. Borgarello[1]

[1]CISE SpA, Environmental Department, via R. Emilia 39, 20090 Segrate, Milan, Italy
[2]ENEL SpA, Environment and Material Research Centre, via Rubattino 54, 20153 Milan, Italy

Summary

ENEL SpA/ CRAM, Environment and Material Research Centre of the Italian Electricity Company, in collaboration with the Environmental Department of CISE, has been developing a long term project since 1992, in order to evaluate the contribution of deposition phenomena in the environmental problems related to atmospheric acidity. This activity is carried out at the meteorological station of San Pietro Capofiume, located in the Po river plain, a rural region in northern Italy affected by anthropogenic pollution events. This project also aims to study dry deposition over a long period in the Mediterranean area. Direct measurements of meteorological and chemical parameters were used to run inferential technique models, in order to quantify dry deposition. The big-leaf and multi-layer approaches were both tested. Concentrations of SO_2, NO_2, NO and O_3 were measured with half an hour resolution, using automatic analysers. During the vegetative season, concentration of gaseous species as SO_2, NO_2, HNO_2, HNO_3, NH_3 and fine particulate as SO_4^{2-} NO_3^- and NH_4^+ were measured weekly, by an annular denuder system. In 1993, during the vegetative season, an intercomparison campaign was performed at S. Pietro Capofiume to test an inferential technique against eddy correlation ozone fluxes. The results of this comparison and the annual dry deposition fluxes are presented.

Aims of the research

There is a general agreement about the fact that the dry deposition can account for up to 30–50 % of total acid deposition [1]. Despite its importance, the dry deposition generally is not monitored on a regular basis by fixed networks on the territory, unlike the wet one. This is due to the complexity of the involved processes, where the interaction between the pollutant and the vegetation cannot be simulated by means of surrogate collection media Hans [2]. A suitable technique for the routine assessment of dry deposition is the so called 'inferential technique', a measurement/modelling approach, developed by US/NOAA. In order to evaluate total deposition (wet + dry + fog) over a long period, ENEL/CRAM, in collaboration with CISE, installed a station to estimate dry deposition by means of the inferential technique. Moreover an intercomparison campaign was realised for a 'site-specific' calibration of the code. Total deposition for sulfur and nitrogen compounds and ozone are evaluated.

Principal scientific results

Direct measurements of dry deposition fluxes are difficult to perform on routine basis. For this purpose, the inferential approach seems to be the most suitable tool, due to its ability to estimate dry deposition fluxes by means of standard meteorological and chemical measurements. The inferential method is based on the assumption that the pollutant dry deposition fluxes are linearly dependent on air concentration via deposition velocity:

$$F = V_d + C$$

F = deposition flux ($mg/m^2\,s^{-1}$)

V_d = deposition velocity (m/s)

C = pollutant concentration (mg/m^3)

Deposition velocity accounts for all the micrometeorological, surface, chemical and sink phenomena involved in deposition. A resistive scheme is used in the model in order to take into account the different contributions. At the moment we are using a code that includes both big-leaf [3, 4] (BL) and multi-layer [4] (ML) model parametrisations. In Fig. 14.1.1 both resistive schemes are shown. The differences between these two approaches are mainly concerned with the interaction with the surface. ML introduces a more detailed description of the interaction phenomena, but this accuracy is coupled with many input requests and parametrisations that need to be carefully calibrated.

The mathematical formulations for the two schemes are:

BL

$$V_d = \frac{1}{R_a + R_b + R_c}$$

ML

$$V_d = \frac{1}{R_a + \frac{1}{\frac{1}{R_s} + \frac{1}{\sum_{i=1}^{N} R_{c(i)}}}}$$

R_a = Aerodynamic resistance, depending on turbulent atmospheric conditions

R_b = Bulk resistance, depending on diffusive and chemical proprieties of the pollutants

R_c = Canopy resistance, which describes the different mechanisms of pollutants absorption by plants (R_s, R_{inc}, R_{ext}, R_{st} and R_{mes})

$R_{c(i)}$ = Canopy resistance (i=1, N vertical level), which describes the different mechanisms of pollutants absorption by plants (R_{inc}, R_{ext}, R_{st} and R_{mes})

R_s = Soil resistance

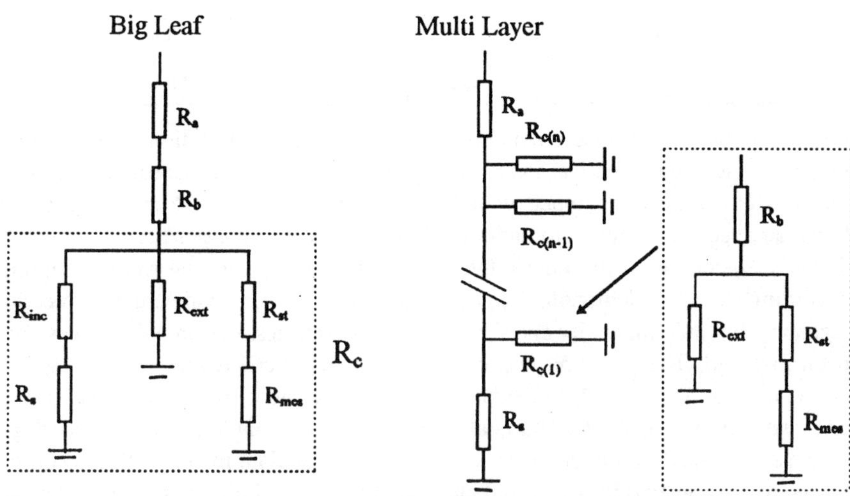

Fig. 14.1.1: Resistive schemes for the two different inferential approaches.

These computation schemes are specific for gaseous compounds, a specific algorithm [4] is included for particle deposition evaluation:

unstable condition

$$V_d = \frac{1}{\frac{1}{0.01\ u^*} + R_a}$$

neutral or stable condition

$$V_d = \frac{1}{\frac{1}{0.002\ u^*} + R_a}$$

The code uses a micrometeorological scheme [5–7] based on similarity theory, in order to compute radiative budget, friction velocity, and to evaluate the turbulence properties of the atmosphere. The model is able to take into account complex vegetative scenarios, as agricultural and forest land, considering up to 21 plant species, defined by plant shape, activation temperatures, threshold active radiation, LAI (leaf area index).

Table 14.1.1: Measured quantities

Meteorological	Chemical	Chemical
half hour	half hour	weekly
air temperature	SO_2	SO_2
relative humidity	NO	SO_4
global radiation	NO_2	NO_2
net radiation	O_3	NH_3
wind velocity		HNO_3
wind direction		HNO_2
wind direction sigma		NO_3
rain		NH_4

For dry deposition measurement in 1992 was installed a station with standard meteorological and chemical instruments. In Table 14.1.1 measured quantities are listed. The station is equipped with Micros meteorological instruments, Monitor Labs analysers and an automatic calibration system. Furthermore an annular denuder system was chosen as the most suitable [8] for one-week sampling; according to the methodologies developed by US Environmental Protection Agency, Harvard University and CNR, Italy Laboratories, two sampling lines have been installed. The first line was composed of one NaCl denuder, two Na_2CO_3, and one citric acid denuder together with a Teflon and a nylon filter; the second line was composed of two charcoal denuders. After sampling, the absorbed gas which is collected on the denuder wall, and the particulate matter collected on the filters, are extracted with water and analysed by means of chromatography. During the intercomparison campaign a Gill sonic anemometer and a GFAS fast ozone analyser were used for eddy correlation flux measurements.

The sampling site is located in the Po Valley (latitude 44°39’13” N and longitude 11°37’27” E). Dry deposition phenomena are affected by the vegetative surface and by the physical characteristics of the soil. The sampling site has no orographic complexity, but it has different cultivation types. A specific investigation on the vegetative cycle in the surroundings ~ 1 km from the sampling site, was performed (Fig. 14.1.2), in order to estimate the deposition fluxes on the area. Some parameters, LAI, plant type, height, *etc.*, of the vegetative species were recorded monthly during the year.

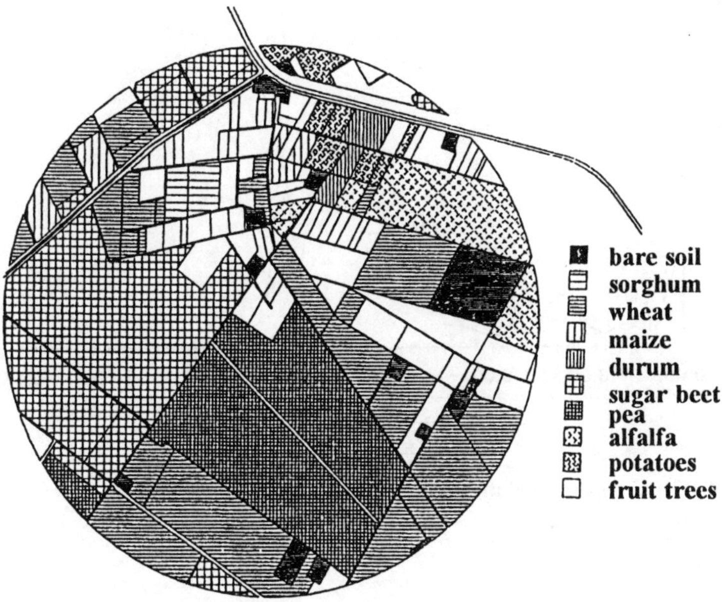

Fig. 14.1.2: Agricultural land use

An intensive field experience was performed, aimed to test the application of the inferential method versus eddy correlation, from 15–29 June, 1993. The intercomparison campaign allowed us to test the computation schemes of O_3 fluxes, and some other routines of the METPRO meteorological pre-processor [6, 7] which derives the micrometeorological characteristics of the atmosphere from conventional measurements. In general we observe a good agreement between model and sonic data (Fig. 14.1.3 and 14.1.4): for example, linear correlation coefficients of 0.96 and 0.93 were computed for sensible heat flux and friction velocity. The ozone deposition velocity is related to the plants stomata activity.

The day/night cycle strongly affects deposition phenomena, in particular we observe maximum values (0.5 cm/s) at noon, and minimum values during the night (0.05 cm/s). Eddy correlation [9] deposition velocity for ozone is compared with ML and BL formulation (Fig. 14.1.5). Both the models show a good agreement, however the BL algorithm seems to derive larger peak values during the day than ML. These differences are due to the fact that ML gives a more detailed description, based on vertical profile and sun/shade algorithms of the stomata activity.

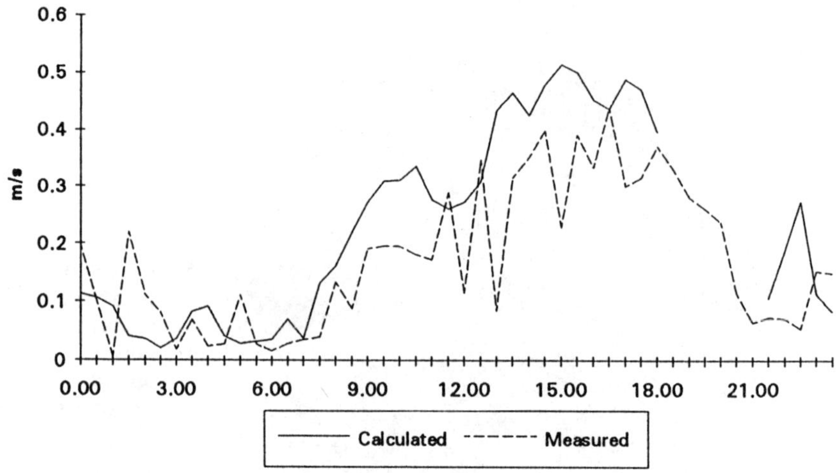

Fig. 14.1.3: Comparison of the friction velocity, on 19 June, 1993.

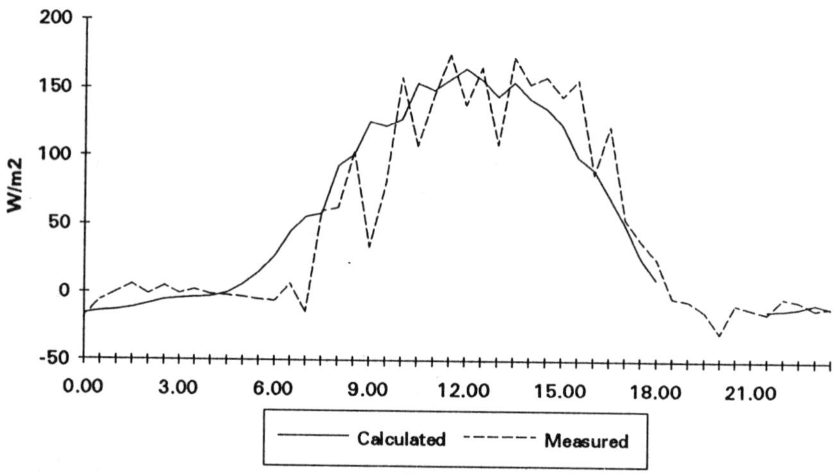

Fig.14.1.4: Comparison of sensible heat flux, on 19 June, 1993.

An estimation of the deposition of total nitrogen and sulfur compounds was performed by means of annular denuder, during the vegetative season of 1993 (Fig. 14.1.6). Table 14.1.2 shows the measured deposition fluxes, velocities and concentrations for the different compounds; for a comparison purpose, in the same table literature [10, 11] typical ranges of these values are listed. The monitored pollutant concentrations are typical of Italian rural sites, with cultivated land. The use of artificial fertiliser forces sensitive N compounds concentrations. Modelled deposition velocities agree quite well with literature reported values, for all the pollutants. Deposition velocities of fine particles are computed in a simplified way, taking in account only the aerodynamic characteristic, so that we cannot

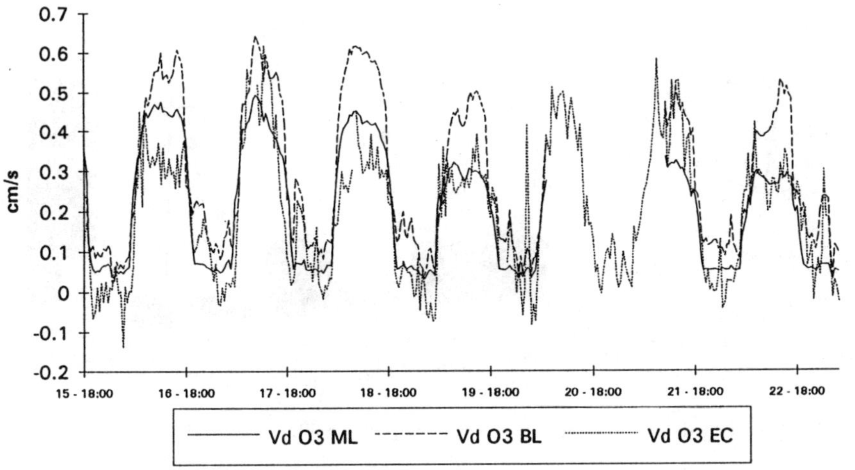

Fig. 14.1.5: Comparison between the inferential and eddy correlation techniques from 15 to 22 June, 1993.

Table 14.1.2: Measured (during the vegetative period in 1993) and literature values for concentration, deposition velocities and fluxes

	Concentration			Deposition Velocity				Deposition Flux			
	Measured	Reference values		Computed daytime		Reference values		Computed summer		Reference values year	
Compounds				BL	ML			BL	ML		
	[μg/m^3] average	[μg/m^3] min	max	[cm/s] average		[cm/s] min	max	[mgS/m^2]		[mgS/m^2] min	max
SO_2	2.9	1	10	0.68	0.36	0.1	1.5	89	48	40	12000
SO_4	6.3	0.2	5	0.19		0.01	2.9	26			
								[mgN/m^2]		[mgN/m^2]	
NO_2	17.9	1	10	0.29	0.16	0.2	0.5	95	47	60	15000
HNO_3	1.0	0.2	0.5	1.86	3.31	1	4	24	44	30	6000
HNO_2	1.0	0.01	0.1	0.71	0.37	0.2	0.5	19	10	0.6	6
NH_3	2.9	0.03	12	0.24	0.12	↑	0.3	72	42	↑	10000
NO_3	3.5	–	–	0.19		0.13	2.25	10		–	–
NH_4	1.0	0.1	5	0.19		0.08	1.4	13		–	–
								[mg/m^2]		[mg/m^2]	
O_3	65.7	1	200	0.32	0.18	0.1	1	1794	1008	–	–

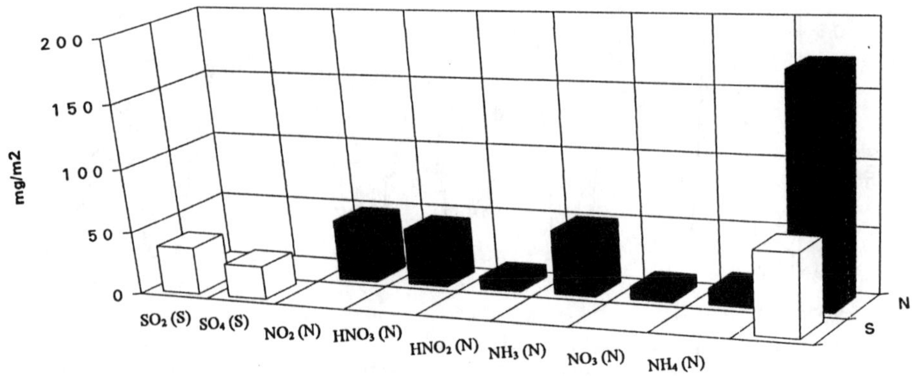

Fig. 14.1.6: Dry deposition integral fluxes during the vegetative period in 1993.

distinguish the different chemical compounds. Dry deposition fluxes are in the literature ranges of variation for all the pollutants, except HNO_2. Significant measured HNO_2 concentrations force high deposition fluxes for this pollutant. As shown in the table, the two computation schemes lead to sizeable differences: the BL approach seems to estimate higher deposition of all compounds (except HNO_3) than the ML method. The differences for HNO_3 depend mainly on the dominant effect of soil resistance in ML scheme. Deposition data of the two approaches reflect the different parametrisations of the radiative and water budgets. This effect is true for all the compounds, even if different removal mechanisms can strongly influence the deposition fluxes. Deposition of SO_2 and HNO_2 are computed by similar schemes [3], for these gases BL algorithm considers a more efficient removing mechanism in high humidity conditions (well wetted soil).

Table 14.1.3: Integral flux ratios during the vegetative period in 1993

Flux ratios	BL	ML
$SO_4(S)/SO_2(S)$	0.29	0.53
$HNO_3(N)/NO_2(N)$	0.25	0.92
$HNO_2(N)/NO_2(N)$	0.20	0.21
$NH_3(N)/NO_2(N)$	0.76	0.89
$NO_3(N)/NO_2(N)$	0.11	0.21
$NH_4(N)/NO_2(N)$	0.14	0.28

In Table 14.1.3 integral flux ratios, between sulfur compounds and SO_2 and between nitrogen compounds and NO_2, for the two models are reported.

Conclusions

Knowledge of dry deposition phenomena and of their controlling factors allowed us to set up a prototype station for dry deposition sampling. This station combines meteorological and chemical measurements with an inferential approach in order to evaluate the dry deposition fluxes for gaseous and particulate pollutants in different vegetative scenarios during the whole vegetative cycle. The main goal for the future studies will be a better comprehension of the removing/emission mechanisms, in particular for HNO_2, HNO_3, and NH_3. Moreover, it is important to understand the influence of chemical reactions in the atmosphere on deposition phenomena in order to quantify the total fluxes of pollutants over a long period. Another goal will be the nesting of the computation schemes into a regional scale model. In the near future we will apply our model and instruments to evaluate dry deposition on complex terrain and in forest areas. This development is particularly relevant for the Italian orographic characteristics.

Acknowledgements

We are grateful to the JRC of Ispra, the Department of Physics of the University of Torino, CNR/ICG Torino and CNR/FISBAT Bologna for the co-operation during the experimental campaign.

References

1. EMEP, Calculated budget for airborne acidifying components in Europe, 1985–1991, *EMEP/MSC-W Report* 1/92.
2. B.B. Hicks, D.D. Baldocchi, T.P. Meyers, R. P. Hosker, D.R. Matt, Preliminary Multiple Resistance Routine for Deriving Dry Deposition Velocities from Measured Quantities, *Water, Air and Soil Pollution*, **36** (1987) 311–330,
3. J Erisman, A. van Pul, P.Wyers, Parametrisation of dry deposition mechanisms for the quantification of atmospheric input to ecosystem – *ECN Report*, Petten, 1992.
4. T. P. Meyers, D.D. Baldocchi, A comparison of models for deriving dry deposition fluxes of O_3 and SO_2 to a forest canopy, *Tellus* **40b** (1988) 270–284.
5. M. L. Wesley, B. B. Hicks, An examination of some micrometeorological methods for measuring dry deposition – EPA 600/7-78–116, 1978.
6. EPA, On site Meteorological Program , Guidance for regulatory modelling application - NC 27711, 1986 (revised 1993).
7. R. Paine, Meteorological Pre-processor program - EPA, 1987.
8. I. Allegrini, F. De Santis, Measurement of Atmospheric Pollutants Relevant to Dry Acid Deposition, *Anal. Chem.* **21** (1989) 237–255.
9. D.D. Baldocchi, B.B. Hicks, T.P. Meyers, Measuring biosphere-atmosphere exchange of biologically related gasses with micrometeorological methods, *Ecology* **69** (1988) 1331–1340.

10. A. van Pul, S.Braun, J. Duyzer, P. Grennfelt, F.X. Meixner, A. Neftel, K. Pilegaard, R. San Jose, Dry deposition of nitrogen dioxide, *Models and Methods for the Quantification of Atmospheric Input to Ecosystems*, Nordiske Seminar, og, Arbejds Rapporter 1993, Nordic Council of Ministers

11. S.E. Lindberg, A.L. Page, S.A. Norton, in: D.C. Adriano and W. Salomons (eds), *Acid Precipitation, Vol.3: Sources, Deposition and Canopy Interactions*, Advances in Environmental Science, Springer Verlag, 1989.

14.2 Improvement and Validation of the Throughfall Technique for Nitrogen Deposition Measurements to Forest Ecosystems

Martin Ferm

Swedish Environmental Research Institute, P.O. Box 47086,
S-402 58 Gothenburg, Sweden

Summary

The throughfall technique cannot be used to estimate the dry deposition of ammonium and nitrate when the dry deposition is small because the two ions are lost in the canopy. A new phenomena partly explaining this loss is the conversion of inorganic nitrogen to organic nitrogen in the canopy and in the throughfall collectors if they are not kept dark. It is difficult to analyse the throughfall deposition of organic nitrogen because it is bound in particles and can also stick to the walls of the collector. The sampling technique will be investigated further using the addition of known amounts of inorganic nitrogen to the canopy.

Aims of the research

The aim of this project is to develop a technique to measure the dry deposition of ammonium and nitrate to a coniferous forest.

Principal scientific results

The nitrogen loss was first interpreted as an uptake by the canopy, because more nitrogen is lost during the warmer seasons. The sampling frequency is often event bases or monthly bases. In the literature nothing could be found on the stability of throughfall samples. Iodine was therefore added as a preservative to half of the throughfall collectors at a forest edge in Sweden. The sampling site is called Anten and is situated 10 km north west of Alingsas and approximately 60 km from the Swedish west coast (58°00' N, 12°26' E) having an altitude of 100 m. The forest stand which is healthy, consists of 50 year old Norway spruce (*Picea abies* [L.] *Karst*) trees 18–20 m tall. When the results from preserved and unpreserved collectors were compared, no difference could be detected during the wintertime. There was, however, a large difference during the summertime. The nitrate and especially the ammonium depositions measured with unpreserved semitransparent samplers were very small compared to present ones [1]. No effect was observed for sulfate or chloride. The nitrogen loss was more pronounced inside the forest where the deposition was lower. There is thus a biological activity in the throughfall that can remove ammonium and nitrate. The intercept precipitation that remains on the tree after a rain event is exposed to light during a long time. The biological activity of the throughfall can thus be an additional explanation to a possible uptake of nitrogen.

In the next phase the disappearance of ammonium and nitrogen in throughfall was further investigated [2]. An increased amount of organic nitrogen was observed when inorganic nitrogen was lost. The conversion rate was increased by the addition of glucose. The efficiency of iodine and another preservative (thymol) was tested by adding glucose to preserved samplers. Both iodine and thymol was found to be efficient despite the presence of glucose. Neither ammonium nor nitrate was transformed to organic nitrogen in the absence of light.

The atmospheric nitrogen deposition to a forest is difficult to estimate. A rough estimate based on the assumption that the relationship between the dry deposition to the forest and the air concentration difference between the inside and outside of the forest is the same for all compounds was used. The relationship was calculated for sulfate and then used to calculate the ammonium and nitrogen depositions. The results are shown in Table 14.2.1. The wet and throughfall depositions are summarised in Table 14.2.2. From Table 14.2.2 it can be seen that there is a deposition gradient in the forest with higher depositions at the edge for all compounds. The estimated ammonium deposition (43 mmol m^{-2} a^{-1}, Table 14.2.l) is almost equal to the net throughfall deposition of NH_4^+ + organic nitrogen (34 mmol m^{-2} a^{-1}, Table 14.2.2)

Table 14.2.1: Measured concentrations in air during October 1990 using a denuder technique (in nmol m^{-3}) and estimated dry depositions during 1990 (in mmol m^{-2} a^{-1}).

Compound	Concentration		Deposition
	field	forest	1990
SO_2	34.5	28.9	29
particulate SO_4^{2-}	33.6	32.2	6
HNO_3	8.96	3.90	25
particulate NO_3^-	25.0	21.9	10
NH_3	7.62	3.51	36
particulate NH_4^+	66.4	66.5	7

Table 14.2.2: Wet and throughfall deposition during 1990 (in mmol m^{-2} a^{-1}) measured using dark and preserved collectors at different distances from the edge.

	WD	TD				NTD
Distance	– 10	0	10	30	60	
mm	1099	862	793	790	890	–278
SO_4^{2-}	41	130	67	78	61	35
Cl^-	95	768	241	247	194	181
NO_3^-	43	162	73	55	41	22
NH_4^+	45	87	46	31	28	– 6
org. N	8	90	42	52	37	40

The measurements made in 1990 were repeated in 1993. The wet and throughfall depositions are shown in Table 14.2.3. As can be seen both the wet and the net throughfall depositions are much smaller during 1993 than during 1990, except for chloride and organic nitrogen. Air concentration was measured at an EMEP station *ca.* 75 km south west of Anten (Rörvik) The concentration of sulfur, but not of nitrogen compounds was lower in 1993 compared to 1990, see Table 14.2.4. Lower nitrogen depositions have also been observed at other sites in this region. The high deposition of organic nitrogen in 1993 was caused by a pollen episode over southern Sweden in May. It was the largest pollen episode in 15 years (measured as wet deposition of K^+). If the organic nitrogen deposition had been normal that month the deposition of organic nitrogen would have been around 33 mmol m^{-2} in 1993, *i.e.* slightly less than during 1990.

Table 14.2.3. Wet and throughfall deposition during 1993 (in mmol m^{-2} a^{-1}) measured using dark and preserved collectors at different distances from the edge.

	WD	TD				NTD
Distance	–10	0	10	30	60	
mm	888	585	596	610	657	–272
H^+	67	161	123	116	131	58
SO_4^{2-}	21	51	38	40	36	18
Cl^-	66	759	218	180	223	182
NO_3^-	27	68	38	21	22	3
NH_4^+	28	25	18	13	13	–13
org. N	9	75	69	61	60	55

Table 14.2.4: Air concentrations (in nmol m^{-3}) at Rörvik.

	1990	1993
SO_2	57	37
particulate SO_4^{2-}	49	37
total NH_4^+	77	93
total NO_3^-	41	46
NO_2	156	161

Denuder measurements were also performed during 1993. There was a lot of problems with the equipment and the data was not consistent. Diffusive samplers, which are much simpler to use was therefore tested. They were only tried for NO_2 (measured outside the forest during the whole year) SO_2 and HNO_3. A diffusive sampler is also available for NH_3 [3]. The results are shown in Table 14.2.5. The concentration difference between the open field and the forest resembles those measured with denuder technique in 1990 (Table 14.2.1).

Table 14.2.5: Air concentrations at Anten from July 1993 measured using diffusive samplers. The concentrations are given in nmole m^{-3}.

month	NO_2 field	SO_2 field	SO_2 forest	HNO_3 field	HNO_3 forest
J	57	13.1	13.3	5.4	0.2
A	32	10.9	6.6	–	–
S	62	13.1	7.8	–	–
O	119	9.0	6.5	6.2	3.1
N	161	75	62	20.0	3.9
D	142	36	23	7.6	1.7
J+F	128	37	34	13.0	2.8

It has earlier been observed that higher Kjeldahl nitrogen (ammonium + organic nitrogen) concentrations are found when a preservative is used or when the whole sample is digested. When a large sample without preservative is taken and a small fraction is used for analysis a lot of nitrogen can remain on the walls. When a preservative is used the walls always seem clean. The whole sample can be analysed if a Kjeldahl tube is used as throughfall collector. The diameter is, however, only 25 mm so the sample is not very representative. A comparison between ordinary collectors (190 mm funnels connected to 51 bottles using 10 collectors per distance from the edge) was compared to 2 Kjeldahl tubes per distance during six months, both preserved with iodine, see Table 14.2.6. As can be seen the throughfall amounts are rather similar with the two techniques, but the ammonium and organic nitrogen concentrations are both higher in the Kjeldahl tubes. No systematic trend in the concentration differences can be observed between the different months (Table 14.2.6) or distances (Table 14.2.7). On the average there was 4 % less throughfall deposition in the Kjeldahl tubes 38% more NH_4^+ and 54 % more organic nitrogen. It is for the moment not clear how representative the measurements with the Kjeldahl tubes were. In another study it was found that the addition of iodine to dark collectors increased the amount of organic nitrogen collected, but not of NH_4^+ or NO_3^-, most likely because losses of organic nitrogen to the walls were avoided.

Table 14.2.6: Throughfall amounts and average concentration (μM) of NH_4^+ and organic nitrogen in Kjeldahl tubes and conventional funnels for different months (Nov. 1991 and Feb.–June 1993). The ratio between the measurements are also given.

month	funnels			Kjeldahl tubes			tubes/funnels		
	mm	NH_4^+	org N	mm	NH_4^+	org N	mm	NH_4^+	org N
N	276	62	14	250	61	64	0.9	1.0	4.6
F	139	38	39	109	49	150	0.8	1.3	3.9
M	59	146	100	59	148	154	1.0	1.0	1.5
A	36	127	237	65	166	382	1.8	1.3	1.6
M	310	60	490	291	110	557	0.9	1.8	1.1
J	139	21	119	133	8	282	1.0	0.4	2.4

Table 14.2.7: Throughfall amount and average concentration (μM) of NH_4^+ and organic nitrogen in Kjeldahl tubes and conventional funnels at different distances from the edge. The ratio between the measurements are also given.

Distance	funnels			Kjeldahl tubes			tubes/funnels		
	mm	NH_4^+	org N	mm	NH_4^+	org N	mm	NH_4^+	org N
–10	267	55	41	250	50	105	0.9	0.9	2.6
0	163	110	298	147	117	512	0.9	1.1	1.7
10	174	57	271	213	83	292	1.2	1.5	1.1
30	167	43	254	156	93	335	0.9	2.2	1.3
60	188	39	229	140	79	353	0.7	2.0	1.5

Conclusions

The wet deposition of NH_4^+, NO_3^- and SO_4^{2-} was much smaller during 1993 than during 1990. The corresponding net throughfall depositions were also smaller for these species, especially for NH_4^+ and NO_3^- which were negative during half the year 1993. The total nitrogen deposition was not negative and seem to be a reasonable measure of the atmospheric deposition of nitrogen (about half of the wet deposition). The sampling technique for organic nitrogen in throughfall has, however, to be further investigated. Diffusive samplers seem to be appropriate for measurements of monthly average concentration gradients in forests.

Acknowledgements

This project has been financed from the National Environmental Protection Board and the Swedish Environmental Research Institute.

References

1. M. Ferm: Throughfall deposition of nitrogen and sulfur compounds, in: P. Borrell, P.M. Borrell, W. Seiler (eds), *Proc. EUROTRAC Symp. '90*, SPB Academic Publishing bv, The Hague 1991, pp. 131–134.
2. M. Ferm: Throughfall measurements of nitrogen and sulfur compounds, *Int. J. Anal. Chem.* **50** (1993) 29–43.
3. L. Svensson, M. Ferm: Mass transfer coefficient and equilibrium concentration as key factors in a new approach to estimate ammonia emission from livestock manure, *J. Agricult. Eng. Res.* **56** (1993) 1–11.

14.3 Modelling the Uptake and Metabolisation of Nitrogen Dioxide and Ozone by Plant Leaves

Gundolf H. Kohlmaier, P. Ramge, M. Plöchl and F. Badeck

Institute for Physical and Theoretical Chemistry, Johann Wolfgang Goethe University, Marie-Curie-Straße 11, D-60439 Frankfurt

Summary

Uptake of NO_2 and ozone via stomata and their elimination in the mesophyll were investigated using mechanistic mathematical models. The common hypotheses, postulating the disproportionation reaction of NO_2 being the predominant path for the transformation of NO_2 into the cellular nitrate and nitrite pools, could be falsified because it fails to reproduce the sorption characteristics as determined experimentally. If, however, the reduction of the pro-oxidative NO_2 by apoplastic ascorbate is taken into account the calculated uptake rates and the effects of changes in immission concentration and vegetation characteristics fit well within corresponding experimental results. The onset of acute injury in case of a fall in apoplastic ascorbate concentration below a critical level is postulated.

In the case of ozone the results of the model calculations corroborate the hypotheses that apoplastic ascorbate is an essential factor controlling the uptake of this chemically reactive trace gas, too. The model developed contains furthermore a comprehensive mathematical description of the dynamic distribution of ascorbic acid/ascorbate between the different plant cell compartments.

Aims of the research

The research work presented here was performed in 1990–1992 as part of the BIATEX project within EUROTRAC. On the basis of plant physiological and biochemical processes mathematical models were to be developed describing uptake and release of atmospheric trace gases like NO_2, O_3 and SO_2 by vegetation and soils. The main issue of our research was the modelling of uptake and metabolisation of atmospheric trace constituents at the leaf and cell level, respectively. With the aid of such models, which consider and describe processes of dissolvation, reaction, and diffusion within leaf and cell, reliable values of the mesophyll resistance of the trace gases taken into consideration were to be evaluated. The mesophyll resistance describes the total of the diffusive and reactive single resistances of the mesophyll and may play an important role within the resistance network.

For reasons of space we are showing in this report only the most important principles and results concerning the modelling of the uptake and metabolisation of NO_2 and ozone into and within plant leaves.

Principal scientific results

Nitrogen dioxide

With respect to the uptake and metabolisation of nitrogen dioxide in leaf tissues many authors formulated the hypotheses that the disproportionation reaction of NO_2 yielding nitrite and nitrate at equal amounts is the rate determining step of the uptake process [1–9]. However, investigating the uptake of NO_2 into surface waters, Lee and Schwartz [10] had already shown that the rate constant of the disproportionation is too low to satisfactorily explain the measured uptake rates. Moreover Norby *et al.* [11] deduced the existence of further elimination reactions for NO_2 from comparisons of uptake and metabolisation rates with those, which were to be expected on the basis of the rate constant of the disproportionation reaction.

Within the frame of our modelling approaches we could show that the very slow disproportionation process as the only elimination reaction would lead to very high NO_2 concentrations inside the cell and therefore to very small concentration gradients between ambient air and mesophyll; so extremely low uptake rates would result [12, 13]. As further metabolisation reactions radical hydrogen abstraction processes at polyunsaturated fatty acids inside cell membranes were considered. On the basis of this hypothesis we simulated uptake rates being somewhat more realistic, but that would lead to severe membrane disintegration processes even under ambient NO_2 mixing ratios. This brought up considerable clues suggesting the existence of fast elimination reactions within the aqueous phase of the cell wall.

Using experimental results from Forni *et al.* [14] who had measured kinetic data of reactions of NO_2 with some organic reduction agents, we could show that hydrogen abstraction reactions with apoplastic ascorbate are most likely the processes determining the uptake rate [12, 13]. With the aid of this model it is now possible to satisfactorily explain the characteristics and the rates of the uptake of NO_2 into plant leaves. In the light of these results the ascorbate content of the apoplast is to be considered as decisive with respect to the detoxification of nitrogen dioxide in plant cells. The induction of nitrate reductase under NO_2 fumigation reported by several authors [6, 15], however, remains an open question, because the elimination process proposed here yields nitrite only.

Fig. 14.3.1 shows some results of NO_2 fumigation experiments by several authors using relatively low NO_2 mixing ratios between 4 and 45 ppb.

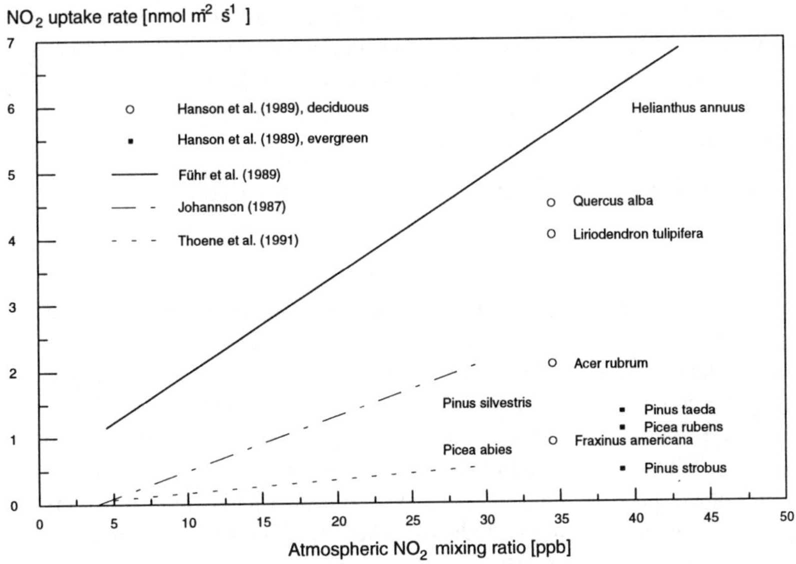

Fig. 14.3.1: Some results of NO_2 fumigation experiments by several authors

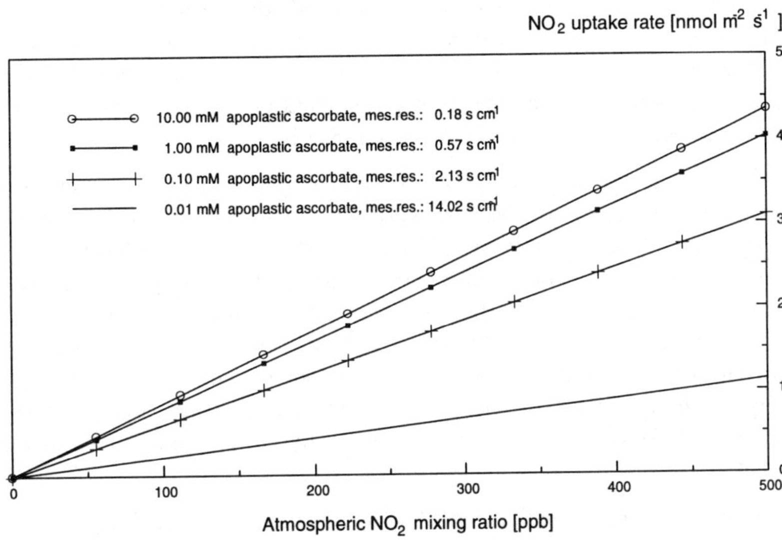

Fig. 14.3.2: Calculated uptake rate and mesophyll resistance as function of the NO_2 mixing ratio

As one can see in Fig. 14.3.2 the model can be validated very well using experimental data. The linear uptake characteristics as well as the uptake rates and mesophyll resistances calculated on the basis of the ascorbate hypothesis do reproduce the measured data within realistic ranges of apoplastic ascorbate concentration. The four straight lines stand for four different concentrations of cellular total ascorbate. The model calculates an uptake rate of 2.01 $nmol \cdot m^{-2} \cdot s^{-1}$ at 25 ppb NO_2 and 1 mM ascorbate in the liquid cell wall phase. The mesophyll resistance is independent of the atmospheric NO_2 mixing ratio but strongly depends on the ascorbate concentration.

For a more detailed description and discussion of the results see [12, 13].

Ozone

Like nitrogen dioxide the oxidative trace gas ozone can be taken up by plants via two different paths. On one hand it may dissolve in the cuticle and react with cuticular material, on the other hand it can be taken up via the stomatal path. However, the cuticular path is not relevant in the case of nitrogen dioxide as well as in the case of ozone, in both cases the stomatal path determines the uptake process [16]. Ozone, which dissolves in the liquid phase of the cell wall reacts with water itself forming further phytotoxic substances like hydrogen peroxide, hydroxyl and superoxide anion radicals [14, 17]. Moreover it can react with organic compounds in the apoplast.

If ozone would not be detoxificated in the apoplast, it would react with the carbon double bonds of the membraneous fatty acids inside the plasmalemma. This would lead to severe disintegration and defunctionalisation and would favour the influx of ozone into the cytosol. In the case of high ozone immission even the chloroplastic membranes and the chloroplasts themselves would be destroyed and photosynthesis processes would be reduced [18, 19].

For a long time, already, ascorbate is known to be an effective antioxidant. Chloroplastic ascorbate has been suggested as an important hydrogen peroxide scavenger [20]. More recently the function of extracellular (apoplastic) ascorbate as a protection agent against atmospheric ozone has been discussed. Castillo and Greppin [21] determined the extracellular content of ascorbate and dehydroascorbic acid in leaves of *Sedium Album* L. under fumigation with ozone. They could show that the concentrations of both compounds rise during fumigation and decline to their original values within 24 hours after ozone stress. The increase of apoplastic dehydroascorbic acid points at the ozone induced oxidation of ascorbate.

The results of the mathematical model of the uptake and metabolisation of ozone strongly corroborate the hypothesis that ascorbate serves as an effective ozone scavenger within the apoplastic cell regions. For a detailed description see [22] who furthermore presents a comprehensive model system of the dynamic distribution of ascorbate between the different cell compartments.

Conclusions

The investigations performed contribute to the research topics pollutant deposition, effects of anthropogenic pollutants and ecotoxicology.

As rate determining steps within the uptake process of both the oxidative trace gases NO_2 and ozone reactions with apoplastic antioxidants like ascorbate could be identified and corroborated, respectively. With respect to NO_2 the widespread hypothesis of the disproportionation of NO_2 being the rate determining step could be discarded.

In the light of our modelling results the apoplastic content of antioxidants is decisively determining the elimination rate of NO_2 as well as of ozone and is therefore an important factor for the protection of cell membranes and other cell compartments against oxidative attack. Consequently, threshold values for an acute injury by oxidative trace gases are determined by the concentration of ascorbate in the apoplast as well as by the rate of its regeneration. Possible toxic effects caused by the increase of the nitrite concentration, by acid effects and disturbance of the metabolism are subject of a discussion, which lies beyond the frame of the project described here [9]. By increasing the rates of ascorbate synthesis and/or reduction of dehydroascorbate plants could increase their capacity for elimination of oxidants in the apoplast and are therefore capable to adapt to oxidant stress within certain limits. Therefore threshold values seem to be dynamic quantities, which depend on the physiological state of the plant [23, 24]. Consequently, it could be expected that plants exposed to sudden high concentrations of oxidative pollutants after long periods of relatively low exposure will show symptoms of acute injury under lower threshold concentrations than plants adapted to a steady mean immission.

Acknowledgements

This work has been done during the EUROTRAC / BIATEX project. We acknowledge the financial support by the Bundesministerium für Forschung und Technologie. The authors have the responsibility for the content of this report.

References

1. A.J. Zeevaart. Some Effects of Fumigating Plants for Short Periods with NO_2. *Environ.l Pollut.* **11** (1976) 97–108.
2. R.L. Heath, Initial Events in Injury to Plants by Air Pollutants. *Ann. Rev. Plant Physiol.* **31** (1980) 395–431.
3. M.L. Wesely, J.A. Eastman, D.H. Stedman, E.D. Yalvac, An Eddy-Correlation Measurement of NO_2 Flux to Vegetation and Comparison to O_3 Flux. *Atmos. Environ.* **16(4)** (1982) 814–820.
4. H.S. Srivastava, D.P. Ormrod, Effects of Nitrogen Dioxide and Nitrate Nutrition on Growth and Nitrate Assimilation in Bean Leaves, *J. Plant Physiol.* **76** (1984) 418–423.

5. A.J. Rowland, M.C. Drew, A.R. Wellburn, Foliar Entry and Incorporation of Atmospheric Nitrogen Dioxide into Barley Plants of Different Nitrogen Status. *New Phytologist* **107** (1987) 357–371.

6. G. Wingsle, T. Näsholm, T. Lundmark, A. Ericsson, Induction of Nitrate Reductase in Needles of Scots Pine Seedlings by NO_x and NO_3^-. *Physiol. Plantarum* **70** (1987) 399–403.

7. R. Tischner, A. Peuke, D.L. Godbold, R. Feig, G. Merg. The Effect of NO_2 Fumigation on Aseptically Grown Spruce Seedlings. *J. Plant Physiol.* **133** (1988) 243–246.

8. C. Latus, Quantitative Untersuchungen zur NO_2-Aufnahme durch Pflanzen aus der Luft. Konstruktion eines Expositionssystems und Ergebnisse an Sonnenblumen (*Helianthus annuus L.*) *Berichte der Kernforschungsanlage Jülich* - Nr. 2326 (1989). Jülich: KFA Jülich.

9. A.R. Wellburn, Why are Atmospheric Oxides of Nitrogen Usually Phytotoxic and not Alternative Fertilizers? *New Phytologist* **115** (1990): 395–429.

10. Y.N. Lee, S.E. Schwartz, Evaluation of the Rate of Uptake of Nitrogen Dioxide by Atmospheric and Surface Liquid Waters. *J. Geophys. Res.* **86** (1981) 11971–11973.

11. R.J. Norby, Y. Weerasuriya, P.J. Hanson. Induction of Nitrate Reductase in Red Spruce Needles by NO_2 and HNO_3 vapour. *Canad. J. Forest Res.* **19** (1989) 889–896.

12. P. Ramge, F.-W. Badeck, M. Plöchl, G.H. Kohlmaier, Apoplastische Antioxidantien als maßgebliche Eliminierungsfaktoren bei der Aufnahme von NO_2 über den Blattpfad. *Berichte. des Zentrums für Umweltforschung* der Johann Wolfgang Goethe-Universität, Frankfurt am Main (1991), No. 18

13. P. Ramge, F.-W. Badeck, M. Plöchl, G.H. Kohlmaier, Apoplastic antioxidants as decisive elimination factors within the uptake of nitrogen dioxide into leaf tissues. *New Phytologist* **125** (1993) 771–785

14. L.G. Forni, V.O. Mora-Arellano, J.E. Packer, R.L Willson, Nitrogen Dioxide and Related Free Radicals: Electron-transfer Reactions with Organic Compounds in Solutions containing Nitrite or Nitrate. *J. Chem. Soc. Perkin Trans. II* (1986) 1–6.

15. B. Thoene, P. Schröder, H. Papen, A. Egger, H. Rennenberg, Absorption of Atmospheric NO_2 by Spruce (*Picea abies L. Karst.*) trees. I. NO_2 Influx and its Correlation with Nitrate Reduction. *New Phytologist* (1991) 1–11.

16. K.J. Lendzian, G. Kerstiens, Interactions Between Plant Cuticles and Gaseous Air Pollutants. *Aspects of Appl. Biol.* **17** (1988) 97–104.

17. M.L. Kilpatrick, C.C. Herrick, M. Kilpatrick, The Decomposition of Ozone in Aqueous Solution. *J. Am. Chem. Soc.* **78** (1955) 1784–1789.

18. R.L. Heath, Ozone. in: Mudd, J.B., Kozlowski, T.T., (eds) *Responses of Plants to Air Pollution.* Academic Press New York (1975) 23–56.

19. D.T. Tingey, Jr G.E. Taylor. Variation in Plant Response to Ozone: A Conceptual Model of Physiological Events, in: Unsworth, M.H., Ormrod D.P.(eds) *Effects of Gaseous Air Pollution in Agriculture and Horticulture* (1982) 113–138.

20. D.J. Gillham, A.D. Dodge. Hydrogen Peroxide Scavenging Systems within Pea Chloroplasts. A Quantitative Study. *Planta* **167** (1986) 246–251.

21. F.J. Castillo, H. Greppin, Extracellular Ascorbic Acid and Enzyme Activities Related to Ascorbic Acid Metabolism in *Sedum Album* L. Leaves after Ozone Exposure. *Environ. and Experimental Bot.* **28(3)** (1988): 231–238.

22. M. Plöchl, Die Bedeutung apoplastischer Ascorbinsäure für die Aufnahme und Reduktion von Ozon und Stickstoffdioxid in Pflanzenblättern: Eine dynamische Modellierung der involvierten metabolischen Prozesse im System Blatt. Ph.D. thesis, Faculty of Chemistry, Johann Wolfgang Goethe University, Frankfurt/Main (1992).

23. A. Polle, K. Chakrabarti, H. Rennenberg. Entgiftung von Peroxiden in Fichtennadeln (*Picea abies,* L.) am Schwerpunktstandort Kalkalpen (Wank). In: *2. Statusseminar der PBWU zum Schwerpunkt "Waldschäden"* (Projektgruppe Bayern zur Erforschung der Wirkung von Umweltschadstoffen (PBWU) (ed) GSF - Forschungszentrum für Umwelt und Gesundheit, Neuherberg (1991) 151–160.

24. A. Polle, H. Rennenberg,. Field studies on Norway spruce trees at high altitudes: II. Defence systems against oxidative stress needles. *New Phytologist* **121** (1992) 635–642.

25. D. Giamalva, D.F. Church, W.A. Pryor, A Comparison of the Rates of Ozonization of Biological Antioxidants and Oleate and Linoleate Esters, *Biochem. Biophys. Res. Comm.* **133** (1985) 773–779.

14.4 Experimental and Modelling Studies of the Deposition Processes in South European Ecosystems

Roberto San José[1], A. García[2] and H. Casado[3]

[1]Group of Environmental Software and Modelling. Computer Science School, Technical University of Madrid, Campus de Montegancedo, Boadilla del Monte, 28660-Madrid Spain
[2]Departamento de Física. Facultad de Ciencias, Universidad de Extremadura, 06071-Badajoz, Spain
[3]Departamento de Física. Facultad de Farmacia, Universidad del Pais Vasco, 01007-Vitoria, Spain

Summary

Since 1988, our contribution to BIATEX research has been focused on modelling and measuring the deposition fluxes of different pollutants, such as SO_2, O_3 and NH_3, under unfertilised dry grassland, which is typical of large parts of inland Spain, as a typical representative of at least part of the Mediterranean area. Results contribute significantly to the knowledge of the deposition velocities and fluxes in a part of Europe where lack of data was very significant. Several deposition field experiments were performed in the Castilla - Leon region and in the final stage in the Madrid area. We investigated the deposition velocities and fluxes of SO_2, O_3 and NH_3. We used the Bowen ratio, profile and eddy correlation techniques to measure the deposition fluxes.

Aims of the research

In our contribution we have measured concentrations and fluxes (in the final stage) of different pollutants over Castilla - Leon and Madrid area Madrid area in Spain. In the first stage we focused on characterising the deposition processes on inland areas of Spain. Particularly, we measured the meteorological variables in a tower at different levels. It allowed us to obtain the turbulent parameters and characterise the micrometeorological conditions of the regions [1]. The profile method and, in the final stage the eddy correlation method, was used to parameterise the turbulent conditions in time and space.

Fig. 14.4.1 shows the location of the two places in Spain where the experiments were held. The Castilla - Leon region is located in the northern part of the Iberian peninsula with typical continental weather, in the Madrid Area the weather is not so extreme. Also, the Madrid area has an ecosystem of its own, as it is surrounded by different mountain ranges which protect the area from the influence of large synoptic fields.

The main objectives were to characterise the canopy resistances and fluxes in the areas. In order to obtain the deposition velocity of the different pollutants. We have used the "resistance" approach to model de deposition fluxes and velocities. Particularly, the Wesely [2] and Baldocchi *et al.* [3] models were used. In the last phase of our contribution over Madrid area, we have used Sutton *et al.* [4], Erisman and Wyers [5] and Schjørring *et al.* [6].

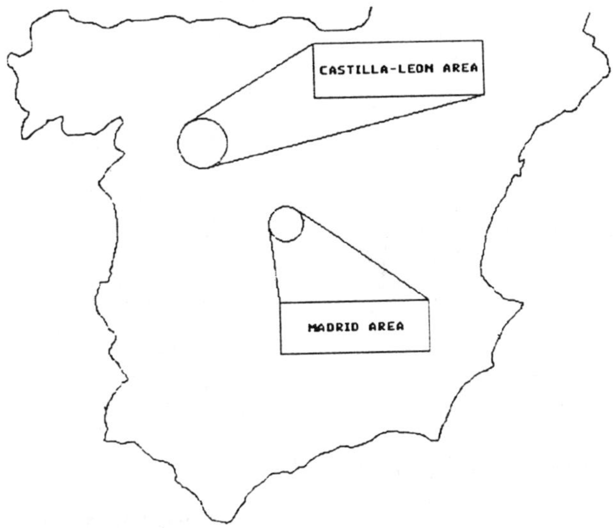

Fig. 14.4.1: Location of measuring sites in Spain.

Principal scientific results

Results of the experiments carried out at CIBA site near Valladolid in the Castilla - León region show that concentrations of SO_2 are low but significant, the gradient method is sometimes not able to produce significant results because of the low values of SO_2 concentrations. Results were used to parameterise the canopy resistance of the surface. Results are very dependent on the land-use classification and the canopy resistance parameterisation. In this sense, a new canopy resistance parameterisation was proposed (modified from Wesley [2] 'Big Leaf' model). Table 14.4.1 shows the modification of the resistance components for sulphur and ozone pollutants. Fig. 14.4.2 shows the SO_2 deposition velocities when we use the agricultural land-use type and the barren land-use type and the comparison with the results from the modified Bowen method.

The results for ozone show that concentrations are found in a range between a few ppb up to 70–80 ppb which for a rural area 32 km away of a medium size city (Valladolid) and during the summer-autumn season are the expected values.

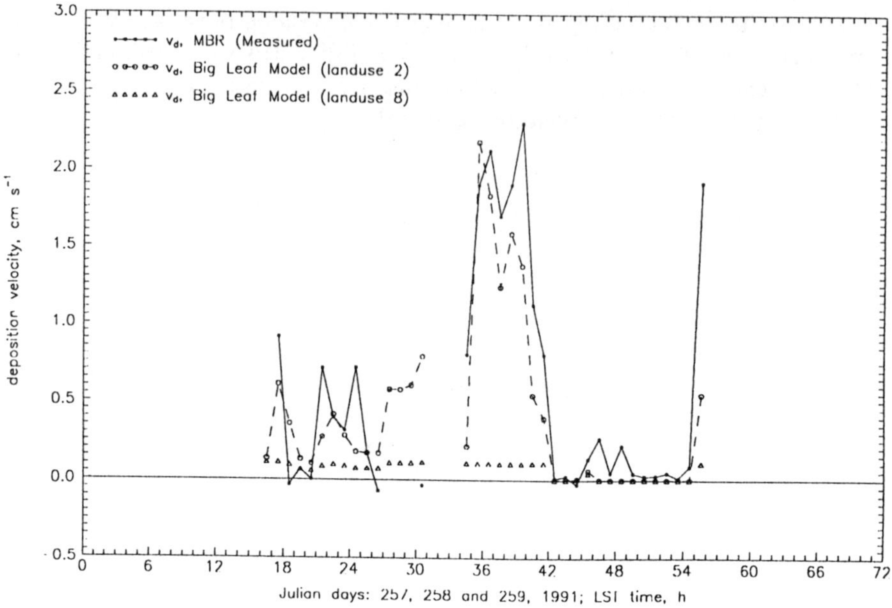

Fig. 14.4.2: Comparison of SO_2 deposition velocities.

Table 14.4.1: Input resistances of r_c (sm^{-1}). Entries of 9999 mean that the effect is too small to affect the total resistance

Resistance	Land-use type	
Component	Agricultural land (2)	Barren land (8)
r_{lu}	2000	9999
r_{ac}	200	0
r_{gsS}	150	1000
r_{gsO}	150	400
r_{clS}	2000	9999
r_{clO}	1000	9999

All the data are applied for the seasonal category corresponding to midsummer with lush vegetation

Fig. 14.4.3 shows ozone concentrations during part of the field campaign. Ozone deposition velocities are found in a range starting in almost zero values to 3.0–4.0 cm s^{-1}. Deposition velocities for ozone and sulphur dioxide seemed to be affected advection effects. Future work should be focused on the investigation of these effects.

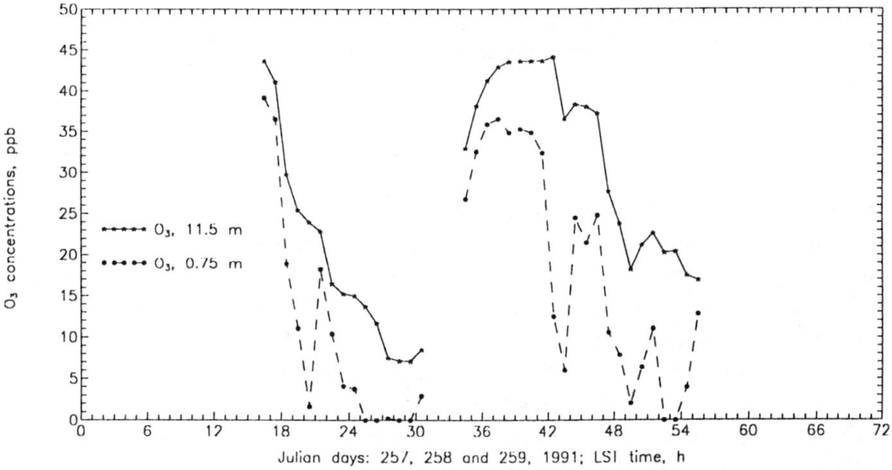

Fig. 14.4.3: Ozone concentration over several days during 1991, at two different heights.

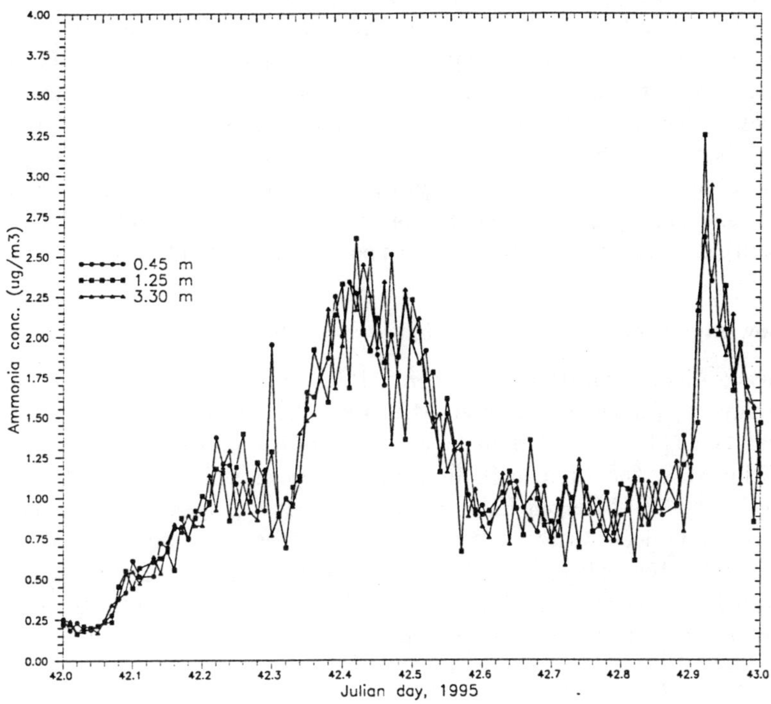

Fig. 14.4.4: Ammonium concentration over several days during 1995, at three different heights.

The last phase of our contribution was focused on obtaining the ozone fluxes by using the eddy correlation technique, SO_2 and NH_3 fluxes by using the modified Bowen ratio method in a place located in the west suburban area of Madrid city (near a not very busy military airport). Fig. 14.4.4 shows results for NH_3 at three different heights by using the denuders developed at ECN by Slanina and Wyers [7]

Conclusions

Results of our contribution can be summarized as follows. We have obtain accurate concentration data for SO_2, O_3 and NH_3 at different inland locations in Spain (typical of the Mediterranean region). We have obtained ozone fluxes by using a modified Bowen method technique and the eddy correlation method. This allowed us to characterise the ozone canopy resistances for a location in Castilla - León and a suburban area in the Madrid region. Results in Madrid show high values for the winter season. Deposition velocities are also high which show the influence of possible advection effects. This effect can also be found for SO_2 in both places and preliminary results for NH_3 reflect a similar effect.

The initial objectives in EUROTRAC project in our contribution has been satisfactory obtained. Although, results open several more questions which should be investigated in the future. Compensation point for NH_3 and studies related to the upper atmosphere and soil layer influence on the deposition processes should be the main research lines for the future of our work. Generalisation of deposition schemes to include emission fluxes and parameterization of the soil flux and characterisation of the flux exchanges between upper parts of the atmospheric boundary layer and surface boundary layer, will be our future research lines. Finally, understanding the influence of the biological processes in the deposition-emission exchanges in the surface will also be studied in our future research work.

Acknowledgements

J. Bilbao; A. de Miguel; V.E. Cachorro; R.E. Viloria and V. Tricio from the University of Valladolid. Cancillo, M.L. from the University of Extremadura. Encinas, D. and Palacios, P. from the University of Basque Country. Moreno, F.J., San Feliu M.A. and Moreno B. from the Technical University of Madrid are acknowledged. IBM-Bergen Environmental Sciences (Bergen, Norway). Project PB88-0338, Project AMB92-0259-CO3-0, Project AMB93-0010-CO4-01 from the CICYT (Technology and Science, Research Spanish Commission) are acknowledged. Project EV5V-CT91-0316, Project EV5V-CT94-0426 from the European Commission are acknowledged.

References

1. San José R., Reguero J., Casanova J., Diabatic wind and temperature profiles obtained from a 100 m tower: a statistical comparison of model performances. *Boundary Layer Meteorology* **37** (1986) 1–15.
2. Wesely M.L., Parameterization of surface resistances to gaseous dry deposition in regional scale numerical models, A*tmos. Environ.* **23** (1989) 1293–1304.
3. Baldocchi D.D., Hicks B.B., Camara P., A canopy stomatal resistance model for gaseous deposition to vegetated surfaces, A*tmos. Environ.* **21** (1987) 91–101.
4. Sutton M.A., Asman W.A.H., Schjørring J.K., Dry deposition of reduced nitrogen. Workshop of atmospheric pollutants, Göteborg 3–6 November 1992, Nordic Council of Ministers, Copenhagen 1993, pp. 125–143.
5. Erisman J.W., Wyers G.P., Continuous measurements of surface exchange of SO_2 and NH_3; implications for their possible interaction in the deposition process, A*tmos. Environ.* **27A** (1993) 1937–1949.
6. Schjørring J.K., Killingsbaek A., Mortensen J.V., Byskov-Nielsen S., Field investigations of ammonia exchange between barley plants and the atmosphere. I. Concentration profiles and flux densities of ammonia, *Plant, cell and environment* **16** (1993) 161–167.
7. Slanina J., Wyers G.P., Monitoring of atmospheric components by automatic denuder systems, *J. Anal. Chem.* **350** (1994) 467–473.

Chapter 15

Biatex Publications: 1988–1995

1988

Höfken, K.D., F. Meixner, D.H. Ehhalt; Dry deposition of NO, NO_2 and HNO_3, in: H. van Dop (ed), *Air Pollution Modelling and its Application VI,* Plenum Press, New York 1988, pp. 15–22.

Keuken, M.P., C.A.M. Schoonebeek, A. van Wensveen-Louter, J. Slanina; Simultaneous sampling of NH_3, HNO_3, HCl, SO_2 and H_2O_2 in ambient air by a wet annular denuder system, *Atmos. Eviron.* **22** (1988) 2541–2548.

Meixner, F.X., F. Slemr, H. Rennenberg; Biosphere-atmosphere exchange of nitrogen oxides over natural and agricultural surfaces, in: S. Beilke, J. Morelli, G. Angeletti (eds), *Field Measurements and Their Interpretation, Air Pollution Research Report* **14**, EUR 11690 EN, CEC DG XII, Brussels 1988, pp. 19–24 (ISBN 2-87263-001-5).

Meixner, H.H. Franken, J.H. Duyzer, R.M. van Aalst; Dry deposition of HNO_3 to a pine forest, in: H. van Dop (ed), *Air Pollution Modelling and its Application VI,* Plenum Press, New York 1988, pp. 23–35.

Riba, M.L., N. Tsiropoulos, B. Clement, A. Golfier, L. Torres; Preconcentration and analysis of atmospheric isoprene and monoterpenes. System automation, *J. Chromatogr.* **456** (1988) 165–173.

1989

Arends, B.G., S. Eenkhoorn; The Influence of manganese on the oxidation of sulfite in dew water, in: H.W. Georgii (ed), *Mechanisms andeffects of pollutants transfer into forests.* Kluwer Academic Publ., Dordrecht 1989, pp. 231–238.

Enders, G., U. Teichmann, G. Kramm; Profiles of Ozone and Surface Layer Parameters over a mature Spruce Stand, in: H.-W. Georgii (ed), *Mechanisms and Eff,ects of Pollutants-Transfer into Forests,* Kluwer Academic Publ., Dordrecht 1989, pp. 21–35.

Keuken, M.P., A. Wayers-Ypelaan, J.J. Mols, R.P. Otjes, J. Slanina; The determination of ammonia in ambient air by an automated thermodenuder system, *Atmos. Environ.* **23** (1989) 2177–2185.

Klockow, D., R. Niessner, M. Malejczyk, H. Kiendl, B. vom Berg, M.P. Keuken, A. Wayers-Yypelaan, J. Slanina; Determination of nitric acid and ammonium nitrate by means of a computer-controlled thermodenuder system, *Atmos. Environ.* **23** (1989) 1131–1138.

Kotzias, D., J.L. Hjorth, H. Skov; A chemical mechanism for dry deposition - The role of biogenic hydrocarbon (terpene) emissions in the dry deposition of O_3, SO_2 and NO_x in forest areas, *Toxicol. Environ. Chem.* **20–21** (1989) 95–99.

Meixner, F.X.; EUROTRAC-BIATEX: An European Program for the investigation of Biosphere-Atmosphere Exchange of Pollutants, in: J. Casanova Colas (ed), *Transporte Atmosferico de Contaminantes a Larga Distancia, Jornadas Celebradas los dias 3 al 5 noviembre de 1988 en al Castillo de la Mota de Medina del Campo (Valladolid)*, Secretario de Publicaciones Universidad de Valladolid, 1989, pp.11–28 (ISBN 84-7762-081-4).

Remde, A., F. Slemr, R. Conrad; Microbial production and uptake of nitric oxide in soil, *FEMS Microbiol. Ecol.* **62** (1989) 221–230.

Rennenberg, H.; Synthesis and emission of hydrogen sulfide by higher plants. In: Biogenic Sulfur in the Environment, in: E.S. Saltzman, W.J. Cooper, (eds) *ACS Symp. Ser.* 393, Amer. Chem. Soc., Washington 1989, pp. 44–57.

Slanina, J., M.P. Keuken, P.A.C. Jongejan; Precision and accuracy of ion chromatography in dry deposition measurements, *J. Chromatog.* **82** (1989) 297–305.

Sutton, M.A, J.B. Moncrieff, D. Fowler; Measurement of atmospheric ammonia and the assessment of its exchange with vegetated surfaces, in: *Changing compostion of the troposphere. Special Environmental Report No. 17.* World Meteorological Organization, Geneva, Switzerland 1989, pp 154–157.

Vitousek, P.M., O.T. Denmead, D. Fowler, C. Johanssen, J. Kesselmeier, L. Klemendtsson, F.X. Meixner, A.R. Mosier, H. Schütz, L.J. Stal, M. Wahlen; What are the relative roles of biological production, micrometeorology, and photochemistry in controlling the flux of trace gases between terrestrial ecosystems and the atmosphere? in: M.O. Andreae, D.S. Schimel (eds), *Exchange of Trace Gases between Terrestrial Ecosystems and the Atmosphere*, John Wiley, Chichester 1989, pp. 249–261.

1990

Abbeel, R. van den, A. Claes, K. Vlassak; Gaseous nitrogen losses from slurry manured land, in: Hansen J. and Hendriksen K. (eds.), *Nitrogen in Organic Wastes Applied to Soil*, A.P. Harcourt Brace Jovanovich Publishers, London 1990, p.213.

Arends, B.G., G.P.A. Kos, R.K.A.M. Mallanr, G.P. Wyers; Nitrogcn compouuds measured in fog and dew in the Netherlands, in: S. Beilke, M. Millan and G. Angeletti (eds) *Field measurements and interpretation of species derived from NO_x, NH_3 and VOC emissions in Europe. Air Pollution Research Report* **25**, CEC DG XII, Brussels 1990, pp. 24–27.

Arends, B.G., G.P.A. Kos, R.K.A.M. Mallant, G.P. Wyers; Nitrogen compounds measured in fog and dew in the Netherlands, in: S. Beilke, M. Millan, G. Angeletti (eds), *Field Measurements and Interpretation of Species derived from* NO , NH_3 *and VOC Emissions in Europe*, *Air Pollution Research Report* **25**, CEC DG XII, Brussels 1990.

Arends, B.G., S. Eenkhoorn; The influence of manganese leached from plants on dew chemistry, *Environ Technol.* **11** (1990) 181–188.

Baumgärtner, M., A. Remde, E. Bock, R. Conrad; Release of nitric oxide from building stones into the atmosphere, *Atmos. Environ.* **24**B (1990) 87–92.

Conrad, R.; Flux of NO_x between soil and atmosphere: importance and soil microbial metabolism, in: J. Soerensen, N.P. Revsbech (eds), *Denitrification in Soil and Sediment,* Plenum Press, New York 1990, pp. 105–128.

Dollard, G.J., B.M.R. Jones, T.J. Davies;The dry deposition of HNO_3 and PAN,*Harwell Report AERE* R 13780 (1990) and *Proc. COST 611 Meeting*, Madrid 1990.

Duyzer, J.H., D. Fowler, F. Meixner, G. Dollard, C. Johansson, M. Gallagher; The Halvergate trace gas experiment on surface exchange of oxides of nitrogen: Preliminary results,in: S. Beilke, M. Millan, G. Angeletti (eds), *Field Measurements and Interpretation of Species derived from* NO , NH_3 *and VOC Emissions in Europe, Air Pollution Research Report* **25**, CEC DG XII, Brussels 1990, pp. 267–276.

Fowler, D., A.P. Morse, M.W. Gallagher, T.W. Choularton; Measurements of cloud water deposition on vegetation using a lysimeter and a flux gradient technique, *Tellus* **42B** (1990) 285–290.

Fowler, D., C. Johansson, J. Duyzer, F.X. Meixner; A comparison between micrometeorological and chamber measurements of nitric oxide emission and ozone depletion from grazed pasture, in: *Proc. SCOPE/IGBP Workshop on Trace Gas Exchange in a Global Perspective*, Sigtuna, Sweden 1990.

Hewitt, C.N., G. L. Kok, R. Fall; Hydroperoxides in plants exposed to ozone mediate air pollution damage to alkene emitters,*Nature* **344** (1990) 56–58.

Hewitt, C.N., P. Lucas, A.R. Welllburn, R. Fall; Chemistry of ozone damage to plants,*Chem. Ind.* **15** (1990) 478–481.

Hewitt, C.N., R.K. Manson, R. Fall; Isoprene emissions from the grass *Arundo donax L* are not linked to photorespiration, *Plant Sci.* **66** (1990) 139–140.

Keuken, M.P., F.P. Bakker, J.J. Möls, B. Broersen, J. Slanina; Atmospheric deposition and conversion of ammonium to nitric acid on a historic building: a pilot study, *Int. J. Environ. Anal. Chem.* **38** (1990) 47–62.

Krämer, W., M. Baugärtner, M. Bender, R. Conrad; Consumption of NO by methoanotrophic bacteria in pure culture and in soil, *FEMS Microbiol. Ecol.* **73** (1990) 345–350.

Meixner, F.X., F. Böswald, C. Buck, K. Eiblmeier, H. Müller; Measurement of HNO_3 dry deposition by gradient techniques, in: S. Beilke, M. Millan, G. Angeletti (eds), *Field Measurements and Interpretation of Species derived from* NO , NH_3 *and VOC Emissions in Europe*, *Air Pollution Research Report* **25**, CEC DG XII, Brussels 1990, pp. 267–276.

Nägele, W., R. Conrad; Influence of pH on the release of NO and N_2O from fertilized and unfertilized soil, *Biol. Fertil. Soil* **10** (1990) 139–144.

Nägele, W., R. Conrad; Influence of soil pH on the nitrate-reducing microbial populations and their potential to reduce nitrate to NO and N_2O, *FEMS Microbiol. Ecol.* **74** (1990) 49–58.

Ramon, H. O. Van Cleemput; Emission of gaseous C_1-C_3 and hydrocarbons out of soils, EUROTRAC, Part 4, (1990) BIATEX, Garmisch-Partenkirchen, Germany, 72–77

Remde, A., R. Conrad; Production of nitric oxide in *Nitrosomonas europaea* by reduction of nitrite, *Archiv. Microbiol.* **154** (1990) 187–191.

Sabljic, A., H. Güsten, J. Schönherr, M. Riederer; Modelling plant uptake of airborne organic chemicals. 1. Plant cuticle/water partitioning and molecular connectivity, *Environ. Sci. Technol.* **24** (1990) 1321–1326.

Siebke, K., F.-W. Badeck, G.H. Kohlmaier, M. Plöchl, C. Wientzek; Modelling pollutant exchange between plant and environment: Uptake and metabolization of sulphur dioxide by different Leaf Cell Compartments, in: Jorgensen (ed), *Modelling in Ecotoxicology*, Elsevier, Amsterdam 1990, pp. 149–172.

Slanina, J., B.G. Arends, M.P. Keuken, A. Veltkamp, G.P. Wyers; Acidification research at ECN: Contribution to the second phase of the Dutch Priority Programme on Acidification *ECN-C--90-064-Report*, Netherlands Energy Research Foundation, Petten 1990.

Slanina, J., G.P. Wyers; Measurements of ammonia in ambient air - Impact of ammonia on ecosystems, *VDI Berichte* **838** (1990), 499–515.

Slanina, J., J.J. Mols and J.H. Baard; The influence of outliers on results of wet deposition measurements as a function of measurement strategy, *Atmos. Environ.* **24** (1990) 1843–1860.

Steinbrecher, R., W. Schürmann, G. Eichstädter, C. Deigele, H. Ziegler; Sources and mechanisms of monoterpene emission in a spruce forest (*Picea abies* [L.] *Karst.*), *Physiol. Plant.* **79** (1990) A117.

Wyers, G.P., A.T. Vermeulen, A. Wayers-Ypelaan, R.P. Otjes, J. Slanina; Determination of NH_3, SO_2, HNO_2, HNO_3, HCl and H_2O_2 in air above a forest by automated thermodenuder and wet annular denuder systems, in: S. Beilke, M. Millan and G. Angeletti (eds) *Field measurements and interpretation of species derived from NO_x, NH_3 and VOC emissions in Europe. Air Pollution Research Report* **25**, CEC DG XII, Brussels 1990, pp. 108–117.

1991

Arends, B.G., G.P. Wyers; The chemistry of dew sampled in Petten, North Holland, in: P. Borrell, P.M. Borrell, W. Seiler (eds), *Proc. EUROTRAC Symp. '90,* SPB Academic Publishing bv, The Hague 1993, pp. 103–106.

Badeck, F.W., M. Plöchl, P. Ramge, G.H. Kohlmaier; Extension and application of the big-leaf model for the deposition of selected trace gases, in: P. Borrell, P.M. Borrell, W. Seiler (eds), *Proc. EUROTRAC Symp. '90,* SPB Academic Publishing bv, The Hague 1993, pp. 107–112.

Baumgärtner, M., F. Sameluck, E. Bock, R. Conrad; Production of nitric oxide by ammonium-oxidizing bacteria colonizing building stones, *FEMS Microbiol. Ecol.* **85** (1991) 95–100.

Beswick, K.M., K.J. Hargreaves, M.W. Gallagher, T.W. Choularton, D. Fowler; Size-resolved measurements of cloud droplet deposition velocity to a forest canopy using an eddy correlation technique, *Q. J. Roy. Met. Soc.* 117 (1991) 623–645.

Bieder, U., A. Schatz; Modeling of the deposition processes of air pollutants during dew formation in complex structures, in: P. Borrell, P.M. Borrell, W. Seiler (eds), *Proc. EUROTRAC Symp. '90,* SPB Academic Publishing bv, The Hague 1993, pp. 113–117.

Bieder, U., A. Schatz; The impact of dew on the deposition of air pollutants, *Annales Geophysicae* **9** Suppl. (1991) C193–C194.

Bingemer, H., L. Ernst, M. Fischer, J. Kesselmeier, M.O. Andreae, G. Helas; Volatile organic acids and ammonia measurements in a spruce forest, in: P. Borrell, P.M. Borrell, W. Seiler (eds), *Proc. EUROTRAC Symp. '90,* SPB Academic Publishing bv, The Hague 1993, pp. 201–205.

Cape, J.N. R.I. Smith, D. Fowler; Modelling dry deposition of SO_2 in Britain, in: D.G. Farmer, M.J. Rycroft (eds) *Computer Modelling in the Environmental Sciences. IMA Conference Series*, Oxford University Press 1991, pp. 285–298.

Chakrabarti, K., H. Rennenberg, A. Polle; Interaction of plants and atmospheric hydrogen peroxide, in: P. Borrell, P.M. Borrell, W. Seiler (eds), *Proc. EUROTRAC Symp. '90,* SPB Academic Publishing bv, The Hague 1993, pp. 119–121.

Cleemput, O. van, H. Ramon, A. Vermoesen; Emission of C_1-C_3 hydrocarbons out of soils, in: P. Borrell, P.M. Borrell, W. Seiler (eds), *Proc. EUROTRAC Symp. '90,* SPB Academic Publishing bv, The Hague 1993, pp. 189–190.

Clement, B., M. Haziza, M.L. Riba, L. Torres; Continuous measurement of atmospheric terpenes by use of an automatic device. BIOVOTEX II, in: P. Borrell, P.M. Borrell, W. Seiler (eds), *Proc. EUROTRAC Symp. '90,* SPB Academic Publishing bv, The Hague 1993, pp. 207–210.

Coe, H.; Report on the BIATEX II experiment : Preliminary data analysis and results, *UMIST APRG Report.* 1991.

Conrad, R., A. Remde, M. Krämer, M. Baumgärtner , W. Nägele; Mechanisms of NO echange between soil and atmosphere, in: P. Borrell, P.M. Borrell, W. Seiler (eds), *Proc. EUROTRAC Symp. '90,* SPB Academic Publishing bv, The Hague 1993, pp. 123–126.

DeKok, L.J., H. Rennenberg, P.C. Kuiper; The internal resistance in spinach leaves to atmospheric H_2S deposition is determined by metabolic processes, *Plant Physiol. Biochem.* **29** (1991) 463–470.

Dollard, G.J., B.M.R. Jones, T.J. Davies; The dry deposition of HNO_3 and PAN, in: S. Beilke, M. Millan, G. Angeletti (eds), *Field Measurements and Interpretation of Species derived from* NO , NH_3 *and VOC Emissions in Europe, Air Pollution Research Report* **25**, CEC, Brussels 1991.

Dore, C.; Report on the BIATEX II experiment : CO_2 flux measurements to a Dutch heathland, *UMIST APRG Report* (1991).

Enders, G., S. Piper, G. Jakobi, U. Teichmann, Mikrometeorologie und Vertikalprofile von Luftschadstoffen in einem Fichtenbestand des Nationalparks, in: *Proc. 2. Statusseminar*, München-Neuherberg, *GSF-Bericht* **26** (1991) 311–320.

Enders, G., U. Teichmann, , S. Piper; Diurnal cycles and vertial profiles of O_3 in a spruce forest. Normal behaviour and irregularities, in: P. Borrell, P.M. Borrell, W. Seiler (eds), *Proc. EUROTRAC Symp. '90,* SPB Academic Publishing bv, The Hague 1993, pp. 211–215.

Enders, G.; Deposition of ozone to a mature spruce forest: measurements and comparison to models, *Environ. Pollut.* **75** (1991) 61–67.

Enders; G.; Schachtenau, National Park Bayerischer Wald (FRG). A European forest site for surface flux measurements, in: P. Borrell, P.M. Borrell, W. Seiler (eds), *Proc. EUROTRAC Symp. '90,* SPB Academic Publishing bv, The Hague 1993, pp. 197–200.

Erisman, J.W., J.H. Duyzer; A micrometeorological investigation of surface exchange parameters over heathland, *Boundary–Layer Meteorol.* **57** (1991) 115–128.

Ernst ,N, H. Güsten, G. Heinrich, E. Monnich, J. Weppner; Measurement of the dry deposition of ozone onto agricultural land, in: P. Borrell, P.M. Borrell, W. Seiler (eds), *Proc. EUROTRAC Symp. '90,* SPB Academic Publishing bv, The Hague 1993, pp. 127–129

Ernst, N., H. Güsten, G. Heinrich, E. Mönnich, J. Weppner; Measurement of the dry deposition of ozone onto agricultural land, in: P. Borrell, P.M. Borrell, W. Seiler (eds), *Proc. EUROTRAC Symp. '90,* SPB Academic Publishing bv, The Hague 1993, pp. 127–129.

Ferm, M.; Throughfall Deposition of Nitrogen and Sulphur Compounds, in: P. Borrell, P.M. Borrell, W. Seiler (eds), *Proc. EUROTRAC Symp. '90,* SPB Academic Publishing bv, The Hague 1993, pp. 131–134.

Förstel, H., B. Jansen; Simultaneous measurements of H_2 and CO deposition velocities into soil, in: P. Borrell, P.M. Borrell, W. Seiler (eds), *Proc. EUROTRAC Symp. '90,* SPB Academic Publishing bv, The Hague 1993, pp. 135–137.

Förstel, H., C. Latus, H.J. Segschneider, A. Neubert, B. Sack; Measurement of NO deposition into leaves using ^{15}N labelled gas, in: P. Borrell, P.M. Borrell, W. Seiler (eds), *Proc. EUROTRAC Symp. '90,* SPB Acedemic Publishing bv, The Hague 1993, pp. 139–141.

Fowler, D., D. Baldocchi, J.H. Duyzer; Inputs of trace gases, particles and cloud droplets to terrestrial surfaces, in: F.T.Last, R. Watling (eds) *Acidic deposition, its nature and the atmosphere, Proc. Roy. Soc. Edinburgh* **97B** (1991) 35–59.

Fowler, D., J.H. Duyzer, F.X. Meixner, G.J. and Dollard; Inputs of nitrogen compounds and photochemical oxidants as components of the pollution climate of Europe, *Progress report.* CEC 1991, 31 pp.

Fowler, D., J.N. Cape, M.A. Sutton, R. Mourne, K.J. Hargreaves, J.H. Duyzer, M.W. Gallagher;
Deposition of acidifying compounds, in: T Schneider (ed) *Acidification Research: evaluation and policy applications*, Maastricht,The Netherlands 1991, pp.553–572.

Fowler, D., K.J. Hargreaves, M. Sutton; Dry deposition of NO, NO_2, O_3 and NH_3 onto vegetation and soil. Modelling dry deposition processes, *Interim report to the Department of the Environment* Contract PECD No. 7/10/79 (1991).

Fowler, D., R.I. Smith; Dry deposition of sulphur dioxide n the United Kingdom, *Ann. Rep. Institute of Terrestrial Ecology* 1991,pp. 34–37.

Hargreaves, K.J., D. Fowler; The exchange of trace nitrogen gases between terrestrial ecosystems and the atmosphere. *Ann. Rep. Institute of Terrestrial Ecology* 1991,pp. 37–41.

Hargreaves, K.J., M.A. Sutton, D. Fowler, D., R.L. Storeton-West; Measurements of NO, NO_2, O_3 and NH_3 exchange with vegetated surfaces, in: *Inputs of Nitrogen Compounds and Photochemical Oxidants as Components of the Pollution Climate of Europe. Progress report.* CEC. ITE, Edinburgh, 1991.

Haziza, M., B. Clement, M.L. Riba, L. Torres; Automatic device for atmospheric terpenes. Determination in a forest area, in: *Proc. 42nd Pittsburgh Conf. and Exposition on Analytical Chemistry and Aapplied Spectroscopy*, Chicago 1991, p. 409.

Hewitt, C.N., G.L. Kok; Formation and occurrence of organic hydroperoxides in the troposhere: laboratory and field observations, *J. Atmos. Chem.* **12** (1991) 181–194.

Hofmann, U., G. Protoschill-Krebs, L. Merk, H. Bingemer, J. Kesselmeier , M.O. Andreae; Field measurements of reduced sulphur compounds and first physiological results on the uptake of COS by higher plants, in: P. Borrell, P.M. Borrell, W. Seiler (eds), *Proc. EUROTRAC Symp. '90,* SPB Academic Publishing bv, The Hague 1993, pp. 149–154.

Kalkowski, I., R. Conrad; Metabolism of nitric oxide in denitrifying *Pseudomonas aeruginosa* and nitrate-respiring *Bacillus cereus*, *FEMS Microbiol. Lett.* **82** (1991) 107–112.

Kesselmeier, J.; Emission of sulfur compounds from vegetation and global-scale extrapolation, in: Th.D. Sharkey, E.A. Holland, H.A. Mooney, (eds), *Trace Gas Emissions from Plants*, Academic Press, San Diego 1991, pp. 261–265.

Kotzias, D., J. van Eijk , R. Daiber; Emission and concentration measurements of monoterpene hydrocarbons in the Bavarian Forest, in: P. Borrell, P.M. Borrell, W. Seiler (eds), *Proc. EUROTRAC Symp. '90,* SPB Academic Publishing bv, The Hague 1993, pp. 217–220.

Krämer, M., R. Conrad; Influence of oxygen on production and consumption of NO in soil, *Biol. Fert. Soils* **11** (1991) 38–42.

Kramm, G., H. Müller, D. Fowler, K.D. Höfken, F.X. Meixner, E. Schaller; A modified profile method for determining the vertical fluxes of NO, NO_2, ozone and HNO_3 in the atmospheric surface layer, *J. Atmos. Chem.* 13 (1991) 265–288.

Kramm, G., H. Müller, D. Fowler, K.D. Hüfken, F.X. Meixner, E. Schaller; A modified profile method for determining the vertical fluxes of NO, NO_2, ozone and HNO_3 in the atmospheric surface layer, *J. Atmos. Chem.* **13** (1991) 265–288.

Kramm, G.; Modellierung des Vertikaltransports von Aerosolpartikeln in der bodennahen Luftschicht,
Meteorol. Rdsch. **43** (1991) 65–80.

Kramm; G.; Numerical investigation of the dry deposition of reactive trace gases, in: P. Borrell, P.M. Borrell, W. Seiler (eds), *Proc. EUROTRAC Symp. '90,* SPB Academic Publishing bv, The Hague 1993, pp. 155–157.

Linne von Berg, K.H., H. Bothe, R. Voßwinkel; Parameterisation and biochemical characterisation of NO_x evolving microorganisms living in association with plants, in: P. Borrell, P.M. Borrell, W. Seiler (eds), *Proc. EUROTRAC Symp. '90,* SPB Academic Publishing bv, The Hague 1993, pp. 159–161.

MacKenzie, A.R., R.M. Harrison, I. Colbeck, C.N. Hewitt; The role of biogenic hydrocarbons in the production of ozone in urban plumes in SE England, *Atmos. Environ.* **25**A (1991) 351–359.

Meixner, F.X., F. Böswald, K. Eiblmeier, G. Kramm, J. Ludwig, H. Müller , M. Nestlen; Vertical Fluxes of NO, NO_2, CO_2, O_3, momentum, sensible and latent heat during the BIATEX-LOVENOX experiment, in: P. Borrell, P.M. Borrell, W. Seiler (eds), *Proc. EUROTRAC Symp. '90,* SPB Academic Publishing bv, The Hague 1993, pp. 163.

Papenbrock, Th., F. Stuhl; Measurement of gaseous nitric acid in Europe and on the Atlantic ocean, in: P. Borrell, P.M. Borrell, W. Seiler (eds), *Proc. EUROTRAC Symp. '90,* SPB Academic Publishing bv, The Hague 1993, pp. 165–166.

Pul, W.A.J. van, J.W. Erisman, J.A. van Jaarsveld, F.A.A.M. de Leeuw; High resolution assessment of acid deposition fluxes, in: T. Schneider (ed). *Acidification Research: Evaluation and Policy Applications* Maastricht,The Netherlands 1991, pp. 277–284.

Ramge P., F.-W. Badeck, M. Plöchl, G.H. Kohlmaier; Apoplastische Antioxidantien als maagebliche Eliminierungsfaktoren bei der Aufnahme von NO_2 über den Blattpfad, *Berichte des Zentrums für Umweltforschung,* Frankfurt am Main, 1991.

Remde, A., R. Conrad; Metabolism of nitric oxide in soil and denitrifying bacteria, *FEMS Microbiol. Ecol.* **85** (1991) 81–94.

Remde, A., R. Conrad; Production and consumption of nitric oxide by denitrifying bacteria under anaerobic and aerobic conditions. *FEMS Microbiol. Lett.* **80** (1991) 329–332.

Remde, A., R. Conrad; Role of nitrification and denitrification for NO metabolism in soil, *Biogeochem.* **12** (1991) 189–205.

Remde, A., R. Conrad; Role of nitrification versus denitrification for production and uptake of nitric oxide in soils, in: P. Borrell, P.M. Borrell, W. Seiler (eds), *Proc. EUROTRAC Symp. '90,* SPB Academic Publishing bv, The Hague 1993, pp. 167.

Rennenberg, H., P. Schröder , B. Huber; Biogenic emission of reduced sulfur compounds, in: P. Borrell, P.M. Borrell, W. Seiler (eds), *Proc. EUROTRAC Symp. '90,* SPB Academic Publishing bv, The Hague 1993, pp. 169–172.

Rennenberg, H., The significance of higher plants in the emission of sulfur compounds from terrestrial ecosystems, in: Th.D. Sharkey, E.A. Holland, H.A. Mooney, (eds), *Trace Gas Emissions from Plants*, Academic Press, San Diego 1991, pp. 261–265.

Riba, M.L., B. Clement, M. Haziza, L. Torres; Trace analysis. Determination of the "breakthrough volume" of atmospheric isoprene, *Toxicol. Environ. Chem.* **31–32** (1991) 235–240.

Rondon, A., C. Johansson , L. Granat; Flux of NO_2 between atmosphere and a pine stand in central Sweden at low NO_2 concentration, in: P. Borrell, P.M. Borrell, W. Seiler (eds), *Proc. EUROTRAC Symp. '90,* SPB Academic Publishing bv, The Hague 1993, pp. 173–174.

Sattler, Th., U. Herrmann , W. Jaeschke; Determination of dry deposition of pollutants - Comparison and correction of different measuring-methods and models, in: P. Borrell, P.M. Borrell, W. Seiler (eds), *Proc. EUROTRAC Symp. '90,* SPB Academic Publishing bv, The Hague 1993, pp. 177–180.

Schaller, E.; Future experiments in BIATEX, in: P. Borrell, P.M. Borrell, W. Seiler (eds), *Proc. EUROTRAC Symp. '90,* SPB Academic Publishing bv, The Hague 1993, pp. 181–182.

Schütz, B., L.J. DeKok, H. Rennenberg; Thiol accumulation and cysteine desulfhydrase activity in H_2S fumigated leaves and leaf homogenates of cucurbit plants, *Plant Cell Physiol.* **32** (1991) 733–736.

Slanina, J., G.P. Wyers; Ammonia: impact and measurement, in: *Proc. Visual Lübecker Umweltsymposium*, Lübeck 1991, pp. 222–237.

Slanina, J.; Forest Die-Back, Pollution Climates and the Role of Ammonia, in: Chadwick, Hudson, (eds), *Acid Depositions in Europe, Environmental Effects, Control Strategies and Policy Options*, Stockholm Environment Institute 1991, pp 48–59.

Slanina, J.; Individual experiments ECN: emissions of tidal flats, influence of water layers on deposition, in: P. Borrell, P.M. Borrell, W. Seiler (eds), *Proc. EUROTRAC Symp. '90,* SPB Acedemic Publishing bv, The Hague 1993, pp. 183–185.

Slanina, J.; Inventory of current tropospheric sampling programs worldwide, in: *Proc. Visual Lübecker Umweltsymposium*, Lübeck 1991, pp. 209–221.

Slanina, J.; Report on the progress in BIATEX, in: P. Borrell, P.M. Borrell, W. Seiler (eds), *Proc. EUROTRAC Symp. '90,* SPB Academic Publishing bv, The Hague 1993, pp. 95–101.

Stahl, K., J. Hahn; Study of the emission of low-molecular-weight organic compounds by various plants, in: P. Borrell, P.M. Borrell, W. Seiler (eds), *Proc. EUROTRAC Symp. '90,* SPB Academic Publishing bv, The Hague 1993, pp. 187–188.

Steinbrecher, R., W. Schürmann, R. Schönwitz, G. Eichstädter, H. Ziegler; Volatile Monoterpenes in a forest of spruce (*Picea abies* [L.] *Karst.*): sources and mechanisms of emission and behavior in the atmosphere, in: P. Borrell, P.M. Borrell, W. Seiler (eds), *Proc. EUROTRAC Symp. '90,* SPB Academic Publishing bv, The Hague 1993, pp. 221 – 224.

Sutton, M.A, D. Fowler, J.B. Moncrieff; The dry deposition of atmospheric ammonia, in: A. Anagnostopoulos, P. Day (eds) *Proc. 2nd Int. Conf. on Environmental Pollution*, University of Thessaloniki, Greece 1991, pp 119–129.

Teichmann, U., S. Piper, G. Enders; Surface Layer Parameter über einem Fichtenbestand, *Bayer. Landw. Jahrbuch 1990* (1991) 55–58.

Van Cleemput, O., H. Ramon, A. Vermoesen; Emission of C_1-C_3 hydrocarbons from soils, in: P. Borrell, P.M. Borrell, W. Seiler (eds), *Proc. EUROTRAC Symp. '90,* SPB Academic Publishing bv, The Hague 1993, pp. 189–190.

Verhage, A.J.L., R.A. Rooth , L.W. Wouters; Photoacoustic measurement of concentration gradients of ammonia in the atmosphere, in: P. Borrell, P.M. Borrell, W. Seiler (eds), *Proc. EUROTRAC Symp. '90,* SPB Academic Publishing bv, The Hague 1993, pp. 191–192.

Vermoesen, A., P. Demeyer, A. Dobbelaere, G. Hofman, O. Van Cleemput; Comparison between the chamber and micrometeorological method for determination of ammonia emission, in: G. Angeletti, S. Beilke, J. Slanina, (eds) *Air Science, Research and Development,*1991, pp. 195–201.

Vermoesen, A., P. Demeyer, G. Hofman, O. Van Cleemput; Field measurement of ammonia volatilisation upon application of different NH_4^+ fertilizers and urea, *Pedologie*, 42 (1991) 119–128.

Vermoesen, A., R. Ramon, O. Van Cleemput; Composition of the soil gas phase. Permanent gases and hydrocarbons, *Pedologie*, 41 (1991) 119–132.

Weston, K., D. Fowler; The importance of orography in spatial patterns of rainfall acidity in Scotland, *Atmos. Environ.* **25A** (1991) 1517–1522.

Wyers, G.P.; Development of instrumentation for the measurement of area integrated deposition fluxes, in: P. Borrell, P.M. Borrell, W. Seiler (eds), *Proc. EUROTRAC Symp. '90,* SPB Academic Publishing bv, The Hague 1993, pp. 193–196.

1992

Anghelus, A., M. Lörinczi, E. Lörinczi, A. Ritivoiu; Dry and wet deposition. Expression of the air pollution in Timis county, in: *Proc. Symp: TEMPUS-Transfer Ro.*(JEP 2760), Timisoara, Romania 1992.

Baumgärtner, M., E. Bock, R. Conrad; Processes involved in uptake and release of nitrogen dioxide from soil and building stones into the atmosphere, *Chemosphere* **24** (1992) 1943–1960.

Baumgärtner, M., R. Conrad; Effects of soil variables and season on the production and consumption of nitric oxide in oxic soils, *Biol. Fertil Soils* **14** (1992) 166–174.

Baumgärtner, M., R. Conrad; Role of nitrate and nitrite for production and consumption of nitric oxide during denitrification in soil, *FEMS Microbiol. Ecol.* **101** (1992) 59–65.

Beswick, K.M., K.J. Hargreaves, M.W. Gallagher, T.W. Choularton, D. Fowler; Size resolved measurements of cloud droplet deposition velocity to a forest canopy using an eddy correlation technique, *Q. J. R. Meteorol. Soc.* **117** (1992) 623–645.

Bieder, U., J. Ludwig, F. Meixner, A. Schatz; On the deposition of air pollutants to wet vegetation surfaces, in: G. Angeletti, S. Beilke, J. Slanina (eds), *Field Measurements and Interpretation of Species related to Photo-oxidants and Acid Deposition, Air Pollution Research Report* **39**, CEC, Brussels 1992, pp. 179–187.

Bingemer, H., M.O. Andreae, T.W. Andreae, P. Artaxo, G. Helas, D.J. Jacob, N. Mihalopoulos, P. Nguyen; Sulfur gases and aerosols in and above the equatorial rainforest, *J. Geophys. Res.* **97** (1992) 6207–6217.

Blösl, M., R. Conrad; Influence of an increased pH on the composition of the nitrate-reducing microbial populations in an anaerobically incubated acidic forest soil, *Syst. Ppl. Microbiol.* **15** (1992) 624–627.

Cape, J.N., D. Fowler (eds); Physical and chemical processes in acidic deposition 2: Networks, models and mechanisms, *Environ. Pollut.* **75(2)** (1991) 2.

Cape, J.N., K.J. Hargreaves, R.L. Storeton-West, D. Fowler, R.N. Colvile, T.W. Choularton, M.W. Gallagher, M.W.; Nitrite in orographic cloud as an indicator of nitrous acid in rural air, *Atmos. Environ.* **26A** (1992) 2301–2307.

Cape, J.N., R.L. Storeton-West, S.F. Devine, R.N. Beatty, A. Murdoch; Acidification research: evaluation and policy applications. The reaction of nitrogen dioxide at low concentrations with natural waters, *Atmos. Environ.* **27A** (1992) 2613–2621.

Cape, J.N.; Direct effects of acid rain and cloud on vegetation, in: *Proc. Egham UNECE Critical Levels* Workshop 1992.

Choularton, T.W., A.J. Wicks, R.M. Downer, M.W. Gallagher, S.A. Penkett, D.J. Bandy, G.J. Dollard, B.H.R. Jones, T.D. Davies, M.J. Gay, B.J. Tyler, D. Fowler, J.N. Cape, K.J. Hargreaves; A field study of the generation of nitrate in a hill cap cloud, *Environ. Pollut.* **75(1)** (1992) 69–74.

Coe, H., M.W. Gallagher; Measurements of dry deposition of NO_2 to a Dutch heathland using the eddy correlation technique,*Q. J. Roy. Meteorol. Soc.* **118** (1992) 767–786.

Crossley, A., D.B. Wilson, R. Milne; Pollution in the upland environment, *Environ. Pollut.* **75(1)** (1992) 81–88.

Demeyer, P., O. Van Cleemput, G. Hofman; Effect of soil moisture content on ammonia volatilisation from surface-applied ammonium fertilisers and urea, in: K. Francois, L. Pitman, Bartiaux-Thill (eds). *Nitrogen cycling and leaching in cool and wet regions of Europe, COST 814,* Commission of the European Communities, Brussels 1992, pp. 22–23.

Dore, A.J, T.W. Choularton, D. Fowler; An improved wet deposition map of the United Kingdom incorporating the seeder-feeder effect over mountainous terrain, *Atmos. Environ.* **26A** (1992) 1375–1381.

Dore, A.J., T.W. Choularton, D. Fowler; Orographic enhancement of snowfall, *Environ. Polut.* **75(1)** (1992) 175–180.

Duyzer, J.H., J.H. Weststrate, H.L.M. Verhagen, M. Gallagher, H.Coe, D. Fowler, M.A. Sutton; The joint experiment on surface exchange of trace gases over the Leende heathland, *IMW-TNO Report* PU 92/015 and *EUROTRAC Newsletter* No. 10 (1992) 30–36.

Enders, G., R. Dlugi, R. Steinbrecher, B. Clement, R. Daiber, J. v. Eijk, S. Gäb, M. Haziza, G. Helas, U. Herrman, M. Kessel, J. Kesselmeier, D. Kotzias, K. Kourtidis, H. H. Kurth, R. T. McMillen, G. Roider, W. Schürmann, L. Torres; Biosphere/atmosphere interactions integrated research in an European coniferous forest ecosystem. *Atmos. Environ.* **26**A (1992) 171–189.

Fontan, J., A. Minga, A. Lopez, A. Druilhet; Vertical ozone profiles in a pine forest, *Atmos. Environ.* **26A** (1992) 863–869.

Fowler, D., J.N. Cape (eds); Physical and chemical processes in acidic deposition 1: Trace gases and clouds, *Environ. Pollut.* **75(1)** (1992).

Fowler, D., K.J. Hargreaves, M.A. Sutton, M.A.; Dry deposition of nitrogen oxides, *Ann. Rep. Department of the Environment* 1992.

Fowler, D., K.J. Hargreaves, M.A. Sutton; Dry deposition of NO, NO_2, O_3 and NH_3 onto vegetation and soil. Modelling dry deposition processes, *Interim Rep. Department of the Environment* 1992.

Fowler, D., M.A. Sutton, K.J. Hargreaves; Surface-atmosphere exchange of ammonia, *Interim Rep. Department of the Environment* 1992.

Fowler, D., M.A. Sutton; Atmospheric budget for fixed nitrogen species in the UK, *Rep. Institute of Terrestrial Ecology 1991–92*, Natural Environment Research Council, Swindon, UK 1992, pp. 36–39.

Fowler, D.; Air pollution transport, deposition and exposure to ecosystems, in: J.R. Barker, D.T. Tingey (eds), *Air pollution effects on biodiversity*, New York: Van Nostrand Reinhold 1992, pp. 31–51.

Fowler, D.; Effects of acidic pollutants on terrestrial ecosystems, in: M. Radojevic, R.M. Harrison (eds), *Atmospheric acidity: sources, consequences and abatement,* Elsevier Applied Science 1992, pp. 341–362

Gallagher, M.W., K. Beswick, T.W. Choularton, H. Coe, D. Fowler, K.J. Hargreaves, K.J.; Measurements and modelling of cloudwater deposition to moorland and forests, *Environ. Pollut.* **75(1)** (1992) 97–107.

Güsten H., G. Heinrich, R.W.H. Schmidt, U. Schurath; A novel ozone sensor for direct eddy flux measurements, *J. Atmos. Chem.* **14** (1992) 73–84.

Güsten, H., N. Ernst, G. Heinrich; Field measurements and interpretation of species related to photooxidants and acid depositions, in: G. Angeletti, S. Beilke, J. Slanina (eds), *Field Measurements and Interpretation of Species related to Photooxidants and Acid Deposition, Air Pollution Research Report* **39**, CEC, Brussels 1992, pp. 219–224.

Güsten, H., N. Ernst, G. Heinrich; Kontinuierliche Bestimmung der turbulenten Ozonflüsse in der bodennahen Luftschicht, *nn. Meteorol.* **27** (1992) 84.

Güsten, H.; Bilanzierungsmethoden des Schadstofftransfers in der Grenzschicht Atmosphäre-Vegetation, *GIT Tachz. Lab.* **36** (1992) 544–550.

Hargreaves, K.J., D. Fowler, R.L. Storeton-West, J.H. Duyzer; The exchange of nitric oxide, nitrogen dioxide and ozone between pasture and the atmosphere, *Environ. Pollut.* **75(1)** (1992) 53–60.

Helas, G., H. Bingemer, M.O. Andreae; Organic acids over equatorial Africa. Results from DECAFE 88. *J. Geophys. Res.* **97** (1992) 6187–6193.

Helas, G., M.O. Andreae, W.R. Hartmann; Behaviour of atmospheric formic and acetic acid in the presence of hydrometeors, *J. Atmos. Chem.* **15** (1992) 101–115.

Hewitt, C.N., R.A. Street; A qualitative assessment of the emission of non-methane hydrocarbon compounds from the biosphere to the atmosphere in the U.K.: present knowledge and uncertainties, *Atmos. Environ.* **26**A (1992) 3069–3077.

Hofmann U., R. Hofmann, J. Kesselmeier; Cryogenic trapping of reduced sulfur compounds under the influence of a NAFION dryer and cotton wadding as an oxidant scavenger, *Atmos. Environ.* **26** (1992) 2445–2449.

Hofmann, U., R. Hofmann, J. Kesselmeier; Field measurements of reduced sulfur compounds over wheat during a growing season, in: S.E. Schwartz, W.G.N. Slinn (eds), *Precipitaion Scavenging and Atmosphere-Surface Exchange. 2-The Semonin Volume: Atmosphere Surface Exchange Processes*, Hemisphere, Washington DC, 1992, pp. 967–977.

Horvath, L., L. Bozo, L. Haszpra, J. Kopacz, A. Molnar, T. Prager, T. Weidinger; Gradient measurement of air-soil exchange of gases, in: S.E. Schwartz, W.G.N. Slinn (eds), *Precipitaion Scavenging and Atmosphere-Surface Exchange. 2-The Semonin Volume: Atmosphere-Surface Exchange Processes*, Hemisphere, Washington DC, 1992.

Joutsenoja, T.; Measurements of the deposition velocity of small atmospheric particles using an eddy correlation technique, *UMISTAPRG Report* (1992).

Kesselmeier J., F.X. Meixner, U. Hofmann, A. Ajavon, St. Leimbach, M.O. Andreae; Distribution of ozone over a tropical rain forest and exchange of reduced sulfur compounds between tropical tree species and the atmosphere: Investigations from the atmospheric boundary layer, at ground level and from the top of the canopy, in: F. Hallé, O. Pascal (eds), *Biologie d'une Canopée de Forêt Équatoriale - II.* Rapport de mission: Radeau des Cimes Octobre / Novembre 1991, Reserve de Campo, Cameroun, 1992, pp 247–263.

Kesselmeier, J.; Plant physiology and the exchange of trace gases between vegetation and the atmosphere, in: S.E. Schwartz, W.G.N. Slinn (eds), *Precipitation Scavenging and Atmosphere–Surface Exchange. 2-The Semonin Volume: Atmosphere-Surface Exchange Processes*, Hemisphere, Washington DC, 1992, pp 949–966.

Kristensen, L.; The NO_x-O_3 triad in the turbulent atmospheric surface layer, in: G. Angeletti, S. Beilke, J. Slanina (eds), *Field Measurements and Interpretation of Species related to Photo-oxidants and Acid Deposition, Air Pollution Research Report* **39**, CEC, Brussels 1992, p. 37.

Ludwig, J., P. Weber, F.X. Meixner, H. Rennenberg; Surface fluxes of NO and NO_2 by dynamic chamber technique - Laboratory studies on wheat, in: G. Angeletti, S. Beilke, J. Slanina (eds), *Field Measurements and Interpretation of Species related to Photo-oxidants and Acid Deposition, Air Pollution Research Report* **39**, CEC, Brussels 1992, pp. 257–265.

Meixner, F.X., J. Ludwig, H. Müller, M. Nestlen; Bestimmung des Austausches von Stickstoffkomponenten zwischen bodennaher Atmosphäre und einer landwirtschaftlich genutzten Fläche, *Ann. Meteor. N. F.* **27** (1992) 403–404.

Meixner, F.X., J. Ludwig; A field experiment on surface exchange of trace gases: Ecosystem "wheatfield", Manndorf/Lower Bavaria, in: G. Angeletti, S. Beilke, J. Slanina (eds), *Field Measurements and Interpretation of Species related to Photo-oxidants and Acid Deposition, Air Pollution Research Report* **39**, CEC, Brussels 1992, pp. 325–337.

Mennen M.G., J.W. Erisman, B.G.van Elzakker, E. Zwart; A monitoring system for continuous measurement of the SO_2 dry deposition flux above low vegetation, in: G. Angeletti, S. Beilke, J. Slanina (eds), *Field Measurements and Interpretation of Species related to Photo-oxidants and Acid Deposition, Air Pollution Research Report* **39**, CEC, Brussels 1992.

Pilegaard, K.; Risoe's integrated environmental project - RIMI, in: G. Angeletti, S. Beilke, J. Slanina (eds), *Field Measurements and Interpretation of Species related to Photo-oxidants and Acid Deposition, Air Pollution Research Report* **39**, CEC, Brussels 1992, p. 253.

Pitcairn, C.E.R., D. Fowler, M.A. Smith; Air quality at St. Fergus North Sea gas terminals and its impact on the environment, *Rep. St. Fergus Consortium.* Institute of Terrestrial Ecology, Edinburgh 1992.

Protoschill-Krebs, G., J. Kesselmeier; Enzymatic pathways for the metabolization of carbonyl sulphide (COS) by higher plants, *Botanica Acta* **105** (1992) 206–212.

Rondón A., L. Granat, C. Johansson; Dry deposition of NO, NO_2 and O_3 to coniferous forests in Sweden, in: G. Angeletti, S. Beilke, J. Slanina (eds), *Field Measurements and Interpretation of Species related to Photo-oxidants and Acid Deposition, Air Pollution Research Report* **39**, CEC, Brussels 1992, pp. 311–318.

Sànches, M.L., A. Calle, J. Domìnguez, A. Rodrìguez; Estudio meteorológico a escala regional en Castilla-León, *Encuentro METEO 92*, Salamanca 1992, pp. 315–317.

Schäfer L., J. Kesselmeier, G. Helas; Formic and acetic acid emission from conifers measured with a "cuvette" technique, in: G. Angeletti, S. Beilke, J. Slanina (eds), *Field Measurements and Interpretation of Species related to Photo-oxidants and Acid Deposition, Air Pollution Research Report* **39**, CEC, Brussels 1992.

Schuster M., R. Conrad; Metabolism of nitric oxide and nitrous oxide during nitrification and denitrification in soil at different incubation conditions. *FEMS Microbiol. Ecol.* **101** (1992) 133–143.

Slanina, J., B.G. Arends, G.P. Wyers; EUROTRAC Projecten: Voortgangsrapportage 1990–1992 (in Dutch), *Contractrapport ECN-C-92-041*, Energieonderzoek Centrum Nederland, Petten 1992.

Slanina, J., G.P. Wyers, P. Kieskamp; The analysis of air pollutants: Development of problem–oriented tools, *Analysis* **20 N8** (1992) M10-M17.

Slanina, J., G.P. Wyers; Development of automated denuder systems for profile measurements, in: *Proc. Fifth International Conference on Precipitation Scavenging and Atmosphere-Surface Exchange Processes*, Richland 1992, pp. 627–635.

Slanina, J., P.J. de Wild, G.P. Wyers; The application of denuder systems to the analysis of atmospheric components, in: J.O. Nriagu (ed.) *Gaseous Pollutants: Characterization and Cycling*. John Wiley & Sons Inc. 1992, pp. 129–154.

Slanina, J.; Environmental analytical chemistry and pollution climates in Europe, *Analysis* **20 N3** (1992) M15–M19.

Slanina, J.; Pollution climate in Europe. Natural and anthropogenic depositions on ecosystems, *Biologi Italiani* **22 (4)** (1992) 27–35.

Steinbrecher, R., W. Schürmann, H. Ziegler; Monoterpene als reaktive Spurenkomponenten in Waldsystemen Wechselwirkungen zwischen Biosphäre und Atmosphäre, *Ann. Meteorol.* **27** (1992) 79–81.

Sutton, M.A., D. Fowler, D.; Dry deposition of ammonia to frozen land surfaces and analysis of the uncertainties in fluxes deriving from measurement errors, in: I. Allegrini (ed) *Development of analytical techniques for air pollution measurement Proc. COST 611 Workshop, Rome*, CEC, Brussels 1992 pp. 9–19..

Sutton, M.A., D. Fowler, K.J. Hargreaves, L.R. Storeton-West; Effects of land-use and environmental conditions on the exchange of ammonia between vegetation and the atmosphere, in: G. Angeletti, S. Beilke, J. Slanina (eds), *Field Measurements and Interpretation of Species related to Photo-oxidants and Acid Deposition, Air Pollution Research Report* **39**, CEC, Brussels 1992, pp. 211–217.

Sutton, M.A., D. Fowler, K.J. Hargreaves, R.L. Storeton-West; Effects of land-use and environmental conditions on the exchange of ammonia between vegetation and the atmosphere, in: G. Angeletti, S. Beilke, J. Slanina (eds), *Field Measurements and Interpretation of Species related to Photo-oxidants and Acid Deposition, Air Pollution Research Report* **39**, CEC, Brussels 1992, pp. 211–217.

Sutton, M.A., J.B. Moncrieff, D. Fowler; Deposition of atmospheric ammonia to moorlands, *Environ. Pollut.* **75** (1992) 15–24.

Sutton, M.A.; D. Fowler; Atmospheric ammonia: acid rain from agriculture, in: *The impact of agriculture on the atmosphere. Proc. SCI/RSC Symp., J. Science Food and Agriculture* 1992.

Van Cleemput, O., J. Van Hoorde, A. Vermoesen; Emission of N_2O under different cropping systems, in: K. Francois, L. Pitman, K. Bartiaux-Thill (eds). *Nitrogen cycling and leaching in cool and wet regions of Europe, COST 814,* CEC, Brussels 1992, pp.20–21.

Vermoesen, A., M. Corre, O. Van Cleemput; Nitrous oxide emission during nitrification, in: K. Francois, L. Pitman, Bartiaux-Thill (eds). *Nitrogen cycling and leaching in cool and wet regions of Europe, COST 814*, Commission of the European Communities, Brussels 1992, pp. 33–34.

Wyers, G.P., A.T. Vermeulen, J. Slanina; Measurement of dry deposition of ammonia on a forest, *Environ. Pollut.* **75** (1992) 25–28.

Wyers, G.P., R.P. Otjes, A.T. Vermeulen, P.J. de Wild, J. Slanina; Measurement of vertical concentration gradients of ammonia by continuous-flow denuders, in: G. Angeletti, S. Beilke, J. Slanina (eds), *Field Measurements and Interpretation of Species related to Photo-oxidants and Acid Deposition, Air Pollution Research Report* **39**, CEC, Brussels 1992, pp. 173–178.

1993

Andersen, H.V., M.F. Hovmand, P. Hummelshøj, N.O. Jensen; Measurements of the NH_3 flux to a spruce stand in Denmark, *Atmos. Environ.* **27**A (1993) 189–202.

Anderson, H.V., H. ApSimon, W.A.H. Asman, K. Barrett, J. Kulwleit, J. Schjorring, M.A. Sutton; Working group report on dry deposition of reduced nitrogen, in: G. Lövblad, J.W. Erisman, D. Fowler (eds) *Models and Methods for the Quantification of Atmospheric Input to Ecosystems.* Nordiske Seminar-og Arbejdsrapporter 573. Nordic Council of Ministers, Copenhagen 1993, pp. 25–32.

Anghelus, A., M. Lörinczi, E. Lörinczi, A. Ritivoiu; Measurements of wet and dry deposition in Timis County, Romania, in: P.M. Borrell, P. Borrell, T. Cvitaš, W. Seiler (eds), *Proc. EUROTRAC Symp. '92*, SPB Academic Publishing bv, The Hague 1993, p. 713.

Arah, J.R.M., K.J. Hargreaves, U.M. Skiba, H. Clayton, D. Fowler, K.A. Smith; Micrometeorological measurement of N_2O flux, *Soil Use and Management* **9** (1993) 124.

Ballmoos, P. von, S. Nussbaum, C. Brunold; Isotopenpraxis, *Environ. Health Stud.* **29** (1993) 59 – 70.

Bartell, U., U. Hofmann, R. Hofmann, B. Kreuzburg, M.O. Andreae, J.Kesselmeier; COS and H_2S fluxes over a wet meadow in relation to photosynthetic activity. An analysis of measurements made on 6 September 1990, *Atmos. Environ.* **27**A (1993) 1851–1864.

Baykal, A., S.G. Tuncel, Ulviye Ozer; Ambient Concentrations of O_3, SO_2, and NO_x in Uludag Mountain, in: *Proc. 7th Symp. Chemistry and Chemical Engineering*, Trabzon 1993, p. 163.

Bieder, U., A. Schatz; Modellierung der Depositionsprozesse während der Taubildung in komplexen Strukturen, *Report IKE* 2-99, Stuttgart 1993.

Bieder, U., A. Schatz; The deposition of air pollutants to wet vegetation, in: P.M. Borrell, P. Borrell, T. Cvitaš, W. Seiler (eds), *Proc. EUROTRAC Symp. '92*, SPB Academic Publishing bv, The Hague 1993, pp. 714–717.

Braun, S., J. Duyzer, P. Grennfeldt, F.X. Meixner, A. Neftel, K Pilegaard, A. van Pul, R. San José; Dry deposition of nitrogen oxides. Working group report, in: G. Lövblad, J.W. Erisman, D. Fowler (eds), *Models and Methods for the Quantification of Atmospheric Input to Ecosystems*, Nordic Council of Ministers, EUROTRAC/BIATEX and EMEP, *Nordiske Seminar og Arbejdsrapporter*, 1993:573, pp. 17–23.

Cao, X-L, C.N. Hewitt; Evaluation of Tenax GR adsorbent for the passive sampling of volatile organic compounds at low concentrations, *Atmos. Environ.* **27**A (1993) 1865-1872.

Cao, X-L, C.N. Hewitt; Thermal desorption characteristics of four adsorbent/adsorbate systems commonly used in air sampling programmes, *Chemosphere* **27** (1993) 695–705.

Cape, J.N. J. Hilton, V.H. Kennedy, A P. Rowland (eds); Chemical analysis of natural waters in environmental research, *Science of the Total Environment* **135 (1–3)** (1993).

Cape, J.N.; Direct damage to vegetation caused by acid rain and polluted cloud: definition of Critical Levels for forest trees, *Environ. Pollut.* **82** (1993) 167–180.

Choularton, T.W., M.W. Gallagher, H. Coe, K. Beswick, D. Fowler, K.J. Hargreaves, M.A. Sutton, G. Follard, T.D. Davies, P. Wyers; Measurements of aerosol exchange to a Dutch heathland, in: P.M. Borrell, P. Borrell, T. Cvitaš, W. Seiler (eds), *Proc. EUROTRAC Symp. '92*, SPB Academic Publishing bv, The Hague 1993, p. 185.

Clement, B., M.L. Riba, V. Simon, L. Torres; Automatic determination of monoterpenes in a forested area, *Int. J. Environ. Anal. Chem.* **50** (1993) 19–27.

Coe, H., M.W. Gallagher, K.J. Hargreaves, T.W. Choularton, D. Fowler; Deposition of NO_2 to a Dutch heathland, in: P.M. Borrell, P. Borrell, T. Cvitaš, W. Seiler (eds), *Proc. EUROTRAC Symp. '92*, SPB Academic Publishing bv, The Hague 1993, p. 186.

Coe, H., M.W. Gallagher, T.W. Choularton, K.H. Hargreaves, D. Fowler, J.H. Duyzer; NO_2 fluxes: measurements and modelling, in: P.M. Borrell, P. Borrell, T. Cvitaš, W. Seiler (eds), *Proc. EUROTRAC Symp. '92*, SPB Academic Publishing bv, The Hague 1993, pp. 699–704.

Conrad, R.; Microbial metabolism of nitric oxide in soil, in: R. Guerrero, C. Pedros-Alio (eds), *Trends in Microbial Ecology,* Spanish Soc. for Microbiology, Barcelona 1933, pp. 67–70.

Dlugi, R.; Interaction of NO_x and VOCs within vegetation, in: P.M. Borrell, P. Borrell, T. Cvitaš, W. Seiler (eds), *Proc. EUROTRAC Symp. '92*, SPB Academic Publishing bv, The Hague 1993, pp. 682–688.

Draaijers, G.P.J., J.W. Erisman; Atmospheric sulphur deposition onto forest stands: throughfall estimates compared to estimates from inference, *Atmos. Environ.* **27**A (1993) 43–55

Duyzer, J., D. Fowler; Dry deposition of nitrogen oxides, in: G. Lövblad, J.W. Erisman, D. Fowler (eds) *Models and Methods for the Quantification of Atmospheric Input to Ecosystems*. Nordiske Seminar-og Arbejdsrapporter 573. Nordic Council of Ministers, Copenhagen 1993, pp. 95–123.

Eichstädter, G., W. Schürmann, R. Steinbrecher, H. Ziegler; Diurnal cycles of soil and needle monoterpene emission rates and simultaneous gradient measurements of monoterpene concentrations in the stem region and above a Norway spruce canopy, in: P.M. Borrell, P. Borrell, T. Cvitaš, W. Seiler (eds), *Proc. EUROTRAC Symp. '92*, SPB Academic Publishing bv, The Hague 1993, pp. 276–282.

Eliassen, A., K. Juda-Rezler, F. de Leeuw, S. Metcalfe, L. Nordberg, U. Pedersen, C. Pressen, H. Sandnes, R. Smith, T. Spranger, J.P. Tuovinen; Long range transport models, in: G. Lövblad, J.W. Erisman, D. Fowler (eds) *Models and Methods for the Quantification of Atmospheric Input to Ecosystems.* Nordiske Seminar-og Arbejdsrapporter 573. Nordic Council of Ministers, Copenhagen 1993, pp. 43–50.

Enders, G., R.T. McMillen; Experimental research on forest air chemistry and deposition modelling, in: P.M. Borrell, P. Borrell, T. Cvitaš, W. Seiler (eds), *Proc. EUROTRAC Symp. '92*, SPB Academic Publishing bv, The Hague 1993, p. 283.

Erisman, J.W., A. Van Pul, G.P. Wyers; Parametrization of dry deposition mechanisms for the quantification of atmospheric input to ecosystems, in: J. Slanina, G. Angeletti, S. Beilke (eds), *General Assessment of Biogenic Emissions and Deposition of Nitrogen Compounds, Sulphur Compounds and Oxidants in Europe*, *Air Pollution Research Report* **47**, EC, Brussels 1993, pp. 215–234.

Erisman, J.W., A.H. Versluis, T.A.J.W. Verplanke, D. de Haan, D. Anink, B.G. van Elzakker, M.G. Mennen, R.M. van Aalst; Monitoring the dry deposition of SO_2 in the Netherlands, *Atmos. Environ.* **27**A (1993) 1153–1161.

Erisman, J.W., B.G. van Elzakker, M. G. Mennen, J. Hogenkamp, E. Zwart, L. van den Beld, F.G. Römer, R. Bobbink, G. Heil, M. Raessen, J.H. Duyzer, H. Verhage, G.P Wyers, R.P. Otjes, J.J. Möls; The Elspeetsche Veld experiment on surface exchange of trace gases: summary of results, *Atmos. Environ.* **28** (1993) 487–496.

Erisman, J.W., G.P. Wyers; Continuous measurements of surface exchange of SO_2 and NH_3: implications for their possible interaction in the deposition process, *Atmos. Environ.* **27A** (1993) 1937–1949.

Erisman, J.W., H. van Jaarsweld, A. van Pul, A. van, D. Fowler, R.I. Smith, G. Lövblad; Comparison between small scale and long range transport modelling, in: G. Lövblad, J.W. Erisman, D. Fowler (eds) *Models and Methods for the Quantification of Atmospheric Input to Ecosystems.* Nordiske Seminar-og Arbejdsrapporter 573. Nordic Council of Ministers, Copenhagen 1993, pp. 185–197.

Erisman, J.W., M. Ferm, D. Fowler, L. Granat, H. Hultberg, J. Kesselmeier, R. Koble, G. Lovett, F. Moldan, J. Padro, G. Runge, C. Simmons, J. Slanina, M. Zapletal; Dry deposition of sulphur dioxide, in: G. Lövblad, J.W. Erisman, D. Fowler (eds) *Models and Methods for the Quantification of Atmospheric Input to Ecosystems.* Nordiske Seminar-og Arbejdsrapporter 573. Nordic Council of Ministers, Copenhagen 1993, pp. 13–15.

Erisman, J.W., M.G. Mennen, J. Hogenkamp, A. Goedhart, A. Van Pul, J. Boermans, J.H. Duyzer, G.P. Wyers; Dry deposition measurements of SO_2 over the Speulder Forest; Estimation of a surface parametrization, in: J. Slanina, G. Angeletti, S. Beilke (eds), *General Assessment of Biogenic Emissions and Deposition of Nitrogen Compounds, Sulphur Compounds and Oxidants in Europe*, *Air Pollution Research Report* **47**, EC, Brussels 1993, pp. 147–156.

Erisman, J.W.; Acid deposition onto nature areas in the Netherlands; Part I. Methods and results, *Water, Soil Air Pollut.* **71** (1993) 51–80.

Erisman, J.W.; Acid deposition onto nature areas in the Netherlands; Part II. Throughfall measurements compared to deposition estimates, *Water, Soil Air Pollut.* **71** (1993) 81–99.

Ferm, M.; Throughfall measurements of nitrogen and sulphur compounds, *Int. J. Anal. Chem.* **50** (1993) 29–43

Fowler, D., J. Duyzer, G. Dollard, F. Meixner, M. Nestlen, T. Choularton; Surface-atmosphere exchange of reactive nitrogen containing trace gases, in: P.M. Borrell, P. Borrell, T. Cvitaš, W. Seiler (eds), *Proc. EUROTRAC Symp. '92*, SPB Academic Publishing bv, The Hague 1993, p. 681.

Fowler, D., L. Granat, R. Koble, G. Lovett, F. Moldan, C. Simmons, S. Slanina, M. Zapletal; Wet, cloud water and fog deposition, in: G. Lövblad, J.W. Erisman, D. Fowler (eds*) Models and Methods for the Quantification of Atmospheric Input to Ecosystems.* Nordiske Seminar-og Arbejdsrapporter 573. Nordic Council of Ministers, Copenhagen 1993, pp. 9–12.

Fowler, D., M.A. Sutton, M.A.; Sources and sinks for atmospheric ammonia in the UK, *Interim Rep. Department of the Environment. Contract PECD 7/12/82*, 1993, pp. 1-9.

Fowler, D., M.A.Sutton, K.J. Hargreaves, R.L. Storeton-West, J.H. Duyzer, J.H. Westrate, H.L.M. Verhagen, F.X. Meixner, J. Ludwig, H. Müller, M. Nestlen, G.J. Dollard, T.D. Davies; Surface-atmosphere exchange of reactive nitrogen compounds in Europe, *Final Rep. Commission for the European Communities.* CEC contract EV4V00299, 1993, pp. 1–480.

Fowler, D., M.W. Gallagher, G.M. Lovett; Fog, cloudwater and wet deposition, in: G. Lövblad, J.W. Erisman, D. Fowler (eds*) Models and Methods for the Quantification of Atmospheric Input to Ecosystems.* Nordiske Seminar-og Arbejdsrapporter 573. Nordic Council of Ministers, Copenhagen 1993, pp. 51–73.

Fowler, D., R.I. Smith, K.J. Weston; Quantifying the spatial distribution of surface ozone exposure at the 1km x 1km scale, in: J. Fuhrer, B. Ackermann (eds) *Critical Levels for Ozone, Rep. UN-ECE Convention on Long-Range Transboundary Air Pollution* 1993, pp. 196–205.

Fowler, D.; Estimating sulphur and nitrogen deposition, in: *Acid Rain and its Impact: the Critical Loads Debate*, Department of Geography, University College London 1993.

Gallagher, M.W., T.W. Choularton, A. Wicks, K.M. Beswick, H. Coe, M.A. Sutton, D. Fowler, J.H. Duyzer, P. Wyers; Measurements of aerosol exchange to a Dutch heather moor, in: P.M. Borrell, P. Borrell, T. Cvitaš, W. Seiler (eds), *Proc. EUROTRAC Symp. '92*, SPB Academic Publishing bv, The Hague 1993, pp. 694–698.

Grieser, J., M. Kessel, W. Wobrock, W. Jaeschke; Exchange of N-species between the atmosphere and a forest ecosystem, in: P.M. Borrell, P. Borrell, T. Cvitaš, W. Seiler (eds), *Proc. EUROTRAC Symp. '92*, SPB Academic Publishing bv, The Hague 1993, pp. 289–292.

Güsten, H., A. Jochum; Meßprinzip und Anwendungsbereiche eines neuartigen Sensors für atmosphärisches Ozon, in: *Proc. AGF Tagung, Atmosphärisches Ozon – Prozesse und Wirkungen,* AGF, Bonn-Bad Godesberg 1993, pp. 21–24.

Güsten, H., G. Heinrich, R.W.H. Schmidt, U. Schurath; A novel ozone sensor for various environmental applications, in: P.M. Borrell, P. Borrell, T. Cvitaš, W. Seiler (eds), *Proc. EUROTRAC Symp. '92*, SPB Academic Publishing bv, The Hague 1993, pp. 743–746.

Güsten, H., N. Ernst, G. Heinrich; Continuous eddy flux measurements of ozone over agricultural land, in: P.M. Borrell, P. Borrell, T. Cvitaš, W. Seiler (eds), *Proc. EUROTRAC Symp. '92*, SPB Academic Publishing bv, The Hague 1993, pp. 739–742.

Helas, G., J. Kesselmeier; Estimates on sinks and sources of formic and acetic acid, in: J. Slanina, G. Angeletti, S. Beilke (eds), *General Assessment of Biogenic Emissions and Deposition of Nitrogen Compounds, Sulphur Compounds and Oxidants in Europe*, *Air Pollution Research Report* **47**, EC, Brussels 1993, pp. 299–304.

Horvath, L., J. Kopacz, Z. Nagy, E. Führer, T. Weidinger; Measurement of dry deposition velocity of different trace compounds above different types of surfaces (in Hungarian), in: *Proc. Exhibition and Conf. on Environmental Protection*, Veszprém 1993, p. 255.

Horvath, L., L. Bozo, L. Haszpra, J. Kopacz, A. Molnar, Z. Nagy, E. Fuhrer, J. Sitkey, T. Weidinger; Gradient measurement of the dry flux of nitrogen dioxide and ozone over a pine forest in Hungary, in: P.M. Borrell, P. Borrell, T. Cvitaš, W. Seiler (eds), *Proc. EUROTRAC Symp. '92*, SPB Academic Publishing bv, The Hague 1993, pp. 736–738.

Horváth, L.; Gradient measurements of the dry deposition flux of nitric oxide, nitrogen dioxide, ozone and sulfur dioxide above forest and low vegetation. Brief summary of the five field campaigns carried out in 1990–1993, in: J. Slanina, G. Angeletti, S. Beilke (eds), *General Assessment of Biogenic Emissions and Deposition of Nitrogen Compounds, Sulphur Compounds and Oxidants in Europe*, *Air Pollution Research Report* **47**, EC, Brussels 1993.

Huber, B., P. Schröder, H. Rennenberg; Emission of reduced sulfur compounds from a wheat field, in: P.M. Borrell, P. Borrell, T. Cvitaš, W. Seiler (eds), *Proc. EUROTRAC Symp. '92*, SPB Academic Publishing bv, The Hague 1993, p. 709.

Hummelshoej, H., W. Ruijgrok, A. Semb, O. Westling, P. Wyers; Dry deposition of particles. W orking group report, in: G. Lövblad, J.W. Erisman, D. Fowler (eds), *Models and Methods for the Quantification of Atmospheric Input to Ecosystems*, Nordic Council of Ministers, EUROTRAC/BIATEX and EMEP, *Nordiske Seminar- og Arbejdsrapporter*, 1993:573, pp. 33–36.

Hummelshoej, P.; Gradient measurements of dry deposition of submicron particles during stable stratification, in: *Physico-Chemical Behaviour of Atmospheric Pollutants*, EC, Brussels 1993, p. 177.

Jaeschke, W., U. Schickedanz, W. Wobrock; Turbulent fluxes of NO_x, and O_3 between a forest ecosystem and the atmosphere under high pollutant levels, in: J. Slanina, G. Angeletti, S. Beilke (eds), *General Assessment of Biogenic Emissions and Deposition of Nitrogen Compounds, Sulphur Compounds and Oxidants in Europe*, *Air Pollution Research Report* **47**, EC, Brussels 1993, pp. 207–215

Kesselmeier, J., F.X. Meixner, U. Hofmann, A. Ajavon, St. Leimbach, M.O. Andreae; Reduced sulfur compound exchange between the atmosphere and tropical tree species in southem Cameroon, *Biogeochemistry* **23** (1993) 23–45.

Kesselmeier, J., L. Merk, M. Bliefernicht, G. Helas; Trace gas exchange between terrestrial plants and atmosphere carbon dioxide, carbonyl sulfide and ammonia under the rule of compensation points, in: J. Slanina, G. Angeletti, S. Beilke (eds), *General Assessment of Biogenic Emissions and Deposition of Nitrogen Compounds, Sulphur Compounds and Oxidants in Europe*, *Air Pollution Research Report* **47**, EC, Brussels 1993, pp. 71–80.

Kesselmeier, J., L. Merk; Exchange of Carbonyl Sulfide (COS) between Agricultural Plants and the Atmosphere Studies on the Deposition of COS to Peas, Com and Rapeseed, *Biogeochem.* **23** (1993) 47–59.

Kesselmeier, J.; Gasaustausch zwischen Pflanzen und Atmosphäre, in: G. Thews, C. Servatius (eds), *Ökosystemanalyse und Umweltforschung*, Band I, Fischer Verlag, Berlin 1993, pp 195–206.

Khlystov, A., H.M. Ten Brink, G.P. Wyers; Hygroscopic growth rates of aerosols at high relative humidity, *Contractrapport ECN-C-93-011*, Energieonderzoek Centrum Nederland, Petten 1993.

Kramm, G., R. Dlugi, H. Müller, E. Schaller, W. Seiler; Modelling of the vertical transport of NO, NO_2, O_3, HNO_3, NH_3 and NH_4NO_3 in the atmospheric surface layer, in: P.M. Borrell, P. Borrell, T. Cvitaš, W. Seiler (eds), *Proc. EUROTRAC Symp. '92*, SPB Academic Publishing bv, The Hague 1993, p. 747.

Linne von Berg, K.-H., H. Bothe; Monitoring the distribution of NO_x evolving and utilizing bacteria by DNA probing, in: P.M. Borrell, P. Borrell, T. Cvitaš, W. Seiler (eds), *Proc. EUROTRAC Symp. '92*, SPB Academic Publishing bv, The Hague 1993, p. 303.

Lopez, A., j. Fontan, A. minga; Analysis of atmospheric ozone measurements over a pine forest, *Atmos. Environ.* **27A** (1993) 555–563.

Lörinczi, M., E. Biro; Some aspects of the air pollution with dry depositions in Timisoara, in. *3rd Natioal Conf. of Atmosphere Protection*, Bucarest, Romania 1993.

Lövblad, G., J.W. Erisman, D. Fowler (eds); Models and methods for the quantification of atmospheric input to ecosystems, *Nordiske Seminar -og Arbejdsrapporter 573*. Nordic Council of Ministers, Copenhagen 1993.

Marnette, E.Cl., H. Houweling, H. van Dam, J.W. Erisman; Effects of decreased atmospheric deposition on the sulfur budgets of two Dutch moorland pools, *Biogeochem.* **23** (1993) 119–144.

Meixner, F.X., M. Possanzini, J. Slanina; Development of automated chemical sensors for dry deposition measurements by the gradient method, *Final Report Project* EV4V-0083-C, CEC, DG XII, Brussels 1993, 74 pp.

Meixner, F.X., S. Braun, J. Duyzer, P. Grennfelt, A. Neftel, K. Pilegaard, A. van Pul, R. San José; Dry deposition of nitrogen oxides, in: G. Lövblad, J.W. Erisman, D. Fowler (eds), *Models and Methods for the Quantification of Atmospheric Input to Ecosystems*, Nordic Council of Ministers, EUROTRAC/BIATEX and EMEP, *Nordiske Seminar- og Arbejdsrapporter*, 1993:573, pp. 17–23

Müller, H., G. Kramm, F.X. Meixner, G.J. Dollard, D. Fowler, M.F. Possanzini; Determination of HNO_3 dry depostion by modified Bowen-ratio and aerodynamic profile techniques, *Tellus* **45B** (1993) 346–367.

Nestlen, M., H. Müller, F. Meixner; Measurements of vertical fluxes of CO_2, H_2O, NO, NO_2 and O_3 by eddy-correlation: First results from the BIATEX/SANA experiment 'Eisdorf', March 1991, in: P.M. Borrell, P. Borrell, T. Cvitaš, W. Seiler (eds), *Proc. EUROTRAC Symp. '92*, SPB Academic Publishing bv, The Hague 1993, pp. 727–731.

Nisbet, T.R., D. Fowler, R.I. Smith; Use of the critical loads approach to quantify the impact of afforestation on surface water acidification, in: *Acid Rain and its Impact: the Critical Loads Debate,* Department of Geography, University College London 1993, pp. 39–40.

Papen, H.; Emission of N-trace gases from terrestrial ecosystems, in: P.M. Borrell, P. Borrell, T. Cvitaš, W. Seiler (eds), *Proc. EUROTRAC Symp. '92*, SPB Academic Publishing bv, The Hague 1993, p. 253.

Pilegaard, K, S. Larsen; Mikrometeorologisk teknik i luftforureningsforskning.Maaling af flukse ved hjaelp af eddy korrelations metoden, *Det strategiske Miljoeforskningsprogram, Nyhedsbrev* (7) (1993) 12–14.

Pilegaard, K., P. Hummelshoej, N.O. Jensen; Deposition of ozone and nitrogen dioxide to open land and forest, in: J. Slanina, G. Angeletti, S. Beilke (eds), *General Assessment of Biogenic Emissions and Deposition of Nitrogen Compounds, Sulphur Compounds and Oxidants in Europe*, *Air Pollution Research Report* **47**, EC, Brussels 1993, pp. 157–164.

Pio, C.A., A. R. Valente; Terpene emissions from aresin tapping apractices, in: J. Slanina, G. Angeletti, S. Beilke (eds), *General Assessment of Biogenic Emissions and Deposition of Nitrogen Compounds, Sulphur Compounds and Oxidants in Europe*, *Air Pollution Research Report* **47**, EC, Brussels 1993, pp. 243–250.

Pio, C.A., T. V. Nunes, S. Brito; Volatile HC emissions from common and negative species of vegetation in Portugal, in: J. Slanina, G. Angeletti, S. Beilke (eds), *General Assessment of Biogenic Emissions and Deposition of Nitrogen Compounds, Sulphur Compounds and Oxidants in Europe*, *Air Pollution Research Report* **47**, EC, Brussels 1993, pp. 291–298.

Pio, C.A.; Measurements of biogenic non-methane hydrocarbon (BNMHC) emissions from vegetation representative of European ecosystems, *Final Report, STEP Contract* CT90–0060.

Pitcairn, C.E.R., D. Fowler; St. Fergus Project II 1992–1993, *Final Rep. HMIPI*, Scottish Development Department, Institute of Terrestrial Ecology, Edinburgh 1993, pp. 1–30.

Plöchl, M., P. Ramge, F.-W. Badeck, G.H. Kohlmaier; Modelling of deposition and detoxification of ozone in the plant leaves, in: P.M. Borrell, P. Borrell, T. Cvitaš, W. Seiler (eds), *Proc. EUROTRAC Symp. '92*, SPB Academic Publishing bv, The Hague 1993, pp. 748–752.

Pul, W.A.J. van, J.W. Erisman, J.A. van Jaarsveld, F.A.A.M. de Leeuw; Mapping small scale deposition of acidifying components over Europe: Methodology and Examples, in: *Proc. CEC/BIATEX workshop: General assessment of biogenic emissions and deposition of nitrogen compounds, sulpher compounds and oxidants in Europe,* Aveiro, Portugal 1993.

Remde, A., J. Ludwig, F.X. Meixner, R. Conrad; A study to explain the emission of nitric oxide from a marsh soil, *J. Atmos. Chem.* **17** (1993) 249–275.

Rondon, A, C. Johansson., L. Granat; Dry deposition of nitrogen dioxide and ozone to coniferous forest, *J. Geophys. Res.* **98** (1993) 5159–5172.

Rondón, A., C. Johansson, L. Granat; Comparison of stomatal and cuticular uptake of ozone by Scots pine and Norway spruce, in: P.M. Borrell, P. Borrell, T. Cvitaš, W. Seiler (eds), *Proc. EUROTRAC Symp. '92*, SPB Academic Publishing bv, The Hague 1993, pp. 723–726.

Rudolph, J., R. Conrad; Turnover of NO in a podsolic soil from the BIATEX site "Grote Heide", in: P.M. Borrell, P. Borrell, T. Cvitaš, W. Seiler (eds), *Proc. EUROTRAC Symp. '92*, SPB Academic Publishing bv, The Hague 1993, p. 718.

Saad, O.A.L.O., R. Conrad; Adaption to temperature of nitric oxide-producing nitrate-reducing bacterial populations in soil, *Syst. Appl. Microbiol.* **16** (1993) 120–125.

Saad, O.A.L.O., R. Conrad; Temperature dependence of nitrification, denitrification, and tumover of nitric oxide in different soils, *Biol. Fertil. Soils* **15** (1993) 21–27.

San José, R., J. Bilbao, A. de Miguel, V.E. Cachorro, R.E. Viloria, V. Tricio, J. Casanova, A. García, M.L. Cancillo, H. Casado, P. Palacios, D. Encinas; Dry deposition fluxes over bare and homogeneous terrain by using the "Inferential" method, in: P.M. Borrell, P. Borrell, T. Cvitaš, W. Seiler (eds), *Proc. EUROTRAC Symp. '92*, SPB Academic Publishing bv, The Hague 1993, pp. 705–708.

Sánches, M.L., J. Domìnguez, F. Sans, R. Rodrìguez; Preliminary study on dry deposition, in: J. Slanina, G. Angeletti, S. Beilke (eds), *General Assessment of Biogenic Emissions and Deposition of Nitrogen Compounds, Sulphur Compounds and Oxidants in Europe*, *Air Pollution Research Report* **47**, EC, Brussels 1993, pp. 65–69.

Sánches, M.L., J. Domìnguez, R. Rodrìguez; Experimental results of acidic aerosol and particle size distribution in a forested area, *C.E.E. Rep. EUR 15609/2 EN 1005-1010*, 1993.

Schäfer, F., R. Conrad; Metabolism of Pseudomonas stutzeri in culture and soil, *FEMS Microbiol. Ecol.* **102** (1993) 119–127.

Schürmann, W., H. Ziegler, D. Kotzias, R. Schönwitz, R. Steinbrecher; Emission of biosynthesized Monoterpenes from needles of Norway Spruce, *Naturwiss.* **80** (1993) 276–278.

Skiba, U., K.A. Smith, D. Fowler; Nitrification and denitrification as sources of nitric oxide and nitrous oxide in a sandy loam soil, *Soil Biol. Biochem.* **25** (1993) 1527–1536.

Slanina, J.; Quality Assurance for Ambient Monitoring, *VDI-Berichte* **1059** (1993) 19–32.

Steinbrecher, R., A.–M. Schreiner, G. Eichstädter, W. Schürmann; VOC-Emissions from some German woodland and crop species, in: K.-H. Becker (ed), *Proc. Third US/FRG/CEC Workshop on Photochemical Ozone Problem and its Control*, Bergische Universität-GH Wuppertal 1993, pp. 125–126.

Steinbrecher, R., I. Karbach, S. Nitz, H. Ziegler; Isoprene emission from Norway spruce (*Picea abies* [L.] *Karst.*) and behavior in the atmosphere of a mature spruce stand, in: P.M. Borrell, P. Borrell, T. Cvitaš, W. Seiler (eds), *Proc. EUROTRAC Symp. '92*, SPB Academic Publishing bv, The Hague 1993, pp. 284–288.

Steinbrecher, R., M. Jocher, M. Wiedemann, J. Slemr, J. Hahn; Low molecular weight compounds above a natural Norway spruce (*Picea abies* [L.] *Karst.*) forest, in: P.M. Borrell, P. Borrell, T. Cvitaš, W. Seiler (eds), *Proc. EUROTRAC Symp. '92*, SPB Academic Publishing bv, The Hague 1993, pp. 299–302.

Steinbrecher, R., W. Schürmann, A.-M. Schreiner, H. Ziegler; Terpenoid emissions from common oak (*Quercus robur L.*) and Norway spruce (*Picea abies L. Karst.*), in: J. Slanina, G. Angeletti, S. Beilke (eds), *General Assessment of Biogenic Emissions and Deposition of Nitrogen Compounds, Sulphur Compounds and Oxidants in Europe, Air Pollution Research Report* **47**, EC, Brussels 1993, pp. 251–261.

Steinbrecher, R.; Uncertainties of biogenic VOC-emisssion databases, in: H. P. Baars et al. (eds), *Proc. TNO/EURASAP workshop on the reliability of VOC emission databases, IMW-TNO Publication* P93/040, TNO, Delft 1993, pp. 89–98.

Stokes, N.J., P.W. Lucas, C.N. Hewitt; Controlled environment fumigation chambers for studies of reactive hydrocarbon effects on plants, *Atmos. Environ.* **27**A (1993) 679–684.

Street, R., J. Wolfenden, C. Duckham, C.N. Hewitt; Field and laboratory measurements of non-metahne hydrocarbons from a range of plant species, in: P.M. Borrell, P. Borrell, T. Cvitaš, W. Seiler (eds), *Proc. EUROTRAC Symp. '92*, SPB Academic Publishing bv, The Hague 1993, pp. 304–308.

Sutton, M.A., C.E.R. Pitcairn, D. Fowler; The exchange of ammonia between the atmosphere and plant communities, *Adv. Ecological Res.* **24** (1993) 202–292.

Sutton, M.A., D. Fowler, J.B. Moncrieff, R.L. Storeton-West; The exchange of atmospheric ammonia with vegetated surfaces. II: Fertilized vegetation, *Q. J. Roy. Met. Soc.* **119** (1993) 1047–1070.

Sutton, M.A., D. Fowler, J.B. Moncrieff; The exchange of atmospheric ammonia with vegetated surfaces. I: Unfertilized vegetation, *Q. J. Roy. Met. Soc.* **119** (1993) 1023–1045.

Sutton, M.A., D. Fowler, K.J. Hargreaves, R.L. Storeton-West; Interactions of NH_3 and SO_2 exchange inferred from simultaneous flux measurements over a wheat canopy, in: J. Slanina, G. Angeletti, S. Beilke (eds) *General Assessment of biogenic emissions and deposition of nitrogen compounds, sulphur compounds and oxidants in Europe. Proc. joint CEC/BIATEX workshop, Air Pollution Research Report* **47**, CEC, Brussels 1993, pp. 173–190.

Sutton, M.A., D. Fowler, R.I. Smith, M. Eager, D.J. Place, W.A.H. Asman; Modelling the net echange of reduced nitrogen, in: J. Slanina, G. Angeletti, S. Beilke (eds) *General Assessment of biogenic emissions and deposition of nitrogen compounds, sulphur compounds and oxidants in Europe. Proc. joint CEC/BIATEX workshop, Air Pollution Research Report* **47**, CEC, Brussels 1993, pp. 117–131.

Sutton, M.A., D. Fowler; A model for inferring bi-directional fluxes of ammonia over plant canopies, in: *Proc. WMO Conf. on the Mesurement and Modelling of Atmospheric Composition Changes including Pollutant Transport*, WMO, Geneva 1993, pp. 179–182.

Sutton, M.A., D. Fowler; Estimating the relative contribution of SO_x, NO_y and NH_x inputs to effects of atmospheric deposition, in: M. Hornung, R.A. Skeffinton (eds), *Critical Loads: Concept and Applications*, ITE Symp. No. 28, HMSO, London 1993, pp. 119–126.

Sutton, M.A., D. Fowler; Interactions of atmospheric NH_3 and SO_2 exchange with plant communities, *Rep. Institute of Terrestrial Ecology 1992–3,* Natural Environment Research Council, Swindon, UK 1993, pp. 59–62.

Tidy, G., J.N. Cape; Ammonia concentrations in houses and public buildings, *Atmos. Environ.* **27A** (1993) 2235–2237.

Torres, L., M.L. Riba, M. Haziza, V. Simon, B. Clement, G. Enders; Emission rates and fluxes of isoprenoic hydrocarbons, in: P.M. Borrell, P. Borrell, T. Cvitaš, W. Seiler (eds), *Proc. EUROTRAC Symp. '92*, SPB Academic Publishing bv, The Hague 1993, pp. 293–295.

Unsworth, M.H., M.R. Ashmore, M.G.R. Cannell, D. Fowler, J.N. Gibbs, M. Hornung, P. Ineson, J.L. Innes, P.W. Lucas, T.A. Mansfield, R.B. Wilson, C.E.R. Pitcairn; Air pollution and tree health in the United Kingdom, *2nd. Rep. Terrestrial Effects Review Group*. London: HMSO 1993.

Vermoesen, A., O. Van Cleemput, G. Hofman; Nitrogen Loss Processes: mechanisms and importance, *Pedologie* 43 (1993) 417–433.

Vermoesen, A., O. Van Cleemput; Emission of methane influenced by different types of harvest residues, in: P.M. Borrell, P. Borrell, T. Cvitaš, W. Seiler (eds), *Proc. EUROTRAC Symp. '92*, SPB Academic Publishing bv, The Hague 1993, pp. 309–312.

Vermoesen, A., O. Van Cleemput; Methane emission influenced by incorporation of different types of harvest residue, in: J. Slanina, G. Angeletti, S. Beilke, S. (eds). *General assessment of biogenic emissions and deposition of nitrogen compounds, sulphur compounds and oxidants in Europe. Air pollution Research Report* **47**, CEC, Brussels 1993, pp. 281–285.

Vermoesen, A., O. Van Cleemput; Methane emission influenced by incorporation of different types of harvest residue, in: J. Slanina, G. Angeletti, S. Beilke (eds), *General Assessment of Biogenic Emissions and Deposition of Nitrogen Compounds, Sulphur Compounds and Oxidants in Europe*, *Air Pollution Research Report* **47**, EC, Brussels 1993, pp. 281–285.

Vermoesen, A., O. van Cleemput; Emission of methane influenced by different types of harvest residues, in: P.M. Borrell, P. Borrell, T. Cvitaš, W. Seiler (eds), *Proc. EUROTRAC Symp. '92*, SPB Academic Publishing bv, The Hague 1993, pp. 309–312.

Weidinger, T.; *Study of Processes of Turbulent Boundary Layer,* University Books for Meteorology, No. 5, Eötvös Lorand University, Budapest 1993, (in Hungarian).

Wyers, G.P., R.P. Otjes, J. Slanina; A continuous-flow denuder for the measurement of ambient concentrations and surfaœ-exchange fluxes of ammonia, *Atmos. Environ.* **27**A (1993) 2085–2090.

Wyers, G.P., R.P. Otjes, J. Slanina; An automated, high-precision denuder for the measurement of ammonia in ambient air, in: J. Allegrini (ed.) *Development of analytical techniques for atmospheric pollutants. Air Pollution Research Report* **41**, EC Brussels 1993, pp. 1–7.

Wyers, G.P., R.P. Otjes, J.W. Erisman; Dry deposition of ammonia onto a coniferous forest, in: J. Slanina, G. Angeletti, S. Beilke (eds), *General Assessment of Biogenic Emissions and Deposition of Nitrogen Compounds, Sulphur Compounds and Oxidants in Europe, Air Pollution Research Report* **47**, EC, Brussels 1993, pp. 139–145.

Wyers, G.P., R.P. Otjes, J. Kuipers, J. Slanina; Sampling of HNO_2 in an automated denuder, in: P.M. Borrell, P. Borrell, T. Cvitaš, W. Seiler (eds), *Proc. EUROTRAC Symp. '92*, SPB Academic Publishing bv, The Hague 1993, pp. 710–712.

1994

Arah, J.R.M., I.J. Crichton, K.A. Smith, H. Clayton, U. Skiba; Automated gas chromotographic analysis system for micrometeorological measurements of trace gas fluxes, *J. Geog. Res.* **99** (1994)16,593–16,598.

Arends, B.G., G.P. Wyers, M.G. Mennen, J.W. Erisman, P. Hofschreuder, J.H. Duyzer; Comparison of concentration measurements for aerosols and gases using different techniques, *Contractrapport ECN-C-94-058*, Energieonderzoek Centrum Nederland, Petten 1994.

Bächmann, K., P. Ebert, I. Haag, T. Prokop, A. Röder, K. Steigenwald; Field measurement for the determination of the main cations, anions and elements as well as some organic species in individual raindrops as a function of their size, in: P.M. Borrell, P. Borrell, T. Cvitaš, W. Seiler (eds), *Proc. EUROTRAC Symp. '94*, SPB Academic Publishing bv, The Hague 1994, pp. 1101–1104.

Bandibas, J., A. Vermoesen, C.-J. De Groot, O. Van Cleemput; Comparative laboratory study of the effect of different moisture regimes and soil characteristics on nitrous oxide emission and consumption by different soils, *Soil Science* **158** (1994) 106–114.

Barthody, J., T. Weidinger, L. Horvath; Nyirjes 1991–1993. Part II: Analysis of data of tower measurements above pine forest; vertical structure and seasonal variation (in Hungarian), in: *Proc. Conf. on Forest and Climate*, Noszvaly 1994.

Baykal, A., S. Tuncel, U. Özer; Measurement of gas phase pollutants in Uludag mountain in northwestern Turkey, in: P.M. Borrell, P. Borrell, T. Cvitaš, W. Seiler (eds), *Proc. EUROTRAC Symp. '94*, SPB Academic Publishing bv, The Hague 1994, pp. 310–312.

Beswick, K.M., M.W. Gallagher, P. Hummelshøj, K. Pilegaard, N. O. Jensen, J. Duyzer; Aerosol exchange to Speulder Forest, in: P.M. Borrell, P. Borrell, T. Cvitaš, W. Seiler (eds), *Proc. EUROTRAC Symp. '94*, SPB Academic Publishing bv, The Hague 1994, pp. 683–688.

Blezinger, S., C. Wilhelm, J. Kesselmeier; Exchange of reduced sulfur compounds between marine phytoplankton and the ocean, in: P.M. Borrell, P. Borrell, T. Cvitaš, W. Seiler (eds), *Proc. EUROTRAC Symp. '94*, SPB Academic Publishing bv, The Hague 1994, pp. 559–562.

Boeckx, P., O. Van Cleemput; Methane oxidation in a landfill cover soil, *Annales Geophysicae, Supplement II.* 12 (1994) C387.

Boeckx, P., O. Van Cleemput; Methane oxidation in a landfill cover soil, in: P.M. Borrell, P. Borrell, T. Cvitaš, W. Seiler (eds), *Proc. EUROTRAC Symp. '94*, SPB Academic Publishing bv, The Hague 1994, pp. 521–524.

Cadariu, C., M. Lorinczi; Some aspects of SO_2 and NO_2 emissions in Timisoara, in: *4th. Thermotechnics National Conf.*, Timisoara, Romania 1994.

Cape, J.N., R.I. Smith, R.I.. D. Fowler; The influence of ozone chemistry and meteorology on plant exposure to photo-oxidants, in: R.M.M. Crawford, G.A.F. Hendry, B.A. Goodman (eds) *Oxygen and Environmental Stress in Plants, Proc. Roy. Soc. Edinburgh (B)* **102** (1994) 11–32.

Choularton, T.W., H. Coe, S. Walton, M.W. Gallagher, K.M. Beswick, C. Dore, J. Duyzer, H. Westrate, K. Pilegaard, N. O. Jensen, P. Hummelshøj; Photochemical modification of ozone and NO_x deposition to forests, in: P.M. Borrell, P. Borrell, T. Cvitaš, W. Seiler (eds), *Proc. EUROTRAC Symp. '94*, SPB Academic Publishing bv, The Hague 1994, pp. 645–650.

Clement, B., J.L. Fugit, M.L. Riba, L. Torres, T. Nunes, C. Pio; Isoprene emission from an *Eucalyptus globulus* forest, *Fressenius Environ. Bull.* **3** (1994) 569–574.

Colvile, R.N., T.W. Choularton, K.N. Bower, M.W. Gallagher, J.N. Cape, D. Fowler, K.J. Hargreaves, G.J. Dollard, T.J. Davies, S.A. Penkett, R.A. Burgess, B.J. Bandy; Field studies of the interaction of oxidised nitrogen compounds with clouds, in: P.M. Borrell, P. Borrell, T. Cvitaš, W. Seiler (eds), *Proc. EUROTRAC Symp. '94*, SPB Academic Publishing bv, The Hague 1994, pp. 1114–1118.

Colvile, R.N., T.W. Choularton, M.W. Gallagher, A.J. Wicks, R.M. Downer, B.J. Tyler, K.J. Hargreaves, R.L. Storeton-West, D. Fowler, J.N. Cape, G.J. Dollard, T.J. Davies, B.R.M. Jones, S.A. Penkett, B.J. Bandy, R.A. Burgess; Observations on Great Dun Fell of the pathway by which oxides of nitrogen are converted to nitrate, *Atmos. Environ.* **28(3)** (1994) 397–408.

De Groot, C.-J., A. Vermoesen, A., O. Van Cleemput; Laboratory study of the emission of NO and N_2O from some Belgian soils, *Environmental Monitoring and Assessment* **31** (1994) 183–189.

De Groot, C.-J., A. Vermoesen, O. Van Cleemput; Laboratory study of the emission of N_2O and CH_4 from a calçareous soil, *Soil Science* **158** (1994) 355–364.

De Groot, C.-J., L. Nollet, A. Vermoesen, P. Boeckx, O. Van Cleemput; The effect of nitrate and ammonium application on the emission of NO and N_2O out of 4 Belgian soils, in. *Proc. 8th Nitrogen Workshop*, University of Ghent 1994.

Dlugi, R., L. Kins, D. Müller, G. Roider, H. Schoers, T. Seiler, M. Weber, M. Zelger, T. Foken, B. Wichura, H. Güsten, G. Kramm; The exchange of trace gases and the surface energy budget: An intercomparison of results from field studies, in: P.M. Borrell, P. Borrell, T. Cvitaš, W. Seiler (eds), *Proc. EUROTRAC Symp. '94*, SPB Academic Publishing bv, The Hague 1994, p. 643.

Domìnguez, J., M.L. Sánches; Metodos analiticos y resultados experimentales de aerosoles de caracter acido en un area rural de Catilla-Leon, in: *Symp. Int. de metodologia analitica en el campö de medio ambiente*, Barcelona 1994.

Domìnguez, J., R. Rodrìguez, M.J. Fernández, J. Sanz, M.A. Garcìa, I. Pérez, M.L. Sánches; Desposicion de sustancias acidificantes en un area rural de Castilla y Leon: Metodologia y resultados experimentales, *XXIV Jornadas de la Asociacion Meteorologica Espanola*, Segovia 1994.

Duyzer, J., D. Fowler; Modelling land atmosphere exchange of gaseous oxides of nitrogen in Europe, *Tellus* **46B** (1994) 353–372.

Erisman, J.W., A. van Pul, P. Wyers; Parameterization of dry deposition mechanisms for the quantification of atmospheric input to ecosystems, *Atmos. Environ.* **28** (1994) 2595–2607.

Erisman, J.W., B.G. Van Elzakker, M.G. Mennen, J. Hogenkamp, E. Zwart, L. Van den Beld, F.G. Römer, R. Bobbink, G. Heil, M. Raessen, J.H. Duyzer, H. Verhage, G.P. Wyers, R.P. Otjes, J.J. Möls; The Elspeetsche Veld experiment on surface exchange of trace gases: Summary of results, *Atmos. Environ.* **28** (1994) 487–496.

Erisman, J.W., C. Beier, G. Draaijers, S. Lindberg; Review of deposition monitoring methods, *Tellus,***46**B (1994) 79–93.

Erisman, J.W., C. Potma, G. Draaijers, E. van Leeuwen, A. van Pul; A generalised description of acidifying pollutants on a small scale in Europe, in: P.M. Borrell, P. Borrell, T. Cvitaš, W. Seiler (eds), *Proc. EUROTRAC Symp. '94*, SPB Academic Publishing bv, The Hague 1994, pp. 588–596.

Erisman, J.W., D.D. Baldocchi; Modelling dry deposition of SO_2, *Tellus* **46**B (1994) 159–171.

Erisman, J.W., G.J.P. Draaijers, J.H. Duyzer, P. Hofschreuder, N. Van Leeuwen, F.G. Römer, W. Ruijgrok, G.P. Wyers; Contribution of aerosol deposition to atmospheric deposition and soil loads onto forests, *Contactrapport ECN-C-94-078*, Energieonderzoek Centrum Nederland, Petten 1994.

Erisman, J.W., M. Mennen, J. Hogenkamp, E. Kemkers, D. Goedhart, A. van Pul, G. Draaijers, J. Duyzer, P. Wyers; Dry deposition monitoring of SO_2, NH_3 and NO_2 over a coniferous forest, in: P.M. Borrell, P. Borrell, T. Cvitaš, W. Seiler (eds), *Proc. EUROTRAC Symp. '94*, SPB Academic Publishing bv, The Hague 1994, pp. 655–659.

Erisman, J.W.; Evaluation of a surface resistance parameterization of SO_2, *Atmos. Environ.* **28** (1994) 2583–2594.

Fowler, D., G.J. Dollard, M.R. Ashmore, R.A. Cox, P.A. Clark, R.G. Derwent, R.m. Harrison, C.N. Hewitt, O. Hov, J.G. Irwin, D.P.H. Laxen, R.L. Maynard, S.A. Penkett, K.J. Weston, M.L. Williams, P.J. Woods; Ozone in the United Kingdom 1993, *3rd.Rep. United Kingdom Photochemical Oxidants Review*, Department of the Environment, London 1994.

Fowler, D., R.I. Smith; Terrestrial inputs of atmospheric nitrogen, in: *The ecological effects of Increased Aerial Deposition of Nitrogen. BES Ecological Issues No. 5.* Field Studies Council, UK 1994.

Gabriel, R., G. Helas, J. Kesselmeier; Determination of the concentration of organic acids and inorganic ions in the apoplastic compartment of beech leaves, in: P.M. Borrell, P. Borrell, T. Cvitaš, W. Seiler (eds), *Proc. EUROTRAC Symp. '94*, SPB Academic Publishing bv, The Hague 1994, pp.525–528.

Gries, C., T.H. Nash III, J. Kesselmeier; Exchange of reduced sulfur gases between lichens and the atmosphere; *Biogeochem.* **26** (1994) 25–39.

Hargreaves, K.J., U. Skiba, D. Fowler, J.R.M. Arah, F.G. Wienhold, L. Klemedtsson, B. Galle; Measurement of nitrous oxide emissions from fertilised grassland using micrometeorological techniques, *J. Geophys. Res.* **99** (1994) 16569–16574.

Hornung, M., K.R. Bull, M. Cresser, A. Davison, D. Fowler, R. Harriman, P. Ineson, J.G. Irwin, C.E.R. Pitcairn, R.B. Wilson; Impact of nitrogen deposition in terrestrial ecosystems, *Rep. UK Review Group on Impacts of Atmospheric Nitrogen* 1994.

Horváth, L., E. Führer, A. Gyuricza, K. Lovas, Z. Nagy, T. Weidinger; Gradient flux measurements of gases above short vegetation and forest, in: P.M. Borrell, P. Borrell, T. Cvitaš, W. Seiler (eds), *Proc. EUROTRAC Symp. '94*, SPB Academic Publishing bv, The Hague 1994, pp. 477–478.

Horvath, L., Z, Nagy, E. Führer, T. Weidinger; Field experiments for estimation of turbulent fluxes over different types of surfaces in Hungary, *Annales Geophysicae Part II. Oceans, Atmosphere, Hydrology and Nonlinear Geophysics* **12 Supp.II** (1994) C335.

Horvath, L., Z, Nagy, E. Führer, T. Weidinger; Nyirjes 1991–1993. Part I: Expedition for tower measurements over pine forest (in Hungarian), in: *Proc. Conf. on Forest and Climate*, Noszvaly 1994.

Ibrom, A., C. Veithen, M. Marques, G. Gravenhorst; The evidence of crown rain interactions in the throughfall solution of a montainous spruce forest, in: P.M. Borrell, P. Borrell, T. Cvitaš, W. Seiler (eds), *Proc. EUROTRAC Symp. '94*, SPB Academic Publishing bv, The Hague 1994, pp. 677–682.

Jaeschke, W., Th. Dietrich, U. Schickedanz; Determination of CO_2-, NO, NO_2 and O_3 fluxes between the atmosphere and a strongly polluted forest ecosystem, in: P.M. Borrell, P. Borrell, T. Cvitaš, W. Seiler (eds), *Proc. EUROTRAC Symp. '94*, SPB Academic Publishing bv, The Hague 1994, p. 549.

Janson, R., L. Granat; Dry deposition of ammonium and nitrate compounds to coniferous forest, in: P.M. Borrell, P. Borrell, T. Cvitaš, W. Seiler (eds), *Proc. EUROTRAC Symp. '94*, SPB Academic Publishing bv, The Hague 1994, pp. 660.

Kramm, G., R. Dlugi, T. Foken, N. Mölders, H. Muller, K. T. Pau U; On the determination of sublayer Stanton numbers of heat and matter for different types of surfaces, in: P.M. Borrell, P. Borrell, T. Cvitaš, W. Seiler (eds), *Proc. EUROTRAC Symp. '94*, SPB Academic Publishing bv, The Hague 1994, p. 644.

Lamaud, E., A. Labatut, J. Fontan, A. Lopez, A. Druilhet, Y. Lopez; Biosphere atmosphere exchanges: ozone and aerosol dry deposition velocities over a pine forest. Non CO2 greenhouse gases. Why and how to control, *Environmental Monitoring Assesment* **31** Kluwer Acad. Pbl. 1994, pp. 175–181.

Lamaud, E., J. Fontan, A. Lopez, A. Druilhet; Pollution control and monitoring, in: *Air Pollution II, Vol. 2,* Computational Mechanics Pbl., Southampton 1994, pp. 433–440.

Lamaud, E., Y. Brunet, A. Labatut, A. Lopez, J. Fontan, A. Druilhet; The Landes experiment: biosphere - atmosphere exchanges of ozone and aerosol particles above a pine forest, *J. Geophys. Res.* **99 D8** (1994) 16511–16521.

Lörinczi, E., A. Ritivoiu, T. Camaroschi; First emission inventory of ammonia in Timis County, in: P.M. Borrell, P. Borrell, T. Cvitaš, W. Seiler (eds), *Proc. EUROTRAC Symp. '94*, SPB Academic Publishing bv, The Hague 1994, pp. 534–540.

Lovas, K., L. Horváth; A model calculation on the possible effect of monoterpenes in the ozone formation above a pine forest, in: P.M. Borrell, P. Borrell, T. Cvitaš, W. Seiler (eds), *Proc. EUROTRAC Symp. '94*, SPB Academic Publishing bv, The Hague 1994, pp. 485–488.

Ludwig, J., F.X. Meixner; Surface exchange of nitric oxide (NO) over three European ecosystems, in: G. Angeletti, G. Restelli (eds), *Physico-Chemical Behaviour of Atmospheric Pollutants, Report* EUR 15609/1 EN, EC, Brussels 1994, pp. 587–593.

Meixner, F.X.; Surface exchange of ammonia. Three different micrometeorological procedures applied, in: G. Angeletti, G. Restelli (eds), *Physico-Chemical Behaviour of Atmospheric Pollutants, Report* EUR 15609/2 EN, EC, Brussels 1994, pp. 749–754.

Meixner, F.X.; Surface exchange of odd nitrogen oxides, *Nova Acta Leopoldina N.F.* **288** (1994) 299–348.

Nunes, T.V., A.R. Valente,S.E. Brito, C.A. Pio; Emissão de hidrocarbonetosvolàteis não metano (HCVNM) pelo *Pinus pinaster* e *Eucalyptus globulus*, in: F. Santana, M.P. Antunes, A.M. Rodrigues, J. Farinha, I. Peres (eds), *4th Conf. Nacional da Qualidade do Ambiente,* Universidade Nova de Lisboa **2** (1994) M209–M219.

Pio, C.A., T.V. Nunes, A.R. Valente; Measurement of VOC emissions from *Pinus pinaster* species by bag enclosure and tracer techniques, in: P.M. Borrell, P. Borrell, T. Cvitaš, W. Seiler (eds), *Proc. EUROTRAC Symp. '94*, SPB Academic Publishing bv, The Hague 1994, pp. 497–501.

Plantaz, M.A.H.G., A.T. Vermeulen, G. P. Wyers; Surface exchange of ammonia over grazed pasture, in: P.M. Borrell, P. Borrell, T. Cvitaš, W. Seiler (eds), *Proc. EUROTRAC Symp. '94*, SPB Academic Publishing bv, The Hague 1994, pp. 661–667.

Pul, W.A.J. van, A.F.G. Jacobs; The conductance of a maize crop and the underlying soil to ozone under various environmental conditions, *Boundary-Layer Meteorology* **69** (1994) 83–99.

Pul, W.A.J. van, C. Potma, E. van Leeuwen, G. Draaijers, J.W. Erisman; EDACS: European Deposition maps of Acidifying Components on a small scale; Model description and preliminary results, *Rep. No. 722401005*, RIVM, Bilthoven, The Netherlands 1994.

Pul, W.A.J. van, C.J.M. Potma, G.P.J. Draaijers, E. van Leeuwen, J.W. Erisman; Small scale deposition fluxes of acidifying components over Europe, in: P.M. Borrell, P. Borrell, T. Cvitaš, W. Seiler (eds), *Proc. EUROTRAC Symp. '94*, SPB Academic Publishing bv, The Hague 1994, pp. 673–676.

Reuder, J., T. Gori, A. Ruggaber, L. Kins, R. Dlugi; Photolysis frequencies of nitrogen dioxide and ozone: Measurements and model calculations, in: P.M. Borrell, P. Borrell, T. Cvitaš, W. Seiler (eds), *Proc. EUROTRAC Symp. '94*, SPB Academic Publishing bv, The Hague 1994, pp. 180–184.

Rondon, A., L. Granat; Studies on the dry deposition of NO_2 to coniferous species at low NO2 concentrations, *Tellus* **46B** (1994) 399–352.

Rudolph, J., R. Conrad; Vertical distribution of nitric oxide production and consumption in soil cores, in: *Proc. EUROTRAC Symp. '94*, p. 533.

San José, R., L. Rodriguez, J. Moreno; An application of the 'big-leaf' deposition approach to the mesoscale meteorological, transport and chemical modelling in a three dimensional context, in: P.M. Borrell, P. Borrell, T. Cvitaš, W. Seiler (eds), *Proc. EUROTRAC Symp. '94*, SPB Academic Publishing bv, The Hague 1994, p. 620.

Sánches, M.L., M.A. Garcia, A. Calle; Characterization of wind fields at a regional scale calculated by means of a diagnostic model using multivariate techniques, in: *Computer Simulation Vol. 1,* UK 1994, pp. 45–54.

Sánches, M.L., M.J. Fernández, J. Domìnguez; VOC and CH_4 observations in a forested area and in a flat terrain in Castille and Leon, in: P.M. Borrell, P. Borrell, T. Cvitaš, W. Seiler (eds), *Proc. EUROTRAC Symp. '94*, SPB Academic Publishing bv, The Hague 1994, pp. 516–520.

Sánchez, M.L., M.J. Fernández, J. Dominguez; VOC and CH_4 observations in a forested area and in flat terrain in Castile and Leon, in: P.M. Borrell, P. Borrell, T. Cvitaš, W. Seiler (eds), *Proc. EUROTRAC Symp. '94*, SPB Academic Publishing bv, The Hague 1994, pp. 516–520.

Simó, R., R. de Wit, J.O. Grimalt, J. Albaigés; Factors controlling the emission of volatile sulphur compounds from a monomictic, sulphide-rich Karstic lake to the atmosphere, in: P.M. Borrell, P. Borrell, T. Cvitaš, W. Seiler (eds), *Proc. EUROTRAC Symp. '94*, SPB Academic Publishing bv, The Hague 1994, pp. 554–558.

Simon, V., B. Clement, M.L. Riba, L. Torres; The landes experiment: Monoterpenes emitted from the maritime pine, *J. Geophys. Res.* **99** (1994) 16501–16510.

Slanina, J., G.P. Wyers; Monitoring of atmospheric components by automatic denuder systems, *Fresenius J. Anal. Chem.* **350** (1994) 467–473.

Slanina, J., M. Plantaz, P. Wyers; Deposition: From local measurements to conclusions on a European scale, in: P.M. Borrell, P. Borrell, T. Cvitaš, W. Seiler (eds), *Proc. EUROTRAC Symp. '94*, SPB Academic Publishing bv, The Hague 1994, pp. 606–613.

Slanina, J.; The Case for Eutrophication, in: P.M. Borrell, P. Borrell, T. Cvitaš, W. Seiler (eds), *Proc. EUROTRAC Symp. '94*, SPB Academic Publishing bv, The Hague 1994, pp. 29–30.

Smith, R.I., C.W. Anderson, D. Fowler; Critical levels of ozone over the United Kingdom: Mapping aggregate exceedances over moderate to high thresholds, *J. Res. National Institute of Standards and Technology* **99** (1994) 353–359.

Spranger, T., E. Hollwurtel, F. Poetzsch-Heffter, A. Branding; Dry deposition estimates from two different inferential models as compared to net throughfall measurements., in: P.M. Borrell, P. Borrell, T. Cvitaš, W. Seiler (eds), *Proc. EUROTRAC Symp. '94*, SPB Academic Publishing bv, The Hague 1994, pp. 615–619.

Steinbrecher, R., G. Eichstädter, L. Torres, B. Clement, V. Simon, W. Schürmann, D. Kotzias, R. Daiber, J. Van Eijk; Monoterpenes in air samples: European intercomparison experiments, *Int. J. Environ. Anal. Chem.* **50** (1994) 283–297.

Steinbrecher, R., R. Rabong; Emission of terpenoids and C_2-C_6 hydrocarbons by Norway spruce: Phytotron experiments and field studies, in: P.M. Borrell, P. Borrell, T. Cvitaš, W. Seiler (eds), *Proc. EUROTRAC Symp. '94*, SPB Academic Publishing bv, The Hague 1994, pp. 502–505.

Steinbrecher, R., W. Junkermann, J. Steinbrecher, J. Slemr; VOCs and peroxides in ambient air above an oak/pine forest in the Mediterranean region, in: P.M. Borrell, P. Borrell, T. Cvitaš, W. Seiler (eds), *Proc. EUROTRAC Symp. '94*, SPB Academic Publishing bv, The Hague 1994, pp. 506–510.

Steinbrecher, R.; Emission of VOCs from selected European ecosystems: The state of the art, in: P.M. Borrell, P. Borrell, T. Cvitaš, W. Seiler (eds), *Proc. EUROTRAC Symp. '94*, SPB Academic Publishing bv, The Hague 1994, pp. 448–454.

Street, R.A., S.C. Duckham, C.N. Hewitt; Biogenic terpeniod emissions from vegetation in Europe, in: P.M. Borrell, P. Borrell, T. Cvitaš, W. Seiler (eds), *Proc. EUROTRAC Symp. '94*, SPB Academic Publishing bv, The Hague 1994, pp. 480–484.

Sutton, M.A., E. Perthue, J.N. Cape, R.L. Storeton-West, D. Fowler, B.G. Arends, H. Muls, T.W. Choularton, M. Wells, K.N. Bower; Measurements of the NH_x budget over Great Dun Fell: Surface atmosphere exchange and cloud chemistry, in: P.M. Borrell, P. Borrell, T. Cvitaš, W. Seiler (eds), *Proc. EUROTRAC Symp. '94*, SPB Academic Publishing bv, The Hague 1994, pp. 1148–1151.

Sutton, M.A., W.A.H. Asman, J.K. Schjorring; Dry deposition of reduced nitrogen, *Tellus* **46B** (1994) 255–273.

Tuncel, M. Baykal, U. Özer; Measurement of NO_x, SO_x, SPM and O_3 at a high altitude site station in north western Turkey, *Israel J. Chem.* **34** (1994).

Valente, A.R., T.V. Nunes, C.A. Pio; Emissão de hidrocarbonetos biogénicos não metano pelo pinheiro (*Pinus pinaster*). Utilização da tècnica de medição do tracer atmosférico, in: F. Santana, M.P. Antunes, A.M. Rodrigues, J. Farinha, I. Peres (eds), *4th Conf. Nacional da Qualidade do Ambiente,* Universidade Nova de Lisboa **3** (1994) 110–115.

Van Cleemput, O., A. Vermoesen, C.-J. De Groot, K. Van Rycheghem; Nitrous oxide emission out of grassland, *Environmental Monitoring and Assessment* **31** (1994) 145–131.

Van Cleemput, O., A. Vermoesen, C.-J. De Groot; Emission of N_2O out of different soils and cropping systems, in: *Proc. 2nd China-EC joint workshop on nitrous oxide and methane productions and emissions from agricultural and forest ecosystems in China.* Shenyang, Sept. 1994.

Van Cleemput, O.; Biogeochemistry of nitrous oxide in wetlands, *Current Topics in Wetland Biogeochemistry* **1** (1994) 3–14.

Weidinger, T., L. Horvath; Analysis of atmospheric deposition of trace materials and the modelling of the surface layer in the frame of international research projects (in Hungarian), *Egytemi Meteorologiai Füzetek* **7** (1994) 33–36.

Wells, M., K.N. Bower, T.W. Choularton, M.A. Sutton, J.N. Cape, D. Fowler, K.J. Hargreaves, G.J. Dollard, T. Davies; A study of the ammonia budget through a hill cap cloud, in: P.M. Borrell, P. Borrell, T. Cvitaš, W. Seiler (eds), *Proc. EUROTRAC Symp. '94*, SPB Academic Publishing bv, The Hague 1994, pp. 1119–1122.

Wyers, G.P., A.C. Veltkamp, A.T. Vermeulen, M. Geusebroek, A. Wayers, J.J. Möls; Deposition of aerosol to coniferous forest, *Contractrapport ECN–C-94-051*, Energieonderzoek Centrum Nederland, Petten 1994.

Wyers, G.P., M. Geusebroek, A. Wayers, J.J. Möls, A.C. Veltkamp; Dry Deposition of submicron aerosol on a coniferous forest, in: P.M. Borrell, P. Borrell, T. Cvitaš, W. Seiler (eds), *Proc. EUROTRAC Symp. '94*, SPB Academic Publishing bv, The Hague 1994, pp. 712–715.

Wyers, G.P., M. Geusebroek, A. Wayers, J.J. Möls, A.C. Veltkamp; Deposition of airbourne radioactivity onto a coniferous forest as tracer for the deposition of submicron aerosol, in: P.M. Borrell, P. Borrell, T. Cvitaš, W. Seiler (eds), *Proc. EUROTRAC Symp. '94*, SPB Academic Publishing bv, The Hague 1994, pp. 712–715.

Zelger, M., J. Schween, B.T. Wichura, Th. Foken, T. Andersson, G. Roider, Th. Seiler, C. Frank, H. Lösslein, R. Dlugi; Micrometeorological measurements for flux determination of trace substances, in: P.M. Borrell, P. Borrell, T. Cvitaš, W. Seiler (eds), *Proc. EUROTRAC Symp. '94*, SPB Academic Publishing bv, The Hague 1994, p. 479.

Ziegler, H., R. Steinbrecher; Biogene Kohlenwasserstoffe als reaktive Luftkomponenten - Messungen im Nationalpark Bayerischer Wald, in:*Rundgespräche der Komission für Ökologie.Bd. 8 'Klimaforschung in Bayern'* , Verlag Dr. Friedrich Pfeil, München 1994, pp. 95–102.

1995 + in press

Boeckx, P., O. Van Cleemput; Methane oxidation in a neutral landfill cover soil: influence of temperature, moisture content and N-turnover, *J. Environmental Quality* (submitted for publication).

Choularton, T.W., M.W. Gallagher, K.N. Bower, D. Fowler, M. Zahneiser, A. Kaye; Trace gas flux measurements at the landscape scale using boundary layer budgets, *Phil. Trans. Roy. Soc.* (in press).

Demeyer, P., G. Hofman, O. Van Cleemput; Fitting NH_3 volatilisation dynamics with a logistic equation, *Soil Science Society America Journal* 59 (1995) 261–265.

Dore, A.J., T.W. Choularton, D. Fowler, R. Storeton-West; Field measurements of wet deposition in an extended region of complex topography, *Q. J. Roy. Met. Soc.* (in press).

Fowler, D., C. Flechard, R.L. Storeton-West, M.A. Sutton, K.J. Hargreaves, R.I. Smith; Long term measurements of SO_2 deposition over vegetation and soil and comparison with models,in: Acid rain research: Do we have all the answers? in: *Proc. Speciality Conf.*, Elsevier Science 1995 (in press).

Fowler, D., I.D. Leith, R.I. Smith, T.W. Choularton, D. Inglis, G. Campbell; Atmospheric inputs of acidity, sulphur and nitrogen in the UK, in: *UCL Conf. on Critical Load of Acidic Deposition.* (in press).

Gallagher, M.W., T.W. Choularton, K.N. Bower, I.M. Stromberg, K.M. Beswick, D. Fowler, K.J. Hargreaves; Measurements of methane fluxes on the landscape scale from a wetland area in north Scotland, *Atmos. Environ.* (in press)

Guenther A., N.C. Hewitt, D. Erickson, R. Fall, C. Geron, T. Graedel, P. Harley, L. Klinger, M. Lerdau, W.A. McKay, T. Peirce, B. Scholes, R. Steinbrecher, R. Tallamraju, J. Taylor, P. Zimmerman; A global model of natural volatile organic compound emissions, *J. Geophys. Res.* (1995) in press.

Horvath, L., Z. Nagy, T. Weidinger, R. Artz, W.T. Luke, R. Waligura, J. Pinto, J. Womack; Measurement of flux of trace gases (O_3, NO_x, SO_2, CO_2, HNO_3), particulate sulfate and nitrate and water vapour over short vegatation by gradient and eddy correlation techniques in Hungary, *Annales Geophysicae* (in press).

Khlystof, A., G.P. Wyers, J. Slanina; The steam-jet aerosol collector, *Atmos. Environ.* (submitted).

Oms, M.T., A.C. Veltkamp, P.A.C. Jongejan, G.P. Wyers, J. Slanina; Continuous monitoring of atmospheric HCl, HNO_2, HNO_3 and SO_2 by wet-annular denuder sampling with on-line chromatographic analysis, *Int. J. Environ. Anal. Res.* (submitted).

Pitcairn, C.E.R., D. Fowler, J. Grace; Deposition of fixed atmospheric nitrogen and foliar nitrogen contents of bryophytes and Calluna vulgaris (L.) Hull, *Environ. Pollut.* **88** (1995) 193–205.

Saad, O.A.L.O., S. Lehmann, R. Conrad; Influence of thiosulfate on nitrification, denitrification and production of nitric oxide and nitrous oxide in soil, *Biol. Fertil. Soils* (1995) in press

Simon, V., M.L. Riba, A. Waldhart, L. Torres; Breakthrough volume (BTV) of monoterpenes on Tenax TA. Influence of temperature and concentration for α-pinene, *J. Chromatogr* (1995) in press.

Simpson D., A. Guenther, C.N. Hewitt, R. Steinbrecher; Biogenic emissions in Europe: Estimates and uncertainties. *J. Geophy. Res.* (submitted).

Smith, R.I., D. Fowler, K.R. Bull; Quantifying the scale dependence in estimates of wet and dry deposition and the implications for critical load exceedances. In: Acid rain research: Do we have all the answers? in: *Proc. Speciality Conf.*, Elsevier Science 1995 (in press).

Sutton, M.A., D. Fowler; A canopy model for inferring bi-directional plant-atmosphere exchange of ammonia, in: *Proc. COST 611 Workshop*, CEC (in press).

Sutton, M.A., D. Fowler; Continuous measurements of ammonia deposition to acidic moorland and application of a new inferential technique to describe ammonia exchange with the plant canopy, in: *Acid Deposition Update: the Critical Loads Debate*, Department of Environment/HMSO, London (in press).

Vermeulen, A.T., G.P. Wyers, F.G. Römer, N.F.M. Van Leeuwen, G.P.J. Draaijers, J.W. Erisman; Fog deposition on douglas fir forest, *Atmos. Environ.* (submitted).

Weidinger, T., Z. Barcza, L. Horvath, R. Meszaros; Applicability and restrictioins of the gradient technique in the analysis of daily and annual courses of heat, momentum and trace gas fluxes, *Annales Geophysicae* (in press).

Weston, K.J., D. Fowler, R.E. Smith; Rural exposure to ozone over the UK, *Barcelona Air Pollution Conf. Proc.*(in press).

Wilson, R.B., K.R. Bull, M.R. Ashmore, R.W. Battarbee, R.N. Butlin, S.J.M. Caporn, M.S. Cresser, R.G. Derwent, D. Fowler, M. Hornung, S.E. Metcalfe, S.J. Ormerod, C.E.R. Pitcairn; *Critical loads of acidity in the United Kingdom. Summary report of the Critical Loads Advisory Group*, Department of the Environment. Penicuik: Institute of Terrestrial Ecology, UK (in press).

Theses

M. Sc. / Diploma

Bakkali, Y.El.; Etude et réalisation d'un analyseur pour la mesure rapide des fluctuations de l'aérosol atmosphérique, Paul Sabatier University, Toulouse, 1991.

Baykal, M.; Measurement and source determimation of air pollutants in Uludag National Park, Technical University of the Middle East, 1994.

Bliefernicht, M.; Untersuchungen zum Ammoniak-Austausch zwischen Pflanzen und Atmosphäre an Fichten, Fachbereich Biologie, University of Mainz, 1992.

Blösl, M.; Bildung von Stickstoffmonoxid durch dissimilatorisch nitratreduzierende Bakterien im Boden, University of Konstanz, 1991.

Buck, C.; Test eines Naß-Denudersystems auf seine Eignung für die Messung von vertikalen Spurengas-Gradienten in der bodennahen Luftschicht, Naturwissenschaftliche Akademie Isny, 1989.

Conze, W.; Aktivitätsbestimmungen der für die <Aufnahme von Carbonylsulfid verantwortlichen Carboanhydrase in Phytoplanktonspezies, Fachbereich Biologie, University of Mainz, 1995.

Dorfner, M.; Die Permeabilität der Cuticula für ausgewählte, niedermolekulare, organische Substanzen, Lehrstuhl für Botanik, Technical University of Munich 1992.

Ernst, N.; Die Bestimmung der Depositionsgeschwindigkeiten von Ozon mit mikrometeorologischen Meßmethoden, University of Karlsruhe, 1991.

Fehsenfeld U.; Untersuchungen zur Monoterpenemission aus Waldböden am Beispiel des Fichtenstandortes Höglwald, Lehrstuhl für Botanik, Technical University of Munich, 1992.

Gabriel, R.; Untersuchungen zur Monoterpenemission aus Zapfen und Stämmen von *Picea abies* (L.) Karst., Lehrstuhl für Botanik, Technical University of Munich 1991.

Haag, I.; An- und Abreicherungsphänomene and Regentropfenoberfächen, TH Darmstadt, 1994.

Hoffmann, B.; Umbau eines kommerziellen Ozonanalysators zur Messung schneller Ozonflüsse nach der Methode der Eddy-Korrelation, University of Karlsruhe, 1992.

Horvath, T.; Analysis of the surface layer on the basis of results of experiments carried out over pine forest, Eotvos Lorand University, Budapest, 1994.

Hubert, A.; Untersuchungen zum Gasaustausch von flüchtigen reduzierten Schwefelverbindungen an Laubstreu, Fachbereich Biologie, University of Mainz, 1992.

Kolb, Ch.; Monoterpenanalytik von Luftproben und Monoterpenemission von Fichtenbaumstämmen im Freiland, Fachhochschule Isny, 1994.

Kuhn, U.; Austausch organischer Säuren zwischen Laubstreu und Atmosphäre, Fachbereich Biologie, University of Mainz, 1993.

Lamaud, E.; Mesure et paramétrisation des flux verticaux de l'aérosol atmosphérique, Paul Sabatier University, Toulouse, 1994.

Lederle, K.; Untersuchungen zum Einfluß von Ozon, NO und Luftfeuchtigkeit auf den Nachweis von Isopren und Monoterpenen in Luftproben, Fachhochschule Isny, 1994.

Lovas, K.; Modelling of atmospheric ozone level above a pine forest, Eotvos Lorand University, Budapest, 1993.

Minga, A.; Etude des transferts d'ozone entre la biosphére et l'atmosphére: Expèriences Atila et Decafé, Paul Sabatier University, Toulouse, 1991.

Müller, M.; Feldmessungen zum Austausch von gasförmigem Ammoniak zwischen Atmosphäre und höheren Pflanzen, Fachbereich Biologie, University of Mainz, 1995.

Nägele, W.; The effect the pH has on the production of nitric oxide in an acid forest soil, University of Konstanz, 1989.

Nascimento, M. Do.; Réalisation et étude d'un analyseur pour le mesure rapide de l'ozone troposphérique, Paul Sabatier University, Toulouse, 1990.

Reichmann, A.; Oberflächentemperaturen von Blättern und Pflanzenbeständen, Institut für Physik, Technical University of Munich, 1993.

Schäfer, F.; Untersuchungen zum Abbau von Stickstoffmonoxid durch Pseudomonas stutzeri im Boden, University of Marburg, 1992.

Schäfer, L.; Untersuchungen zur Abgabe von organischen Säuren durch Pflanzen an die Atmosphäre.; Fachbereich Biologie, University of Mainz, 1992.

Schreiner, A.-M.; Isoprenemission und CO_2-H_2O-Gaswechsel von häufig in Mitteleuropa vorkommenden Laubbaum-, Nadelbaum- und Straucharten, Lehrstuhl für Botanik, Technical University of Munich, 1992.

Schuster, M.; Die Rolle der Nitrifikation und Denitrifikation bei der Umsetzung von NO und N_2O in Böden, University of Konstanz, 1991.

Seubold, T.; Feldversuch zur Bestimmung der Trockenen Deposition von gasförmiger Salpetersäure und Ammoniak auf der Esterbergalm (1265 m üb.NN), Naturwissenschaftliche Akademie Isny, 1990.

Sprung, D.; Kontinuierliche Bestimmung der turbulenten Ozonflüsse in der bodennahen Grenzschicht mittels der Eddy-Korrelation, University of Karlsruhe, 1993.

Terrey, B.; Aufbau und Test eines dynamischen, feldtauglichen Expositionskammersystems zur Bestimmung von Emission und Deposition von Stickstoffmonoxid, Ozon und Kohlendioxid an Weizenpflanzen, Georg-Simon-Ohm-Fachhohchschule, Nürnberg 1991.

Velmeke, F.; Versuche zur Korrelation von Photosynthese und dem Austausch flüchtiger Schwefelverbindungen zwischen Bäumen und Atmosphäre, Fachbereich Biologie, University of Mainz, 1993.

Winterhalter, M.; Spektralanalyse meteorologischer und luftchemischer Grüssen - Theorie und Anwendung auf Eddy Korrelationsmessungen über einem Waldbestand, University of Munich, 1991.

Ph. D.

Baumgärtner, M.; Umsetzung von Stickoxiden (NO_x) in Böden, auf Gebäudeoberflächen und in Mikro–organismen, University of Konstanz, 1991.

Blezinger, S.; Untersuchungen zum Austausch von DMS und COS zwischen marinen Phytoplankton Spezies und der Atmosphäre, Fachbereich Biologie, University of Mainz, 1995.

Bode, K.; Die Angabe von organischen Säuren an die Atmosphäre durch Pflanzen verschiedener Entwicklungsstufen, Fachbereich Biologie, University of Mainz, 1994.

Clement, B.; Dispositif automatique de dosage des composés organiques volatils (COV) biogéniques, Institut National Polytechnique de Toulouse, 1990.

Eichstädter G.; Monoterpenemission des Bodenkompartiments in einem Fichtenwaldökosystem und einiger in Mitteleuropa häufig vertretener Laubbaum-, Nadelbaum- und Straucharten sowie Verhalten der Monoterpene in der Atmosphäre, Technical University of Munich, 1993.

Erisman, J.W.; Atmospheric deposition of acidifying compounds in the Netherlands, University of Utrecht, 1992.

Haziza, M.; Contribution to the determination of isoprenic compounds emission rates and fluxes by forested ecosystems, Polytechnical National Institute, Toulouse 1991.

Hofmann, U.; Vergleichende Untersuchungen zum Austausch von reduzierten Schwefelverbindungen zwischen Vegetation und Atmosphäre im Feldversuch und unter kontrollierten Bedingungen, Fachbereich Biologie, University of Mainz, 1993.

Hummelshøj, P.; Dry deposition of particles and gases, Risø National Laboratory, Roskilde 1992.

Lehmann, S.; Umsetzung von reduzierten Schwefelverbindungen, insbesondere von Carbonylsulfid (OCS), in Böden, University of Konstanz, 1994.

Ludwig, J.; Untersuchungen zum Austausch von NO und NO_2 zwischen Atmosphäre und Biosphäre, University of Beyreuth, 1995.

Minga, A.; Etude de transport d'ozone entre la biosphere et l'atmosphere. Experience ATILA et DECAFE, University Paul Sabatier, 1991.

Plochl, M.; Die Bedeutung apoplastischer Ascorbinsäure für die Aufnahme und Reduktion von Ozon und Stickstoffdioxid in Pflanzenblättern: Eine dynamische Modellierung der involvierten metabolischen Prozesse im System Blatt, Johann Wolfgang Goethe University of Frankfurt, 1992.

Protoschill-Krebs, G.; Untersuchungen zur Aufnahme und zum Metabolismus von Carbonylsulfid in Pflanzen, Fachbereich Biologie, University of Mainz, 1991.

Pul, W.A.J. van The flux of ozone to a maize crop and the underlying soil during a growing season, Wageningen Agricultural University, 1992.

Remde, A.; Umsetzung von NOx in Böden und Bodenmikroorganismen, University of Konstanz, 1989.

Riba, M.L.; Isoprenoids in the plant and in the atmosphere. A study of their evolution in forested media, Polytechnical National Institute, Toulouse 1991.

Rondon, A.; Atmosphere-surface exchange of nitrogen oxides and ozone, Dept. of Meteorology, Stockholm University, 1993.

Rudolph, J.; Umsetzung und Transport von Stickstoffmonoxid (NO) in Böden, University of Konstanz, 1994.

Saad, O.A.L.O.; Influence of soil temperature on the microbial population metabolizing nitric oxide, University of Konstanz/Minia-Egypt, 1991.

Schäfer, L.; Emissionen organischer Säuren und Aldehyde aus der mediteranen Vegetation, Fachbereich Biologie, University of Mainz, 1995.

Schürmann W.; Emission von Monoterpenen aus Nadeln von *Picea abies* (L.) *Karst.* sowie deren Verhalten in der Atmosphäre, Technical University of Munich, 1993.

Simon, V.; Les hydrocarbures biogéniques monoterpéniques. Taux et flux d'eémission par les couverts végétaux;Institut National Polytechnique de Toulouse, 1993.

Stahl, K.; Untersuchungen zum Austausch niedermolekularer organischer Verbindungen zwischen Atmosphäre und Pflanzen - Freiland- und Gewächshausversuche an Fichten (*Picea abies* [L.] *Karst.*) und Buchen (*Fagus sylvatica* L.).; Technical University of Munich, 1990.

Sutton, M.A.; The surface/atmosphere exchange of ammonia, University of Edinburgh, 1990.

Teichmann, U.; Anreicherungseinheiten für SO und NO - Wirkungsweise, Funktionskontrolle, Kalibrierung und Einsatz zur Messung von Vertikalprofilen in und über Wald, University of Munich, 1991.

Valente, A.R.; Hidrocarbonetos naturais na atmosfera. Contribuição para a compreensão dos fluxos de emissão de hidrocarbonetos não metano pelo *Pinus pinaster* em Portugal, University of Aveiro, 1995.

Weber, P.; Aufnahme und Metabolisierung von atmosphärischen Stickstoffdioxid (NO_2) in Weizen (*Triticum aestivum* L.), Technical University of Munich, 1993.

Subject Index

E

F

G

H

I

J

K

L

M

S

T

U

V

W

X

Printing: Mercedesdruck, Berlin
Binding: Buchbinderei Lüderitz & Bauer, Berlin